SMALL AND MICRO HYDROELECTRIC POWER PLANTS

SMALL AND MICRO HYDROELECTRIC POWER PLANTS

Technology and Feasibility

Edited by Robert Noyes

NOYES DATA CORPORATION
Park Ridge, New Jersey, U.S.A.
1980

Library of Congress Catalog Card Number: 80-19099
ISBN: 0-8155-0819-0
Printed in the United States

Published in the United States of America by
Noyes Data Corporation
Noyes Building, Park Ridge, New Jersey 07656

Library of Congress Cataloging in Publication Data
Main entry under title:

Small and micro hydroelectric power plants.

(Energy technology review ; 60)
Bibliography: p.
Includes index.
1. Hydroelectric power plants. I. Noyes, Robert.
II. Series.
TK1081.S64 333.91'4 80-19099
ISBN 0-8155-0819-0

FOREWORD

The need to search for alternative sources of energy, as well as new capacity, is becoming more urgent as fossil fuel becomes less available and more expensive; thus there is increasing interest in the development of small scale or low-head hydroelectric power plants. Such plants have been defined to have an electrical capacity of less than 15 megawatts (small scale) or a dam height of less than 20 meters (low head).

Small scale hydroelectric power generation has been developed extensively in Europe, but, because of relatively low fuel costs in the past for electric thermal generation, many old plants in the United States have been abandoned. Today, with fuel costs escalating, it would seem that future economic conditions will favor a return to small scale hydroelectric power generation for at least some part of the national power supply. The fundamental concepts in the design and construction of a small scale plant are that the impacts of implementation are modest; the project is likely to be noncontroversial; and, therefore, license and permit granting should be possible within a relatively short time frame.

This book provides a general survey of the feasibility and technology for small scale (less than 15 megawatts) and micro (less than 100 kilowatts) hydroelectric power production. The information here is intended for use by public and private agencies and utilities, private investors, engineers, researchers, and educational institutions. The subject matter is divided into four parts. Part 1 is a brief discussion of the resource potential for small scale hydroelectric power in six regions comprising the United States. Part 2 details the feasibility of a small scale facility, including discussions on economic and financial analysis, hydrologic studies, facility integrity, electromechanical systems, and civil features. Part 3 gives data on available equipment, and Part 4 presents an information package for the individual investor or property owner who is considering micro hydropower—from the initial idea to the final production of power.

The book is based on the following reports:

> *Preliminary Inventory of Hydropower Resources,* prepared by the U.S. Army Corps of Engineers' Institute for Water Resources and Hydrologic Engineering Center as part of the National Hydroelectric Power Resources Study, July 1979.
>
> *Feasibility Studies for Small Scale Hydropower Additions—A Guide Manual,* prepared by the U.S. Army Corps of Engineers' Hydrologic Engineering Center and Institute for Water Resources, with funding from the U.S. Department of Energy, June 1979.
>
> *Proceedings—Small/Low-Head Hydropower PRDA-1706 Contractors' Symposium* (held in Albany, New York), prepared for the U.S. Department of Energy, Idaho Operations Office under DOE Contract DE-AM07-76IDO1570, May 8-10, 1979.

Micro-Hydropower: Reviewing an Old Concept, by R. Alward, S. Eisenbart, and J. Volkman of the Technical Research Staff of The National Center for Appropriate Technology, prepared for the U.S. Department of Energy under DOE Contract ET-78-S-07-1752, January 1979.

The table of contents is arranged so as to provide easy access to the information contained in the book and is organized in such a way as to serve as a subject index.

In order to keep the price of this large and extensive book to a reasonable level, it has been reproduced by photo-offset directly from the original reports and the cost savings passed on to the reader. Due to this method of publishing, certain portions of the reports may be less legible than desired.

CONTENTS AND SUBJECT INDEX

Part 1
Resource Potential for Small Scale Hydroelectric Power

The information in this part is from *Preliminary Inventory of Hydropower Resources*, prepared by the U.S. Army Corps of Engineers' Institute for Water Resources and Hydrologic Engineering Center as part of the National Hydroelectric Power Resources Study, July 1979.

ACKNOWLEDGEMENTS

The preparation of these reports was a coordinated effort accomplished with the assistance of many individuals in the U.S. Army Corps of Engineers. The primary responsibility for these reports was assigned to the U.S. Army Corps of Engineers, Institute for Water Resources (IWR), under the direction of Mr. A. J. Fredrich. The Preliminary Inventory of Hydropower Resources was developed as a major component of the Corps' National Hydropower Study. Supplemental funding was provided by the United States Department of Energy (DOE) through the DOE Small-Scale Hydropower Development Program. Both of these studies are under the direction of Mr. James R. Hanchey, Deputy Director for Special Studies at the Institute for Water Resources.

The manuscript herein was written and prepared by Dr. Wayne R. Sigleo, Mr. James R. Hanchey and Mr. Darrell G. Nolton of the Corps' Institute for Water Resources. The text had the benefit of informal review and comment by the staff of the National Hydropower Study group at the Institute. The data presented in these reports were collected by the Corps' Division and District field offices. The presentation of these data, particularly the tables and computer format, were made possible through the concentrated efforts of Mr. Gary Franc of the Corps' Hydrologic Engineering Center (HEC) who, based on instructions from Mr. Jim Dalton of the Corps' Southwestern Division (SWD), developed the computer software to summarize the data from the inventory and made all necessary computer runs. HEC arranged for the printing of these reports and is responsible for their distribution.

Some of the major responsibilities associated with the National Hydropower Study were assigned to the Corps' Hydrologic Engineering Center, under the supervision of Mr. Bill S. Eichert, the Center's Director. HEC was assigned the tasks of developing the data management software, the editing and analysis programs required in the screening studies and in making the computer runs required in the screening process. Mr. Jim Dalton (SWD) was instrumental in formulating the computational techniques used and was assigned the responsibility of technical management. Mr. Dale R. Burnett was HEC's overall coordinator; Mr. Tom White and Mr. Orval Bruton of the Corps' North Pacific Division (NPD) developed the cost-estimating procedures; Messrs. Arthur Pabst and Mark Lewis (HEC) developed the file management software; and Ms. Marilyn Hurst (HEC) did most of HEC's computer production runs for the National Hydropower Study.

Grateful acknowledgements are extended to the support staff of IWR and HEC for their patience and endurance in the overall effort to complete these reports. In particular, Ms. Sharon Blake and Ms. Denise Henderson of IWR and Ms. Penni Baker of HEC should be recognized. Finally, since it is not possible, because of the scope of these reports, to mention all participants by name, acknowledgements are extended to all, especially the National Hydropower Study coordinators and other Division and District personnel who devoted many hours to the organization and data collection activities necessary to provide this preliminary inventory of hydroelectric power resources in the United States.

INTRODUCTION

Since completion of the world's first central hydroelectric generating facility at Appleton, Wisconsin in 1882, hydropower has played a major role in our nation's social and economic development. Although this first installation was comparatively small (providing only enough power to light 250 light bulbs), it had a large impact, and streams and rivers across the country were rapidly developed to generate electricity. Today, hydropower provides about 13 percent of the nation's total electric power with a conventional installed capacity of about 64,000 megawatts and an average annual energy generation of some 280 thousand gigawatt-hours.

Hydroelectric power development was rapid during the first half of the twentieth century, but by the mid-1960's many factors had combined to diminish its contribution to electrical utility systems. First, the most favorable sites were developed early, and the undeveloped potential simply did not look as attractive when compared to other available energy sources. Second, demand for electricity increased rapidly during the 50's and 60's, and even with the continued development of new sites, hydropower's "share of the load" steadily decreased. Finally, the low cost of fossil fuels and optimistic forecasts concerning nuclear technology and its public acceptability led many planners to believe that the nation's energy future was secure.

During the past decade, a number of interacting factors, including rising fuel prices, rapid escalation of the costs in constructing thermal generating facilities, and increased public concern over the safety of nuclear plants have prompted not only a search for new energy alternatives, but also a reexamination of previously ignored or discounted alternatives. Because of the immediate need to develop new sources of energy, planners at all levels of organization have significantly increased their efforts to assess the most feasible alternatives to meet present and future energy demands. Hydroelectric power development, particularly incremental or new capacity at existing facilities, could provide an important contribution to our nation's growing energy needs.

The U.S. Army Corps of Engineers is currently conducting a detailed assessment of the nation's hydroelectric resources as part of the National Hydroelectric Power Study authorized by Section 167 of the Water Resources Development Act of 1976 (P.L. 94-587). The study is designed to provide a current and comprehensive estimate of the potential for incremental or new generation at existing dams and other water resource projects, as well as for undeveloped sites in the United States. In addition, the study will address the demand for hydroelectric power, and will investigate various related policy and technical considerations to determine the incentives, constraints and impacts of developing hydropower to meet a portion of our future energy demands. When complete in 1981, the effort will provide a more detailed evaluation of the nation's hydroelectric resources, and will serve as a framework for future planning and development of this important renewable energy source.

The National Hydropower Study addresses all conventional hydroelectric power potential at Federal and non-federal installations, and considers both large and small-scale dams and other water resource projects. The Corps of Engineers involvement in studying the nation's small-scale potential dates from President Carter's Energy Plan of 1977. This program specifically recognized the opportunity for redeveloping small-scale hydropower as an alternative source of energy and the President directed the Corps to produce summary estimates of the potential at existing small dams in the country.

The directive led to the Corps' preliminary 90-day hydropower study which was published in 1977[1]. This study was the first to provide comprehensive estimates of the small-scale potential at existing dams and also identified key areas of the country where small-scale hydropower development could potentially reduce dependence on fossil fuels as a source of energy generation. It is important to note that these estimates were based largely on theoretical potentials calculated for the river basins in the United States and were not the product of site-specific investigations.

During the initial planning stages of the National Hydropower Study, the U.S. Department of Energy requested that a more detailed assessment be made of the nation's small-scale hydroelectric resources. Because of the wide public interest in this potentially valuable alternative energy resource, the small-scale assessment has been integrated into the overall National Hydropower Study and is included in this series of reports.

PURPOSE AND SCOPE

Site-specific information on the physical hydroelectric power potential is essential in determining the social, economic, institutional and environmental feasibility of developing this resource. Because of the immediate need for wide dissemination of state, regional and national hydropower data, the Corps' Institute for Water Resources has prepared

---

[1] R. J. McDonald, Estimate of National Hydroelectric Power Potential at Existing Sites, Institute for Water Resources, Ft. Belvoir, Virginia, July 1977.

this series of regional reports, Preliminary Inventory of Hydropower Resources. The inventory is the result of a comprehensive data collection effort conducted by the Corps of Engineers and is based on site-specific analysis and evaluation of 6 regional areas (Figure 1).

The purpose of these reports is to provide preliminary estimates of the existing and potentially feasible hydroelectric power resources in the United States, and to briefly evaluate their regional significance. The estimates of existing, incremental and undeveloped hydropower potential have been grouped in three categories which are based on megawatt (MW) capacity. These include small-scale (.05-15 MW); intermediate (15-25 MW); and large-scale (greater than 25 MW).

METHODS OF STUDY

The preliminary inventory of potentially feasible hydropower resources includes an estimate of the capacity and energy available at both existing dams and undeveloped sites in the United States. The major source of data on existing hydropower facilities was the National Inventory of Dams developed by the Corps of Engineers as part of the National Dam Safety Program.[2] This inventory contains geographic, physical, and ownership data on approximately 50,000 dams in the nation. Identification and data collection on undeveloped sites was more limited since only about 5,000 sites had been identified or previously studied by the Corps of Engineers and other local, state and Federal water resource agencies. In addition, no attempt was made to include pumped storage sites in the inventory.

The data in the original national inventory of dams were supplemented as necessary to develop preliminary estimates of the hydroelectric power potential at each site. Computer routines which utilized head, storage and streamflow estimates were developed to compute the capacity and energy potential of each existing dam and undeveloped site. A screening routine was used to eliminate those sites without sufficient storage, head or streamflow to generate a significant amount of electrical energy. Generally, the existing dams and undeveloped site locations listed in the inventory are those with a capacity of 50 kilowatts or greater. In most cases, the current installed capacity at existing dams was derived from the nameplate capability. This initial screening procedure reduced the number of sites in the active inventory from approximately 55,000 to about 17,500.

During the second stage of the preliminary screening, additional physical data were collected for all sites remaining in the inventory. In particular, the supplemental data included the designation of a U.S. Geological Survey (U.S.G.S.) reference gaging station; a refined estimate of the available net power head; and an estimate of the drainage area associated with each site. Computer routines developed by

---

[2] U.S. Army Corps of Engineers, National Program of Inspection of Dams, in 5 Volumes, Office of the Chief of Engineers, Washington, D.C., May 1975.

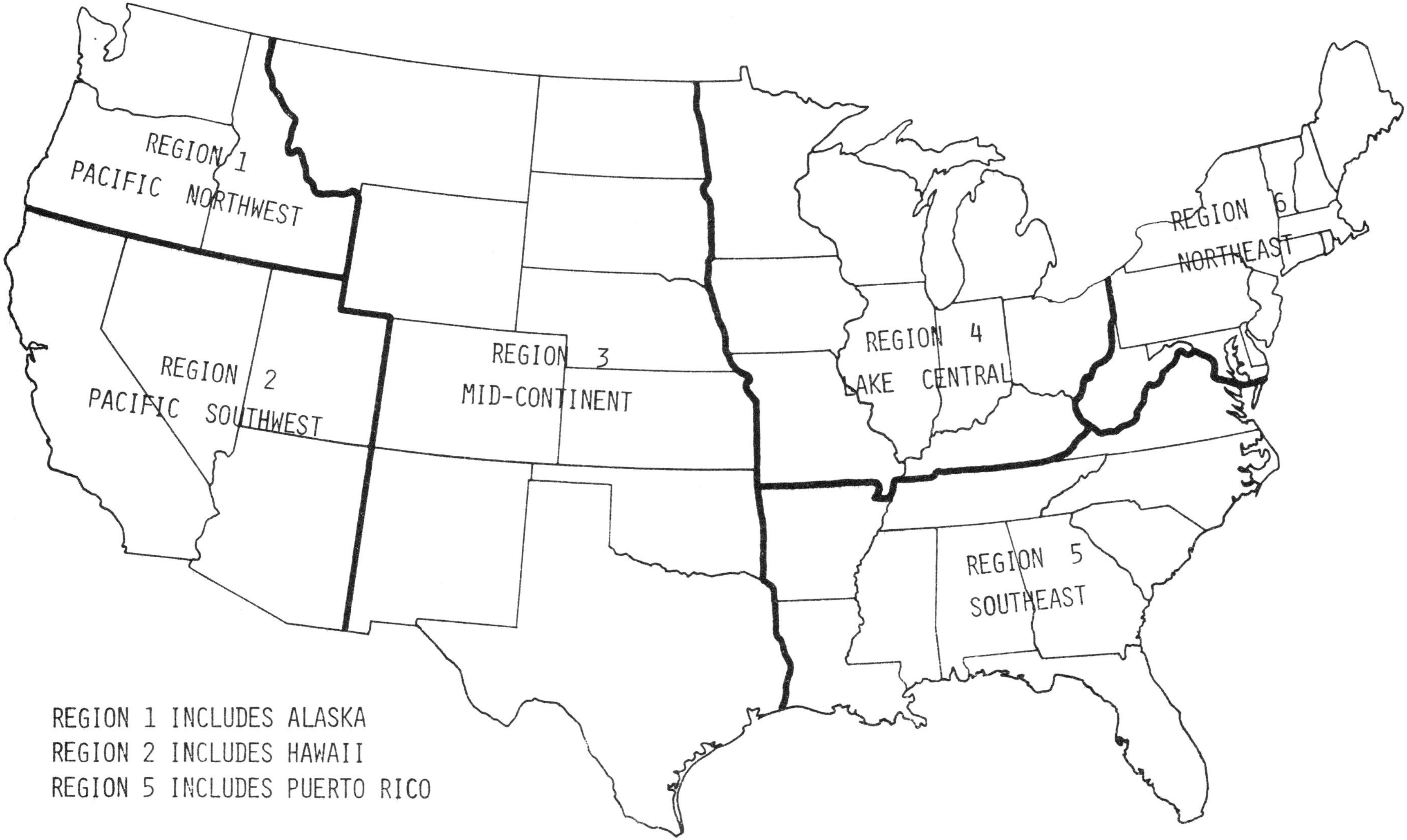

FIGURE 1: REGIONS AS DEFINED FOR THE PRELIMINARY INVENTORY OF HYDROPOWER RESOURCES

the Hydrologic Engineering Center and the Corps' Southwestern Division were utilized with USGS streamflow data and drainage area measurements to produce a synthetic flow-duration curve at each site. Conventional flow-duration analysis was used to estimate the capacity and energy available at each site for a range of plant factors.

Generalized cost estimates were developed by the Corps' North Pacific Division to approximate the cost of turbines, generators, and other powerhouse costs associated with the representative capacity selected for each site in the inventory. Generalized regional power values, developed for the study by the Federal Energy Regulatory Commission (FERC), were used to provide a preliminary estimate of the value of the potential capacity and energy at each site. Each site was then sized at the capacity and energy which gave a maximum net benefit. A second screening, comparing the estimated powerhouse cost with the value of power to be produced, eliminated those sites which had doubtful economic feasibility. This screening process reduced the active inventory to approximately 11,000 sites which are contained in these regional reports.

The basic objective of the preliminary inventory and analysis procedures is to provide a comprehensive assessment of the undeveloped hydroelectric power potential in the United States and to determine which sites merit more thorough investigation. Accordingly, conservative assumptions have been made in the screening and analysis process to avoid eliminating any potentially feasible sites. The current summary tables provide the best estimates to date, but to some degree, may overstate the actual capacity and energy which could be developed. The estimates for individual sites may be overstated for the following reasons:

a. A reduction of net power head due to rising tailwater conditions during high flows was not computed.

b. The analysis technique of maximum net benefits, using incomplete project cost resulted in a low plant factor operation. This type of operation could require more reservoir storage than is available for regulating power flows or could cause fluctuations in the surface elevation of the reservoir or downstream flow that would not be acceptable.

c. Computations ignored diversion of water for other uses, as well as losses due to evaporation.

d. Turbines were assumed to be 100 percent efficient, and head losses through penstocks were not estimated.

e. During periods of high flow, it was calculated that streamflow would pass through the turbines at the design discharge rate when in fact, during excessively high flows, the plant may be shut down because of high tailwater and reduced head.

f. Summary tables include estimates of the potential capacity and energy at each site in the inventory. In some cases, individual projects may be site alternatives to others in the same general location, when only one can be considered for hydropower development.

g. Detailed consideration of the social, economic, institutional and environmental constraints associated with hydropower development were not specifically included in the analysis.

All of the issues listed above will be addressed during future stages of the National Hydropower Study through the addition of more detailed site-specific information, and by refinements in the computer routines used in assessing the data.

RESOURCE ASSESSMENT

Estimates of the existing, incremental and undeveloped conventional hydroelectric power potential for the various regions of the United States are presented in Table 1. The total physical resource for all regions is estimated to exceed 512,000 MW of capacity with an average annual energy generation greater than 1.4 million GWH. At the present time, the Corps has identified 1,251 existing hydropower facilities currently generating power with a total installed capacity of some 64,000 MW producing over 280,000 GWH of average annual energy. There are over 5,400 existing dams which have the potential for new incremental power development. Some of these are currently generating power, and full development of the incremental potential could yield an additional capacity of some 94,000 MW with an average annual energy generation exceeding 223,000 GWH. There are also some 4,500 potentially feasible, undeveloped sites which, if fully developed for hydropower, could produce another 354,000 MW with an estimated average annual energy greater than 935,000 GWH.

The distribution of the overall hydroelectric power resource in the nation is shown in Figure 2. The Pacific Northwest has the largest proportion of the nation's installed capacity and currently generates some 48 percent of the conventional hydroelectric energy produced in the United States. Other areas with a significant, but smaller proportion of the total installed capacity and energy generation include the Southeast, Northeast, and Pacific Southwest regions. Nearly all existing hydroelectric facilities and other water resource projects in the country have the capability for incremental energy generation with the Northeast, Lake Central and Pacific Northwest having a large share of this potential. The undeveloped hydroelectric resource is widely distributed, but appears greatest in the Pacific Northwest, Mid-Continent and Southeast regions, particularly at large-scale sites.

There are over 5,600 small-scale dams in the country which are either generating power, or have the potential for incremental development. The installed capacity at existing small-scale facilities is estimated to be some 3,000 MW with an average annual energy generation exceeding 15,000 GWH. These values represent about 5 percent of the nation's current installed hydroelectric capacity and energy generation. Approximately 5,400 MW of new incremental capacity could be installed at a large percentage of the existing small-scale dams for an estimated energy generation of about 17,000 GWH annually. In addition, some 2,600 potentially feasible, undeveloped sites have been identified which could provide an estimated capacity of 8,000 MW and more than 28,000 GWH of average annual energy generation.

As shown in Figure 3, the amount and regional distribution of the small-scale resource potential varies considerably, as these patterns closely reflect an interaction between climate, landforms and settlement history. The greatest number and density of small-scale facilities with installed capacity are found in the Northeast and Lake Central regions of the country. When considered together, these two regions generate more than 53 percent of the total energy produced from all small-scale facilities in the United States. All regions have the potential for incremental power development at existing sites, especially the Northeast, Lake Central and Mid-Continent regions. Significantly, many of the small dams with incremental potential in these regions are located near smaller population and industrial centers where existing transmission interties are well developed. The undeveloped hydroelectric potential at small-scale sites is widely distributed, but appears greatest in the Pacific Northwest, Lake Central, and the Northeast regions of the country.

The estimates of existing, incremental and the undeveloped hydropower potential for all states in the various regions of the country are presented in Table 2.

SUMMARY

Over 5,400 existing structures have been identified as having the physical potential to add hydropower plants or increase hydropower output thereby increasing our present hydropower capacity from a total of 64,000 MW to 158,000 MW and our energy from 280,000 GWH to 503,000 GWH. While the physical potential for this increase is clearly available, some of these projects will undoubtly not satisfy more detailed economical analysis as well as the institutional and environmental criteria which will be imposed upon them.

More than 4,500 undeveloped sites have been identified as having the physical potential to increase our capacity by 354,000 MW and our energy by 936,000 GWH. Many of these have less chance of acceptance than the modifications to the existing projects because of the more adverse environmental and institutional effects. Unfortunately, 47 percent (166,700 MW) of this undeveloped potential is located in Alaska where it would be economically difficult to transmit the power to the potential user.

For the nation's existing hydroelectric power sites, large-scale facilities, 25 MW and greater, account for approximately 92 percent of the capacity and energy generation, particularly those located in the Pacific Northwest and Southeast regions. Small-scale facilities account for about 5 percent of the nation's installed capacity and hydroelectric energy, but incremental development of other potentially feasible, existing small-scale projects could more than double this output by adding another 5,400 MW of capacity and 17,000 GWH of energy to the total. The distribution of the existing small-scale resource is extremely variable, but nearly all regions of the country have the potential for incremental energy development. The undeveloped potential for all sites and capacity ranges is also widely distributed, and appears greatest in the Pacific Northwest, Southeast and Mid-Continent regions of the country.

As stated earlier, these data are preliminary; the capacity and energy estimates represent the maximum physical hydroelectric potential which could be developed in each state and region. The incremental potential and that estimated for undeveloped sites do not include detailed consideration of the engineering, economic, financial and environmental constraints; nor do they include an assessment of the competitive use of water at existing impoundments, or consideration of the complex social, legal and institutional feasibility, all of which could preclude full development of the hydroelectric potential. Future investigations by the Corps of Engineers and other local, state and federal agencies will consider these factors in more detail, and further refine the actual feasibility of the most favorable sites in the inventory.

Publication of preliminary resource information involves the risk that errors and omissions may exist, and this inventory is no exception. At present, the Corps' inventory of hydroelectric power resources is an active screening tool; its primary function and widest utility is to present a viable list of existing and potentially feasible hydroelectric power sites, and to provide reasonably accurate estimates of the aggregrate state, regional and national development potential.

TABLE 1. PRELIMINARY INVENTORY OF HYDROELECTRIC POWER RESOURCES

REGIONAL SUMMARIES

REGION	EXISTING,[1] POTENTIAL INCREMENTAL[2] AND UNDEVELOPED[3] CAPACITY RANGES												TOTAL			
	Small-Scale (.05-15 MW)				Intermediate (15-25 MW)				Large-Scale (Greater Than 25 MW)				(All Sizes)			
	Exist	Incre	Undev	Total	Exist	Incre	Undev	Total	Exist	Incre	Undev	Total	Exist	Incre	Undev	Total
Vol. 1																
Pacific N. West																
No. of Sites	93	282	745	1,120	13	36	208	257	73	83	896	1,052	179	401	1,849	2,429
Cap. (MW)	430	642	3,702	4,774	234	700	4,069	5,003	26,141	31,919	259,709	317,769	26,804	33,262	267,480	327,546
Ener (GWH)	2,441	2,234	16,390	21,065	1,216	1,943	14,738	17,897	130,365	33,999	673,918	838,282	134,022	38,175	705,045	877,242
Vol. 2																
Pacific S. West																
No. of Sites	111	354	272	737	9	17	26	52	69	43	110	222	189	414	408	1,011
Cap. (MW)	410	574	632	1,616	171	345	509	1,025	9,347	5,109	16,043	30,499	9,928	6,028	17,184	33,140
Ener (GWH)	2,176	1,569	1,640	5,385	837	550	1,059	2,446	37,311	8,729	31,877	77,917	40,325	10,849	34,577	85,751
Vol. 3																
Mid-Continent																
No. of Sites	54	779	666	1,499	11	15	63	89	44	59	234	337	109	853	963	1,925
Cap. (MW)	184	850	1,182	2,216	218	317	1,311	1,846	6,087	6,589	27,376	40,052	6,488	7,758	29,868	44,114
Ener (GWH)	1,372	2,138	3,074	6,584	1,006	524	3,142	4,672	22,403	12,481	64,274	99,158	24,781	15,144	70,491	110,416
Vol. 4																
Lake Central																
No. of Sites	204	601	551	1,356	10	43	16	69	17	88	59	164	231	732	626	1,589
Cap. (MW)	734	914	926	2,574	180	875	319	1,374	1,689	14,038	6,552	22,279	2,602	15,830	7,799	26,231
Ener (GWH)	3,439	3,128	2,859	9,426	940	2,124	763	3,827	5,475	39,514	17,380	62,369	9,854	44,766	21,004	75,624
Vol. 5																
Southeast																
No. of Sites	110	566	265	941	19	29	54	102	98	87	146	331	227	682	465	1,374
Cap. MW)	285	704	1,077	2,066	360	559	1,114	2,033	11,182	11,758	20,969	43,909	11,827	13,021	23,160	48,008
Ener (GWH)	1,000	2,189	3,349	6,538	1,105	1,185	2,863	5,153	36,409	21,466	67,460	125,335	38,514	24,840	73,672	137,026

TABLE 1. PRELIMINARY INVENTORY OF HYDROELECTRIC POWER RESOURCES

REGIONAL SUMMARIES (CONTINUED)

REGION	EXISTING,[1] POTENTIAL INCREMENTAL[2] AND UNDEVELOPED[3] CAPACITY RANGES												TOTAL			
	Small-Scale (.05-15 MW)				Intermediate (15-25 MW)				Large-Scale (Greater Than 25 MW)				(All Sizes)			
	Exist	Incre	Undev	Total	Exist	Incre	Undev	Total	Exist	Incre	Undev	Total	Exist	Incre	Undev	Total
Vol. 6*																
Northeast																
No. of Sites	270	2,231	143	2,644	19	26	20	65	27	85	58	170	316	2,342	221	2,879
Cap. (MW)	914	1,771	491	3,176	354	524	400	1,278	4,784	16,446	7,568	28,798	6,053	18,737	8,457	33,247
Ener (GWH)	4,620	6,009	1,531	12,160	1,613	1,533	938	4,084	26,276	81,898	28,610	136,784	32,508	89,440	31,078	153,026
NATIONAL TOTAL																
No. of Sites	842	4,813	2,642	8,297	81	166	387	634	328	445	1,503	2,276	1,251	5,424	4,532	11,207
Cap. (MW)	2,957	5,455	8,010	16,422	1,517	3,320	7,722	12,559	59,230	85,859	338,217	483,306	63,702	94,636	353,948	512,286
Ener (GWH)	15,048	17,267	28,843	61,158	6,717	7,859	23,503	38,079	258,239	198,087	883,519	1,339,845	280,004	223,214	935,867	1,439,085

[1] Existing hydroelectric power facilities currently generating power.

[2] Existing dams and/or other water resource projects with the potential for new and/or additional hydroelectric capacity.

[3] Undeveloped sites where no dam or other engineering structure presently exists.

*Data on undeveloped sites in the New England states are not available (NA).

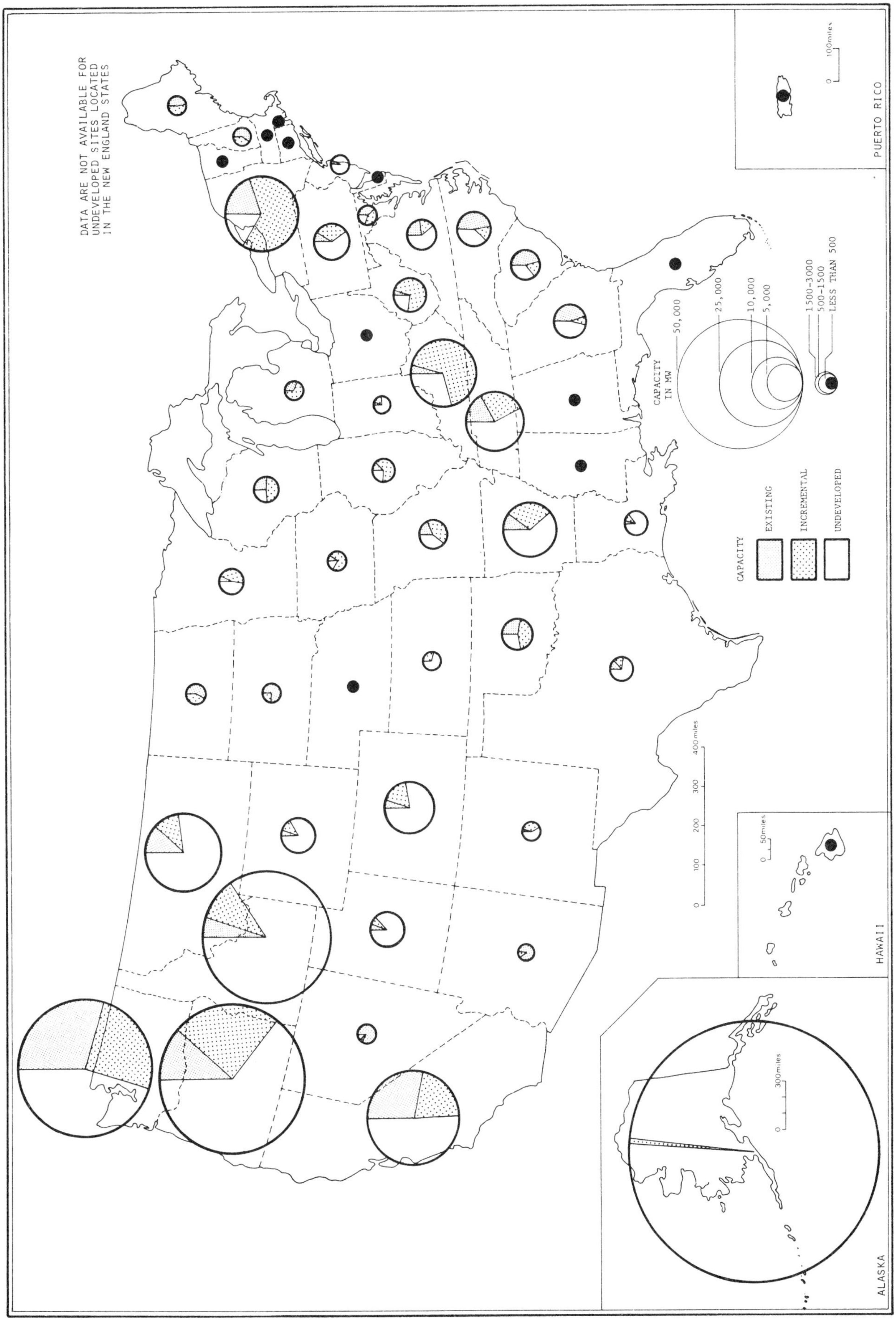

Figure 2: NATIONAL HYDROELECTRIC POWER RESOURCES, (ALL SITES)

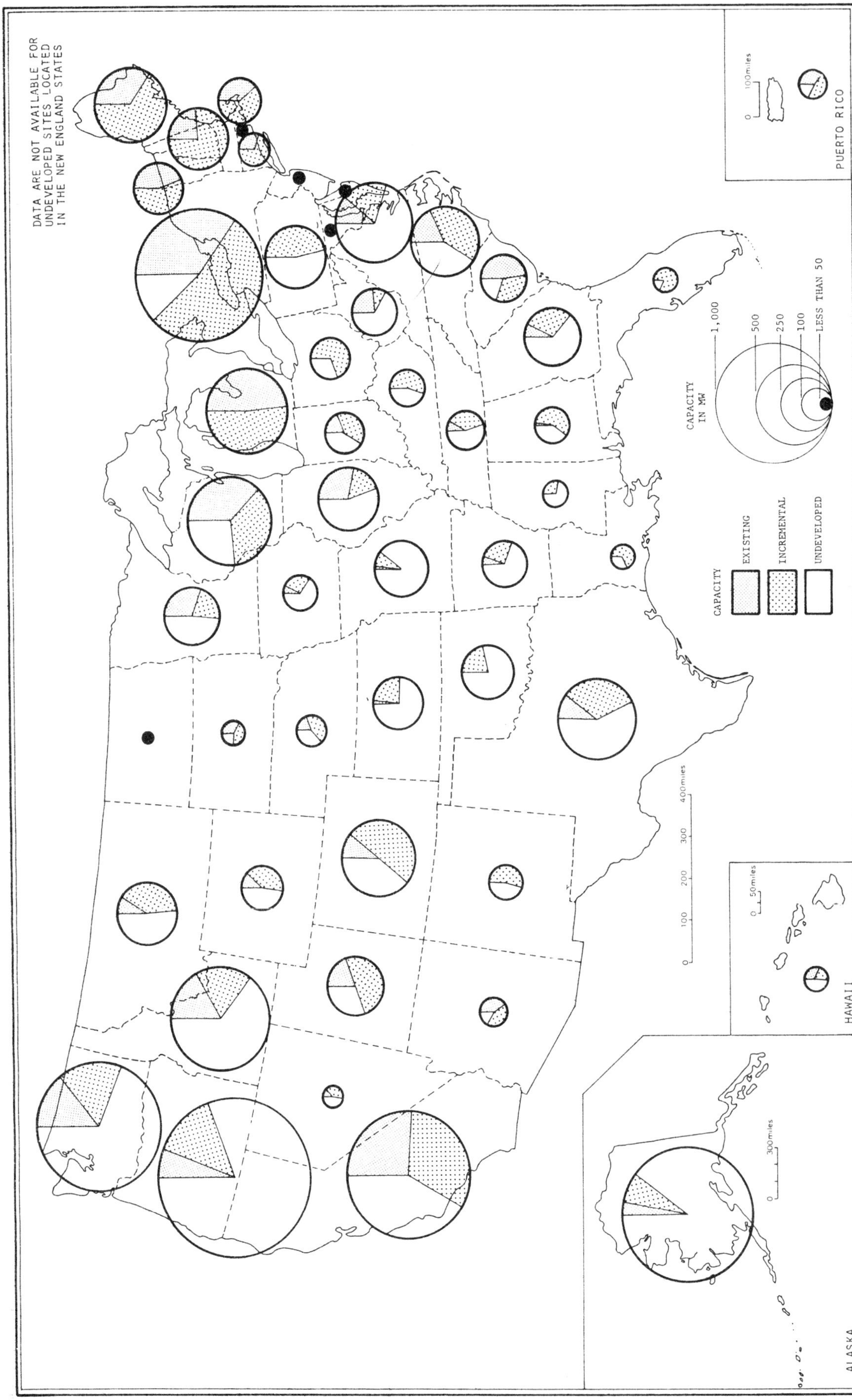

Figure 3: NATIONAL HYDROELECTRIC POWER RESOURCES, (SMALL-SCALE SITES)

TABLE 2. PRELIMINARY INVENTORY OF HYDROELECTRIC POWER RESOURCES
REGIONAL STATE SUMMARIES

PACIFIC NORTHWEST

STATE	EXISTING,[1] POTENTIAL INCREMENTAL[2] AND UNDEVELOPED[3] CAPACITY RANGES												TOTAL			
	Small-Scale (.05-15 MW)				Intermediate (15-25 MW)				Large-Scale (Greater Than 25 MW)				(All Sizes)			
	Exist	Incre	Undev	Total	Exist	Incre	Undev	Total	Exist	Incre	Undev	Total	Exist	Incre	Undev	Total
Alaska																
No. of Sites	16	27	184	227	1	6	53	60	2	5	190	197	19	38	427	484
Cap. (MW)	37	86	1,053	1,176	15	120	1,014	1,149	77	212	164,709	164,998	129	418	166,775	167,322
Ener (GWH)	146	362	4,754	5,262	41	309	4,158	4,508	333	626	432,995	433,954	520	1,297	441,907	443,724
Idaho																
No. of Sites	24	80	68	172	1	5	39	45	15	24	213	252	40	109	320	469
Cap. (MW)	131	140	497	768	16	101	787	904	2,301	4,931	39,252	46,484	2,448	5,172	40,536	48,156
Ener (GWH)	818	435	1,904	3,157	142	195	2,218	2,555	11,130	5,522	82,398	99,050	12,089	6,152	86,520	104,761
Oregon																
No. of Sites	30	96	388	514	9	18	66	93	21	16	253	290	60	130	707	897
Cap. (MW)	105	231	1,390	1,726	157	349	1,291	1,797	6,591	13,609	34,771	54,971	6,853	14,190	37,453	58,496
Ener (GWH)	630	751	6,426	7,807	841	993	4,770	6,604	35,404	8,352	90,039	133,795	36,875	10,095	101,235	148,205
Washington																
No. of Sites	23	79	105	207	2	7	50	59	35	38	240	313	60	124	395	579
Cap. (MW)	157	185	762	1,104	46	130	977	1,153	17,172	13,167	20,977	51,316	17,374	13,482	22,716	53,572
Ener (GWH)	847	686	3,306	4,839	192	446	3,592	4,230	83,498	19,499	68,486	171,483	84,538	20,631	75,383	180,552
Region Total																
No. of Sites	93	282	745	1,120	13	36	208	257	73	83	896	1,052	135	401	1,849	2,429
Cap. (MW)	430	642	3,702	4,774	234	700	4,069	5,003	26,141	31,919	259,709	317,769	26,804	33,262	267,480	327,546
Ener (GWH)	2,441	2,234	16,390	21,065	1,216	1,943	14,738	17,897	130,365	33,999	673,918	838,282	134,022	38,175	705,045	877,242

TABLE 2. PRELIMINARY INVENTORY OF HYDROELECTRIC POWER RESOURCES
REGIONAL STATE SUMMARIES

PACIFIC SOUTHWEST

STATE	EXISTING,[1] POTENTIAL INCREMENTAL[2] AND UNDEVELOPED[3] CAPACITY RANGES												TOTAL			
	Small-Scale (.05-15 MW)				Intermediate (15-25 MW)				Large-Scale (Greater Than 25 MW)				(All Sizes)			
	Exist	Incre	Undev	Total	Exist	Incre	Undev	Total	Exist	Incre	Undev	Total	Exist	Incre	Undev	Total
Arizona																
No. of Sites	4	27	37	68	0	0	0	0	5	3	0	8	9	30	37	76
Cap. (MW)	32	34	13	79	0	0	0	0	1,374	122	0	1,496	1,406	156	13	1,575
Ener (GWH)	105	134	19	258	0	0	0	0	5,959	261	0	6,220	6,064	395	19	6,478
California																
No. of Sites	50	216	185	451	9	12	20	41	61	38	90	189	120	266	295	681
Cap. (MW)	298	365	474	1,137	171	242	387	800	7,167	4,840	12,192	24,199	7,636	5,447	13,053	26,136
Ener (GWH)	1,647	990	1,227	3,864	837	342	789	1,968	28,621	8,421	22,993	60,035	31,106	9,753	25,009	65,868
Hawaii																
No. of Sites	14	11	7	32	0	1	0	1	0	0	0	0	14	12	7	33
Cap. (MW)	19	12	30	61	0	19	0	19	0	0	0	0	19	31	30	80
Ener (GWH)	102	26	77	205	0	39	0	39	0	0	0	0	102	65	77	244
Nevada																
No. of Sites	5	21	19	45	0	1	2	3	1	0	0	1	6	22	21	49
Cap. (MW)	9	28	34	71	0	18	40	58	668	0	0	668	677	46	74	797
Ener (GWH)	68	55	97	220	0	26	116	142	2,056	0	0	2,056	2,124	82	213	2,419
Utah																
No. of Sites	38	79	24	141	0	3	4	7	2	2	20	24	40	84	48	172
Cap. (MW)	52	135	81	268	0	66	82	148	138	147	3,851	4,136	190	348	4,014	4,552
Ener (GWH)	254	364	220	838	0	143	154	297	675	47	8,884	9,606	929	554	9,259	10,742
Region Total																
No. of Sites	111	354	272	737	9	17	26	52	69	43	110	222	189	414	408	1,011
Cap. (MW)	410	574	632	1,616	171	345	509	1,025	9,347	5,109	16,043	30,499	9,928	6,028	17,184	33,140
Ener (GWH)	2,176	1,569	1,640	5,385	837	550	1,059	2,446	37,311	8,729	31,877	77,917	40,325	10,849	34,577	85,751

TABLE 2. PRELIMINARY INVENTORY OF HYDROELECTRIC POWER RESOURCES
REGIONAL STATE SUMMARIES
MID-CONTINENT

STATE	EXISTING,[1] POTENTIAL INCREMENTAL[2] AND UNDEVELOPED[3] CAPACITY RANGES												TOTAL			
	Small-Scale (.05-15 MW)				Intermediate (15-25 MW)				Large-Scale (Greater Than 25 MW)				(All Sizes)			
	Exist	Incre	Undev	Total	Exist	Incre	Undev	Total	Exist	Incre	Undev	Total	Exist	Incre	Undev	Total
Colorado																
No. of Sites	10	167	53	230	1	2	19	22	5	4	79	88	16	173	151	340
Cap. (MW)	49	229	177	455	22	39	419	480	330	1,325	6,477	8,132	401	1,593	7,072	9,066
Ener (GWH)	275	660	423	1,358	70	79	889	1,038	1,264	2,644	13,515	17,423	1,609	3,383	14,827	19,819
Kansas																
No. of Sites	1	64	184	249	0	1	0	1	0	3	6	9	1	68	190	259
Cap. (MW)	2	61	183	246	0	18	0	18	0	141	296	437	2	220	480	702
Ener (GWH)	10	117	382	509	0	38	0	38	0	229	508	737	10	384	890	1,284
Montana																
No. of Sites	7	69	43	119	1	2	10	13	12	17	81	110	20	88	134	242
Cap. (MW)	29	140	176	345	17	43	189	249	2,372	2,148	14,948	19,468	2,418	2,332	15,313	20,063
Ener (GWH)	642	350	500	1,492	111	83	528	722	8,969	4,761	38,321	52,051	9,722	5,195	39,348	54,265
Nebraska																
No. of Sites	11	39	19	69	3	1	4	8	2	1	0	3	16	41	23	80
Cap. (MW)	16	37	30	83	54	21	82	157	66	37	0	103	136	94	112	342
Ener (GWH)	50	121	139	310	300	43	320	663	216	160	0	376	566	323	459	1,348
New Mexico																
No. of Sites	0	26	44	70	1	1	0	2	0	4	3	7	1	31	47	79
Cap. (MW)	0	55	46	101	24	24	0	48	0	207	359	566	24	286	404	714
Ener (GWH)	0	144	120	264	96	49	0	145	0	469	1,101	1,570	96	662	1,221	1,979
N. Dakota																
No. of Sites	0	44	2	46	0	0	0	0	1	1	0	2	1	45	2	48
Cap. (MW)	0	21	10	31	0	0	0	0	430	303	0	733	430	324	10	764
Ener (GWH)	0	45	18	63	0	0	0	0	2,400	568	0	2,968	2,400	612	18	3,030

TABLE 2. PRELIMINARY INVENTORY OF HYDROELECTRIC POWER RESOURCES
REGIONAL STATE SUMMARIES
MID-CONTINENT (CONTINUED)

STATE	EXISTING,[1] POTENTIAL INCREMENTAL[2] AND UNDEVELOPED[3] CAPACITY RANGES												TOTAL			
	Small-Scale (.05-15 MW)				Intermediate (15-25 MW)				Large-Scale (Greater Than 25 MW)				(All Sizes)			
	Exist	Incre	Undev	Total	Exist	Incre	Undev	Total	Exist	Incre	Undev	Total	Exist	Incre	Undev	Total
Oklahoma																
No. of Sites	0	98	170	268	0	4	2	6	11	13	12	36	11	115	184	310
Cap. (MW)	0	49	178	227	0	87	44	131	1,029	1,494	797	3,320	1,029	1,630	1,019	3,678
Ener (GWH)	0	86	346	432	0	133	77	210	2,350	1,991	1,270	5,611	2,350	2,210	1,693	6,253
S. Dakota																
No. of Sites	8	23	4	35	0	0	0	0	4	3	1	8	12	26	5	43
Cap. (MW)	17	22	12	51	0	0	0	0	1,483	397	25	1,905	1,500	420	37	1,957
Ener (GWH)	69	65	33	167	0	0	0	0	6,056	832	38	6,926	6,125	898	72	7,095
Texas																
No. of Sites	9	196	129	334	2	1	8	11	5	4	22	31	16	201	159	376
Cap. (MW)	52	165	288	505	45	22	167	234	225	185	1,420	1,830	321	372	1,875	2,568
Ener (GWH)	212	372	854	1,438	149	7	457	613	542	240	3,149	3,931	903	619	4,461	5,983
Wyoming																
No. of Sites	8	53	18	79	3	3	20	26	4	9	30	43	15	65	68	148
Cap. (MW)	19	71	82	172	56	63	410	529	152	352	3,054	3,558	227	487	3,546	4,260
Ener (GWH)	114	178	259	551	280	92	871	1,243	606	587	6,372	7,565	1,000	858	7,502	9,360
Region Total																
No. of Sites	54	779	666	1,499	11	15	63	89	44	59	234	337	109	853	963	1,925
Cap. (MW)	184	850	1,182	2,216	218	317	1,311	1,846	6,087	6,589	27,376	40,052	6,488	7,758	29,868	44,114
Ener (GWH)	1,372	2,138	3,074	6,584	1,006	524	3,142	4,672	22,403	12,481	64,274	99,158	24,781	15,144	70,491	110,416

TABLE 2. PRELIMINARY INVENTORY OF HYDROELECTRIC POWER RESOURCES
REGIONAL STATE SUMMARIES
LAKE CENTRAL

STATE	EXISTING,[1] POTENTIAL INCREMENTAL[2] AND UNDEVELOPED[3] CAPACITY RANGES												TOTAL			
	Small-Scale (.05-15 MW)				Intermediate (15-25 MW)				Large-Scale (Greater Than 25 MW)				(All Sizes)			
	Exist	Incre	Undev	Total	Exist	Incre	Undev	Total	Exist	Incre	Undev	Total	Exist	Incre	Undev	Total
Illinois																
No. of Sites	16	39	230	285	0	8	0	8	1	7	2	10	17	54	232	303
Cap. (MW)	100	52	169	321	0	145	0	145	32	533	89	654	132	730	259	1121
Ener (GWH)	569	109	411	1,089	0	347	0	347	15	1,750	178	1943	584	2,206	589	3,379
Indiana																
No. of Sites	4	30	45	79	0	2	0	2	0	0	3	3	4	32	48	84
Cap. (MW)	28	58	61	147	0	37	0	37	0	0	383	383	28	96	444	568
Ener (GWH)	98	189	162	449	0	90	0	90	0	0	816	816	98	279	978	1,355
Iowa																
No. of Sites	3	25	37	65	0	1	0	1	1	12	3	16	4	38	40	82
Cap. (MW)	7	28	67	102	0	21	0	21	128	1,068	190	1,386	135	1,117	257	1,509
Ener (GWH)	36	81	200	317	0	39	0	39	805	3,468	408	4,681	841	3,588	608	5,037
Kentucky																
No. of Sites	0	52	23	75	0	2	0	2	4	30	10	44	4	84	33	121
Cap. (MW)	0	64	51	115	0	48	0	48	636	9,159	3,985	13,780	636	9,271	4,036	13,943
Ener (GWH)	0	183	121	304	0	88	0	88	2,259	24,547	11,697	38,503	2,259	24,818	11,819	38,896
Michigan																
No. of Sites	86	136	0	222	3	6	0	9	3	4	0	7	92	146	0	238
Cap. (MW)	283	303	0	586	52	121	0	173	151	709	0	860	486	1,133	0	1,619
Ener (GWH)	1,145	1,238	0	2,383	312	399	0	711	438	2,735	0	3,173	1,895	4,371	0	6,266
Minnesota																
No. of Sites	18	97	45	160	0	5	6	11	1	12	17	30	19	114	68	201
Cap. (MW)	91	63	146	300	0	100	125	225	67	825	755	1,647	158	989	1,027	2,174
Ener (GWH)	536	191	492	1,219	0	288	314	602	318	1,868	1,602	3,788	854	2,346	2,408	5,608

TABLE 2. PRELIMINARY INVENTORY OF HYDROELECTRIC POWER RESOURCES
REGIONAL STATE SUMMARIES
LAKE CENTRAL (Continued)

STATE	EXISTING,[1] POTENTIAL INCREMENTAL[2] AND UNDEVELOPED[3] CAPACITY RANGES												TOTAL			
	Small-Scale (.05-15 MW)				Intermediate (15-25 MW)				Large-Scale (Greater Than 25 MW)				(All Sizes)			
	Exist	Incre	Undev	Total	Exist	Incre	Undev	Total	Exist	Incre	Undev	Total	Exist	Incre	Undev	Total
Missouri																
No. of Sites	2	31	93	126	1	2	8	11	4	9	17	30	7	42	118	167
Cap. (MW)	5	22	227	254	16	45	154	215	577	1,301	868	2,746	598	1,368	1,249	3,215
Ener (GWH)	17	61	643	721	94	88	357	539	1,272	4,154	1,739	7,165	1,383	4,303	2,740	8,426
Ohio																
No. of Sites	0	68	18	86	0	7	0	7	0	2	1	3	0	77	19	96
Cap. (MW)	0	105	47	152	0	153	0	153	0	56	43	99	0	314	90	404
Ener (GWH)	0	308	131	439	0	323	0	323	0	134	70	204	0	768	201	969
Wisconsin																
No. of Sites	75	123	60	258	6	10	2	18	3	12	6	21	84	145	68	297
Cap. (MW)	220	219	158	597	112	205	40	357	98	387	239	724	429	812	437	1,678
Ener (GWH)	1,038	768	699	2,505	534	462	92	1,088	368	858	870	2,096	1,940	2,087	1,661	5,688
Region Total																
No. of Sites	204	601	551	1,356	10	43	16	69	17	88	59	164	231	732	626	1,589
Cap. (MW)	734	914	926	2,574	180	875	319	1,374	1,689	14,038	6,552	22,279	2,602	15,830	7,799	26,231
Ener (GWH)	3,439	3,128	2,859	9,426	940	2,124	763	3,827	5,475	39,514	17,380	62,369	9,854	44,766	21,004	75,624

TABLE 2. PRELIMINARY INVENTORY OF HYDROELECTRIC POWER RESOURCES
REGIONAL STATE SUMMARIES
SOUTHEAST

STATE	EXISTING,[1] POTENTIAL INCREMENTAL[2] AND UNDEVELOPED[3] CAPACITY RANGES												TOTAL			
	Small-Scale (.05-15 MW)				Intermediate (15-25 MW)				Large-Scale (Greater Than 25 MW)				(All Sizes)			
	Exist	Incre	Undev	Total	Exist	Incre	Undev	Total	Exist	Incre	Undev	Total	Exist	Incre	Undev	Total
Alabama																
No. of Sites	1	52	8	61	0	2	5	7	15	19	8	42	16	73	21	110
Cap. (MW)	2	70	49	121	0	41	108	149	2,269	4,010	424	6,703	2,271	4,121	581	6,973
Ener (GWH)	6	190	137	333	0	91	244	335	9,710	7,141	995	17,846	9,716	7,422	1,376	18,514
Arkansas																
No. of Sites	1	89	50	140	0	3	11	14	10	13	17	40	11	105	78	194
Cap. (MW)	11	51	143	205	0	67	218	285	1,069	2,768	5,874	9,711	1,080	2,886	6,235	10,201
Ener (GWH)	43	145	412	600	0	105	393	498	2,756	5,239	19,824	27,819	2,799	5,489	20,629	28,917
Florida																
No. of Sites	1	17	2	20	0	0	1	1	1	0	0	1	2	17	3	22
Cap. (MW)	0	45	10	55	0	0	20	20	30	0	0	30	30	45	30	105
Ener (GWH)	0	151	30	181	0	0	66	66	232	0	0	232	232	151	96	479
Georgia																
No. of Sites	5	61	31	97	6	1	9	16	15	6	33	54	26	68	73	167
Cap. (MW)	20	79	182	281	106	23	188	317	1,924	304	1,690	3,918	2,050	406	2,060	4,516
Ener (GWH)	87	316	538	941	311	52	518	881	3,825	501	4,892	9,218	4,223	869	5,948	11,040
Louisiana																
No. of Sites	0	19	5	24	0	0	0	0	1	4	6	11	1	23	11	35
Cap. (MW)	0	38	17	55	0	0	0	0	81	253	2,336	2,670	81	291	2,353	2,725
Ener (GWH)	0	110	55	165	0	0	0	0	215	618	7,141	7,974	215	728	7,196	8,139
Mississippi																
No. of Sites	0	50	38	88	0	1	1	2	0	2	1	3	0	53	40	93
Cap. (MW)	0	20	51	71	0	16	23	39	0	97	45	142	0	133	119	252
Ener (GWH)	0	71	137	208	0	65	54	119	0	192	87	279	0	328	278	606

TABLE 2. PRELIMINARY INVENTORY OF HYDROELECTRIC POWER RESOURCES
REGIONAL STATE SUMMARIES
SOUTHEAST (Continued)

STATE	EXISTING,[1] POTENTIAL INCREMENTAL[2] AND UNDEVELOPED[3] CAPACITY RANGES												TOTAL			
	Small-Scale (.05-15 MW)				Intermediate (15-25 MW)				Large-Scale (Greater Than 25 MW)				(All Sizes)			
	Exist	Incre	Undev	Total	Exist	Incre	Undev	Total	Exist	Incre	Undev	Total	Exist	Incre	Undev	Total
North Carolina																
No. of Sites	53	117	28	198	5	5	12	22	18	9	22	49	76	131	62	269
Cap. (MW)	72	162	160	394	103	86	259	448	1,762	405	1,134	3,301	1,937	653	1,553	4,143
Ener (GWH)	248	429	546	1,223	396	244	744	1,384	5,958	760	3,387	10,105	6,602	1,433	4,677	12,712
Puerto Rico																
No. of Sites	5	10	6	21	2	3	0	5	0	0	0	0	7	13	6	26
Cap. (MW)	28	37	13	78	36	55	0	91	0	0	0	0	64	92	13	169
Ener (GWH)	64	48	63	175	54	78	0	132	0	0	0	0	118	126	63	307
South Carolina																
No. of Sites	29	49	5	83	4	3	4	11	10	13	13	36	43	65	22	130
Cap. (MW)	88	61	34	183	76	54	80	210	1,368	513	1,061	2,942	1,532	628	1,175	3,335
Ener (GWH)	390	354	130	874	233	145	280	658	2,117	1,201	3,093	6,411	2,740	1,700	3,503	7,943
Tennessee																
No. of Sites	1	31	9	41	2	4	2	8	24	14	23	61	27	49	34	110
Cap. (MW)	11	47	70	128	39	80	45	164	2,046	3,142	7,149	12,337	2,096	3,269	7,264	12,629
Ener (GWH)	33	57	207	297	111	56	145	312	11,064	5,113	25,004	41,181	11,208	5,226	25,356	41,790
Virginia																
No. of Sites	14	71	83	168	0	7	9	16	4	7	23	34	18	85	115	218
Cap. (MW)	53	94	348	495	0	137	173	310	633	266	1,256	2,155	686	497	1,777	2,960
Emer (GWH)	129	318	1,094	1,541	0	349	419	768	532	701	3,037	4,270	661	1,368	4,550	6579
Region Total																
No. of Sites	110	566	265	941	19	29	54	102	98	87	146	331	227	682	465	1,374
Cap. (MW)	285	704	1,077	2,066	360	559	1,114	2,033	11,182	11,758	20,969	43,909	11,827	13,021	23,160	48,008
Emer (GWH)	1,000	2,189	3,349	6,538	1,105	1,185	2,863	5,153	36,409	21,466	67,460	125,335	38,514	24,840	73,672	137,026

TABLE 2. PRELIMINARY INVENTORY OF HYDROELECTRIC POWER RESOURCES
REGIONAL STATE SUMMARIES
NORTHEAST

STATE	EXISTING,[1] POTENTIAL INCREMENTAL[2] AND UNDEVELOPED[3] CAPACITY RANGES												TOTAL			
	Small-Scale (.05-15 MW)				Intermediate (15-25 MW)				Large-Scale (Greater Than 25 MW)				(All Sizes)			
	Exist	Incre	Undev	Total	Exist	Incre	Undev	Total	Exist	Incre	Undev	Total	Exist	Incre	Undev	Total
Connecticut*																
No. of Sites	13	205	NA	218	0	0	NA	0	2	0	NA	2	15	205	NA	220
Cap. (MW)	36	88	NA	124	0	0	NA	0	68	0	NA	68	103	88	NA	191
Ener (GWH)	156	308	NA	464	0	0	NA	0	216	0	NA	216	372	308	NA	680
Delaware																
No. of Sites	0	0	2	2	0	0	0	0	0	0	0	0	0	0	2	2
Cap. (MW)	0	0	2	2	0	0	0	0	0	0	0	0	0	0	2	2
Ener (GWH)	0	0	6	6	0	0	0	0	0	0	0	0	0	0	6	6
Maine*																
No. of Sites	33	469	NA	502	3	1	NA	4	2	2	NA	4	38	472	NA	510
Cap. (MW)	147	284	NA	431	58	20	NA	78	148	64	NA	212	354	369	NA	723
Ener (GWH)	881	992	NA	1,873	388	67	NA	455	507	226	NA	733	1,776	1,285	NA	3,061
Maryland																
No. of Sites	2	15	7	24	0	1	0	1	1	4	2	7	3	20	9	32
Cap. (MW)	2	18	20	40	0	19	0	19	474	496	232	1,202	476	532	252	1,260
Ener (GWH)	14	50	58	122	0	41	0	41	1,719	650	550	2,919	1,733	741	608	3,082
Massachusetts*																
No. of Sites	23	301	NA	324	2	0	NA	2	4	0	NA	4	29	301	NA	330
Cap. (MW)	73	115	NA	188	33	0	NA	33	131	0	NA	131	237	115	NA	352
Ener (GWH)	313	403	NA	716	176	0	NA	176	154	0	NA	154	643	403	NA	1,045
New Hampshire*																
No. of Sites	24	541	NA	565	2	1	NA	3	2	0	NA	2	28	542	NA	570
Cap. (MW)	74	238	NA	312	31	23	NA	54	281	0	NA	281	386	261	NA	647
Ener (GWH)	359	836	NA	1,195	180	82	NA	262	558	0	NA	558	1,097	918	NA	2,015
New Jersey																
No. of Sites	2	36	0	38	0	1	0	1	0	0	5	5	2	37	5	44
Cap. (MW)	6	21	0	27	0	23	0	23	0	0	647	647	6	40	647	693
Ener (GWH)	18	58	0	76	0	56	0	56	0	0	1,821	1,821	18	114	1,821	1,953

TABLE 2. PRELIMINARY INVENTORY OF HYDROELECTRIC POWER RESOURCES
REGIONAL STATE SUMMARIES
NORTHEAST (CONTINUED)

STATE	EXISTING,[1] POTENTIAL INCREMENTAL[2] AND UNDEVELOPED[3] CAPACITY RANGES												TOTAL			
	Small-Scale (.05-15 MW)				Intermediate (15-25 MW)				Large-Scale (Greater Than 25 MW)				(All Sizes)			
	Exist	Incre	Undev	Total	Exist	Incre	Undev	Total	Exist	Incre	Undev	Total	Exist	Incre	Undev	Total
New York																
No. of Sites	123	251	43	417	11	15	11	37	9	40	11	60	143	306	65	514
Cap. (MW)	422	657	148	1,227	216	309	226	751	3,103	11,491	2,754	17,348	3,741	12,458	3,127	19,326
Ener (GWH)	2,155	2,250	539	4,944	799	976	563	2,338	20,581	70,227	17,211	108,019	23,535	73,453	18,313	115,301
Pennsylvania																
No. of Sites	0	138	58	196	0	6	4	10	4	19	26	49	4	163	88	255
Cap. (MW)	0	158	189	347	0	107	79	186	403	1,466	2,977	4,846	403	1,731	3,245	5,379
Ener (GWH)	0	452	567	1,019	0	252	170	422	1,681	3,618	6,969	12,268	1,681	4,322	7,706	13,709
Rhode Island*																
No. of Sites	2	105	NA	107	0	0	NA	0	0	0	NA	0	2	105	NA	107
Cap. (MW)	2	40	NA	42	0	0	NA	0	0	0	NA	0	2	40	NA	42
Ener (GWH)	6	139	NA	145	0	0	NA	0	0	0	NA	0	6	139	NA	145
Vermont*																
No. of Sites	44	155	NA	199	1	0	NA	1	2	0	NA	2	47	155	NA	202
Cap. (MW)	106	134	NA	240	16	0	NA	16	74	0	NA	74	197	134	NA	331
Ener GWH)	436	472	NA	908	70	0	NA	70	317	0	NA	317	822	472	NA	1,294
W. Virginia																
No. of Sites	4	15	33	52	0	1	5	6	1	20	14	35	5	36	52	93
Cap. (MW)	46	18	132	196	0	23	95	118	102	2,929	958	3,989	148	2,969	1,184	4,301
Ener (GWH)	282	49	361	692	0	59	205	264	543	7,177	2,059	9,779	825	7,285	2,624	10,734
Region Total																
No. of Sites	270	2,231	143	2,644	19	26	20	65	27	85	58	170	316	2,342	221	2,879
Cap. (MW)	914	1,771	491	3,176	354	524	400	1,278	4,784	16,446	7,568	28,798	6,053	18,737	8,457	33,250
Ener (GWH)	4,620	6,009	1,531	12,160	1,613	1,533	938	4,084	26,276	81,898	28,610	136,784	32,508	89,440	31,078	153,025

[1]Existing hydroelectric power facilities currently generating power.

[2]Existing dams and/or other water resource projects with the potential for new and/or additional hydroelectric capacity.

[3]Undeveloped sites where no dam or other engineering structure presently exists.

*Data on undeveloped sites in the New England states are not available (NA).

Part 2
Feasibility of Small Scale Hydroelectric Power Generation

The information in this part is from *Feasibility Studies for Small Scale Hydropower Additions—A Guide Manual*, prepared by the U.S. Army Corps of Engineers' Hydrologic Engineering Center and Institute for Water Resources with funding from the U.S. Department of Energy, July 1979.

ACKNOWLEDGEMENTS

The manual preparation was the responsibility of the Hydrologic Engineering Center, Bill S. Eichert, Director. The U.S. Army Corps of Engineers Institute for Water Resources sponsored the manual preparation as a complementary task to the management of the National Hydropower Plan activities for which they are responsible. The Department of Energy provided funding support under their small scale hydro commericalization program. The preparation of the manual was a joint effort by staff of the Hydrologic Engineering Center and several private contractors. Mr. Darryl W. Davis of the Hydrologic Engineering Center was the principal-in-charge. The Technical Guide, Volume I, was written by Mr. Davis aided by Mr. Brian W. Smith of his staff. The Hydrologic Studies, Volume III, was written by Mr. Dale R. Burnett of the Hydrologic Engineering Center aided by his staff. The remaining volumes were prepared under contract to the Hydrologic Engineering Center. The Economic and Financial Analysis, Volume II, was prepared by Development and Resources Corporation, Sacramento, CA. Mr. David C. Auslam, Jr. was the project manager, Mr. Mark Henwood was the principal author, and Mr. James Gibbs and Mr. Norman Sturn served as consultants. Also prepared by Development and Resources Corporation was the Great Falls Hydroelectric Project Case Study, appended to Volume I, with major technical contributions by Mr. Clarence Korhonen. The Existing Facility Integrity, Volume IV, was prepared by W. A. Wahler & Associates, Palo Alto, CA. Mr. Forrest W. Gifford was the project manager and Mr. Clifford S. Cortright served as a consultant. The Electromechanical Features, Volume V, and the Civil Features, Volume VI, were prepared by Tudor Engineering Company, San Francisco, CA. Mr. David C. Willer was the project manager for both volumes, and Mr. Donald J. Guild and Mr. Horace E. Burrier were the principal investigators for Volume V and Volume VI, respectively. Also prepared by Tudor Engineering Company was the Rollins Power Project Case Study, appended to Volume I.

VOLUME I: TECHNICAL GUIDE

SECTION 1: INTRODUCTION AND OVERVIEW

Scope and Purpose of Manual

The recent focus on our national energy resources has generated significant renewed attention in hydroelectric power development. In particular, recent investigations (Federal Power Commission, 1976; Trisco, 1975) that analyze the undeveloped hydroelectric potential at existing reservoir sites indicate that detailed studies of many sites are warranted. An attractive feature of these sites is that many of the difficulties in developing new power sites have already been dealt with (e.g., an impoundment exists). Another finding (McDonald, 1977) was that the need exists for updating and refining analysis data and methods, especially for small power additions of 15,000 kilowatts or less. This manual, referred to hereafter as the "guide manual" or simply "manual" has been prepared to meet this need.

The guide manual is designed for use by public agencies (federal, state, local, and special districts), public and private utilities, private investors, and research and educational institutions. It is a procedural guide that includes technical data and methods suitable for the systematic appraisal of potential small hydropower additions to existing facilities. It focuses upon the concepts, technology, and economic and financial issues unique to small hydropower additions.

The manual is comprised of six volumes: Volume I, "Technical Guide," overviews the investigation process, provides implementation guidance, and documents case study applications; Volume II, "Economic and Financial Analysis," includes criteria and procedures for marketing and valuing power output, determining economic feasibility, and analyzing financial requirements and issues critical to implementation; Volume III, "Hydrologic Studies," describes investigations necessary to evaluate the hydrologic integrity of the dam and to estimate the power output of plant additions; Volume IV, "Existing Facility Integrity," provides guidance for assessing the ability of a site to safely accommodate a power addition; Volume V, "Electromechanical Features," describes selection criteria and performance characteristics of small hydro generation and ancillary equipment; and Volume VI, "Civil Features," provides preliminary design and cost guidelines for the civil features of power additions.

A glossary of hydropower terms is included as an appendix. The terms and definitions were derived from hydropower industry sources, and textbooks. Where conflicts and uncertainty in definitions were found, the prevailing common usage was adopted.

Overview - Guide Manual Volumes

Volume I - Technical Guide. This volume defines small hydropower and discusses the issues and technology associated with power additions to existing impoundments. The volume provides an overview of the manual, presents the purpose, concept, and configuration of the manual, and describes the components of a feasibility study and their interrelationships. Feasibility investigations are characterized as a continuum that begins with generalized resource assessments, such as the many resource assessments underway across the U.S., and concludes when construction is initiated. Decision points exist at several critical stages. The guide manual provides guidance for the reconnaissance stage (should a feasibility study be initiated?) and the feasibility stage (should an implementation commitment be made?) decision points. It is recognized that subsequent events could alter the implementation decisions; such as undiscovered site problems of integrity, foundation competence; financing difficulties (problems in bond marketing for instance); or unfavorable bid openings. The manual is quite comprehensive and following its guidance should significantly minimize the likelihood of unforeseen problems in late implementation stages.

The volume includes a description of major task elements needed to perform the reconnaissance and feasibility studies. Emphasis is placed on the facts that the planning studies need to be performed in considerable detail, site specific conditions are important, and investigation costs must be kept to a minimum. The contents of the other five volumes are described and their use conceptually integrated into the analysis process.

Included as exhibits to the "Technical Guide" are two case studies of existing projects, one from the far west area that is nearing construction completion and start up and one from the northeast that is in the licensing stage. The case studies reformulate the two projects following the data and guidance in the manual and serve both as a test of the manual and illustrated examples of manual use.

Volume II - Economic and Financial Analysis. This volume provides a documented procedure for performing the economic and financial studies necessary for a feasibility determination. The three major subjects covered in the volume are the market analysis, the economic feasibility determination, and the financial feasibility determination. The perspectives appropriate for public and private utilities and private investors are considered.

The market analysis section discusses the factors governing marketability of capacity and energy as related to the unique nature of small hydropower, the procedure used to determine the energy and capacity values for small hydropower, and the marketing arrangements applicable to small hydropower. The market analysis takes the stance of an owner/project sponsor performing the analysis so that the benefits and costs of the sale of power from a small hydro project can be evaluated.

The economic feasibility section clearly distinguishes economic feasibility from financial analysis. Economic feasibility is defined as positive when project benefits exceed project costs. Included is a discussion as to the appropriate perspective of the project evaluations that are performed, procedures and guidelines for arranging cost and benefit data, suggested presentation format, and a description of economic feasibility determination. Concepts for including cost escalation (e.g., fuel costs) in the analysis are discussed.

Financial feasibility is defined as positive when it can be demonstrated that the project can secure the needed financing for implementation and that the revenue receipt pattern will provide debt service at a reasonable rate of return for loans that may be incurred. Included are procedures and guidelines for revenue and cash flow analysis, opportunities for innovation in financial and revenue arrangements with utilities and other energy institutions, alternative construction financing possibilities and financial implication of those useful in small hydropower development, and a description of financial feasibility determination. The important role of a financial advisor in project studies is presented.

Volume III - Hydrologic Studies. This volume describes the studies needed to determine the integrity of the existing structure during the passage of major flood events and to determine the capacity and energy potential at the site. The topics of spillway adequacy, basic streamflow development methods, and capacity and energy calculations are discussed in major sections. The spillway is the safety valve of a dam and is the primary facility protecting it from failing by overtopping due to flooding. The current criteria for spillway performance as a function of reservoir capacity, dam height, and vulnerability of downstream areas that has emerged from dam safety studies by the Corps of Engineers are described. The hydraulic characteristics of spillways and outlet works are described and technical references for analysis procedures included. Flow-exceedance frequency and hydrograph analysis techniques to enable calculation of the range of events needed for spillway evaluation are presented.

The degree to which streamflow records are short, contain gaps, are poorly recorded, or to which changes in operating policy have occurred or are possible in the future, determines the complexity of the task and effort needed to assemble a representative record. Reconstruction of a long period of record by simulation of the hydrologic process and operation of the project is the most accurate and time consuming analysis technique, and adaptation of processed synthetic data from generalized studies such as flow duration curves requiring minimal effort, can be used but could be of poor quality. The appropriate strategy for a small hydro study will certainly vary from site to site but is likely to be found somewhere between the two extremes. A typical situation is likely to require the use of one or more simple approaches initially and the eventual adoption of a likely representative record for more detailed analysis.

Power analysis procedures including duration curve analysis, mass inflow curves, low flow frequency, and sequential period of record routing are described and examples included. Duration curve analysis is characterized as the least exact but easiest to perform (and many times is entirely adequate) and sequential period of record routing as the most accurate (depending on the quality of the available record) but requiring the most effort. Computational aids in the form of references and computer programs are described.

Volume IV - Existing Facility Integrity. The volume adopts the posture that a prerequisite to serious consideration of a site for a small hydro addition is that it be capable of meeting current dam safety standards. The small hydro addition could be expected to make modest improvements to meet integrity deficiencies but would not often generate adequate benefits to "carry" significant remedial work. This observation changes if alternative financing for safety related remedial work is separately provided. The integrity volume is designed to identify early in the feasibility study, any deficiencies that might exist and thus provide a decision point for study termination. Guidance for formulating a range of suitable remedial measures is included.

The volume can by no means provide inexperienced engineers with the capability to perform definitive safety studies. The intent is to provide a strategy that will alert investigators to potential problems. Should the problems appear critical, the volume recommends terminating the power addition feasibility study and notifying appropriate state and federal authorities of the existence of the identified integrity deficiencies.

The volume classifies and describes the principal dam types (concrete, masonry, and earth and rockfill) likely to be encountered in a small hydro addition feasibility study. The appurtenant works associated with dams (spillway, outlet works, power plants, locks, and fish ladders) are described by type and function. The typical deficiencies and failure modes of dam overtopping, uncontrolled or excessive seepage, foundation instability, embankment slope instability, slope protection deterioration on embankment dams, concrete deterioration, excessive uplift pressures, spillway/outlet works failure, and erosion are described and the principal mechanism causing the deficiencies are discussed. Potential adverse effects of power additions are highlighted to alert investigators to problems that may be created by the modification of existing facilities to accommodate a power plant.

The integrity investigation is outlined as a three staged process: (1) records collection and examination, (2) supplemental data collection and analysis to support conclusions relative to integrity, and (3) formulation of repair schemes, if they prove necessary, for rehabilitation work. The elements of each stage and strategies for their performance are outlined.

Volume V - Electromechanical Features. The volume defines electromechanical equipment as the features and systems needed to harness the energy, both potential and kinetic, available in impounded or flowing water, to convert it to electrical energy, to control it, and to transmit it to a regional power grid. The major equipment items are the hydraulic turbine, the electric generator, and a switchyard consisting of a transformer, circuit breaker, and switch gear. Included are supporting systems which control and protect these major equipment items. Maintenance facilities such as a crane for lifting are also included in a broad definition of electromechanical equipment.

Several domestic and foreign equipment manufacturers have historically provided small turbines and are active in standardizing unit sizes and packaging relatively complete generating sets for marketing. These current trends are defined. Relaxing the need for some control and protection equipment is becoming accepted as the scaling down to small facilities takes root within the industry. Simpler low cost governors and similar items are appearing on the market. Smaller hydroelectric plants can also be designed with less flow control than larger plants. The flow of water to most turbines is controlled by a set of wicket gates. These gates are regulated by signals from the governor to control the amount of power produced. Where power control is not needed (many small plants) the gates can be eliminated and the cost of the turbine reduced by as much as 10 percent.

The volume outlines a procedural strategy for selecting and sizing the generating equipment, and includes description, cost, and performance data for Francis, Crossflow, Propeller, Tube, Bulb, Slant, and Rim turbines suitable for the range of heads and power outputs for a small plant. The common indexing parameter used among data and relationships within the electromechanical volume is the turbine throat diameter. This parameter is carried forward to the Civil Features volume (discussed next) as the indexing parameter to determine powerhouse layout dimensions and costs.

A section describing generators suitable for small hydro is included and data on dimensions and weights tabulated. Descriptive data, performance curves, and costs are likewise included for generation control and protection equipment, and switching, transmission and miscellaneous equipment.

Volume VI - Civil Features. The civil features of small hydropower additions are defined as site preparation works, hydraulic conveyance facilities, and powerhouse and appurtenant facilities. Site preparation includes grading, foundation excavation, drainage and erosion control, access roads and parking facilities, and construction noise abatement and dust control. Hydraulic conveyance facilities include penstocks, tunnels, canals, valves and gates, inlet and outlet works, and tailraces. Powerhouse and appurtenant facilities include all structures for the powerhouse and equipment handling facilities, foundations for both the powerhouse and switchyard, and fencing around the project area.

Civil features can at times comprise a significant component of construction cost of small hydro additions. Since major elements of the site are fixed (e.g., embankment, outlet works, spillway) it is important to approach the layout task with an open and innovative attitude. The difference between a feasible and infeasible project may be determined by the cleverness with which use is made of the existing site arrangement and features. The civil features differ from those of major hydropower plants both in scale and in substance. It is appropriate to design adequate outdoor type plants for small units and often portable lifting equipment will suffice for maintenance obviating the need for enclosing structures and fixed gantry cranes. Protection equipment can likewise often be minimized. Layout guidance, dimensions, and cost functions for the several categories of civil features are included. Descriptive text is included to alert the project investigator to circumstances in which the generalized relationships that are included are unreliable and guidance is given for developing alternative data when necessary.

Cost escalation indicies are included so that the cost data (cost data in all volumes are in July 1978 dollars) may be scaled to the base period used for the feasibility analysis. Both this volume and volume V include cost summary sheets keyed to the Federal Energy Regulatory Commission (FERC) account numbers.

Use of Guide Manual

The manual is designed for use by the variety of organizations and private individuals that might study small hydropower projects for feasibility. The document is a guide; not a cookbook. A structure is presented within which the majority of studies are expected to fall. A feasibility investigation of sufficient quality to provide a basis for investment decisions requires the services of qualified professional engineers.

The technical data of selection criteria and charts, physical feature layouts, and performance charts are considered adequate for both reconnaissance and feasibility studies. The cost charts are expected to be adequate for the majority (perhaps 80%) of project settings and configurations likely to be encountered. Notes alerting analysts to special conditions for which the charts would be less accurate are included. For those instances, a specific layout, preliminary design and cost estimate would probably be necessary even at the feasibility level of study.

The material in the manual should be informative to those interested in small hydro (e.g., engineers, administrators, and private enterprenuers). Most of the material is presented in common narrative terms but this should not be construed to suggest that unqualified individuals can thus perform quality studies. Several scenarios of use are envisioned. Institutions/organizations with small technical staffs are expected to find the manual adequate to guide them in preparation of a reconnaissance study and then provide an information base that would be helpful in proceeding to procure the services of qualified consultants, should the reconnaissance finding be positive. Institutions/organizations with technical staffs not experienced in small hydropower but experienced in the several technical areas involved are expected to find the manual helpful in developing capability to perform the feasibility level studies, by having available an organized set of material and guidance (including references) on small hydro. Institutions/organizations experienced in hydropower development (but perhaps not small hydro) should find the manual to be a useful reference that documents many important concepts and represents a compilation of the current state-of-the-art in small hydro.

SECTION 2
SMALL HYDROPOWER

Definition

Small hydropower projects include installations that have 15,000 kilowatts (kW) or less capacity. Although the concept is not limited to additions to existing impoundments, most activities by federal, state, local agencies, and private organizations are so focused. This manual is concerned exclusively with hydropower additions to existing facilities. "Small hydro" and "low head hydro" are not synonomous even though the tendency in public statements and published literature and documents is to blur the distinction. Small hydro as defined above has been an informal breaking point used for various federal and other agency statistical tabulations and informal communications. The concept has now been defined by the Public Utility Regulatory Policies Act (PL 95-617, November 1978) to be 15 megawatts (MW) for purposes of special handling for licensing, loans, incentives, and other promotional programs. Provisions of the law specifically related to small hydro are limited to additions to existing facilities. Low head hydro is a term associated with a research and development program managed by the Department of Energy that is designed to advance the technology for generating hydropower from sites with heads of less than 20 meters (66 feet). A large number of the presently identified small hydro addition sites fall within the low head criteria. This distinction between small and low head hydro will be preserved herein for convenience in communication and consistency with existing and emerging federal and state programs.

The fundamental thesis for small hydro as a concept (apart from hydropower in general) is that the impacts of implementation (especially for an addition to an existing impoundment) are likely to be modest; thus, projects will be essentially non-controversial so that simpler license and permit granting programs are appropriate, and physical facilities can be kept simple and functional. Implementation will therefore be possible in relatively short time frames.

Existing and Potential Development

A significant number of existing hydropower installations in the United States could be classified as small hydro. Current installed hydropower capacity is near 60 million kW in about 1,400 plants, which results in an average installed capacity of about 40 MW per plant. The latest published inventory (Federal Power Commission, 1976) lists 142 plants as having installed capacities greater than 100 MW. Deducting the sum of the capacities for plants in excess of 100 MW from the total results in the average plant size for the remaining 1,260 plants dropping to 12 MW. There are, therefore, a great number of existing plants that meet the small hydro criteria. It would seem that the U.S. should have a considerable body of technology and technical expertise, but on the other hand, the smaller plants tend to be older plants and were specifically designed for the site. It should be noted as well that 385 MW (McDonald, 1977) of hydropower, mostly small plants, have been retired from service in the last 15 years, a trend that recent events are likely to reverse.

Initial estimates of power potential at existing non-hydropower dams indicated that about 30,000 MW and 95 billion kilowatt-hours per year (McDonald, 1977) exist at several thousand sites. These sites are among the some 50,000 dams identified in the national dam inventory prepared by the Corps of Engineers and range in size from retired small hydro plants in the New England area to major federal reclamation projects in the west. Other potential sites not identified in previous studies include irrigation canal drops (significant in the west), municipal water supply delivery systems such as in Southern California and the North Atlantic, and waste management systems such as the Chicago tunnel plan (Gladwell, Warnick, 1978; Macaitis, Schonsett, 1979). An improved resource assessment of small hydro potential and sites will be generated as a component of the Corps of Engineers National Hydroelectric Power Study activities (Institute for Water Resources, 1979). Preliminary results indicate the gross potential at existing dams lies in the 6,000 to 10,000 MW range at upwards of 5,000 sites. Since a significant portion of small hydro development is likely to have no dependable capacity, the annual energy potential is a more meaningful index of the contribution to the nations energy needs than is capacity. The gross potential annual energy at existing dams lies in the 18 billion to 25 billion kWh range, which is equivalent to a savings of 80,000 to 140,000 barrels of oil per day.

Analysis of the national dam inventory data (50,000 dams) indicates that about 1/3 of the sites have heads in the 6 to 20 foot range (considered extremely low in the "low head" literature) and about 2/3 of the dam sites have intermittent flow (inflow ceases some time during the year). Also, a number of significant physical, economic, and institutional obstacles exist that presently inhibit development of a large number of the sites. The economically attractive sites under present conditions would total significantly less than the 30,000 MW reported potential, but it is generally agreed that several hundred sites are likely to be found economically attractive for immediate development. The cost of fossil fuels is expected to continue to grow and thus increase the economic attractiveness of hydropower in general, and in particular small hydropower, such that within the next ten years, upwards of 2,000 sites could be considered as a reasonable count for the number of small hydro sites warranting serious study for implementation.

Implementation Issues

A significant major positive feature of small hydro is that many of the important environmental issues have been previously resolved (e.g., the impoundment site exists and is presently in service). This suggests that it should be substantially less complex to plan developments, marshal support, acquire needed permits, and construct small hydro additions than to develop other new hydro projects or alternative thermal power generation plants. The lag time from conception to implementation could be as little as 3 years (Figure 2-1) compared to the often 10-15 years for major projects. The current trend in small hydro is to take advantage of the head and existing flow release patterns to avoid the environmental and legal complexities that would ensue from altering water use, modifying release patterns, and adding storage (thus increasing pools level). The inferred judgment seems to be that the complexities induced by altering existing use and release patterns to enable generation of more power and perhaps development of some dependable capacity (see glossary for definition) are not worth the time delays and added implementation complexity that would result. In effect the thrust is "let's develop what's presently lost through energy dissipation structures and get it on line quickly, since we are at least aiding in meeting near term energy requirements."

The belief that there will not be instances of important environmental issues is not realistic, however. Any alteration of the flow pattern and released water quality will require careful documentation and analysis. Also, past mitigation omissions will likely be surfaced during studies and will need to be corrected. A specific case in point is that fish passage facilities (especially for anadromous fisheries) are likely to be insisted upon for sites from which they were omitted in a prior era, and preliminary indications are that precedents exist to backup the insistence. Small hydro offers an opportunity for engineers to provide the leadership early in project development to identify and formulate solutions to potential environmental problems. The key point is to define issues early in investigations so that they may be included as a normal component of project feature planning.

Factors Important for Feasibility

Several important issues that can be inferred from the previous discussion are pertinent to establishing the conceptual base for the feasibility guide manual. One is understanding the reasons underlying the major national attention that is focused on small hydro, an admittedly small element of the national energy array. Simply stated they seem to be the national desire to move to energy independence, the current national concern for resource conservation and use of renewable resources, the potential for quick results from public and private efforts (an increasingly rare commodity in today's world), and most assuredly, the demand for nonfirm energy (previously referred to with the tainted label "dump energy") presently valued in many areas at 15 to upwards of 40 mills per kilowatt-hour as compared to 1 to 2 mills per kilowatt-hour several years ago.

The greatest potential seems to be at existing sites with the major civil works already in place. The sites typically are in non-federal ownership (about one-half of existing hydropower plants are in non-federal ownership). The sites are often in the low head range (under 20 meters), with a significant number falling in the head ranges of less than 30 feet. The marketable output will most often only be energy with little, if any, dependable capacity. This means the value of small hydro output will be primarily due to fuel and other operating cost savings and not due to offsetting the need for new power plants to supply capacity.

The feasibility of projects is expected to be quite sensitive to site specific conditions. The value of power produced will not likely support an extensive array of ancillary features such as long transmission lines, access roads, or significant site preparation. The nature of the market area load characteristics and present generating facilities servicing the load are critical elements in valuing power output. Areas served with major fossil fuel plants, or systems with high operating cost plants operating at the margin will be more attractive for small hydro development. A significant issue of project feasibility is that investigation, design, construction management, and administration (the non-hardware elements of a project) are a major project cost burden. Figure 2-2 schematically illustrates the cost elements in small hydro projects. Contingencies, which are not shown, are normally considered as a percentage of all of the items listed, and range from 10 to 20 percent. The feasibility study itself is likely to be viewed as a significant financial burden warranting an investment type decision by the project sponsor prior to initiation of the study.

Small hydro is therefore a unique set of hydroelectric power developments with potential that exists at a relatively large number of existing sites that are mostly in non-federal ownership, primarily of low head, likely to generate "non-essential" power, and sensitive to site specific conditions, and will require investigations whose costs are a significant issue. The guide manual has been formulated to be responsive to these characteristics and to provide a foundation to encourage relatively quick, efficient formulation and assessment of attractive projects.

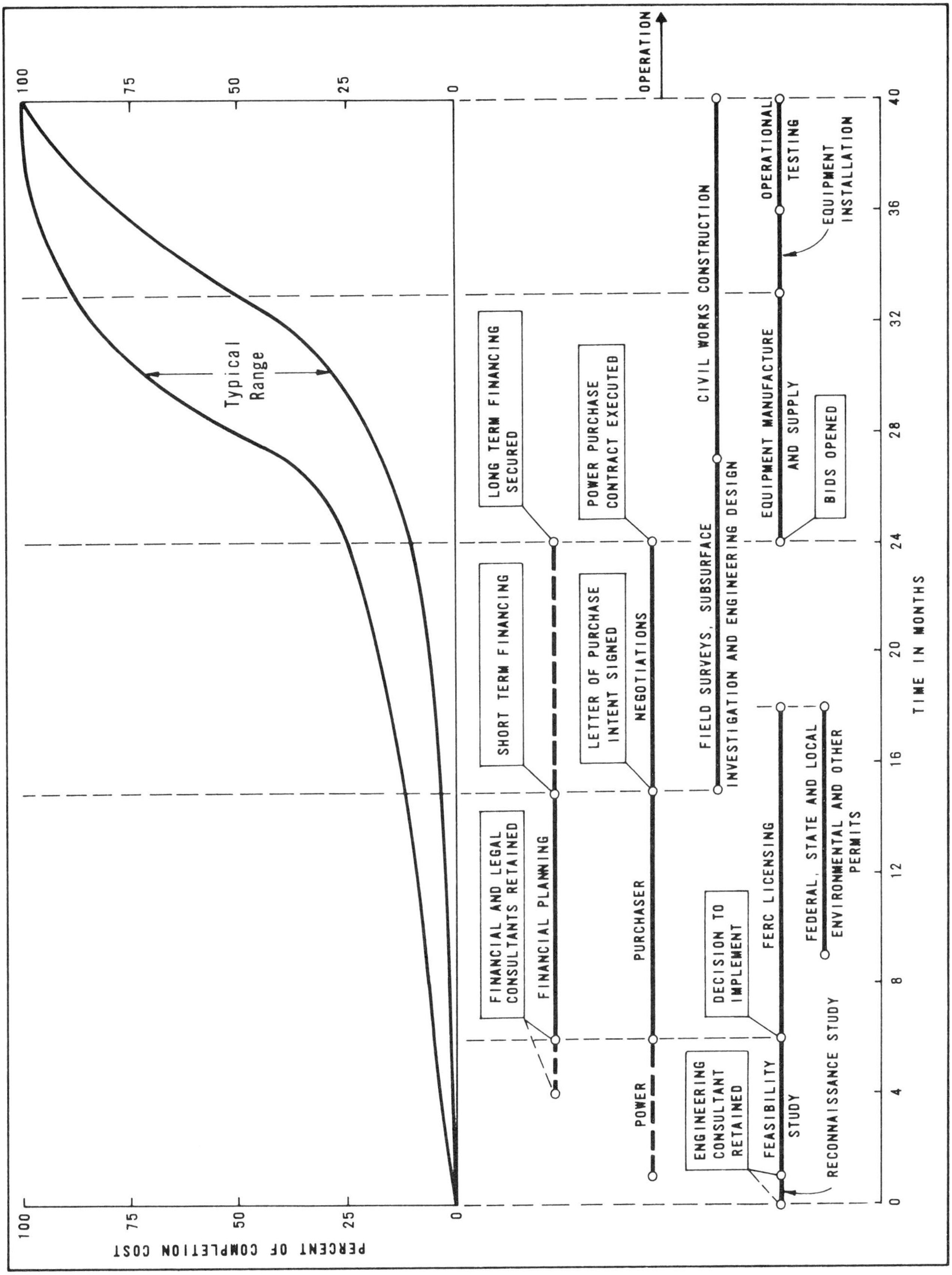

Figure 2-1. Typical project implementation schedule and expenditure pattern (From Volume II-Economic and Financial Feasibility)

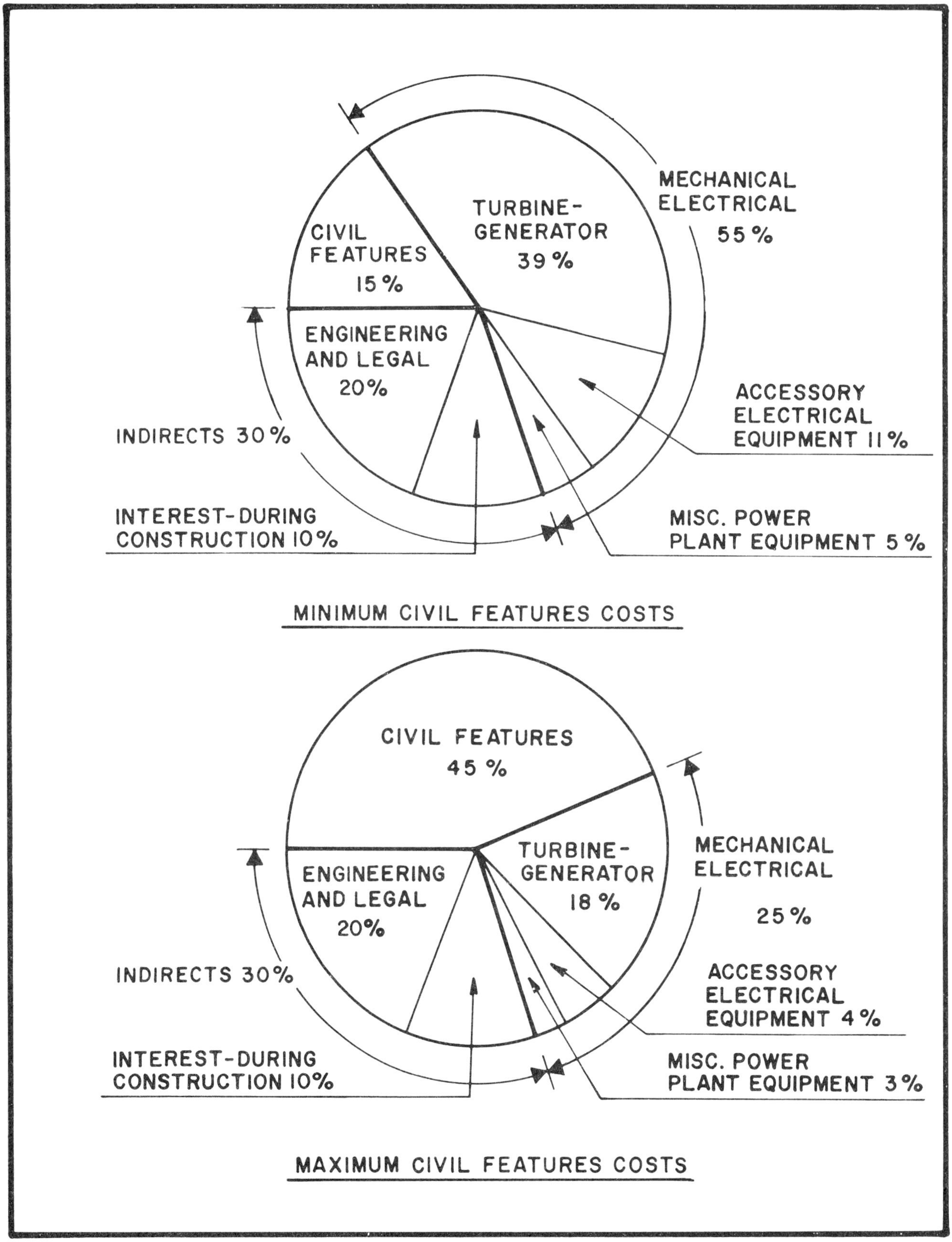

Figure 2-2. Cost elements of small hydro projects. (From Volume VI-Civil Features)

SECTION 3
PLANNING INVESTIGATIONS

Definitions

Several types of investigations varying in scope, detail, and intended client are performed to determine the desirability of public and private implementation of hydropower proposals. These investigations are usually referred to as planning studies. The end points of these studies are relatively easy to define; the formation of an idea or concept for a project at the beginning, and the initiation of construction at the end. Varying degrees of decisions and commitments occur in a continuous sense but normally are formally adopted at discrete decision points throughout this period. Government agencies and private organizations normally have specific steps (study types) that are standard for their purposes.

The general collective term for most of the planning studies performed prior to an implementation decision are "feasibility" studies. Other studies take place between the implementation decision point and construction initiation. This guide manual has adopted the standard sequence of preconstruction studies commonly followed in private and international practice (United Nations, 1964) and several Federal agencies. They are "reconnaissance study" (should a feasibility study be performed?), "feasibility study" (should an investment commitment be made?), and "definite plan studies" (the collective group of studies that are performed between the time of an implementation commitment and initiation of construction that result in permit applications, preparation of marketing agreements and financial arrangements, determination of design parameters, etc). Figure 2-1 schematically identified these studies in the project development and implementation sequence. The guide manual is designed to aid in the performance of the reconnaissance and feasibility studies.

The Glossary defines a reconnaissance study as ... "a preliminary feasibility study designed to ascertain whether a feasibility study is warranted" and feasibility study as ... "an investigation performed to formulate a hydropower project and definitively assess its desirability for implementation."

Objective of Reconnaissance Study

The performance of a feasibility study can be a significant investment in time and resources suggesting that a decision to proceed with a study should be based on a finding that a potentially viable project proposal will be forthcoming. The reconnaissance study is designed to reduce the chance of a subsequent unfavorable feasibility finding and maximize the potential for identifying and moving forward attractive projects. The reconnaissance study is therefore a relatively complete small scale feasibility investigation in which the issues expected to be important at the feasibility stage are raised (the intent is to appraise the critical issues, not formulate approaches and solutions), and to perform a first cut economic analysis. A favorable economic feasibility finding is a strong indicator that further detailed study (a feasibility study) is warranted subject to assessment of potentially critical negative issues. The finding of a reconnaissance study should be either a positive recommendation to proceed with a feasibility study and then also include a study plan and method of accomplishment, or a recommendation to terminate further investigations. The strategy for performing a reconnaissance study is first to perform a preliminary economic analysis and then identify and assess the issues that may be critical to implementation. Section 4 describes the components that are likely to be important in a reconnaissance assessment and suggests appropriate levels of work efforts.

Objective of Feasibility Study

The feasibility study is designed to formulate a viable small hydro project, design an implementation strategy, and provide the bases for an implementation commitment. The significant legal, institutional, engineering, environmental, marketing, economic, and financial aspects are to be defined, investigated, and definitively assessed in support of an investment decision. The feasibility study is a decision document that defines and recommends a course of action. The findings of a feasibility investigation should be whether or not a commitment to implementation is warranted, and should the finding be positive, define the steps needed to assure implemention. A positive economic feasibility finding is normally necessary for further implementation to be initiated. However, other concerns can be equally important in serving the broad public interest and the feasibility study should be performed in the modern spirit of wise natural resource management and multi-objective planning principles. Section 5 provides strategic guidance on performance of feasibility studies and suggests appropriate levels of work efforts.

SECTION 4
RECONNAISSANCE STUDIES

Reconnaissance Study Tasks

The components identified as important in reconnaissance studies are shown on Figure 4-1. The tasks include those required to perform the economic feasibility (power potential, value, cost, and site capabilities) and those that should aid in defining and assessing critical issues (authority and legal issues, site issues, facility integrity, and financial and revenue issues). Subsequent paragraphs briefly discuss the tasks shown on Figure 4-1. Table 4-1 summarizes the pertinent reference sections in the supporting volumes of this manual.

Plan Reconnaissance Study. The specific scope and purpose of the study should be defined and needed output products identified. The scope and purpose have been generally identified in this section of the volume, but variations in emphasis may exist, depending on project proponent (private, public) and prior studies (national, regional screenings), which should be defined. A study plan should be formulated identifying the important work tasks (e.g., refining the suggestions of this section). It is suggested that by this point at the initiation of a reconnaissance study that all volumes of the guide manual be read by the responsible participants.

Contact Principal Agencies. This task has been identified a bit out of context because it would logically be an element within each of several tasks. However, activities by various institutions have developed valuable information that is presently, or soon will be, available that warranted highlighting in the guide manual.

The U. S. Department of Energy (DOE) has specific programs designed to encourage the development of small hydro. The local regional office (for phone number see government section of the phone book) can provide information on up to date activities within that agency. The Idaho Operations Office, DOE, (550 Second Street, Idaho Falls, Idaho 83401) is the action office in small hydro and is active in developing an information referral service and compiling data on small hydro projects.

TABLE 4-1
RECONNAISSANCE STUDY TASKS*/
MANUAL REFERENCE SECTIONS

Study Tasks	Volume	Manual Reference Section	Description
Plan Reconnaissance Study	I	4	Par. of same title.
Contact Principal Agencies	I	4	Par. of same title.
Scope Economic Evaluation	II	4	
Define Power Potential	III	3	
Assess Market Potential	II	3	
Estimate Power Output	III	3	
Develop Spillway Hydrology	III	4	Early paragraphs.
Identify Physical Works	V, VI	1, 2	Fig. 2-2, Vol. V.
Formulate and Cost Project	I	4	Fig. 4-2, Table 4-2.
Develop Cost Stream	I	4	Par. of Same Title.
Adopt Power Values	II	3	
Develop Power Benefit Stream	II	2	
Determine Economic Feasibility	I	4	Par. of same title.
Identify Critical Issues	I	4	Par. of same title.
Assess Legal/Institutional Issues	I	4	Par. of same title.
Assess Site Issues	I	4	Par. of same title.
Assess Facility Integrity	IV	3	Stage 1 discussion.
Assess Financial Issues	II	6	Early pages.
Document Reconnaissance Findings	I	4	Par. of same title.

*Tasks identified are those shown on Figure 4-1 and are discussed in this section.

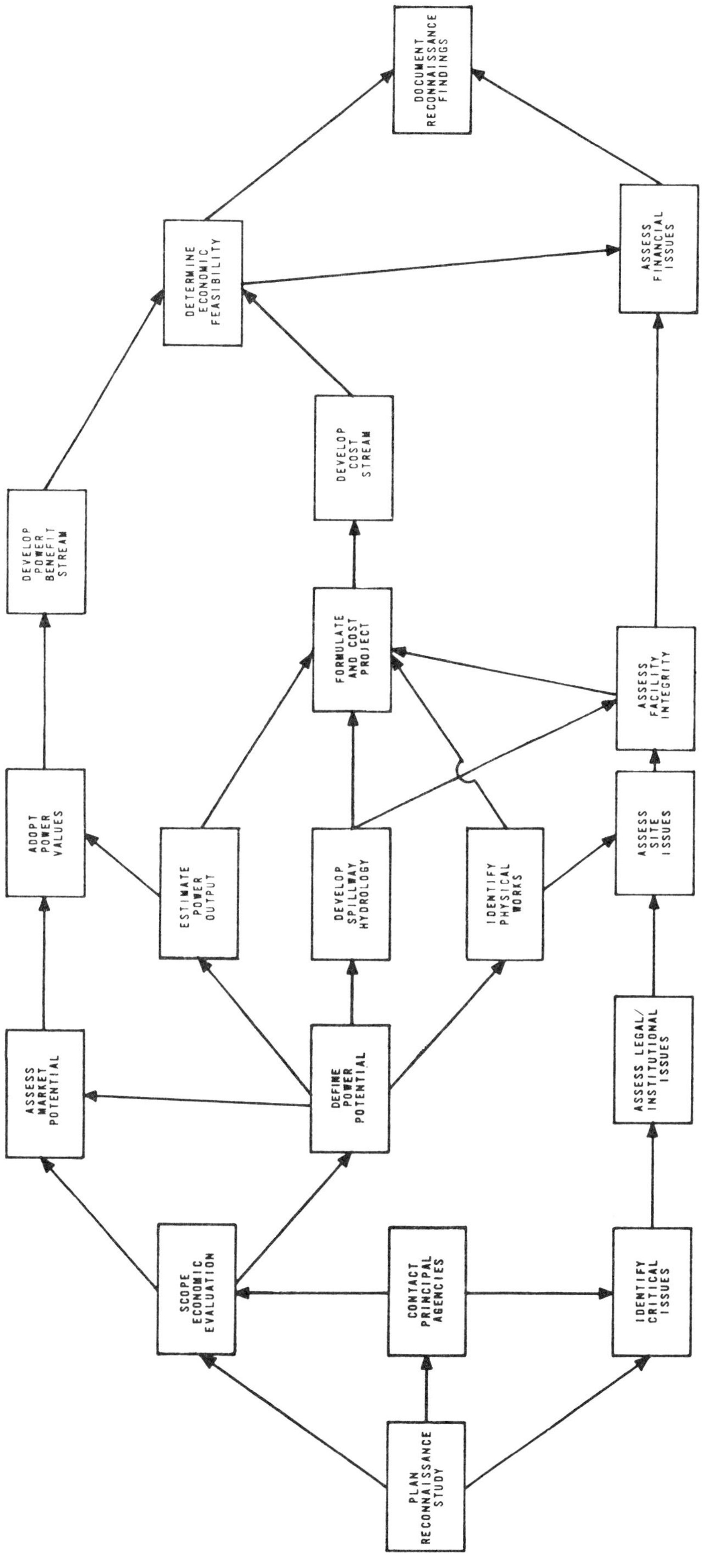

Figure 4-1. Reconnaisance study components

Several agencies have performed assessments of the potential for small hydro in their geographic areas - it is possible the site under investigation might exist in one of these inventories. Agencies to contact include state water (or natural resources agency) offices, regional river basin commissions, and local offices of the U.S. Bureau of Reclamation and U.S. Army Corps of Engineers. A nationwide hydropower resources assessment has been compiled as a feature of the U.S. Army Corps of Engineers National Hydropower Investigation (Institute for Water Resources, 1979). All potential sites that could be identified from reports or are in the national dam inventory are included in a computerized inventory that could provide valuable reconnaissance data. A summary of the file contents has been made available to the public. The responsible agency is the Corps of Engineers Institute for Water Resources, Kingman Building, Ft. Belvoir, VA 22060.

Scope Economic Evaluation. Small hydro projects are generally single purpose power projects. As such, the economic justification is based on the value of power that can be generated. If other project features are to be considered in the economic evaluation such as recreation, fish and wildlife, etc., they should be defined at this point and tasks related to their quantification formulated. See Section 4, Volume II for further discussion.

Define Power Potential. The value of power output from a proposed project, and the appropriate physical facilities are sensitive to the nature of the power potential. Is the plant likely to produce only energy or does it have potential for dependable capacity value as well? About how much output is likely and what is its variability? These are information items that are needed to assess market potential and provide formulation data. See Section 3, Volume III.

Assess Market Potential. Potential buyers of power output should be identified so that estimates of the value of power may be determined. Information important to determining the value of power includes: who is presently generating and selling power in the area, what types of generating equipment are in operation, and who are major customers. Purchasers could include utilities, cooperatives, private industry and other institutions. See Section 3, Volume II.

Estimate Power Output. The value of power output and the cost of works to produce the power are functions of the magnitude and character of output. Several project installed capacities should be investigated to estimate power potential, covering a range of likely installed capacities. Three potential sizes would seem appropriate. A mid value of installed capacity chosen to correspond to the 25% flow-exceedance value is a reasonable starting point with the other two selected at say 15% and 35% exceedance values. Computation methods described as Reconnaissance Sizing Procedures in Section 3 of Volume III provide suggested guidance. The desired product of this task is an array of installed capacities and corresponding annual energy output, indicators of the range of likely output by seasons and years (high and low flow periods), and an assessment as to the amount of capacity (if any) that might be credited as dependable. The head and flow ranges of the array are likewise needed to size and cost the power features.

Develop Spillway Hydrology. The flood flows that must be passed, and the present spillway capability to pass the flood events of rare occurrences are important indicators of the integrity of the existing facility. Reconnaissance estimate methods of flow determination and spillway performance analysis are contained in Section 4 of Volume III, particularly early paragraphs of the section.

Identify Physical Works. The power generation and appurtenant works must be suitable to the intended installation and site. A specific preliminary design is not required but sufficient formulation to define likely machine type and possible configurations are needed to assess site issues, and provide a basis for cost estimates. Introductory sections of Volume V and Volume VI provide general information; note particularly Figure 2-2, Volume V.

Formulate and Cost Project. Cost estimates for construction, site acquisition, operation and maintenance, and engineering and administration are needed to assess economic feasibility. To facilitate reconnaissance estimates, the charts contained in Volumes V and VI have been analyzed to develop the chart and tables contained in this section. Figure 4-2 provides a basis for estimating the major share of construction costs for items that are governed by capacity and head, e.g., turbine, generator, and supporting electrical/mechanical equipment. The chart was developed by studying the generator and powerhouse costs for a variety of turbine types for a complete set of head/capacity values. The chart is therefore the locus of least cost points for head/capacity values shown. The reader is cautioned that this chart is based on the figures contained in other manual volumes and least construction cost criteria governed so that site issues of space and configuration, and generation issues of performance ranges were not used. The chart should be adequate, however, for reconnaissance estimates. Installation of multiple units can be considered using these charts although the refinement of analysis might be questionable at this level of study. A recent paper (O'Brien, George, Purdy, 1979) suggests that multiple units may be critical to small hydro feasibility because of the goal of generating as much energy as possible from the available flow regime. Projects approaching the upper limits of small hydro capacity (15 MW) probably warrant using the charts of Volumes V and VI at the reconnaissance level of study. The remaining components needed for preparing construction cost estimates are included in Table 4-2. Other cost items that may have surfaced during study of the critical issues (access, fish passage, integrity, etc.) should be estimated at this

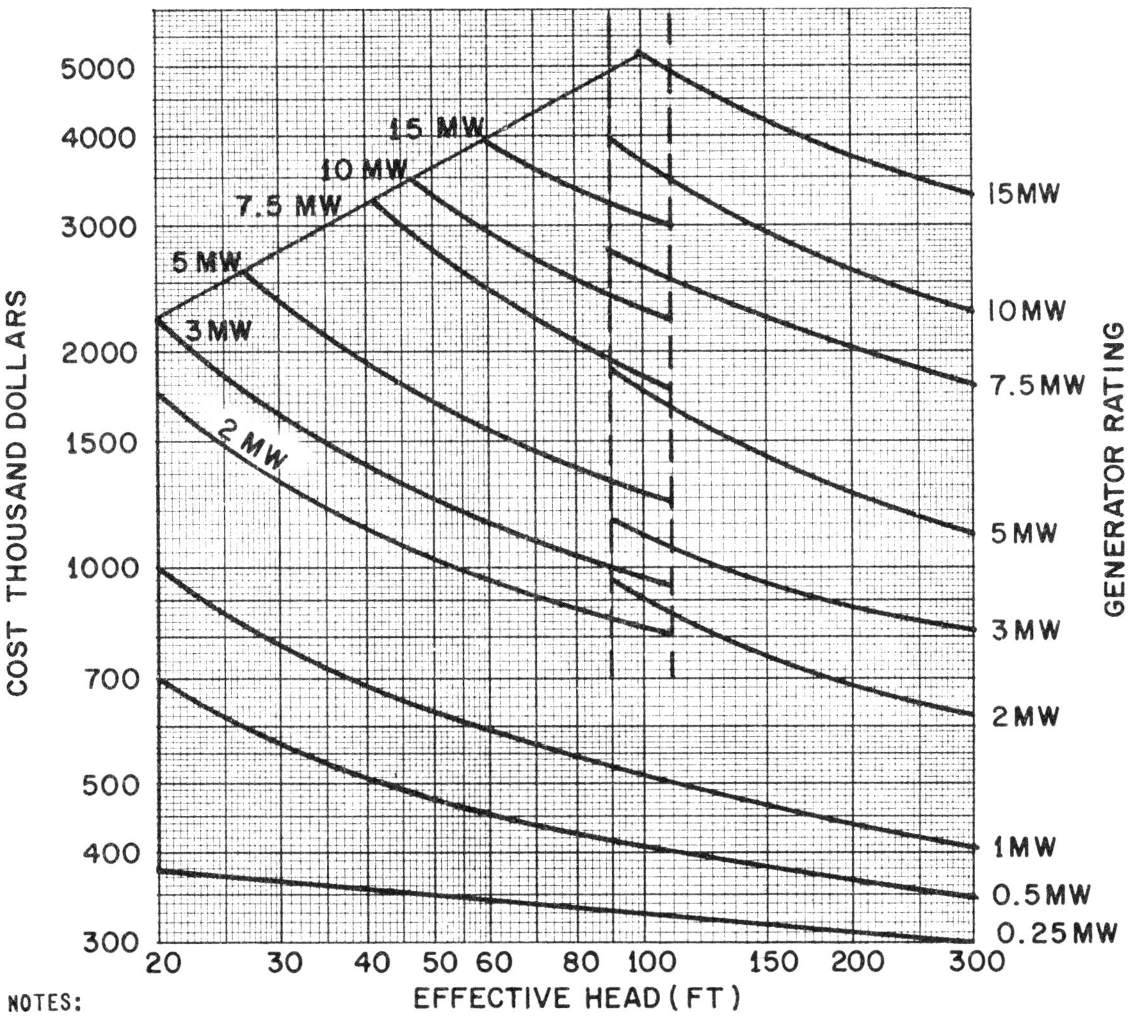

NOTES:

1. Estimated costs are based upon a typical or standardized turbine coupled to a generator either directly or through a speed increaser, depending on the type turbine used.
2. Costs include turbine/generator and appurtenant equipment, station electric equipment, miscellaneous powerplant equipment, powerhouse, powerhouse excavation, switchyard civil works, an upstream slide gate, and construction and installation.
3. Costs not included are transmission line, penstock, tailrace construction and switchyard equipment.
4. Cost base July 1978.
5. The transition zone occurs as unit types change due to increased head.
6. For a Multiple Unit powerhouse, additional station equipment costs are \$20,000 + \$58,000x(n-1) where n is the total number of units.
7. Data for this figure was obtained from figures and tables in Volumes V and VI.

Figure 4-2. Power features cost - reconnaissance

stage as well. In the absence of specific estimates for these additional items, a reconnaissance omissions allowance of up to 20% would be appropriate. The products of this task should be an array of costs for the range of installed capacities for which power estimates were prepared.

Develop Cost Stream. The construction cost values developed in the previous paragraph need to be gathered, organized, and arrayed to permit expeditious performance of the economic feasibility calculations. The construction costs should be escalated to the study date. Section 6 of Volume VI presents a strategy for escalating costs of civil features.

It is recommended that the cost index for "structures" be used as a composite value for all construction items for the reconnaissance cost estimates. Cost estimates are also needed for the nonphysical works cost items. An allowance for unforeseen contingencies ranging from 10% to 20% should be added to the sum of the construction costs, the value depending upon a judgment as to the uncertainties. A mid value of 15% for contingencies is appropriate in the absence of more detailed analysis. All investigation, management, engineering and administration costs that are needed to implement the project and continue its service are appropriately included in the feasibility determination. It is suggested that indirect costs for administration, engineering, interest during construction, etc., of 25% be added. Total indirect costs to be added will therefore vary between 35% and 45%.

Adopt Power Values. The power values needed are the value of energy that the project proponent could reasonably expect to receive for the sale of output, and if any dependable capacity is likely to be present, the value of the dependable capacity of the project. It is suggested that reconnaissance values be adopted from values solicited from the local Federal Energy Regulatory Commission (FERC) office in the case of potential sale to utilities,

TABLE 4-2
MISCELLANEOUS RECONNAISSANCE ESTIMATE COSTS*
(Cost Base July 1978)

PENSTOCK COST

Effective Head (Ft)	10	20	50	100	200	300
Cost Index (CI)	960	480	200	110	55	35

Installed cost = CI x Penstock Length (ft) x Installed Capacity (MW)
Minimum Penstock Cost is $50 per linear foot.

TAILRACE COST

Construction Cost = $15,000 fixed plus $200 per linear foot

SWITCHYARD EQUIPMENT COST
(Thousand Dollars)

Plant Capacity	Transmission Voltage 13.8	34.5	69	115
1 MW	50	60	110	160
3 MW	85	100	120	175
5 MW	110	125	150	210
10 MW	150	170	210	280
15 MW	185	220	250	320

TRANSMISSION LINE COST
(Thousand Dollars)

Plant Capacity	Miles of transmission line 1	2	5	10	15
0.5 MW	30	60	150	—	—
5 MW	45	80	160	320	500
10 MW	60	100	180	380	600
15 MW	80	140	230	460	700

*Data derived from Volume V (Figures 6-4 and 6-5) and Volume VI (Figure 3-1 and Table 4-2).

municipal organizations and cooperatives, or be extracted from existing rate schedules (available from the local utility office) in the case of potential sale to a private industrial buyer. A benchmark value that can often be used as the minimum value for energy is the fuel replacement cost that is obtainable from the nearest FERC regional office. A generous value seems appropriate in light of presently escalating fuel and operations costs. Generalized power values are expected to be published as part of the Corps National Hydropower Investigation (Institute for Water Resources, 1979). Current values (1979) for energy in the range of 20 to 40 mills per kilowatt-hour are considerable reasonable. Section 3, Volume II discusses power values in detail.

Develop Power Benefit Stream. The power generation benefits from the proposed project are the sum of the energy value times the energy production and the capacity value times the estimated dependable capacity (if any). In the instance of a private purchaser, the difference in their power bill with and without the proposed project is the benefit. The project benefit stream is the annual array of power benefits (plus other project benefits if determined to be appropriate). Project benefit streams should be prepared for the several installed capacities under study. See Section 2, Volume II.

Determine Economic Feasibility. Economic feasibility is positive when the stream of benefits exceeds the stream of costs. It is suggested that the Internal Rate of Return method of characterizing project feasibility be employed. The Internal Rate of Return is the discount rate at which the benefits and costs are equal, e.g., the discount rate at which the benefit to cost ratio is unity. This avoids the need at the reconnaissance stage to adopt a discount rate and thus provides an array of economic feasibility results. See Economic Analysis Cost Needs paragraph of Section 5 for additional commentary on costs, benefits, discount rates, evaluation periods, and cost escalation. The analysis should be performed for each of the several installed capacities under study. The alternative is to compute a benefit cost ratio using the discount rate that represents the minimum attractive rate of return for the project proponent. A value in the 9% range has been used in many studies for special districts and agencies in the public sector and a value of 17% in the private sector.

An example computation and display is included in Figure 4-3. Should the outcome of the economic feasibility test appear uncertain, simple sensitivity analysis based on the important variables (power values/fuel costs, amount of energy/capacity, etc.) could significantly contribute to narrowing the band of uncertainty. Use of values contained in Table 5-2 of Section 5 greatly facilitates study of the effect of cost and value escalation on project feasibility.

Identify Critical Issues. The potentially critical issues should be identified and actions required to clarify their importance defined. The issues have been generally identified in this section but important variations may exist depending on project proponent, prior studies, location, etc. The issues that are likely to emerge are primarily related to legal and institutional factors, physical factors focused on the site, integrity of the existing facilities, and financial and revenue capabilities.

Assess Legal/Institutional Issues. An assessment is needed at the reconnaissance stage to define the mechanisms that are likely to be needed to implement a project (e.g., site ownership, legal authority to develop/sell power, access to power grids) and to appraise the actions needed to overcome obstacles, should they exist. Several studies are nearing completion by the Department of Energy (Brown, 1979) that will aid in issue definition. The finding required here is whether and to what degree (qualitatively) impediments to development exist so they may be planned for in the feasibility investigation, should the reconnaissance findings prove to be positive.

Assess Site Issues. A site visit should be considered essential at this stage for (rare exceptions excluded) all reconnaissance investigations of projects. Sketches and drawings may be made and/or existing ones verified defining space for plant siting, terrain and construction features, access, operational status of facilities, and other items pertinent to the physical arrangement of the site, construction of the needed works, and transmission of the power to distribution facilities. The site visit by responsible professionals should be coordinated to provide for a reconnaissance stage integrity assessment as well.

Assess Facility Integrity. The integrity of the site to satisfactorily serve as a power facility and be safe from failure could be a major issue in power addition proposals for many old existing impoundments. Volume IV, especially the discussion of Stage 1 investigations described in Section 3, provides guidance on the needed assessment. See also previous paragraph entitled Spillway Hydrology. The Corps of Engineers (U.S. Army Corps of Engineers, 1975) has been charged with preparing an inventory of existing dams (estimated at 50,000) and performing preliminary assessments of the integrity of certain sites classified as critical. The local offices of the Corps of Engineers can provide information on the current status of integrity investigations, and if a study has been completed, may provide a copy of the report. The fact that a facility exists and continues to function (e.g., has not yet failed) is not conclusive evidence that the dam is safe. The potential impacts of increased stresses from constructing a powerhouse addition should be identified and appraised.

Assess Financial Issues. Sufficient funds must be raised to construct the plant, and adequate flow of revenues generated to provide for maintaining the plant in service, retiring loans, and producing a profit to the developer. The nature of likely financing needs to be defined, potential marketing and revenue arrangements described, and perhaps most important at this recon-

PLANT CHARACTERISTICS:

RUN OF RIVER

Head	= 90 feet	Penstock	= 115 feet
Capacity	= 8 MW	Transmission Line	= 2.5 miles @ 34.5 kV
Efficiency	= 90%	Economic Life	= 50 years
Dependable Capacity	= 0 MW	Evaluation Date	= July 1979
Tailrace	= 250 feet	Average Yearly Energy Generated	= 35 x 10^6 kWh

INVESTMENT COST: ($1,000)

Turbine, Generator and Civil (Figure 4-2)		2,000
Additional Station Equipment (Multi-Unit)		None Required
Penstock (Table 4-2)	(128 x 115 x 8)	118
Tailrace (Table 4-2)	(15,000) + (200 x 250)	65
Switchyard Equipment (Table 4-2)	(8 MW @ 34.5 kV)	152
Transmission Line	(8 MW @ 2.5 miles)	105
Dam Rehabilitation (Integrity)		None Required
Other (Access, Fish Passage, Miscellaneous Site Construction)		None Required
SUBTOTAL		2,440
Escalation (July 78 to July 79 - Figure 6-1, Vol. VI - Ratio: 2.52/2.28)		2,697
Contingencies at 10%-20%	(Used 15%)	405
SUBTOTAL		3,102
Indirect @ 25%		776
TOTAL INVESTMENT COST		3,877

ANNUAL COST: ($1,000)

Annualized Investment Cost is a function of discount rate and economic life of a project and is computed by multiplying the Total Investment Cost by the Capital Recovery Factor for the discount rate and economic life selected. See Table Below

Operation and Maintenance (O&M) Cost = ($20,000 Minimum or 1.5%-4%) (Used 3%) 116

TOTAL ANNUAL COST (Sum of Annualized Investment Cost and O&M Cost) = See Table Below

BENEFIT ESTIMATE:

Capacity Benefit (Dependable Capacity x Value of Capacity) = None

Energy Benefit (Average Annual Energy Generated x Value of Energy) = See Table Below

TOTAL ANNUAL BENEFIT (Sum of Capacity Benefit and Energy Benefit) = See Table Below

COST AND BENEFIT COMPUTATION TABLE

DISCOUNT (INTEREST) RATE (%)	CAPITAL RECOVERY FACTOR	ANNUALIZED INVESTMENT COST[1] ($1,000)	TOTAL ANNUAL COST[2] ($1,000)	BREAK EVEN ENERGY VALUE[3] (Mills/kWh)	TOTAL ANNUAL BENEFIT[4] ($1,000)	NET BENEFIT[5] ($1,000)	B/C RATIO[6]
12	.12042	467	583	16.7	770	187	1.32
14	.14020	544	660	18.9	770	110	1.17
16	.16010	621	737	21.1	770	33	1.04
18	.18005	698	814	23.3	770	-44	0.95
20	.20002	775	891	25.5	770	-121	0.86

NOTES:

[1] Capital Recovery Factor x Total Investment Cost ($3,877).

[2] Annualized Investment Cost + O&M Cost ($116).

[3] Total Annual Cost ÷ Average Annual Energy Generated (35x10^6kWh).

[4] Average Annual Energy Generated (35x10^6kWh) x Value of Energy (taken as 22 mills/kWh) plus the Capacity Benefit (equal to zero for this example).

[5] Total Annual Benefit ($770) - Total Annual Cost.

[6] Total Annual Benefit ($770) ÷ Total Annual Cost.

Figure 4-3. Reconnaissance economic feasibility example

INTERNAL RATE OF RETURN:

The Rate of Return on Investment is the interest rate at which the present worth of annual benefits equals the present worth of annual costs (Net Benefits equal to zero or Benefit/Cost Ratio equal to unity). The internal Rate of Return is 16.8%.

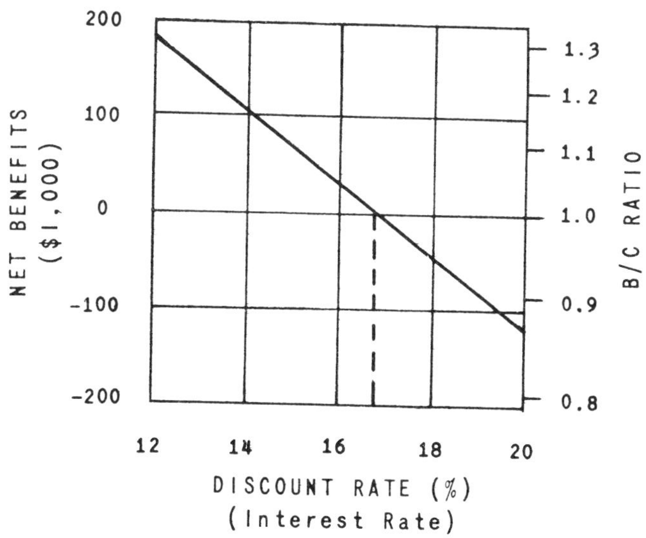

BREAK EVEN ENERGY VALUE:

A similar alternative return type graph is presented here based on the concept of the Break Even Energy Value. This is the value of energy (mills/kWh) which makes annual costs equivalent to the annual return. It is determined by dividing the Average Yearly Generation (kWh) into the Total Annual Cost ($) for each discount rate selected as shown in the table above. At 22 mills/kWh, the Rate of Return is identical to that derived above.

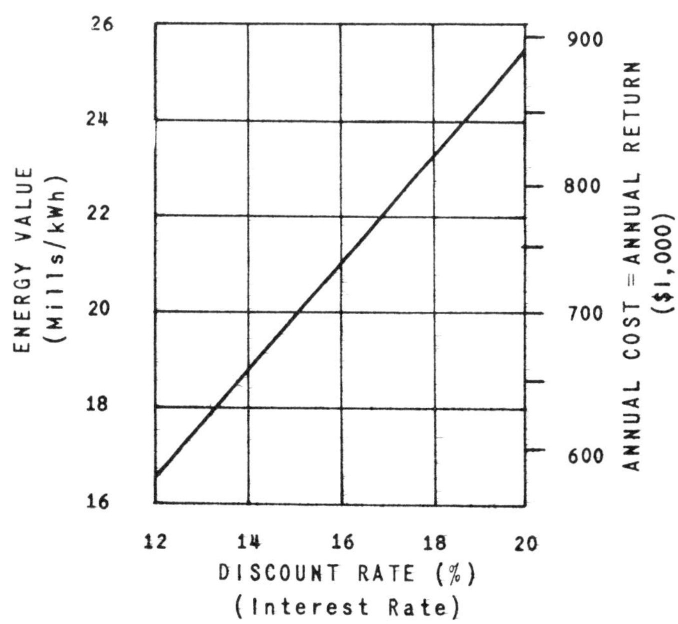

Figure 4-3 continued. Reconnaissance economic feasibility example

naissance stage, the probable cost of capital (interest rate on financing) determined. The early pages of Section 6, Volume II provide guidance on defining financial issues.

Document Reconnaissance Findings. The findings of the reconnaissance investigation should be documented for study by responsible authorities (public officials, boards of directors, private investors, etc.); supporting studies, facts, and references described and codified to expedite performance of further studies; and should the finding be positive, a plan of action for the next steps outlined for execution by the project proponent. The decision to either proceed with a feasibility investigation or terminate further serious study of the potential project concludes the reconnaissance stage of project investigations.

Time, Cost, and Resources for Reconnaissance Studies

The time, costs, and manpower resources required to perform reconnaissance studies for small hydroelectric power plants will vary depending on expected plant size, site conditions, specific scope and depth of study, and availability of information (prior resource assessments and screening studies).

The paragraph of the above title in Section 5 provides general guidance on the expected range of costs for feasibility studies. It concludes that a multiplier of 2.5% of estimated construction cost is a reasonable value for planning purposes. Since reconnaissance studies are in fact mini-feasibility studies, a value of 10% of feasibility cost seems reasonable. Reconnaissance study costs should therefore fall in the range of .15% to .3% of estimated construction cost. A reconnaissance study for a 1 MW plant might cost approximately $3,000 (or about 10-15 man-days) and require 15 to 30 days to complete, and for a 15 MW plant, perhaps $12,000 (45 to 60 man-days) and require 45 to 90 days. The participating professionals would likely include civil, mechanical, and electrical engineers, and power economist for larger proposed projects. Reconnaissance investigations of smaller projects may require more versatility in fewer professionals such as an experienced engineer and economist.

SECTION 5
FEASIBILITY STUDIES

Overview

The tasks identified on Figure 4-1 for reconnaissance studies are applicable for feasibility investigations as well. The emphasis changes from the performance of a preliminary economic analysis and identification of critical issues to study of the full range of issues necessary to support decisions. The work sequence will be similar and the guidance provided in the supporting guide manual volumes is directly applicable to the component investigations. This section presents a general strategy for performing the feasibility study and provides guidance on several topics, the most significant being project formulation.

Strategy

The addition of small hydropower generation to an existing facility is, with few exceptions, a single purpose project planning task. The overriding objective is to formulate a power addition project that is economically attractive and consistent with modern concepts of resource planning and management. Opportunities to enhance other purposes, such as recreation, water quality, and fish and wildlife, should be exploited where possible and where equitable cost sharing arrangements are feasible. Any adverse impacts must be mitigated in accordance with existing statutes. The planning should therefore focus on power addition requirements and impacts, and accommodate other resource management issues as they become evident during studies.

The planning strategy adopted by several federal agencies is conceptually suitable to the small hydro planning task. See for example *Planning Process-Multiobjective Planning Framework* (Corps of Engineers, 1975). The basic thrust is to proceed through several stages of planning increasing in detail and narrowing in focus. The feasibility study strategy can be characterized as successive performance of the tasks shown in Figure 4-1, increasing in specificity on each pass. With no prior studies, 3 passes (stages) would be likely with the final two stages perhaps blurred. A prior reconnaissance study performed as suggested in this manual reduces the successive passes (stages) to 2 maximum and quite likely only one (issues identified at the reconnaissance stage may need no further study). The substantive formulation/evaluation tasks will likely be performed successively to explore the range of project opportunities. Paragraphs following describe the project formulation activities in more detail.

Project Formulation

The selection of the installed capacity, the number of units, and the supporting ancillary physical works are the specific objectives of project formulation. The target in small hydro project formulation is to develop one or more proposals that have the greatest economic value consistent with the array of constraints that may modify the attractiveness of a purely economic formulation. Financial, legal, environmental, and public interest issues may significantly influence the final proposal or even prevent a hydro project from being developed. Performing the project formulation as is suggested herein in an open style and with sensitivity to the significant interfaces depicted on Figure 4-1 should assure that an economically attractive and acceptable project is produced by the formulation efforts.

A strategy for performing the power project formulation is depicted in Figure 5-1. Table 5-1 summarizes the pertinent reference sections in the supporting volumes of this manual. The chart is an expansion of the project formulation tasks that were described for reconnaissance studies. The significant interacting factors in the formulation are the nature of flow/head availability, the performance characteristics of the turbine equipment, and the configuration of the powerhouse structure needed to accommodate the specific generating equipment. The amount of energy that can be generated is dependent upon the range of flow that can be passed through the turbine and upon the head variation. The range of flow that can be utilized is therefore a function of the installed capacity, type of turbine (operating range and efficiency characteristics), and the number of units. Each of these variables affects the size and shape of the powerhouse. The strategy suggested in Figure 5-1 is designed to pragmatically accommodate the set of interacting variables in arriving at the formulated project features.

The strategy shown progresses through three stages of project feature sizing and selection. The first stage (ending with Select Installed Capacity) yields an estimate of the project installed capacity. The second (ending with Select Project Power Features) yields a selection of the number and type of turbine units, considering site conditions and trade offs between unit performance and energy generated. The final stage (ending with Refine Power Features) concludes the project formulation for power facilities. Note that information flow (from other elements of the feasibility study) to specific formulation tasks occurs as the formulation process proceeds. Although not shown, it should be evident that information flow to other than formulation tasks likewise takes place. The following paragraphs discuss the tasks in detail.

Initial Tasks. The first several tasks of the formulation strategy are basically repeats of formulation elements of the reconnaissance study discussed in Section 4. The amount of effort and significance of performance of these initial tasks will depend on whether or not a reconnaissance study was previously performed, the level of detail of the study, and whether the data that was used remains current. Note that prior reconnaissance findings and early feasibility level information flow to the tasks and therefore are assumed to provide the bases for improved estimates. The formulation benefit criteria or values may reflect, if available, additional (to reconnaissance) market studies, and the estimated power output may make use of improved data (e.g., adjusted flow-duration data), if available. A range of project installed capacities should be studied. Selection of installed capacities near a mid value corresponding to the installed capacity at 25% flow-exceedance (15%, 25%, and 35%, are good choices) should provide a reasonable initial array for analysis. Flow-duration analysis techniques described in Volume III are adequate at this stage and optimistic turbine performance criteria are appropriate.

The project benefit stream is developed in the same fashion as the reconnaissance estimates and the project cost estimate can be prepared using the functions and procedures presented and discussed in Section 4. Only costs associated with power features or directly affected by power features are needed. The capacity selection is performed by arraying the costs and benefits of each of the installed capacities investigated, and selecting the one that yields the highest net present value. Plotting capacity versus net present value (present worth benefits minus costs) is a simple and practical means of arraying the date to define the installed capacity to be subjected to additional study. Rate of return or annual cost computations could likewise be used to aid in the selection of the installed capacity.

Subsequent formulation tasks of Figure 5-1 are designed to develop refined estimates of capacity and output by progressively considering site conditions and constraints, turbine performance characteristics, and flow/head variability.

Formulate Power Features. The objective of this task is to formulate an array of project features to allow refinement of estimates of installed capacity, energy output, and project power costs. Specific site assessments and constraint information should be available from other concurrent studies and used for this task. The turbine selection methodology presented in Volume V provides overview guidance (Figure 2-1) and supporting charts and data.

Should only a single turbine type appear suitable, the significant remaining issue is that of the number and size of the units. More units of lesser capacity will result in higher cost but may be justified if performance characteristics and flow regime result in significantly more energy being generated. Several (at least three) proposals of capacity/number of units should be formulated for additional study. The total installed capacity, (e.g., sum of the units) of each alternative should most likely fall near the capacity selected in the previously completed task (say plus or minus 25%).

TABLE 5-1
PROJECT FORMULATION TASKS*/
MANUAL REFERENCE SECTIONS

Formulation Tasks	Volume	Manual Reference Section	Description
Initial Tasks	I	5	Par. of same title.
Formulate Power Features	V	2	Figure 2-1.
Refine Power Output Estimate	III	6	
Recompute Benefit Stream	II	3	
Cost Project Power Features	V, VI	ALL	
Select Project Power Features	I	5	Par. of same title.
Perform Sequential Routing	III	3	
Refine Power Features and Performance Characteristics	I	5	Par. of same title.
Finalize Project Costs/Benefits	II	2, 4	Tables 2-1, 4-3
	I	5	Par. "Project Cost Estimates".
	II	3	Par. "Hydroelectric Capacity and Energy".
Remaining Tasks	I	5	Par. of same title.

*Tasks identified are those shown on Figure 5-1 and are discussed in this section.

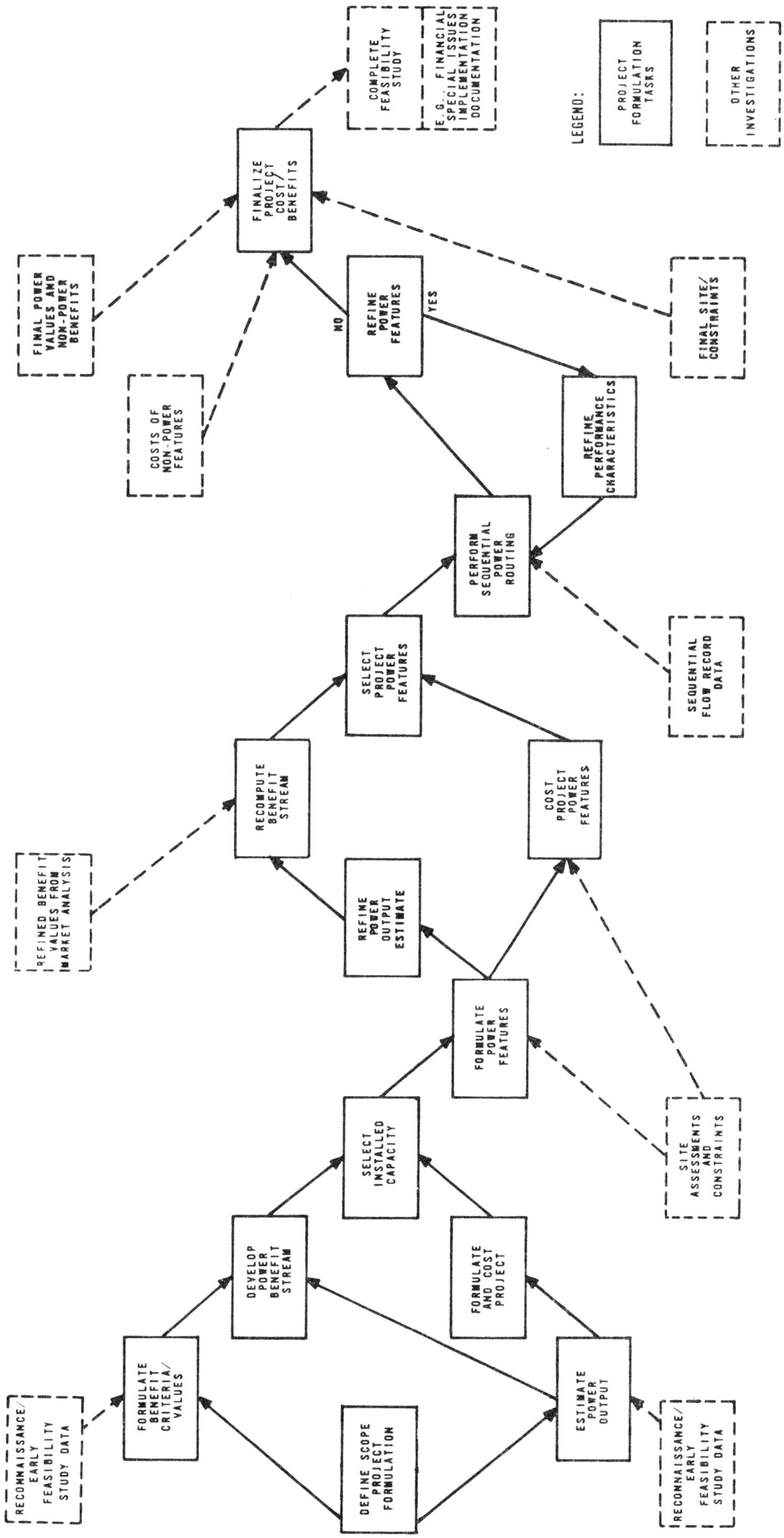

Figure 5-1. Project formulation components - feasibility

If more than a single turbine type seems suitable, and their performance characteristics are quite similar, the least costly is likely to be the best selection. If their performances are different (efficiency over operating range and limits of flow), alternatives for each turbine, and perhaps alternatives of multiple units, should be formulated for further analysis following the guidance of the previous paragraph.

Refine Power Output Estimate. A revised set (from the general data used in the initial tasks) of performance parameters (weighted efficiency and flow range) are to be used in computing refined capacity and energy values for each of the alternatives that were formulated. Flow duration techniques may continue to be adequate for this task. The alternative strategy of developing a continuous record of streamflow and performing sequential routing may be required for those instances in which significant water level fluctuations (e.g., changing head on turbines) are in evidence. See Section 6, Volume III for additional discussion.

Recompute Benefit Streams. Power values or power benefit criteria specific to the proposed project output should now be available. Capacity and energy values based on prevailing alternative power costs are the appropriate criteria. See Section 3, Volume II. A preliminary alternative set of values reflecting analysis of price shift trends should also be available for use (later) in testing the sensitivity of the project to price level changes. The power benefit stream for each alternative set of power features is computed and arrayed for further processing as the final output of this task.

Cost Project Power Features. The complete set of cost estimating charts, tables, and guidelines contained in Volumes V and VI are applicable. Care should be taken to make use of site assessment data and constraints to assure that the features for which costs are being estimated are physically feasible and sensible for the site. The cautions noted on the charts and tables of Volumes V and VI should be particularly noted so that specific layout and cost analysis will be performed if warranted. The output from this task is the initial construction cost, and annual operation, maintenance, repair, and replacement costs for each alternative set of power features.

Select Project Power Features. The power features selection is performed by arraying the cost and benefit streams for each of the alternative sets of power features and computing the net value of each. All other constraints being equal, the alternative exhibiting the highest net value should be selected. If a clear choice is not evident, reanalysis of the leading candidates using alternative power benefit values that include price shifts (representing for example rising fuel costs) should aid in narrowing the choice. The one or more (if still close) alternatives selected should be advanced to the next step in project formulation analysis.

The remaining tasks shown on Figure 5-1 provide for finalizing the power features, power output, and cost and benefit streams. Should the power output estimates from the refined sequential analysis not differ significantly from the prior estimates, additional refinement in the power features is unnecessary.

Perform Sequential Power Routing. The power output for use in completing the feasibility analysis should generally be developed by sequential power routing studies. If sequential routings were used in the previous analysis step, this task and the following task may be omitted. This added refinement assures that important sequential issues of fluctuating upstream and downstream water levels and flow passage by the site and the proper efficiency is selected for the turbine for partial turbine flows are properly incorporated in the analysis. Guidance for developing data and performing the sequential analysis is provided in Section 3 of Volume III. The sequential analysis should incorporate the performance (flow and efficiency) characteristics of the selected generating equipment. The analysis may be required for one or more of the alternatives that remain in contention.

Refine Power Features and Performance Characteristics. Sequential power analysis could yield information that would suggest refinement of turbine capacity/performance might be advantageous. Previous duration curve analysis necessarily required use of a single value (weighted) for head and a single value (average) for efficiency. The more complete simulation will accurately trace the turbine performance and may result in slightly higher or lower power and energy output estimates. The degree of variability (say plus or minus 10%) will suggest whether additional power feature refinement is warranted. The power output values developed at this stage will provide the basis for initiating development of power sales agreement should the feasibility findings be positive.

Although it is possible to perform the sequential power routing by hand methods, several of the computer programs mentioned in Volume III are available to public and private requestors and can be used to efficiently perform the analysis.

Finalize Project Cost/Benefits. The feasibility study findings will normally be presented in complete detail for the selected alternatives. Additional analysis and data (over that developed within the project formulation investigations) are needed to complete the economic feasibility assessment. If uncertainty has prohibited the selection of a single alternative, it may be necessary to present two or at most three alternatives in detail. Tables 2-1 and 4-3 of Volume II tabulate the categories of complete information needed for the feasibility assessment.

Construction cost estimates must be finalized for the power features and cost estimates for non-power features, such as integrity corrective actions, environmental enhancement and mitigation, and acquisition of water rights, lands, easements, and rights-of-way must be prepared. Studies performed to yield these latter estimates do not necessarily directly affect the power features selection and therefore can be performed concurrently with late stage formulation analysis. The integrity of the facility could well be adversely affected by the power features selected and should have been coordinated when performing the Formulate Power Features task. See paragraph Economic Analysis Cost Needs (later in this section) for additional comments on costs, benefits, discount rates, evaluation period, and cost escalation.

Project benefit estimates must also be finalized. Power benefits will be comprised of the product of the values of capacity and energy concluded from the marketing analysis and the dependable capacity (if any) and energy estimates derived from the sequential power routing analysis. Refinements of credits for dependable capacity and firm energy (see paragraph Hydroelectric Capacity and Energy, Section 3, Volume II, for amplification) should be determined and incorporated. A firm decision as to the incorporation of price escalation in the feasibility assessment is needed. It is suggested that if price escalation concepts are incorporated, the feasibility assessment also be performed and presented using price levels in existence at the time of study completion (e.g., a non-escalated project benefit analysis). Non-power project benefits should be estimated and incorporated as well at this stage. The non-power benefits that may be included should be carefully formulated so as to avoid discrediting the economic analysis. It seems prudent that only benefits that could be directly attributable to the project features be included. If a specific category (such as recreation, fisheries enhancement, etc.) is significant, a small scale analysis to separate costs for an incremental justification may be warranted.

Remaining Tasks. The other important elements of the feasibility analysis (e.g., financial, special issues, implementation, documentation) are directly influenced by the physical space and layout requirements of the specific power features selected and the resulting benefits and implementation costs. These assessments proceeded concurrently with project formulation tasks, receiving important inputs from the investigations. These other studies are now to be completed following the finalization of costs and benefits. The detail appropriate for concluding the remaining feasibility assessment tasks will depend on the economic feasibility finding. A positive finding will generally indicate implementation decision level detail is needed; a negative finding should probably result in terminating remaining studies. If a carefully staged study strategy, as suggested herein, has been followed, it should be the rare exception wherein the study has progressed to this point and a negative finding results.

Project Cost Estimates

Time streams of cash flow for both cost and income items are needed for economic and financial analysis. Time streams of cost are assembled from estimates of construction (physical facility) cost estimates, recurring costs, and indirect costs. Table 5-2 tabulates the array of cost items commonly needed to provide cost data for performance of economic and financial analysis. The following paragraphs discuss these items and suggest a systematic framework for dealing with cost issues.

Economic Analysis Cost Needs. Economic and financial analysis have been carefully defined as having distinctly different purposes, and consequently distinctly different (although very much similar) cost data. Economic feasibility analysis compares economic costs with project economic benefits. The comparison is properly made using a common value base. It is normal practice that costs and benefits be stated in the value terms existing at the time of feasibility study completion (e.g., stated in dollar values as of the study year). Federal government policies have generally also resulted in fixing price levels for valuing future costs and benefits in value terms as of the study date as well. The time frame commonly used for cost/benefit analysis begins the first year of project operation and extends through the project economic life. For example: a feasibility report may be completed in January 1980 (the dollar and price level year) with the project to begin operation in 1984 (the year the project benefits begin) and have an economic life extending until 2033 (50 years). The cost/benefit comparison would therefore be performed for the year 1984 using 1980 dollars and price levels. Project cost estimates for economic feasibility analysis using tables and charts presented in July 1978 dollars would be indexed upward to January 1980 dollar costs for use in the economic analysis. Recurring costs

TABLE 5-2
PROJECT COST ITEMS

Construction
- Power and Site Facilities
  - Electromechanical Features
  - Civil Features
- Facility Integrity Works
- Environmental Mitigation/Enhancement Works
- Licenses

Site Acquisition/Rental
- Existing Works
- Lands, Easements, and Rights-of-Way

Recurring
- Operation and Maintenance
- Repair and Replacement
- Water Rights/Use Fee
- Headwater Benefits (Federal Power Act)

Indirect
- Engineering, Construction Management, and Other Studies
- Interest During Construction
- Administration and Management
- Insurance

such as annual operation and maintenance would be forecast in 1980 dollars considering such issues as increased equipment needs and facilities age. Similar adjustment of the expected project benefits to assure they are likewise stated in 1980 dollars may be required. The alternative convention often adopted in the private sector is to state all project costs and benefits in dollar values as of the initial year of operation (e.g. escalate cost and benefit value for our example to represent 1984). Since small hydro projects are expected to be implemented in short time frames, the time and year statement of dollar values should usually not be critical.

The project evaluation period can vary among project proponents. Federal agencies often use 100-years, 50-years, on special occasions, (Corps of Engineers, 1975), as the evaluation period (economic life). Public agencies, and private as well, often use the expected useful FERC license period of about 45 years (license period of 50 years less start-up time). Another commonly used period, most consistent among private investors, is the loan repayment period of 30 to 40 years. In the absence of specific guidance to the contrary, an economic life of 50 years is suggested.

The inclusion of cost and value changes in economic feasibility analysis must be handled with care. If all items in the economic comparison are changing at the same rate, inclusion of these changes in the feasibility assessment would affect the findings because the cost and benefit streams are different in time. Careful treatment of real and inflation affected discount rates, theoretically (Howe, 1971; Hanke et al., 1975), would result in identical conclusions with and without general price escalation (inflation) being considered. This is normally not performed and in practical fact is quite difficult. The usual result of including cost and value escalation in projects such as small hydro (large initial cost followed by a small operation and maintenance cost, and a long stream of project benefits) is to make them appear economically more attractive, e.g., benefits grow with time while costs increase slightly based on operation and maintenance. The impetus for including value changes is the conviction that benefits will continue to rise knowing that some benefit elements are increasing more rapidly than the general inflation rate, e.g., fossil fuel. The argument is that ignoring these value shifts leads to incorrect decisions, e.g., the project may appear infeasible when it should be found to be feasible.

In principle, a price level change economic analysis should forecast the change in value of all aspects of the feasibility assessment, both the cost side and its several components, and the benefit side (e.g., alternative fuel costs) and its several components. The cost and benefit streams are then constructed from these forecasts and the feasibility assessment performed. An alternative is to forecast only the relative difference (from the general inflation trend) for the critical items such as fuel and construction costs.

The argument against including price level change or general cost escalation in economic feasibility analysis is that change in price forecasting is fraught with pitfalls that are both institutionally and technologically dependent. The resulting analyses thus often becomes suspect and a candidate for subjective manipulation, i.e., a means of justifying projects. This criticism is most often levied against public projects rather than private investments. If cost and value change analysis are adopted for the economic analysis, considerable care should be taken to rigorously observe the basic principles and to document the critical value change forecasts.

Table 5-3 has been prepared to aide in computations that consider escalation of project annual costs and benefits over the life of the project. The reason for the caution against indiscriminate use of escalation in benefit analysis is evident from examination of values in the table. For example, using a project evaluation period of 40 years, general escalation rate of 6% and discount rate of 9% (values commonly used in investment decisions for non-federal public agencies), would result in multiplying the average annual benefits by 2.21. In effect more than doubling the value of the benefits!

Financial Analysis Cost Needs. Financial feasibility analysis develops the specific cash flow (dollars in and out of the accounts of the project) characteristics of the project. The need is therefore to forecast the amount and timing of cash outflow and revenue income as accurately as possible. It is common practice for the cash flow analysis to be constructed for the project implementation period; the first year of operation often being critical to project cash reserves. See Section 6, Volume II. Construction costs are therefore indexed to the actual date of contract award, interest during construction added to bring the base to the project initial operation date, and the revenue stream adjusted based on anticipated power sale contract provisions for payment of project output. Recurring costs (operations and maintenance) are frequently escalated based on increased costs to service aging equipment and on anticipated general cost inflation. Private sector economic analysis often is very near to a financial cash flow analysis because of the tendency to classify economic costs as the cash flow from project accounts and benefits as strictly contract revenues. In effect the scope of project costs and benefits are the "cash" impacts on the private developer.

If there were no cost inflation, no borrowing required, and if project revenues captured all project benefits exactly, the economic cost and benefit streams for the economic analysis would be identical to the cost and revenue cash flow streams for the financial analysis.

Construction Costs. Cost estimating charts and tables are included in Volume V and VI that encompass virtually all aspects of the civil and electromechanical features of power additions. The information is presented in July 1978 dollars and a method for indexing to future dates is included. Unusual site conditions, use of

TABLE 5-3
PLANNING PERIOD ESCALATION ADJUSTMENT RATIOS

DISCOUNT RATE (%)	ESCALATION RATE (%) 2	3	4	5	6	7	8
PLANNING PERIOD - 30 YEARS							
2	1.34	1.56	1.84	2.17	2.57	3.06	3.66
4	1.30	1.50	1.73	2.02	2.36	2.78	3.28
6	1.27	1.44	1.64	1.89	2.18	2.53	2.95
8	1.24	1.39	1.57	1.77	2.02	2.31	2.66
9	1.22	1.37	1.53	1.72	1.95	2.22	2.54
10	1.21	1.34	1.50	1.68	1.89	2.13	2.42
12	1.19	1.31	1.44	1.59	1.77	1.98	2.23
14	1.17	1.27	1.39	1.52	1.68	1.86	2.06
16	1.15	1.25	1.35	1.47	1.60	1.75	1.93
18	1.14	1.22	1.32	1.42	1.54	1.67	1.82
20	1.13	1.20	1.29	1.38	1.48	1.60	1.73
25	1.11	1.17	1.23	1.31	1.39	1.47	1.57
PLANNING PERIOD - 40 YEARS							
2	1.46	1.80	2.23	2.80	3.54	4.52	5.82
4	1.39	1.67	2.02	2.47	3.06	3.82	4.81
6	1.33	1.56	1.84	2.20	2.66	3.24	3.99
8	1.28	1.47	1.70	1.98	2.34	2.79	3.35
9	1.26	1.43	1.64	1.89	2.21	2.60	3.10
10	1.24	1.40	1.58	1.81	2.09	2.44	2.87
12	1.21	1.34	1.50	1.68	1.91	2.18	2.51
14	1.18	1.30	1.43	1.58	1.76	1.98	2.24
16	1.16	1.26	1.37	1.50	1.65	1.83	2.04
18	1.15	1.23	1.33	1.44	1.57	1.72	1.89
20	1.13	1.21	1.30	1.39	1.50	1.63	1.77
25	1.11	1.17	1.24	1.31	1.39	1.48	1.58
PLANNING PERIOD - 50 YEARS							
2	1.59	2.06	2.71	3.63	4.93	6.77	9.41
4	1.47	1.84	2.33	3.00	3.93	5.22	7.04
6	1.38	1.66	2.03	2.51	3.17	4.07	5.30
8	1.31	1.53	1.80	2.16	2.63	3.25	4.09
9	1.28	1.47	1.72	2.03	2.42	2.95	3.64
10	1.26	1.43	1.64	1.91	2.25	2.69	3.27
12	1.22	1.36	1.53	1.73	1.99	2.31	2.72
14	1.19	1.30	1.44	1.61	1.81	2.05	2.35
16	1.16	1.27	1.38	1.52	1.68	1.87	2.10
18	1.15	1.23	1.34	1.45	1.58	1.74	1.92
20	1.13	1.21	1.30	1.40	1.51	1.64	1.79
25	1.11	1.17	1.24	1.31	1.39	1.49	1.59
	2	3	4	5	6	7	8
	ESCALATION RATE (%)						

ADJUSTMENT RATIO EXAMPLE

GIVEN: Annual Energy Generation – 10×10^6 kWh
Value of Energy – 25 mills/kWh
Investment Cost* – $2,000,000
Annual O & M Cost – $30,000
Growth in Power Value – 6% per year
Growth in O & M Cost – 4% per year
Discount Rate – 9%
Planning Period – 30 years

* Already Escalated to Construction Date Using Cost Indices

NO ESCALATION

ANNUAL COST

Investment	= $2. x 10^6 x 0.0973	=	$194,600
O & M		=	30,000
		TOTAL	$224,600

ANNUAL BENEFITS

Energy	= 10. x 10^6 x $0.025	=	$250,000

ESCALATION CONSIDERED

Ratio (6%, 9%) = 1.95
Ratio (4%, 9%) = 1.53

ANNUAL COST

Investment	= $2. x 10^6 x 0.0973	=	$194,600
O & M	= $30,000 x 1.53	=	45,900
		TOTAL	$240,500

ANNUAL BENEFITS

Energy	= $250,000 x 1.95	=	$487,500

REVISED OCTOBER 1979

an existing abandoned powerhouse, refurbishing equipment, etc., could result in the requirement to perform feasibility layouts and design, computing construction material quantities, and preparing a specific cost estimate. Prevailing industry cost estimating methods would be employed (see Case Studies). A common practice in estimating turbines and generators when costs are a critical issue, is to solicit preliminary quotes from equipment suppliers. Care should be taken to recognize the values as only estimates, not firm price bids. Supplier lists are included in Volume V.

Cost estimates for facility remedial work (integrity rehabilitation) are not particularly amendable to generalization and therefore the feasibility design layout approach as described above is usually necessary. Guidance on major elements of cost for rehabilitation is included in Volume IV. Data contained in Volume VI for gates, valves, and penstocks may be helpful.

Cost estimating guides for environmental enhancement and mitigation works (such as fish hatcheries and ladders) are not included in this manual. The range of potential mitigation alternatives prohibits formulation of generalized data at this time. Specialists in such issues should be consulted if such features are determined to be a critical item in project development.

It is common practice to add a contingency to construction costs to allow for uncertainties and minor ommissions. Contingencies are often in the range of the 10% to 20% depending on project complexity. The construction cost components could each have a separate contingency applied if warranted. Normally a single contingeny value is applied to the sum.

Several acquisiton/rental fee type costs may need to be estimated. Land acquisition for siting power and other features may be required. Temporary and permanent easements and rights-of-way could likewise be needed.

Recurring Costs. The recurring costs include such items as operation and maintenance, repairs, replacements, and insurance (for private developers). The discussion in Section 4 is pertinent and repeated here. "Operation and maintenance costs can vary considerably depending on present staff resources of the project proponent, the site proximity to other sites, and the intended degree of on site operation requirements. The value used should not be less than a base (suggested as $20,000/year) and may range upwards to 4% if the project proponent cannot efficiently integrate the plant into their work program." Specific guidance is contained in the last section of Volumes V and VI.

Fees may be payable for use of water to generate power. Private developers at federal sites are likely to be required to pay an upstream storage fee. FERC also requires private developers (other than federal) to pay for any storage and re-regulating of the water supply above the project, provided that the upstream entity either holds a FERC license or permit, or is a federal agency. This is the so-called "headwater benefit." Other financial arrangements depending on the owner and project proponent may be needed. The purpose of the analysis (economic/financial) and the perspective of the proponent (federal, public, private) will determine the need and influence the degree to which the dollar transfers between the project development parties are included in project analyses.

Indirect Costs. The discussion in subsection Develop Cost Stream, Section 4 of this volume, is pertinent and is repeated here. "All investigations, management, engineering and administrative costs that are needed to implement the project and continue it in service are appropriately included in the project feasibility analysis." These indirect costs may be estimated directly (e.g., the analysis of the component factors) or included as a multiplier of the investment costs. Volumes V and VI suggest a multiplier of 20% of the total construction cost plus contingencies as a mid value. A table documenting the elements of this multiplier is included in the last section of both volumes.

Licenses, Permits, and Approvals

The feasibility report is the primary source of the information needed to secure the necessary government approvals to proceed with project implementation. A discussion of these issues is included here to alert project investigators to their requirements with the view that parts of the feasibility investigation may be made to efficiently serve these information needs as well.

Federal, state and local governments all have certain requirements that must be satisfied prior to construction and operation of a hydropower plant. Some agencies within these governments only require notification while others require specific data about the project and issue licenses or permits for the construction and operation of the plant. Realizing that a list of all the local, state and federal agencies would be difficult if not impossible to create, a general discussion is provided about local, state, and federal responsibilities and types of agencies on the local and state level that are usually interested in a hydroelectric project. The federal agencies are coordinated for the most part through the federal licensing process. The Rollins Power Project case study (Exhibit II) includes a listing of the permits that were necessary to implement that project.

State and Local Requirements. States operate in several different ways. Some states have resources agencies which are comprised of most of the departments which need to be contacted. In this case coordination is generally straight forward. States that have separate agencies without a main coordination office require the applicant to contact each office individually to initiate compliance with state regulations. Agencies most often contacted are listed in Table 5-4. Many of these state agencies will also be contacted by federal agencies which have similar responsibilities but on a national level. Some state agencies may defer comment by pointing out that a federal license is required and they they will make comments and recommendations on the application for federal license. If comments are deferred compliance with state laws still apply and it would be useful to obtain the laws, regulations, and guidelines the agency will use to evaluate the application so that these concerns are addressed in the application. Some of the major state concerns are water rights, fish and wildlife habitat, water quality, compliance with environmental laws, and dam safety.

TABLE 5-4
STATE CONTACT AGENCIES

Department of Dam Safety
State Energy Office/Commission
Department of Fish and Game/Wildlife
Flood Control/Reclamation Board
Governor's Office
State Historical Preservation Officer
Department of Planning and Research
Public Utilities Commission
Resources Agency
Water Quality Control Board
Department of Water Resources
Division/Board of Water Rights

In most instances local governments, county or city planning department, will be the lead agency with respect to coordination within the state and compliance with state environmental laws. They may also have ordinances and laws concerning construction, employment, road weight limits, and possibly generation, to name a few, which should be complied with.

Federal Energy Regulatory Commission. The Federal Energy Regulatory Commission (FERC), Department of Energy, formerly the Federal Power Commission (FPC), is the lead federal agency and issues licenses for all non-federal hydroelectric projects which fall under their jurisdiction (Code of Federal Regulations, Title 18). Very few projects are exempt from FERC licensing requirements. Being the lead federal agency the FERC coordinates all comments on environmental statements, contacts all other federal agencies that require coordination, coordinates with the appropriate state governors offices and agencies, holds hearings with Administrative Law Judges presiding to settle legal and jurisdictional disputes, and issues a federal license for the construction and operation of the project. Other federal agencies which issue permits or approval which must be contacted individually are discussed later in this section.

Projects requiring a FERC license are divided into two classes based on installed capacity. Minor projects have an installed capacity of 2000 horsepower (1500 kW) or less while major projects have an installed capacity of more than 2000 horsepower. Applications for license are submitted directly to the FERC for processing and approval. Forms, procedures, and requirements for filing may be obtained from the FERC, Washington, D.C. office or any of their regional offices (see Exhibit I, Volume II). An application for a FERC major license for an unconstructed project must contain, in general, the following information:

Applicants name and address.
Applicants business status.
Description of the project (civil features).
Location of the project.
Lands and reservations of the U.S. affected by the project.
Description of ultimate scheme of development (electromechanical features).
Proposed use or market for the power.
Location and capacity of other electric facilities owned or operated by the applicant.
Description of any historical or archeological properties.
Detailed statement of environmental factors.
Other data which the applicant may consider pertinent.

This information is presented in the application in the form of Exhibits. Contents of an application for a minor license, plants with 2000 horsepower (1500 kW) or less installed capacity, are similar but do not require as much detail on most subjects (FERC, 1978). Also applications for proposed or existing plants at existing impoundments have slightly different requirements with respect to the detail required for some exhibits. In general, use of an existing impoundment does not create the same magnitude of environmental impacts as construction of a dam and new reservoir, thereby reducing the time, effort, and coordination required to evaluate the project. Small hydropower developments at existing impoundments are included in this last analysis and, therefore, applications can usually be processed in a shorter amount of time and with less expense than those projects proposing construction of a dam and reservoir.

The FERC also issues preliminary permits for projects of more than 2000 horsepower (1500 kW) installed capacity for the purpose of enabling the applicant to secure the data and perform the acts required by law for filing an application for the issuance of a license (Code of Federal Regulations, Title 18). The preliminary permit retains the application right of the applicant with respect to the site so that his application for license may not be preempted by another applicant's application. It would seem prudent for a developer to apply for a preliminary permit on completion of a positive reconnaissance study so as to establish his application right. The maximum duration for which a preliminary permit may be issued is three years and it may not be renewed.

U.S. Army Corps of Engineers. A permit must be obtained from the U.S. Army Corps of Engineers (or a negative determination that no permit is needed) to locate a structure, excavate, or discharge dredged or fill material in waters of the United States (Corps of Engineers, 1977) . Since most hydroelectric power plants are located in or adjacent to a river and require excavation, a permit must be obtained. The reference, U.S. Army Corps of Engineers Pamphlet (EP) 1145-2-1, provides the procedure for filing and the requirements for a permit. To initiate the process, contact the District Engineer who has jurisdiction over the area where the structure will be built. Request a copy of EP 1145-2-1, an application form (ENG Form 4345), and any special instructions that may not be furnished in the pamphlet.

The permit investigation process requires furnishing a detailed description of the location and nature of the proposed activity, including the purpose, use, type of structures, types of vessels (if any) that will use the facility, facilities for handling wastes, and the type, composition, and quantity of dredged or fill material.

Other Federal Agencies. Several other federal agencies become involved at the time of project implementation. Radio communication permits (for remote operation) are required by the Federal Communications Commission and construction that might obstruct airspace (transmission towers) must be reported to the Federal Aviation Administration. A Water Quality Certificate issued in accordance with Section 401 of the Federal Water Pollution Control Act is generally required. State organizations such as Regional Water Quality Boards are normally the administering agency.

Time, Cost, and Resources for Feasibility Studies

The time, cost, and manpower resources required to perform feasiblity studies for small hydroelectric power plant additions varies depending on expected plant size, site conditions, specific scope and depth of study, and availability of information (basic data and prior reconnaissance assessment). Each of the five support manual volumes provides general guidance on this topic in their respective subject areas. The following paragraphs discuss the range of costs and resources that are likely to be needed for the studies as a whole. The unique characteristics of each project should, however, be evaluated in scheduling use of in-house personnel or in procuring professional services for specific feasibility investigations.

The American Society of Civil Engineers has published general guidelines for the performance of engineering services (ASCE, 1972). The guidelines suggest that professional services for projects in the small hydro category may cost from 6% to 10% as a proportion of construction cost. "Preliminary Phase" studies (those prior to final design) may require up to 40% of the basic compensation yielding total preliminary phase professional services costs of 2.5% to 4.0% of construction cost. Feasibility studies are generally acknowledged as comprising 1/3 to 1/2 of "Preliminary Phase" costs. Noting that marketing, financial, and increased special studies needed for the feasibility study are likely, the range of 1.5% to 3% of estimated construction cost seems appropriate.

Using 2.5% as a conservative estimate, feasibility study costs could range from $25,000 (80 to 110 man-days) for a 1 MW plant to $150,000 (600 to 750 man-days) for the larger plants. The time required to perform the feasibility study could range from 60 days for the small, relatively simple power addition to upwards of 6 to 9 months for larger more complex projects.

The participating professionals include civil, electrical, and mechanical engineers, power economists, and especially for private proponent projects, the services of financial specialists. Projects that significantly alter the flow regime or physical environment will likely need the services of water quality and fish and wildlife specialists.

REFERENCES

American Society of Civil Engineers, ASCE Manuals and Reports on *Engineering Practice, No-45, Consulting Engineer . . . . A Guide for the Engagement of Engineering Services*, 1972.

Brown, Peter, "Social, Legal and Institutional Considerations in Planning for Small-Scale Hydroelectric Development," presented at Water Systems '79 American Society of Civil Engineers Specialty Conference, Houston, Texas, 1979.

Code of Federal Regulations (CFR), Title 18, Conservation of Power and Water Resources, Parts 1 to 149, 1978.

95th Congress, 2d Session, PL 95-617, *Public Utilities Regulatory Policies Act*, November 1978.

Federal Energy Regulatory Commission, (FERC), "Short-Form License (Minor)," Docket No. RM78-9, Order No. 11, September 1978.

Federal Power Commission, *Hydroelectric Power Resources of the United States, Developed and Undeveloped*, January 1976.

Gladwell, John S. and Warnick, Calvin C., *Low Head Hydro, An Examination of an Alternative Energy Source*, Idaho Water Resources Research Institute, September 1978.

Hanke, Steve H.; Carrier, Philip H.; and Bugg, Paul, "Project Evaluation During Inflation," *Water Resources Research* Vol. II, No. 4, August 1975.

Howe, Charles W., "Benefit Cost Analysis for Water System Planning," *American Geophysical Union - Water Resources Monograph 2*, 1971.

Institute for Water Resources, *National Hydroelectric Power Study - Plan of Study*, January 1979.

Macaitis, Bill, and Schonsett, Edward, "Energy Production from the Chicago Tunnel and Reservoir Plan," Presented at Water Systems '79 American Society of Civil Engineers Specialty Conference, Houston, Texas, February 1979.

McDonald, Richard J., *Estimate of National Hydroelectric Power Potential at Existing Dams*, Institute for Water Resources, U.S. Army Corps of Engineers, July 1977.

O'Brien, Eugene; George, Alexander C.; and Purdy, Clayton C., *Evaluation of Small Hydroelectric Potential*, Tippets-Abbett-McCarthy-Stratton, April 1979.

Trisko, Ralph L., *Hydroelectric Power Potential at Corps of Engineers Projects*, IWR Report 75 R1, Institute for Water Resources, U.S. Army Corps of Engineers, July 1975.

U.S. Army Corps of Engineers, *National Program of Inspection of Dams*, May 1979.

U.S. Army Corps of Engineers, *Permit Program . . . . A Guide for Applicants*, ER 1145-2-1, 1 November 1977.

U.S. Army Corps of Engineers, *Planning Process-Multiobjective Planning Framework*, ER 1105-2-200, November 1975.

U.S. Army Corps of Engineers, *Digest of Water Resources Policies*, EP 1165-2-1, January 1975.

United Nations, *Manual of Standards and Criteria for Planning Water Resource Projects*, Water Resources Series No. 26, 1964.

EXHIBIT I
GREAT FALLS HYDROELECTRIC PROJECT
CASE STUDY

CONTENTS

SECTION 1
INTRODUCTION

This case study describes the feasibility investigation of the Great Falls Hydroelectric Project, located in and owned by the City of Paterson, New Jersey. It applies the methodology for preparing a feasibility study for small hydro power projects presented in this manual.

The project feasibility of the Great Falls project has already been determined by a feasibility report (Development and Resources Corporation, 1978) prepared for the City of Paterson. This case study provides a basis for comparing the procedures and methods described in the manual to the results obtained by in-depth feasibility study.

Overview of Findings

The following overview of the feasibility case study findings are categorized according to the five manual components, followed by a summary.

Hydrologic Studies. The hydrologic studies were based on daily average flow conditions for the period 1950-1960. These 10 years of data were assumed to be representative of the longer data period available for the period 1897-1976. The daily records for the 10-year representative period were used to simulate runoff and calculate the resulting potential energy production of between 22.1 million kWh and 32.3 million kWh on an annual basis with an installed capacity of between 5,100 kW and 7,875 kW.

Existing Facility Integrity. The Great Falls dam was built in the period 1838-1840 of large blocks of masonry stone with a total length of 315 feet and a height varying from 8 to 15 feet, and is of the gravity overflow design type. Field inspection of the dam showed there is significant deterioration and erosion of the existing stone masonry section to the point where about 10 percent of the stone section requires replacement. Several alternatives were examined in lieu of restoration of the dam and restoration of existing structure was chosen for historical reasons. The total cost of $1,056,700 was close to other alternatives. The powerhouse and appurtenant structures were found to be in good condition and could be utilized for the project after being refurbished.

Electromechanical Equipment. This investigation studied 17 alternatives involving four manufacturers of hydroturbine equipment. Of the 17, four were chosen for detailed comparisons as alternatives and are presented in this case study. The four manufacturers considered were Allis-Chalmers, Leffel, Ossberger, and Tampella. The estimated installed equipment costs in 1978 ranged from $2,933,850 to $5,074,100. It was determined that only after firm bids for turbine and generation equipment, guaranteed performance data, delivery times, and complete dimensional data had been obtained, could the final equipment selection be made.

Civil Features. The total costs of the civil works for this project, not including the dam restoration cost, were estimated at between $639,200 to $976,200, representing the four alternatives analyzed in the case study. These costs represent an average of 21 percent of the total project costs. This is consistent with the range of civil feature costs identified on Figure 1-1 of Volume 4 of the manual which placed the minimum civil features costs at 15 percent and the maximum at 45 percent.

Economic and Financial Analysis. The financing required to construct the project would vary from between 5.9 and 7.9 million dollars. This further breaks down into a first year (1981) annual cost ranging between $607,000 and $808,000 which includes debt amortization based on a 40-year project life, seven percent interest money, annual operating costs, and repair and replacement costs. The corresponding value of the energy produced would range from between $726,000 and $962,000 on an average production basis for the first year of operation.

The cost of service in 1981 dollars (the first year of project operation) would vary from 21 to 25 mills per kilowatt hour. This compares to a value of energy of around three cents per kilowatt hour in 1981, based on the energy generated at the Great Falls site replacing the fuel costs for oil fired generations.

Summary. The results of the feasibility study show that installed capacities between 5,400 and 10,500 kilowatts are possible for new equipment and that with the rehabilitation and upgrading of existing turbine and generation equipment 5,100 kilowatts could be realized. The average annual production would range between 22,000,000 and 37,000,000 kWh. The project would be run-of-the river. The feasibility study includes 17 alternatives, while this case selected four alternatives to cover the range of turbine equipment.

Project Description

The Great Falls Hydroelectric Project is located in the City of Paterson, New Jersey. The location of the existing powerhouse and diversion dam is indicated on Figure 1-1. The drainage area above the project site as measured at Little Falls is 762 square miles. The mean annual flow is 730 cubic feet per second. The facilities that make up the Great Falls Hydroelectric Generating Facility consist of a masonry stone diversion dam, concrete intake and forebay structure, gated concrete control structure, steel-lined penstocks, and powerhouse constructed of concrete and brick. The powerhouse is located immediately downstream of Great Falls, a natural rock barrier created by a massive basalt sill.

The site is owned by the City of Paterson, New Jersey, and has significant historical importance. The

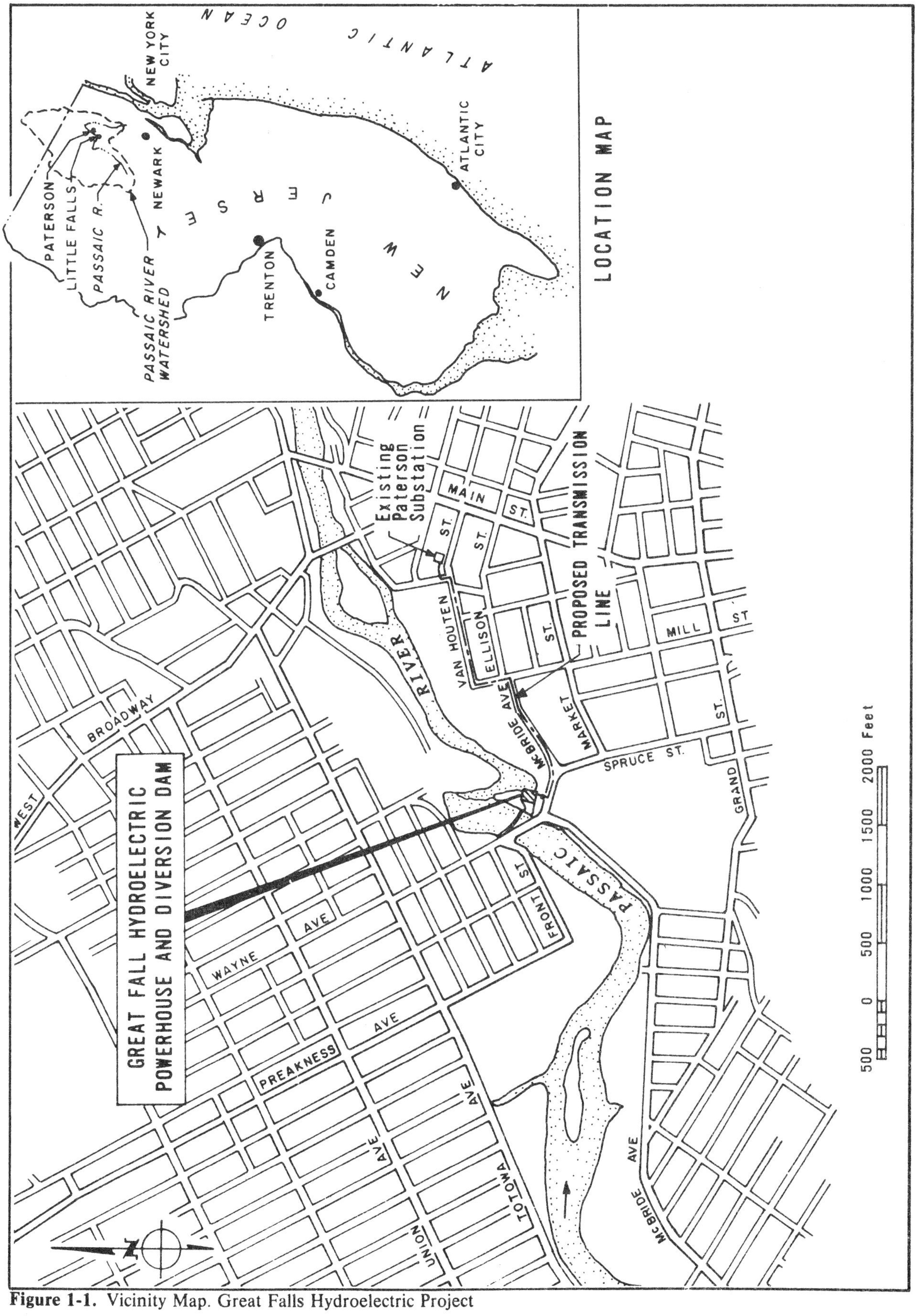

Figure 1-1. Vicinity Map. Great Falls Hydroelectric Project

water power from the site was developed as early as 1794 through a series of three raceways which promoted the establishment of many manufacturing plants. In 1912, waterwheels gave way to a hydroelectric plant. In 1914, the plant was completed and conversion to electrical power was begun by the mills in the area. The plant was decommissioned in 1969 after it was determined that the facilities were in need of major repairs. The raceways are still used in a limited way for water supply and for processing water for manufacturers.

In 1971, Congress declared the Great Falls site a National Historical Landmark and the City has since created a park in the area surrounding the Falls. The view of the Great Falls, located below the diversion dam, is considered to be a tourist attraction and release of approximately 200 cfs of water during the low flow summer months is required to maintain the Falls aesthetic appearance.

The project qualifies for a tax-exempt status since the total financing required is less than $10 million. This tax-exempt status has had some impact on the economic feasibility of the project.

A license to construct and operate the project has been filed with the Federal Energy Regulatory Commission (FERC) and is under review as of January 1979.

Project Formulation and Case Study Data

In August 1978 the Department of Community Development of the City of Paterson authorized consultant services for the preparation of a feasibility study for reactivating hydroelectric power at the Great Falls site. Earlier, in 1976, a reconnaissance level study was made that addressed itself to the Restoration of the Diversion Dam and Power Plant for the Great Falls Historic District. This previous study, coupled with data contained in the Passaic River Survey Report for Water Resource Development (U.S. Army Corps of Engineers, 1971) and the Flood Insurance Study of the Passaic River (U.S. Department of Housing and Urban Development, Federal Insurance Administration, 1975), as well as independent data collection, served as the basis for the case study.

The data and information presented in past reports have been put into the analysis framework as presented in the manual and all results were recalculated then compared.

SECTION 2
HYDROLOGIC STUDIES

This section describes studies performed to determine the adequacy of the facility to pass flood flows and to calculate energy production at the site. Uses of the guidelines contained in Volume III of the manual are indicated.

Passage of Flood Flows

Data. Adequate daily flow records are available for the Passaic River to allow flood frequency analyses to be performed. USGS daily average flow records are available for the Passaic River gage (USGS 01389500) from 1897 through 1976.

Topographic maps and river cross sections from the New Jersey State Riparian Streams and Waterways Survey of 1935 were used in assessing river hydraulics. Previous studies were utilized to obtain information on Passaic River flood flows and water surface profiles (U.S. Army Corps of Engineers, 1971) (U.S. Department of Housing and Urban Development, Flood Insurance Administration, 1975).

Flood Flow and Water Surface Elevation. Flood discharge frequency relationships included in the Passaic River Survey Report were used to establish the design flood flow of 23,500 cfs for an average return period of 100 years. The 100-year flood event provides a water surface elevation at the diversion dam that produces a loading condition appropriate for analysis of the dam's structural integrity under flood flow conditions. This report used the log-Pearson Type III distribution to establish the peak flow-exceedance interval relationship, as is recommended in Volume III of the manual.

River cross sections and the hydraulic characteristics of the current overflow dam structure were used to calculate headwater and tailwater rating curves. Some upstream flooding occurs for the 100-year flood event.

Analysis showed that the current overflow diversion structure is capable of passing the selected design flow. The structural integrity of the dam under flood conditions is examined in the Integrity Section. An analysis was also made to determine the flooding limits that would result from a breaching of the diversion dam. Results show that no downstream flooding would be caused by a dam breach.

One of the dam options considered was construction of a new concrete dam just downstream from the existing dam with a higher crest elevation of 120 feet. The structure was designed with gates so it would be hydraulically equivalent to the current structure. The required gate structure would be approximately 150 feet long and 10 feet high. This option has not been ruled out but for the purposes of this case study only repair of the existing structure was considered.

Power Production

Power production for all options was computed on a detailed level by sequential power routing using daily flow records and a detailed model of power generation. The simulation accounted for turbine and other equipment efficiencies, net head available to the turbines, multiple turbine scheduling, and scenic diversion over Great Falls in the summer months. Sequential power routing is the technique recommended in Volume III of the manual for use during the feasibility level investigation when the increased accuracy over flow-duration analysis is desirable.

Data and Assumptions. After examining the historical record from 1897-1976, project power output was calculated using the records for water years 1950-1960, a representative decade. The project was simulated as a run-of-the-river project because of the very small amount of working storage available. Consequently, flow was used as it occurred at the gage. Daily average flow was used since monthly average flows would tend to overstate power production in this case. The Passaic River has a fairly large flow variation, particularly in the fall and spring. To preserve the scenic value of Great Falls during the low flow months, 200 cfs for the hours between 10:00 a.m. and 8:00 p.m. during June, July, and August were planned for direction over the Falls, thus bypassing the powerhouse. Headwater and tailwater rating curves developed from river cross sections were used in calculating the net head availability to the turbines. All the options considered use of multi-turbines.

The turbine efficiencies were supplied by the manufacturers as a function of the specified flow and head availability. See Section V for a detailed comparison of turbine efficiencies. Other efficiencies and losses were used as shown below:

Item	Percent Loss
Single stage speed increaser	2.5
Double stage speed increaser	4.0
Generators over 1000 kW	5.0
Step-up transformers	2.0
Forced outages	3.0

Results. Energy production for the four options considered are shown on Figure 2-1. Also shown is the minimum energy production as a percentage of average annual production. These results show that on an annual basis substantial fluctuation occurs in energy production. For planning purposes, a worst case analysis was based on energy production at no more than 65 percent of average.

ENERGY PRODUCTION FOR FOUR ALTERNATIVES

	Alternative 1 Rehabilitation (Allis-Chalmers)	Alternative 2 New Horiz. Runners (Leffel)	Alternative 3 Cross Flow (Ossberger)	Alternative 4 Tube Turbines (Tampella)
Annual Energy Production				
Assumed Installed Capacity (kw)	5,100	7,500	6,800	7,875
Average (Millions of kwh)	24.4	30.8	27.9	32.3
Maximum	34.2	45.2	40.3	47.6
Minimum	17.1	20.1	18.7	21.0
Plant Factor (%) 1/	56%	49%	51%	48%
Minimum Production as % of Average	70%	65%	67%	65%

1/ Based on actual production and maximum possible production after accounting for all losses except forced outages.

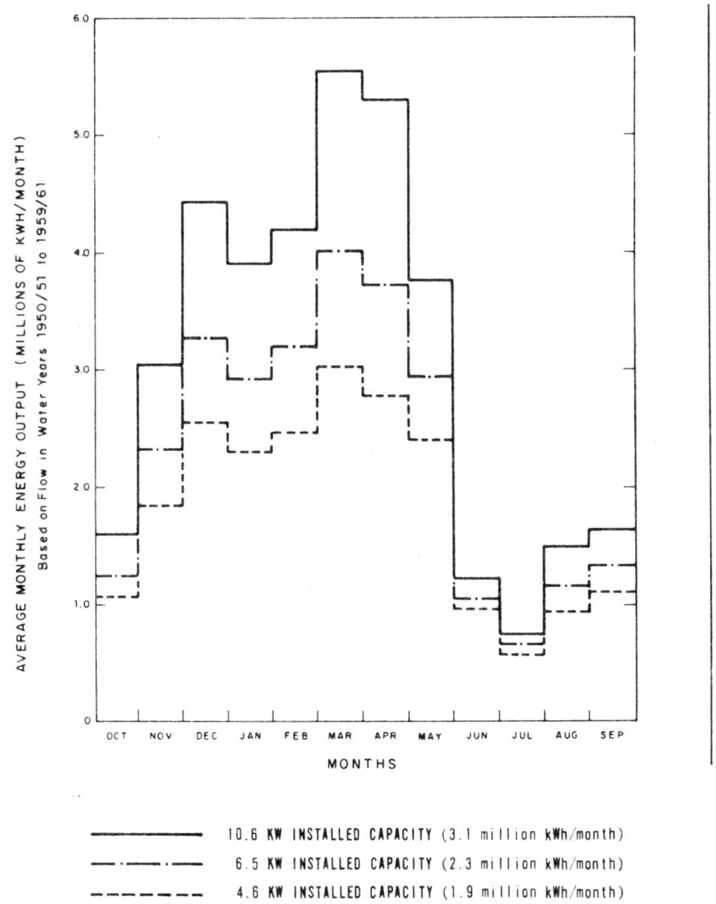

AVERAGE MONTHLY ENERGY OUTPUT FOR THREE SIZES OF INSTALLED CAPACITY

Figure 2-1. Energy Production

The analysis allowed monthly average energy production to be compared for different sized installations. Figure 2-1 displays these results for three different installed capacities. As shown, additional capacity adds little to summer energy production.

The use of daily flow also determines whether periods of non-generation occurred. For all of the options considered, extended periods of no production occur in the summer and, to a lesser extent, in spring and fall months. Consequently, the project has no firm capacity or energy and is strictly run-of-the-river.

SECTION 3
INTEGRITY ASSESSMENT

This section describes the investigation performed to assess the structural integrity of the existing diversion dam and to estimate the cost of rehabilitation of the dam. The lack of engineering records showing the diversion dam's dimensions or methods of construction required making the following assumptions in assessing the dam's structural integrity:

1. Assuming a representative cross section based on field observations and experience gained on similar structures.
2. Assuming the strength parameters of the dam's foundation based on a reconnaissance level engineering geologic investigation and engineering experience.
3. Assuming the strength properties of the granite stone building material for the dam and the cement mortar used to bond the granite stone together.

Loading Criteria

Loading criteria for use in analyzing the dam's structural integrity were developed from the 79 years of daily flow records for the Passaic River at the dam site. Flow frequency curves developed by use of the log-Pearson Type III analysis were used to establish the expected flow for a given frequency storm event. This information, when combined with the developed diversion dam's headwater and tailwater rating curves, allowed selection of appropriate water surface elevations for use in establishing the loading cases.

The design and loading criteria adopted to assess the dam's structural adequacy were based on three cases. These were 1) normal operating conditions, 2) normal flow conditions with .1 g horizontal seismic loading, and 3) flood conditions with the flow being increased from a normal 200 cfs to 23,500 cfs. The adopted criteria follows guidelines as suggested in Section 3, Volume 4 of the manual.

Results

Table 3-1 displays the results of the evaluation of the dam's structural integrity.

These results show that the existing dam has factors of safety below those generally regarded as acceptable for sliding and overturning. Historical records indicate that the original dam section was anchored "to the rocky bed with powerful clamps of iron." The condition of these "clamps" is unknown and to assure the safety of the restored dam for the full anticipated project life, it was decided to provide anchorage by means of a concrete slab placed on the upstream face of the dam. The concrete slab would be reinforced and dowelled to the dam section, and secured to the bedrock by steel anchors grouted into the foundation.

Restoration Costs

The estimated costs for restoration of the diversion dam were based on the preliminary designs, estimated construction quantities, unit costs from cost estimating guides and costs from other similar projects in the engineers' files (Dodge Guide to Public Works and Heavy Construction Costs, 1978 and Engineering News Record Quarterly Cost Roundup, 1978). These reference sources are identified in the manuals. Table 3-2 displays the estimated costs for restoration of the dam including contingencies, engineering and administration.

TABLE 3-1
STRUCTURAL INTEGRITY EVALUATION - DIVERSION DAM

Loading Condition	Uplift Req'd	Uplift Actual	Sliding Req'd	Sliding Actual	Overturning *Req'd	Overturning Actual	Stresses (psi) Toe	Stresses (psi) Heel
Case I Normal	1.50	2.9	1.50	1.66	2.0	1.55	21.6	4.2
Case 2 Seismic	1.25	2.9	1.25	0.81	1.5	1.15	29.9	5.3
Case 3 Flood Flow	1.25	2.2	1.25	1.16	1.5	1.15	29.7	11.4

*The factor of safety against sliding was calculated as being the difference between the summation of the horizontal and uplift forces multiplied by a sliding factor of 0.7 divided by the summation of the Vertical forces (USBR Design of Small Dams, 1965, p. 240).

TABLE 3-2
RESTORATION COSTS

Item	Unit	Quantity	Unit Cost $	Cost $
Cofferdam - first stage	LF	400	700	280,000
Cofferdam - second stage	LF	400	300	120,000
Dewatering	LS			75,000
Excavation - Earth	CY	1,800	10	18,000
Concrete - Reinforced	CY	275	200	55,000
Rock Anchors	LF	1,500	20	30,000
Reinforcing Steel	LBS	40,000	.40	16,000
Cofferdam Removal	LF	400	50	20,000
Replace Stone	CY	140	350	49,000
Reconstruct Stone	CY	350	250	87,500
Grouting Masonry	LS			18,000
Subtotal				768,500
Contingencies at 25%				192,125
Subtotal				960,625
Engineering and Administration at 10%				96,000
TOTAL				1,056,700

SECTION 4
CIVIL FEATURES

This section describes and estimates the cost of the project civil works, excluding the dam, which is presented in the Integrity Section. The Great Falls site has been designated a National Historic Landmark and has certain features and facilities that have been maintained and are suitable for use without additional repair or replacement. In addition, by having the site declared a National Historic Landmark, reuse of the site and facilities carries with it the responsibility of maintaining the exterior appearance of the existing facilities in an "as is" condition.

The civil features of the Great Falls Hydroelectric Project fall into the following categories in accordance with suggested guidelines contained in Volume VI of the manual, Section 1. These are: site preparation, hydraulic conveyance facilities, and powerhouse and appurtenant facilities.

The powerhouse configuration is fixed and therefore the turbine generator equipment selected was based on its being compatible with the existing powerhouse space.

Figure 1-3, Volume VI, graphically displays the steps that should be followed in determining the civil costs for a potential hydroelectric power project. Volume VI does not cover the civil costs associated with repair and rehabilitation or alteration of the impounding or diversion structure. This is covered in Volume IV of the manual. This is a civil cost and must be included to arrive at a total civil cost. In the following estimates, the steps in Figure 1-3, Volume VI, are followed where applicable.

Site Preparation

Since the site now has adequate parking, access, and drainage control, no site preparation costs are included.

Hydraulic Conveyance Facilities

These facilities include:

1. Repair of forebay, gatehouse and penstock inlet
2. Replacement of penstocks
3. Replacement of draft tubes, repair of tailrace, and installation of draft tube bulkheads
4. Cofferdamming

Forebay, Gatehouse and Penstock Inlet. Due to standing water in the forebay area, it was necessary to estimate the extent of repairs that will be required to the forebay intake structure as well as the forebay walls and penstock inlet gate structure. This estimate was based on visual observation, use of engineering drawings, engineering experience, use of vendor supplied estimates, and engineering calculations. It is important that on-site inspections and evaluations be made to complement any office calculations.

Cofferdamming. In order to perform repairs or undertake new construction in the dry, it is necessary that the work area be in a dewatered condition. Therefore, cofferdamming will be required to insure that the work area from the forebay inlet to the tailrace outlet be maintained in a dewatered condition. Cofferdamming cost estimates were developed from engineering experience on similar projects and use of cost estimating guides such as Dodge and Engineering News Record.

Penstocks. The existing steel riveted penstocks have deteriorated to the point where replacement is required. This was determined by site inspections and from discussions with personnel familiar with the plant's condition when it was in operation. Therefore, new penstocks will have to be fabricated, the old penstocks removed, and the new ones installed. The estimated cost for installing new penstocks was compared with the cost as determined by the use of Figure 3-1 in Volume VI.

In the case of the Great Falls power plant, costs in addition to those obtained by use of Figure 3-1 need to be included. These additional costs consist of removal of the existing penstocks and use of a higher unit price for the steel due to its special fabrication. There are four penstocks, each 8 feet 6 inches in diameter, and approximately 60 feet long.

Draft Tubes, Tailrace, Draft Tube Bulkheads. The amount of remedial or new construction work required is dependent on the type of turbine selected. Section 5 covering the Electromechanical Features presents the types of turbines investigated.

For the Allis-Chalmers and Leffel alternatives the draft tubes will require replacement; whereas the Ossberger and Tampella alternatives are complete packages which include the draft tube. The costs for the draft tube replacement alternatives were estimated by use of cost estimating guides (Dodge Guide to Public Works and Heavy Construction Costs, 1978 and Engineering News Record Quarterly Cost Roundup, 1978), engineering experience, and cost information in the engineers' files.

Bulkheads will be required at the discharge end of the powerhouse. Cost for the bulkheads was estimated from costs for similar facilities designed by the engineer.

Powerhouse and Appurtenant Facilities

The powerhouse and appurtenant facilities include:

1. Repair of water supply and sanitary facilities
2. Repair and replacement of broken windows, roof tiles, box gutters
3. Cleaning and repainting of all exposed metal work (stairs, piping, doors, etc.)

4. Cleaning of concrete surfaces in the interior of the powerhouse

5. Inspection and repair as needed to the powerhouse interior back wall

6. Rehabilitation of overhead traveling crane

7. Modification of existing powerhouse floor to accommodate turbine generator equipment.

The existing powerhouse is constructed of brick and reinforced concrete. Engineering drawings were located which show most details of the powerhouse and were utilized to the maximum extent possible.

Field inspection and building code requirements formed the basis for determining what types of repairs or replacements may be required. On-the-site inspections are needed to make reasonable estimates for existing powerhouses in which conditions vary considerably from site to site. The guidelines contained in Section 4, Volume VI of the manual, can only make one aware of the items that need to be considered. Therefore, no comparisons are made with the cost guidelines shown in Section 4, Volume VI.

Cost Estimates

Table 4-1 displays the estimated cost for three alternatives for repairing, altering, or constructing required civil features at the Great Falls Hydroelectric Project, not including the diversion dam rehabilitation.

TABLE 4-1
CIVIL FEATURES - COST ESTIMATE (DOLLARS)

FERC Account Number	Description	ALTERNATIVE 1 Manual Procedures	ALTERNATIVE 1 Feasibility Study	ALTERNATIVE 2 Manual Procedures	ALTERNATIVE 2 Feasibility Study	ALTERNATIVE 3 Manual Procedures	ALTERNATIVE 3 Feasibility Study
331	**STRUCTURES & IMPROVEMENTS:**						
	SITE PREPARATION:						
	Drainage System						
	Erosion Control						
	Final Grading						
	Access Road	--	N/A	--	N/A	--	N/A
	Parking & Miscellaneous Site Features						
	Environmental Construction Controls						
	POWERHOUSE:						
	Structural	Manual not		Manual not		Manual not	
	Excavation	applicable	Lump Sum	applicable	Lump Sum	applicable	Lump Sum
	Foundation	to site.	170,000	to site.	170,000	to site.	132,000
	Switchyard	17,500	Included in Acct. No. 350	23,000	Included in Acct. No. 350	21,500	Included in Acct. No. 350
332	**RESERVOIRS, DAMS & WATERWAYS**[1/]	Not included		Not included		Not included	
	Intake Trashracks	in manual	75,000	in manual	75,000	in manual	75,000
	Intake Slidegates	80,000	112,000	80,000	112,000	80,000	112,000
	Penstocks	107,520	360,000	107,520	360,000	107,520	360,000
	Valves	--	N/A	--	N/A	--	N/A
	Bifurcation	--	N/A	--	N/A	--	N/A
	Tailrace	--	N/A	--	N/A	--	N/A
	Outlet Bulkheads (Slidegates)	20,000	24,000	20,000	24,000	20,000	24,000
	Cofferdamming & Pumping	Not included in manual	139,000	Not included in manual	139,000	Not included in manual	139,000
	TOTAL ACCOUNT 331, 332		880,000		880,000		842,000
	TOTAL CIVIL COSTS	Manual	880,000	Manual	880,000	Manual	842,000
		Procedures		Procedures		Procedures	
	CONTINGENCIES (25%)	Costs	220,000	Costs	220,000	Costs	210,500
	REGIONAL CORRECTION FACTOR	not	1.0	Not	1.0	Not	1.0
	CORRECTED CIVIL COSTS	Complete	1,100,000	Complete	1,000,000	Complete	1,052,500
	ENGINEERING, CONSTRUCTION	For		For		For	
	MANAGEMENT & OTHER COSTS	This	110,000[2/]	This	110,000[2/]	This	105,250[2/]
	GRAND TOTAL	Project	1,210,000	Project	1,210,000	Project	1,157,750

1/ Diversion Dam Rehibilitation Cost Not Included.

2/ Feasibility study used 10% for Engineering, Construction Management and Other Costs.

SECTION 5
ELECTROMECHANICAL EQUIPMENT

The Great Falls Study considered a full range of alternative turbine-generator equipment types. For this case study, four of the 17 alternatives examined and the costs of two compared with manual procedures contained in Volume V.

The four alternatives considered here represent equipment supplied by four different vendors and are summarized in Table 5-1. The turbine types and sizes selected were based on the following factors: available head in feet (gross head 70 feet); available flow in cubic feet per second on a daily basis (range 50 over 3000); use of available powerhouse space without alteration of its exterior (inside dimensions approximately 40 by 102 feet) due to historical considerations; rehabilitation of the existing four S. Morgan Smith Francis turbines, and installation of new turbine-generator equipment. The determination of turbine efficiency was made by using Figure 3-5 in Volume V of the manual and comparing it with vendor-supplied information. In the case of Alternative 1 it was found that the vendor-supplied information resulted in somewhat lower efficiencies than those obtained by use of manual curves.

Table 5-2 displays the comparison between the manual procedures and vendor supplied information of the turbine efficiencies for Alternative 1.

Description of the turbine units for the four alternatives contained in this case study are described below.

Alternative 1 - Allis-Chalmers (Rehabilitated Units)

This alternative investigated the rehabilitation of the four existing in-place S. Morgan Smith turbines. These units are Twin Francis horizontal units installed in 1923 and operated until 1969. Three of the units are rated at 1340 kilowatts and one is rated at 1080 kilowatts.

Alternative 2 - Leffel (Uprating Existing Units)

This alternative investigated the uprating of the existing four Francis-type units. The work required would be similar to Alternative 1 with the exception that all new parts would be provided. Only the middle portion of the existing pressure cases would be used along with the existing or replaced penstocks and draft tubes. To accommodate the new Francis-type runners and wicket gates it will be necessary to extend the pressure cases on each end. This extension can be accommodated without apparent need for structural modification. As a result of the uprating, new higher capacity generators will be needed, thereby necessitating some modification to the existing floor at the generator.

Based on vendor-supplied information the smaller units will operate over a flow range of 120 to 282 cubic feet per second with a net head of 67 feet. Its corresponding efficiencies would be 78 percent at 2/5 load to 90 percent at 4/5 load. The three larger units will operate over a flow range of 236 to 457 cubic feet per second. Their corresponding efficiencies would be 80 percent at 1/2 load to 90 percent at 9/10 load.

Alternative 3 - Ossberger (New Units)

The alternative investigated the installation of four new cross flow turbines manufactured by F.W.E. Stapenhorst, Inc. These units are modified impulse-type turbines with cylindrical runners. The turbines are low speed (136 rpm) and therefore speed increasers are provided to permit use of high speed (1200 rpm) standard generators.

The four cross flow generating set units would operate over a flow range of 76 to 378 cubic feet per second with their corresponding efficiencies being 80 percent at 1/5 load to 84 percent at 3/4 load.

Alternative 4 - Tampella (New Units)

The Tampella units investigated would be low specific speed adjustable blade propeller. The units can be set at a higher elevation than similar Allis-Chalmers units, which permits the use of vertical, conical-shaped draft tubes.

This arrangement results in significantly reduced structural modifications in the tailrace. However, the lower speed results in more costly generators. The generators would be supported integrally with the turbine, which also reduces the required structural modification but would necessitate removal of the generator when removal of the turbine is necessary. The Tampella unit includes an upstream butterfly valve to be used for shutoff, thus eliminating the need for the penstock headgates.

The four Tampella-supplied turbines would operate over a flow range of 106 to 530 cubic feet per second with their corresponding efficiencies being 70 percent at 1/5 load to 90 percent at 4/5 load.

Electromechanical Cost Comparisons

Retrofitting or rehabilitation of existing equipment is unique to itself and therefore use of guidelines contained in Volume V for determining costs is of limited assistance. Procedures illustrated by Figure 2-1 of Volume V were utilized to determine the electrical/mechanical equipment costs for comparison with those obtained by in-depth study.

Electrical/mechanical costs determined by use of the procedures and guidelines contained in Volume V were grouped into the following categories:

1. Turbine-generator equipment
2. Station electrical equipment
3. Switchyard equipment

TABLE 5-1
GENERATING UNIT ALTERNATIVES

Alternative	No's and Capacity (kW) of Units	Installed Capacity kW	Orientation	Turbine Type	Speeds-RPM Turbine	Speeds-RPM Generator	Generator Output Voltage	Turbine Manufacturer
1	1-1080 3-1340	5100 1/	H	DRF	1-514 3-450	1-514 3-450	2400	Allis-Chalmers
2	1-1290 3-2070	7500	H	DRF	1-400 3-327	1200 2/	4160	Leffel
3	4-1700	6800	H	CF	136	1200 2/	4160	Ossberger
4	3-2625	7875	V	AP	300	300	6300	Tampella

Legend:
H = Horizontal
V = Vertical
DRF = Double Runner Francis
CF = Cross Flow
AP = Adjustable Blade Propeller

Notes:
1/ Generator Uprated
2/ Speech Increaser Provided

TABLE 5-2
ALTERNATIVE 1 - TURBINE EFFICIENCIES

Rated Capacity	MANUAL PROCEDURE (FIGURE 3-5) Expected Efficiency	Add Correction For Throat Dia. (1340 kW Unit)	Add Correction For Throat Dia. (1080 kW Unit)	Expected Efficiency (1340 kW Unit)	Expected Efficiency (1080 kW Unit)	VENDOR SUPPLIED INFORMATION[1] Expected Efficiency (1340 kW Unit)	VENDOR SUPPLIED INFORMATION[1] Expected Efficiency (1080 kW Unit)
100%	88%	2.4%	2.1%	90.4%	90.1%	80%	79%
75%	89%	2.4%	2.1%	91.4%	91.1%	82%	83%
60%						77%	77%
50%	83%	2.4%	2.1%	85.4%	85.1%		
25%	73%	2.4%	2.1%	75.4%	75.1%		

1/ These efficiencies are based on rehabilitated turbine.

4. Miscellaneous power plant equipment
5. Special equipment

Cost comparisons between the manual and feasibility results for Alternatives 1 and 3 are shown in Table 5-3. Alternative 1 is a comparison of the rehabilitated Allis-Chalmers turbine and Alternative 3 compares results for the Ossberger turbine.

It should be noted that the total installed costs are higher using manual procedures than those found by the feasibility study. The costs were 10 percent higher for Alternative 1 (rehabilitated equipment) and 25 percent higher for Alternative 3 (new equipment). Vendor-supplied equipment quotes were assumed to have contingencies included. An item where there is a large cost difference is the transmission line cost. Part of this line will be overhead and a portion in underground conduit. The local utility, Public Service Electric and Gas Company (PSE&G), furnished the cost for this work. The cost difference for this item is in excess of 200,000 dollars.

TABLE 5-3
COMPARISON OF ELECTROMECHANICAL COSTS FOR ALTERNATIVES 1 and 3

FERC Account Number	Description	ALTERNATIVE 1 Manual Procedures	ALTERNATIVE 1 Feasibility Study	ALTERNATIVE 3 Manual Procedures	ALTERNATIVE 3 Feasibility Study
333	Water Wheels, Turbines and Generators Installed Cost (Figures 3-12 to 3-16 of the Manual)	$1,793,100	$1,433,500	$2,840,000	$2,120,900
334	Station Electrical Equipment: Transformer, Lightning Arrestor, Air Breaker Swt. Gen. Breaker and Line OCB (Figure 6-3 of the Manual)	$ 105,000	Cost included in transmission line cost.	$ 120,000	Cost included in transmission line cost.
	Battery Sys., Sta. Swt. Gear, Sta. Ser. Trans., Bus, Cable Conduit, Grd., Control Bd., Lighting Sys., Freight and Installation (Figure 5-4 of the Manual)	$ 180,000	$ 213,000	$ 195,000	$ 213,200
335	Misc. Power Plant Equipment: Ventilation, Fire Protection, Cooling Water, Communication Sys., Freight and Installation (Figure 6-5 of the Manual)	$ 55,000	$ 115,500	$ 60,000	$ 123,700
350	Transmission Line (Figure 6-4 of the Manual)	$ 6,000	$ 305,000[1/]	$ 46,000	$ 316,000[1/]
Control System		Not covered in Manual	$ 260,000	Not covered in Manual	$ 260,000
Elevator Rehabilitation, Stilling Well, Crane Rebuilding and Misc.		--	$ 62,000	--	$ 62,000
	Subtotal	$2,179,100	$2,389,200	$3,261,000	$3,095,800
	Contingency (25%)	544,775	277,936[2/]	815,250	251,970[2/]
	Subtotal	$2,723,875	$2,667,136	$4,076,250	$3,347,770
	Engineering, Constr. Mg. & Other Costs (20%)	544,775	266,714[3/]	815,250	334,777[3/]
	TOTAL INSTALLED COST	$3,268,650	$2,933,850[4/]	$4,891,500	$3,682,547[4/]

1/ Cost estimate furnished by local utility (PSE&G).
2/ Contingencies not included for vendor furnished costs.
3/ Feasibility study used 10% for Engineering Costs.
4/ Total installed costs do not include contingencies for vendor supplied equipment.

SECTION 6
POWER MARKETING ANALYSIS

General

The value of the output from the Great Falls plant depends on the project's electric production characteristics and the economics of the power purchaser. The production characteristics determine the type of power the project can displace, and the potential users, and the purchaser's economics determine the value of this class of power. This section closely follows the guidelines contained in Volume II of the manual.

Production Characteristics

Previous studies have shown that no firm generation capacity can be provided by the Great Falls project. Periods of flow below levels required for the hydrogeneration equipment studied occur between June and November annually and flow fluctuates substantially throughout the year. Since plant storage is limited to a small amount of daily pondage, the project is a run-of-the-river project with no firm capacity. In the case of a utility purchaser, the project value will be the energy cost of electricity displaced. For other users, the project value is based on reducing purchased electricity.

Previous investigation explored the possibility of raising the dam to achieve increased energy production. It was shown that the dam could safely be raised 5.7 feet, thereby increasing annual energy output by approximately six percent, but no firm capacity is gained. However, the increased dam height with accompanying gates for flow control would increase the pondage available and could affect the power value estimate.

Power Value

Sale of the Great Falls electrical output to the local utility (Public Service Electric and Gas (PSE&G)) and to an end user were considered.

Sale to Public Service Electric & Gas (PSE&G). PSE&G is New Jersey's biggest utility and the one serving the project area. PSE&G is a member of the Pennsylvania-New Jersey-Maryland (PJM) Interconnection, a power pool with centralized dispatch and free flowing power exchange. PSE&G has tentatively agreed to purchase the project energy at a price related to the cost of energy purchased through the PJM Interconnection. In 1976, this value was put at between 20 and 25 mills/kWh.

Since this offer prices the project output based on the marginal value of energy in the interconnected system, it fairly represents the economic value of the Great Falls project. However, because of the long-term nature of hydroelectric facilities, the future value of energy displacement in the PSE&G system was investigated.

PSE&G's current sources of energy and how they are used to meet demand are shown in Figure 6-1. As this figure shows, PSE&G is burning oil as a baseload fuel. The energy cost of oil firing in the system (based on the weighted average oil-fired heat rate of 11,000 Btu/kWh and oil cost at $2.35/MMBtu) is 2.54¢/kWh. This value will escalate at least as fast as inflation.

It is possible that PSE&G's aggressive nuclear expansion program could result in oil no longer being a baseload fuel. Figure 6-1 also shows the expected growth in energy sales and the timing and capacity of future nuclear addition. Future baseload production was investigated by projecting a series of load duration curves into the 1980's and superimposing energy production by source. Energy was assumed to be produced based on 45 percent annual capacity factors for nuclear and coal generation. (1977 capacity factors were 40.4 percent for nuclear and 44.5 percent for coal.) This analysis showed that oil will still be a baseload fuel through 1989.

It can therefore be concluded that through 1989, the minimum value of energy produced by the Great Falls plant will be based on the energy cost of oil-fired generation in the PSE&G system. In 1977, this value was 25.4 mills per kWh and over this period the minimum *escalation* rate should be the general inflation rate. Most observers predict the real cost of oil will rise, hence leading to a faster escalation than the general inflation rate.

Sale to End User. Power sales to an end user were evaluated and it was concluded that this is an infeasible method of selling the project output. This is so because transporting the energy to the user's site could prove very difficult and expensive. The two options are to construct a separate transmission line or to wheel the power over PSE&G lines. Construction of a separate line in this urbanized area would pose serious right-of-way problems.

The Director of the Office of Technical Assistance, New Jersey State Energy Department, was contacted in regard to wheeling. To his knowledge there are no current wheeling arrangements that would allow an industrial or other non-resale purchaser to wheel power over utility lines. He thought such an arrangement would be very difficult to obtain because the project is nonfirm and significant standby charges would be levied; the energy value of the power displaced would be related to the average energy cost of PSE&G, which is considerably less than the marginal cost; also only small pondage is available, causing energy to be lost during low usage hours. This is in contrast to a situation where the utility takes all project output.

The combination of these four factors makes it unlikely a nonutility would find the purchase of Great Falls power to be beneficial.

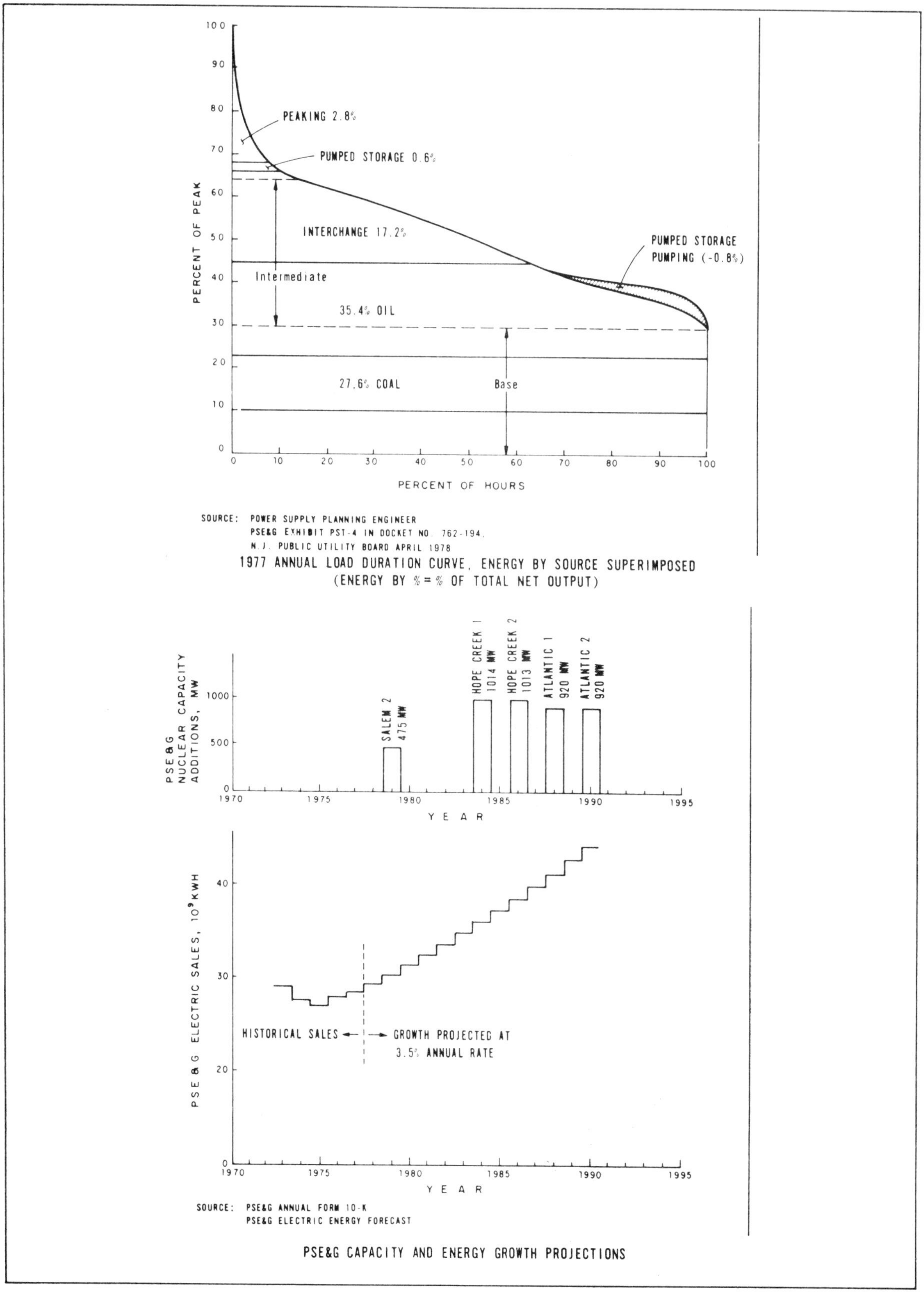

Figure 6-1. PSE&G Energy Source and Growth Projection Curves

SECTION 7
ECONOMIC AND FINANCIAL ANALYSIS

Introduction

The cost and power value information developed in previous sections allows the economic and financial feasibility of the Great Falls project to be evaluated. For this analysis, two major criteria were used:

1. The project was analyzed as a stand alone venture receiving the full economic value of the energy produced. This perspective results in the true economic merits of the project being established.
2. The project has been assumed to be both owned and financed with tax exempt revenue bonds by the City of Paterson. With municipal ownership, no local or income taxes are levied against the project. For financial feasibility, 40 year, seven percent bonds were assumed.

In addition, in this section a sinking fund has been calculated which will provide sufficient funds, in future dollars, to perform major repairs and replacements. These expenditures will be necessary to maintain the facility in functional order through the financing period.

The steps followed in analyzing the plant are discussed below. The actual computations were performed by several computer programs developed for this purpose and described in the manual. By design, the cost of service (financial feasibility) and the internal rate of return (economic feasibility) were calculated in one program and consequently separate calculations are not presented here.

Economic and Financial Analysis

Economic feasibility is the evaluation of project costs and benefits with the project deemed feasible when benefits exceed costs. Financial feasibility is the evaluation of the ability of the project to provide debt service from the capital required to construct and operate the project.

The financial calculations of receipts and disbursements determine the expected "cash flow" for the project. For Great Falls, cash flow represented all quantified costs and benefits so that the financial analysis provided the costs (disbursements) and benefits (receipts) for the economic analysis. The economic criteria used was the internal rate of return (IRR).

The following analysis of the economic evaluation procedure presented in Table 4-3 of Volume II utilizes the Economic and Financial Analysis Manual. Financial calculations are made, then become the quantitative inputs for the economic analysis.

Escalation. It was first determined that inflation would be explicitly included in the analysis. A general escalation rate of six percent was used as representative of expectations of the long-run inflation rate. This rate was used for all costs and revenues.

Economic Life. The project economic life was established at 40 years, the same as the financing period. Since major repairs and replacements are periodically required for the project to remain operational, the period when these repairs are not made determines the project life. In this case, provisions were made for a 40-year operation.

Unescalated Costs. Construction and annual costs in 1978 dollars for the alternatives have been established in previous sections. These are reproduced in summary form in Table 7-1 for use in the economic and financial analysis.

The construction period was estimated to last three years. Capital expenditures were estimated to be 20 percent in the first year and 40 percent in each of the following two years.

The electrical/mechanical investigation determined that repair and replacement of major equipment components are periodically necessary for continued operation of the plant. The costs were estimated as percentages of the original cost of several major asset classes. The procedure described here was used to convert these percentages into a constant annual cost that will provide sufficient funds, in future dollars, to make the required expenditures. In this analysis, provisions were made for a 40-year project.

The first step was to use the replacement schedules and the 1978 value of the asset classes to determine the total replacement (in 1978 dollars) required in the 20th and 30th years of operation. These values were then escalated to the year of occurrence accounting for the construction period. Next, using the city's cost of borrowing (seven percent) as the discount rate, the present value of these future replacements was calculated in 1981, the first year of project operations. This amount and the equivalent 30-year, seven percent sinking fund are shown in Table 7-1. Note that this annual cost (about 40 percent of other annual operating costs) is significant and must be incorporated in the financing plan to assure project operations through the financing period.

Unescalated Benefits. The only project benefit considered in this analysis is power production since no other monetary benefits could be identified. The power marketing analysis established the value of the output at a minimum of 25 mills per kWh in 1977. For this analysis, the valve of power was set at 25 mills per kWh in 1978. This value was also escalated.

Discount Rate. The City of Paterson's cost of bond financing is the appropriate discount rate to use in the analysis. The tax status of revenue bonds used for this purpose has a major impact on their cost. Since the total

TABLE 7-1
COST SUMMARY (1978 DOLLARS)

	Alternative 1 Rehabilitation (Allis-Chalmers)	Alternative 2 New Horiz. Runners (Leffel)	Alternative 3 Cross Flow (Ossberger)	Alternative 4 Tube Turbines (Tampella)
Construction Cost				
Dam[1/]	1,056,700	1,056,700	1,056,700	1,056,700
Civil Features	1,210,000	1,210,000	1,157,750	878,900
Elec/Mech	2,933,850	3,112,000	3,682,547	5,074,100
Total	5,200,550	5,378,000	5,896,997	7,009,000
Annual Cost				
O&M	73,000	90,000	83,000	89,000
Admin (20% of O&M)	14,600	18,000	16,600	17,800
Insurance (.2% of Const.)	10,400	11,505	11,794	14,019
License Fee	2,000	2,000	2,000	2,000
Major Repairs and Replacement[2/]				
Present Value in 1981 of R&R through 30 years	577,580	569,230	605,860	833,770
Constant Annual Sinking Fund Payments (30 years @ 7%)	46,540	45,870	48,820	67,190

1/ Restoration of existing structure
2/ Described in the text

bonding required for all the options is less than $10 million, the established limit for tax exemption of small issues which are not otherwise exempt, the issue was assumed to be tax exempt. See Section 6 of the Economic and Financial Analysis Manual for more detail in this regard. Since the cost of financing can have a major impact on the financial feasibility, an opinion from a bond counsel should be obtained on the tax status prior to further major commitments of funds.

After reviewing *Moody's Bond Record*, a seven percent cost of bonding was used.

Results

Summary results for the four alternatives are contained in Table 7-2. The internal rate of return (IRR) was the economic evaluation criteria used to evaluate this project. IRR is defined and its method of calculation explained in Section 4 of the Economic and Financial Analysis Manual.

The project's IRR was calculated for a range of initial energy values to investigate the project's sensitivity to this major parameter. Over the range of 20 to 30 mills per kWh of initial value, the project's IRR for Alternative 2 was at least twice the client's discount rate, indicating an economically feasible project given the assumptions concerning escalation. The other three alternatives were also shown to be economically feasible.

A number of important financial quantities were determined for each alternative. These were cost escalation and interest during construction and cash receipts and disbursements. Cost escalation and interest during construction increase the Leffel alternative's completed cost by approximately $700,000 over the lump sum estimate of $5.4 million. The constant annual debt service on the bonds required to finance the project will be approximately $460,000 per year. This may vary depending on the exact structure of the bond issue.

Impact of Low Flow. If the output from this project is sold on a per kWh basis, the revenue impact of low flow must be determined. The first year of operation will be examined since this is the most critical period.

Table 7-3 shows the first year financial results of low flow. As shown, all the options have cash flow deficits under these conditions. Provisions for this possibility must be provided in the marketing agreement for each option or a reserve fund must be established for contingencies.

TABLE 7-2
SUMMARY OF TECHNICAL, ECONOMIC AND FINANCIAL DATA

	Alternative 1 Rehabilitation (Allis-Chalmers)	Alternative 2 New Horiz. Runners (Leffel)	Alternative 3 Cross Flow (Ossberger)	Alternative 4 Tube Turbines (Tampella)
Installed Capacity (kW)	5,100	7,500	6,800	7,875
Average Annual Energy Production in millions of kWh[1/]	24.4	30.8	27.9	32.3
Value of Energy Produced:				
1978 Value in ¢/kWh	2.50	2.50	2.50	2.50
1981 Value in ¢/kWh (First year of operation)[2/]	2.98	2.98	2.98	2.98
Capital Costs:				
Civil[3/]	2,266,700	2,266,700	2,214,450	1,935,600
Electrical/Mechanical	2,933,850	3,112,000	3,682,547	5,074,100
TOTAL (1978 Dollars)	5,200,550	5,378,000	5,846,997	7,009,700
$/kW	1,020	717	867	890
Completed Project Cost[4/]				
TOTAL (1981 Dollars)	5,887,500	6,089,200	6,676,000	7,935,700
Annual Project Costs (First year of service - 1981):				
Debt Repayment (Dollars)[5/]	441,600	456,700	500,800	595,200
Repairs and Replacement Sinking Fund (Dollars)[6/]	46,500	46,800	48,800	67,200
Operating Costs (Dollars)[7/]	119,100	143,900	135,100	146,300
Cost of Service (First Year of Operation - 1981) (¢/kWh)	2.49	2.10	2.45	2.50
Internal Rate of Return (%)[8/]	15.4	17.5	15.5	15.4

1/ Based on Water Years 1950/51 through 1959/60 (Representative Decade)
2/ Energy Value escalated at 6.0% per year
3/ Includes restoration of the dam. Dam cost is for most expensive of various dam alternatives studied.
4/ Capital costs are escalated to the Year of Occurrence at 6%, then Construction Interest charged at 7.0% per year
5/ Fully amortized for 40-year life at 7%
6/ Provides sufficient funds for replacements, in Future $'s, to allow 40 years of operation
7/ Includes O&M, Administrative Overhead, Insurance, and License Fees
8/ Assumes 6% General Price Escalation over the Project Life of 40 years

TABLE 7-3
FIRST YEAR RECEIPTS AND DISBURSEMENTS
LOW FLOW CONDITIONS
PER kWh SALE
(In Thousands of Dollars)

	Alternative 1 Rehabilitation (Allis-Chalmers)	Alternative 2 New Horiz. Runners (Leffel)	Alternative 3 Cross Flow (Ossberger)	Alternative 4 Tube Turbines (Tampella)
Percent of Normal year Energy Production	70%	65%	67%	65%
Low Flow Revenue (Per kWh Sale)	$508.6	$596.1	$556.6	$625.1
Less:				
Operations	119.1	143.8	135.1	146.3
Bond Amortization	441.6	456.7	500.8	595.2
Replacement Sinking Fund	46.5	46.8	48.8	67.2
Net Funds	-98.6	-51.2	-128.1	-183.6

REFERENCES

Development and Resources Corportion, *Great Falls Hydroelectric Power Plant Energy Generation Study*, October 1977.

Development and Resources Corporation, *Great Falls Hydroelectric Project Feasibility Report*, October 1978.

Dodge Guide to Public Works and Heavy Construction Cost, 1978.

Engineering News Record-Quarterly Cost Roundup, 1978.

Justin and Courtney, *Restoration of S.U.M. Dam and Hydro Plant for the Great Falls Development Corporation*, 1976.

U.S. Army Corps of Engineers, *Passaic River Survey Report for Water Resource Development*, 1971.

U.S. Department of Housing and Urban Development, Federal Insurance Administration, *Flood Insurance Study of the Passaic River*, 1975.

EXHIBIT II
ROLLINS POWER PROJECT
CASE STUDY
SMALL HYDROPOWER ADDITIONS

Contents

SECTION I
INTRODUCTION

Scope

This case study document describes the application of the guidance and technical data presented in the draft guide manual. Cost and design information for the Rollins Power Project, Bear River, California, is presented as an illustrative example of use of the manual materials. Also, the validity of the data and guidance provided therein is evaluated. This information is presented "after the fact", since the construction of the Rollins Power Project (Project) began in the fall of 1978. It is anticipated that the Project will begin generation in the spring of 1980. The Project was formulated and executed by the Nevada Irrigation District (District) with Tudor Engineering Company as consultants.

Existing Project

The Rollins Power Project is located at Rollins Dam on the Bear River, about 16 miles north of Auburn in the Sierra Nevada mountains of Central California. The dam was completed in 1966 as part of the Yuba-Bear River Development Project, constructed by the District. The Yuba-Bear Project stores and diverts water from the upper Yuba River watershed into the Bear River watershed for irrigation and domestic use in Nevada and Placer Counties. Above Rollins Dam, in addition to other Yuba-Bear Project facilities, the District owns and operates two hydroelectric plants, Dutch Flat No. 2 and Chicago Park. The energy from the power plants, both located on the Bear River, is sold to Pacific Gas and Electric Company (PG&E).

Rollins Dam is a 220-foot high rockfill dam with an impervious core. The concrete ogee spillway in the right abutment was designed for a maximum flow of approximately 60,000 cubic feet per second. The diversion and outlet works for the reservoir were constructed together. A single 18-foot diameter horseshoe shaped conduit was excavated through the left abutment from the reservoir for about 300 feet. At that point, a bifurcation leads into two smaller tunnels. One is a 16-foot flat invert, partially-lined tunnel which was used as the diversion during construction. The other is a 12-foot horseshoe-shaped, concrete-lined tunnel with a 60-inch Howell- Bunger valve which is currently used for water deliveries to downstream users. After construction, the diversion tunnel was plugged with 50 feet of mass concrete. This plug was pierced for the Project penstock.

The intake tower is located within the reservoir near the upstream toe of the dam. It is an ungated structure, equipped with a large trash-rack cage. Within the outlet works, there are no control gates upstream of the bifurcation. Downstream of the dam is a small afterbay and a diversion dam with head-works for the PG&E Bear River Canal. Discharges from the outlet works also flow down the Bear River to Combie Dam and are diverted at that point for use in Placer and Nevada Counties by the District.

Power Plant Addition

The Rollins Power Plant will include the following:

1. A semi-outdoor powerhouse with an installed capacity of 12,700 kilowatts will be constructed near the toe of the dam and the existing outlet portal. A switchyard, enclosed by fencing, will be built adjacent to the powerhouse.
2. A steel penstock approximately 550-feet long, will rest on concrete piers placed in the existing 16-foot diversion tunnel with an emergency control butterfly valve at the upstream end near the existing tunnel plug. The tunnel plug was pierced during a previous work phase and a steel liner was inserted to convey water to the penstock. Control equipment will be provided to allow for synchronous passage of water either from the existing outlet valve in the adjacent outlet tunnel or hydraulic turbine.
3. A tailrace channel downstream of the proposed powerhouse will be excavated in the rock between the tunnel outlet and the existing diversion dam.
4. Supplemental site development features will be built, including an apron adjacent to the power house for parking and the staging of maintenance activities. A storage and office building will be constructed for the accomodation of operation and maintenance personnel and the storage of spare parts and maintenance materials which cannot be stored within the powerhouse. An access road will be developed, by upgrading the existing service road, to accomodate the vehicular traffic to the powerhouse.
5. A transmission line will be constructed by PG&E from the power plant switchyard in a westerly direction to an existing PG&E transmission line. This feature is not considered as part of the Rollins Power Project.

The existing project features and new power facilities are shown on Figure 1-1.

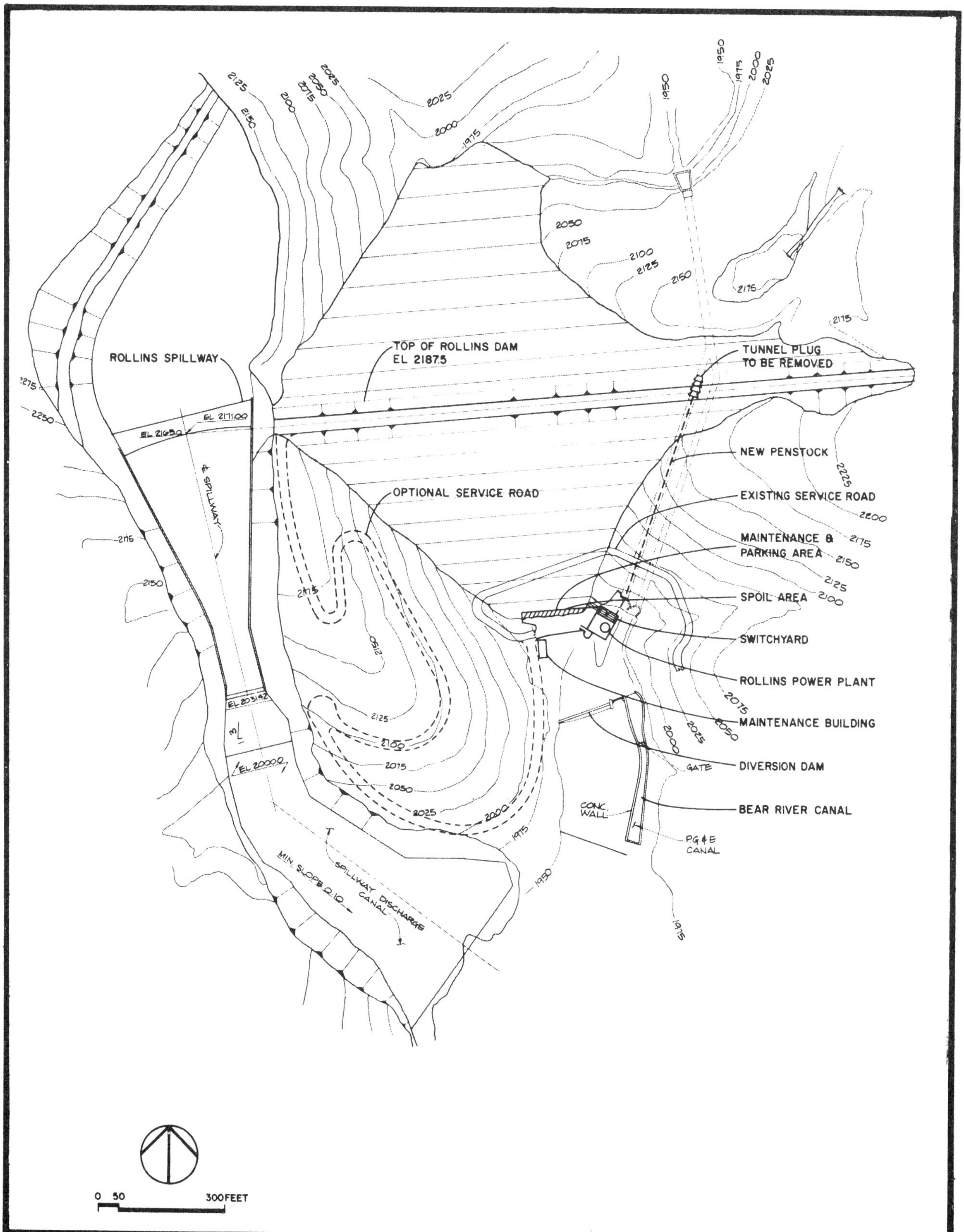

Figure 1-1. Principal Features - Rollins Power Project

SECTION 2
PROJECT FORMULATION

Formulation and initiation of the Project was accomplished by the preparation of a feasibility study, the marketing of the power to be generated, and the preparation of the necessary applications and permits. Other activities which then followed and are described in the next section on implementation of the Project, included the final design, bidding and award of construction contract. Construction of the Project is now proceeding.

Feasibility Studies

Work on the feasibility study was authorized by the District Directors in the spring of 1974 and was completed in August 1974. The main study items consisted of the review of existing studies, the formulation of four alternative project developments, the preparation of operation studies for the alternatives and cost and benefit studies of the alternatives. Conclusions and recommendations were made, along with a proposed time schedule for further action.

Four Project alternatives were formulated as follows:

1. Add power plant to existing dam with no change in present operating agreement with PG&E.
2. Add power plant to existing dam, raise maximum water surface from elevation 2171 to 2185, continue present operating agreement.
3. Add power plant to existing dam, change present operating agreement to maximize water and power output.
4. Add power plant to existing dam, raise maximum water surface from elevation 2171 to 2185, change present operating agreement to maximize water and power output.

The study period for the reservoir operation studies was taken to be 1928 through 1937. This is the same period previously used by the District water supply studies and it was considered important to be able to compare results. This study period included an extreme drought period and the average annual energy from this period was lower than would be realized if a longer-term more representative record was used. An example of the systematic routing operation studies used for this Project is shown in Figure 2-1.

Due to the uncertainty at that time in future cost of fuel oil (circa Spring 1974) on which a traditional benefit and cost analysis should be based, the report included the calculation of the cost of energy from the four project alternatives in mills per kWh. That cost was then converted to a cost for an equivalent barrel of fuel oil. It was assumed that fuel oil would be the source of replacement energy if the project was not constructed. The lower the equivalent fuel oil cost, the greater the benefit of the project. Table 2-1 shows these equivalent fuel oil costs for the four alternatives ranked in order of benefit. It can be seen that the costs range from $5.04 to $8.55 per barrel. At approximately the time of the report, it was reported by PG&E that the cost of imported low sulfur fuel oil rose from $7.75 to $13.00 per barrel. From this information, it was concluded that all of the alternatives considered would be economically feasible.

After evaluating the economic and institutional aspects of each alternative, alternative 1 was selected. A 12,700 kW turbine/generator unit would be installed, with no increase in the height of the dam or addition of spillway gates. The plant would be operated as a run-of-the-river plant with no change in the release pattern. The raising of the water surface entailed by alternatives 2 and 4 was not selected because of the impact on the environment, disruption to the existing recreational facilities next to the reservoir and the added cost of the relocation of old Highway 40 where it crossed an arm of the reservoir. Alternative 3 was not selected since it was decided by the district not to attempt a renegotition of the operating agreement with PG&E.

Since the present operation requires the reservoir to be occasionally lowered to an elevation below the minimum head for power generation, no dependable capacity was credited to the installation. A peaking operation was not considered as an alternative because of the lack of a suitable afterbay site.

Several constraints on the Project were found during the feasibility study. Financially, the District had 7.8 million dollars in authorized but unissued revenue bonds remaining from the construction of the Yuba-Bear Project. These could be used for the Rollins Project, but if that amount was exceeded, other forms of financing the overrun, including possible additional authorization by the electorate to sell more bonds, would be necessary. Also, if the power was sold to an investor-owned utility, the bonds would take the form of Industrial Development Bonds (IDBs) and would lose their tax exempt status. (Revenue bond financing and IDBs are discussed further on page 6-8 of Volume II.) The District's financial consultant indicated that the IDBs would carry an interest rate greater than the maximum allowed by California Irrigation District's law, i.e., eight percent. Therefore, it was proposed, and later accomplished, that the District's law be amended to permit a higher interest rate, not to exceed 10 percent.

The most difficult physical constraint discovered was necessity to pierce the plug in the original diversion tunnel for the penstock. There was no valve or control gate with which to close off the upstream side of the plug so that the work could be performed without draining the reservoir. Several unique and challenging

RESERVOIR OPERATION STUDY

CASE III - MODIFIED OPERATION ONLY

CASE III
Sheet 1 of 4

Water Year (month)	Total Inflow (1000 a-ft)	Releases (1000 a-ft) Demand[1/]	Evap.	Total	Inflow Minus Release (1000 a-ft)	Storage @ End of Month (1000 a-ft)	Reservoir Stage @ End of Month	Spill (1000 a-ft)	Deficiency (1000 a-ft)	Avg. Head on Turbine (feet)	Energy[2/] M-KWH	Average Power[3/] MW
1928												
Oct.	16.9	33.3	.2	33.5	-16.6	49.3	2150			194	6.40	8.6
Nov.	44.2	29.0	.1	28.1	16.1	65.3	2170			214	5.83	8.1
Dec.	30.1	25.8	.1	25.9	4.2	65.9	2171	3.6		215	6.13	8.2
Jan.	33.7	25.8	.1	25.9	7.8	65.9	2171	7.8		215	7.01	9.7
Feb.	35.3	23.4	.1	23.5	11.8	65.9	2171	11.8		215	7.34	10.9
Mar.	124.8	14.3	.2	14.5	110.3	65.9	2171	110.3		215	4.25	12.7
Apr.	65.2	30.3	.3	30.6	34.6	65.9	2171	34.7		215	9.11	12.7
May	33.9	37.9	.3	38.2	-4.3	61.6	2165			209	7.58	10.2
June	23.0	33.2	.6	33.8	-10.8	50.8	2152			196	6.26	8.7
July	34.1	37.9	.8	38.7	-4.6	46.2	2145			189	6.89	9.3
Aug.	36.3	37.8	.6	38.4	-2.1	44.1	2141			185	6.75	9.1
Sep.	19.9	37.0	.4	37.4	-17.5	26.5	2117			161	5.44	7.6
											78.99	
1929												
Oct.	12.8	23.0	.2	23.2	-10.4	16.2	2079			123	2.42	3.3
Nov.	13.2	12.6	.1	12.7	.5	16.7	2080			124	1.33	1.8
Dec.	15.8	11.6	.1	11.7	4.1	20.8	2092			136	1.36	1.8
Jan.	14.5	10.5	.1	10.6	3.9	24.7	2103			147	1.44	2.0
Feb.	17.5	14.1	.1	14.2	3.3	28.0	2110			154	2.01	3.0
Mar.	23.7	19.1	.1	19.2	4.5	32.5	2119			163	2.94	8.8
Apr.	29.6	18.8	.2	19.0	10.6	43.1	2139			183	3.42	4.8
May	24.7	37.3	.3	37.6	-12.9	30.2	2114			158	5.33	7.2
June	27.0	36.9	.4	37.3	-10.3	20.0	2090			134	4.34	6.0
July	34.7	37.9	.4	38.3	-3.6	16.5	2079			123	4.12	5.5
Aug.	36.6	37.8	.3	38.1	-1.5	15.0	2074			118	3.78	5.1
Sep.	19.7	34.6	.3	34.9	-15.2	4.6	2025		4.7	---	--	2.2
											32.49	
1930												
Oct.	12.4	22.0	.1	22.1	-9.7	4.6	2025		9.7	---	--	---
Nov.	12.2	11.9	.1	12.0	.2	4.8	2026				--	---
Dec.	56.0	25.8	.1	25.9	30.1	34.9	2123				3.97	5.3
Jan.	40.1	25.8	.1	25.9	14.2	49.1	2150				4.87	6.8
Feb.	48.3	23.4	.1	23.5	24.8	65.9	2171	8.0			8.54	12.7
Mar.	49.5	14.3	.2	14.5	35.0	65.9	2171	35.0			4.25	12.7
Apr.	47.2	30.3	.2	30.5	16.7	65.9	2171	16.7			9.10	12.7
May	37.1	37.9	.3	38.2	-1.1	64.8	2170				7.90	10.6
June	24.0	34.2	.6	34.8	-10.8	54.0	2157				6.84	9.5
July	34.0	37.9	.7	38.6	-4.6	49.4	2151				7.29	9.8
Aug.	36.2	27.8	.6	38.4	-2.2	47.2	2147				6.87	9.2
Sep.	19.9	37.0	.5	37.5	-17.6	29.7	2115				5.29	7.3
											64.92	

1/ Assume No Demand for 15 days each March.
2/ Includes use of spills when available up to maximum turbine capacity.
3/ No power is generated when head is less than 100 feet.

Figure 2-1. Example - Systematic Routing Study

TABLE 2-1
SUMMARY OF STUDIES

Alternative	Average Generation (Year 1927-1937) kWh × 10^6	Energy Cost Mills/kWh	Equivalent Fuel Oil Cost $/barrel
1	53.0	7.96	5.04
2	55.0	11.45	8.55
3	56.8	7.43	5.55
4	60.4	10.42	7.78

proposals were made for accomplishing the work under those adverse conditions. However, during the unprecedented drought in the summer of 1977, the reservoir water surface was lowered below the level of intake structure and the proposed work schedule was accelerated to take advantage of the unexpected opportunity to pierce the plug in the dry. This work was approved and performed under the supervision of the Division of Safety of Dams, Department of Water Resources, State of California.

Spillway Flood Studies

Two of the alternative project formulations investigated included the maintenance of the existing maximum reservoir water level (Alternatives 1 and 3) and two others entailed the increase of the spillway crest elevation from 2171 to 2185 in order to increase the power and water conservation yield (Alternatives 2 and 4). The raising of the spillway crest would have required a similar raise in the dam crest to facilitate the passage of the spillway design flood. This dam would be classified as a large dam (over 50,000 acre-feet) in a "significant" hazard area (see Table 4-3 Volume III); therefore, the spillway would be required to pass the total probable maximum flood (PMF).

In order to investigate the adequacy of the spillway, the PMF hydrograph and the criteria used to establish the maximum probable preciptiation were obtained from the Division of Safety of Dams in Sacramento, California. The source of the probable maximum precipitation data was found to be Hydromet Report 36 and was judged by the Consultant and Safety of Dams to be adequate. The PMF hydrograph was routed by computer model over the existing spillway crest and the resulting maximum water surface was contained by the dam with about two feet of free board. The spillway was judged to be adequate. The inflow hydrograph and the routed outflow hydrograph are shown on Figure 2-2.

Integrity Investigation

The investigation of the integrity of the existing dam was minimal. The dam has been reviewed for safety each year by the engineering staff of the Division of Safety of Dams and once each five years by staff of the Federal Energy Regulatory Commission. The FERC requires as a part of their five-year review that the owner furnish to FERC a report prepared by a Consultant on the safety of the dam. During these investigations, no conditions have been observed that required remedial measures.

The State of California Water Code, Section 6225, requires that any additions or alterations to a dam receive the approval of the Division of Saftey of Dams prior to construction. An application was made to cover the removal of the tunnel plug. The design and construction criteria, plans and specifications were provided to Safety of Dams and approval was granted. Since blasting of the concrete plug would take place under the dam within 20 to 30 feet of the existing outlet, the consultant proposed and the State agreed that the wave velocity of the explosion be limited to less than three inches per second. During construction, the wave velocity was monitored by instruments and did not prove to be an unreasonable constraint on the blasting operation.

Representatives from Safety of Dams have continued to review and to monitor the construction and will provide final approval upon completion.

Selection of Turbine/Generator

The two turbine options considered for the Rollins Project were Francis and Crossflow. These are the appropriate options for head conditions of between 150 and 200 feet (from Figure 2-2, Volume V). The Crossflow turbine was not considered in detail because of limited available unit capabilities, as described in the manual.

The design turbine flow of 610 cfs was determined by the contractual commitments to PG&E for release and by the District's requirements for irrigation and domestic releases and low flow augmentation. Controlled flows are not released in excess of this demand condition. Uncontrolled flows spill over the spillway, and could be routed through the turbine up to the maximum hydraulic capacity.

The average weighted gross head on the turbine was calculated by multiplying the measured outflow from the reservoir in cfs-days times the daily gross head and dividing by the summation of measured outflow. This computation was accomplished as a part of the computer program used for the systematic routing. The results of

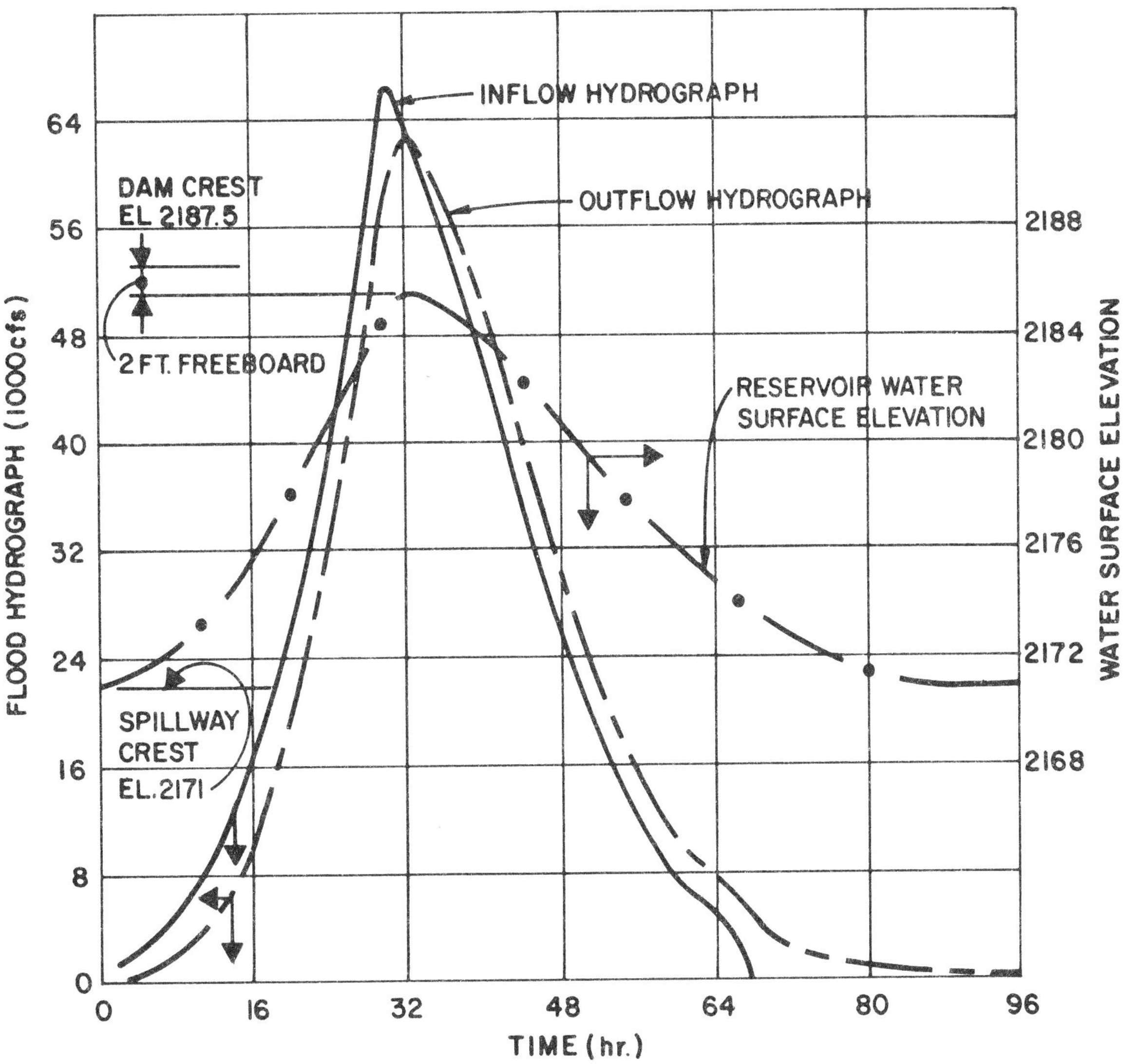

Figure 2-2. PMF Flood - Inflow and Outflow Hydrographs and Water Surface Elevation

this study indicated that the average weighted head was 175 feet.

From the maximum flow of 610 cfs, and with a head of 175 feet and an efficiency of 87 percent, the output of the generator would be about 7800 kW. The generator and related electrical equipment must be designed, however, to receive the maximum hydraulic output of the turbine at the maximum reservoir water surface elevation. This corresponds to a gross head of 204 feet, a flow of 845 cfs, and a plant capacity of about 12,700 kW.

Power Operation Studies

Systematic routing operation studies were performed by computer to estimate the amount of energy to be generated by the power plant. The studies were based on the assumption that Rollins Reservoir will continue to be operated under rules set forth in the Yuba-Bear Water Operation Contract dated July 12, 1963. All discharges will be dictated by the downstream requirements of the Bear River Canal, as operated by PG&E, diversion at Combie Reservoir, as operated by the District and minimum fish flow requirements, as set forth in Article 33 of Federal Energy Regulatory Commission License 2266. Discharges solely for the purpose of generating energy will not be made.

The studies were based on the assumption that excess flows above the capacity of the turbine would be spilled and that flows too small to drive the turbine or flows released when the turbine head is below the safe operating head would be passed through the existing outlet works. The latter case occurs when the water elevation falls below approximately elevation 2040 corresponding to a head of 82 feet on the unit. Below this stage turbine cavitation and rough operation would make power generation undesirable. Operation limitations for a Francis turbine are shown in Volume V.

The tailwater elevation will be controlled by the diversion dam downstream at the Bear River Canal headworks. The normal tailwater elevation was assumed to be at elevation 1958. Spills from Rollins Reservoir will cause no increase in tailwater because of the diversion dam.

For the purposes of estimating energy, inflows for two cases were evaluated: the hypothetical conditions and the actual flows since the dam was completed. The hypothetical study was based on an assumed operation scheme from October 1928 to September 1947 and was derived from the 1960 Ebasco "Yuba-Bear River Project Report". For the purposes of estimating a probable average of the energy to be generated, the years 1939-1947 appear to be most representative. These years have average runoff characteristics, similar to the 65 year average of all years for which flow records have been kept. From this study it is estimated that the average annual energy generated would be 71.1×10^6 kWh and that the average annual capacity factor would be 74 percent. The minimum and maximum generation for this period was 58.5 and 88.5×10^6 kWh. Figure 2-3 illustrates the reservoir elevations, flow duration and plant capability for this study. Note that about 10 percent of the flows discharged from Rollins Reservoir would be at flow rates in excess of the maximum possible turbine outflow. Also note that the capacity of the power plant fluctuates with head. During approximately five percent of the time, no energy could be produced.

The second study, with historical data, was based on the records of inflow and outflow of Rollins Reservoir since operation began. The period of study is from October 1964 to September 1976. During this period, the average annual energy would be 85.4×10^6 kWh and the average annual capacity factor would be 89 percent. Figure 2-4 shows the reservoir elevations and flow duration for this study.

Power Marketing

The procedure followed to market the power consisted of distribution of the project report to interested power purchasers, discussions with the prospective power purchasers, review and ranking of offers received and the negotiation of a memorandum of understanding with the selected power purchaser. This marketing procedure is generally described on page 122, Volume II as "Cost Plus a Royalty Subject to Escalation". Offers to purchase the power were received from PG&E, the California Department of Water Resources and the Northern California Power Agency. The Sacramento Municipal Utility District and the U.S. Bureau of Reclamation did not submit an offer. After study and review, the District's Directors voted to negotiate first with PG&E, an investor-owned utility

The main points of the offer as made by PG&E were as follows:

1. District will own and operate the power plant.
2. District will finance the Project through sale of revenue bonds, the total debt service to be guaranteed by the power purchase agreement from PG&E.
3. PG&E will receive all of the power from the Project.
4. PG&E will pay for debt service on bonds, and annual operation and maintenance costs; PG&E will advance "development costs", to be paid back from the sale of revenue bonds.
5. PG&E will pay to the District an added incentive payment or royalty equal to at least 4 mills per kWh.
6. PG&E will escalate the added incentive payment based upon the change in cost of wholesale price of energy in Northern California.

The offer by PG&E was judged to be reasonable. At the time of the offer, December 1975, the cost of the fuel oil being used to generate power in California resulted in a cost of electrical power of about 20 mills per kWh. This cost was considered the highest replacement value of energy in the PG&E system. The cost to develop power at Rollins was estimated to result in a cost of about 12 mills per kWh. Therefore, payment of 4

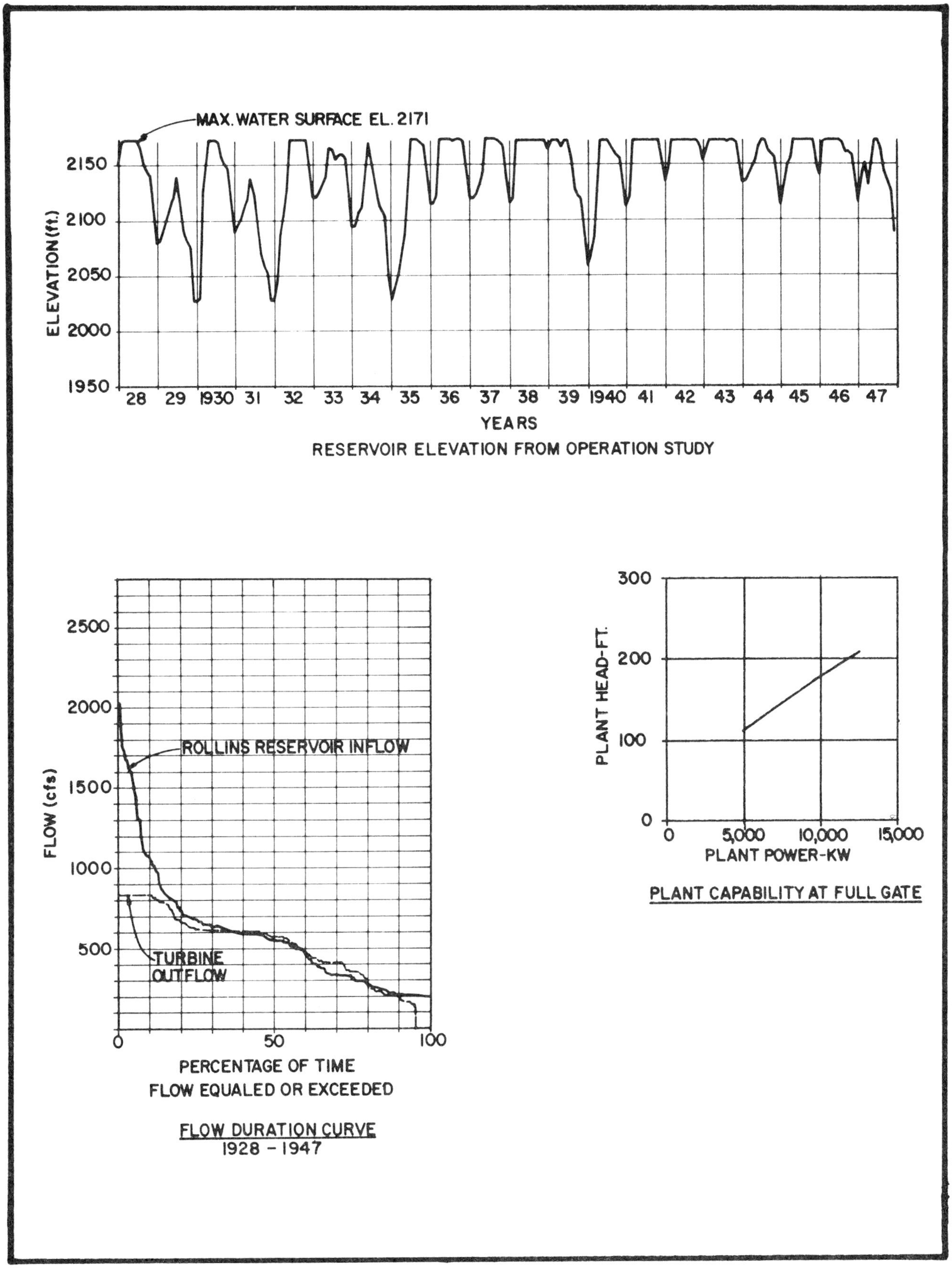

Figure 2-3. EBASCO Operation Study

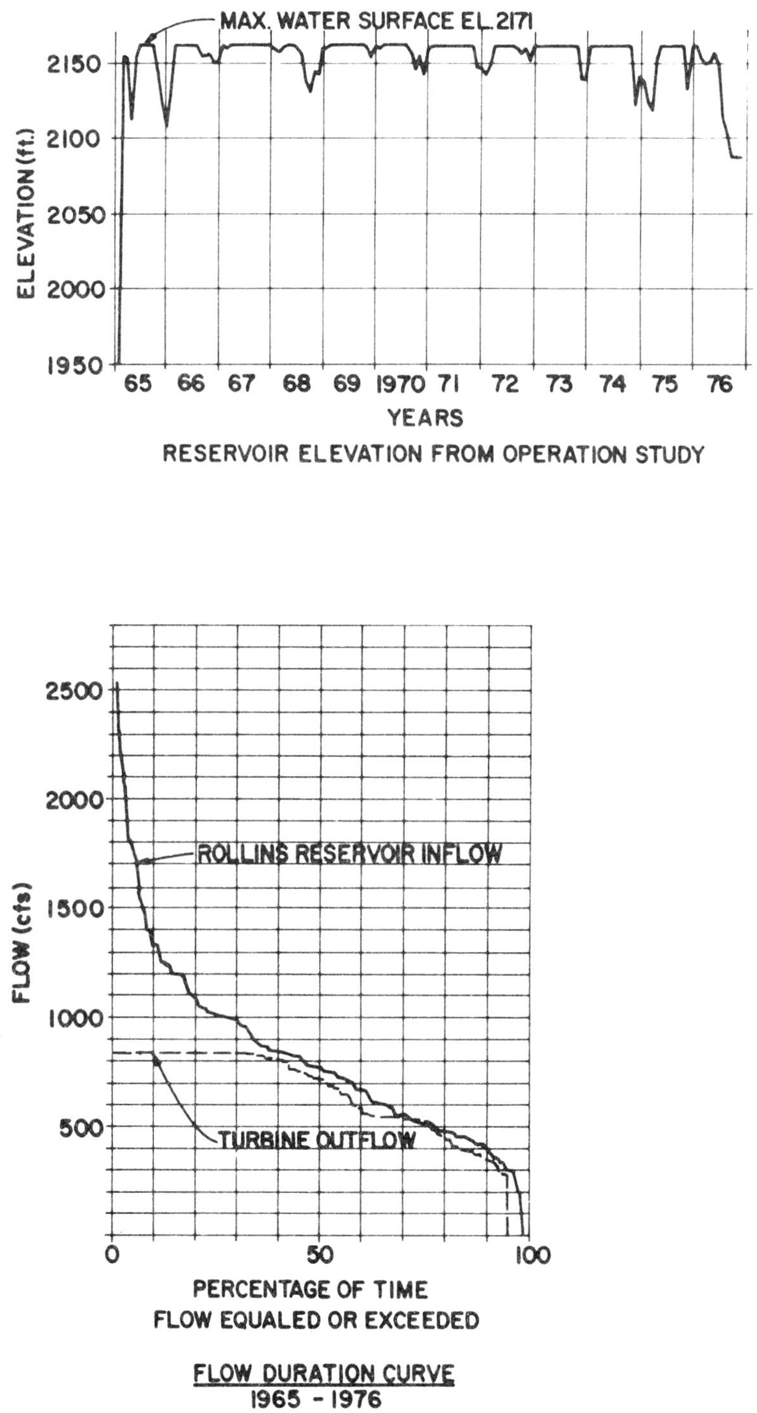

Figure 2-4. Historical Operation Study

mills per kWh, one-half of the difference between the replacement value of energy and the cost to produce the energy, was approved as a fair royalty to the District.

The last step in concluding the marketing arrangement was the preparation of a memorandum of understanding. The memorandum encompassed all the major points of the offer. In addition, since only 7.8 million dollars were available for the project, provision was made to permit short-term warrants to be used for any cost overrun. These warrants could be authorized by a majority vote of the Board of Directors under the Irrigation District law.

The revenue bonds which were issued by the District for construction of the Project were sold with an interest rate of 9 7/8 percent (taxable IDBs). The term of the bonds, 32 years, coincided with the years remaining on the District's FERC license for the Yuba-Bear Project. The total annual cost to PG&E including debt service on bonds, estimated operation and maintenance cost, and added incentive payment amounts to the sum of \$810,000, \$75,000 and \$284,400, respectively for a total of \$1,169,400. With an annual energy production of 71.1×10^6 kWh, the cost of energy, delivered at the bus bar, is 16.5 mills per kWh.

Application and Permits

The applications and permits which were prepared and received are as shown on Table 2-2. The table indicates several significant points. The actual experience shows that, with the exception of the time required to obtain water rights from the State of California, the schedule for project implementation provided in the manual can be achieved. The time required for water rights was due in part to slow processing by the State and to the intervention of a downstream irrigation district. This intervener was eventually satisfied by the execution of a supplemental agreement between the two parties which primarily reiterated each party's intent not to cause harm to the other.

Another significant aspect of the application process was the determination by the District that no significant adverse environmental impact would be caused by the construction. A negative declaration was therefore submitted by the District's Directors. This determination was considered by the Federal Energy Regulatory Commission and, after further review and consultation, indicated to the Council on Environmental Quality that no adverse impact would occur and a negative declaration should be issued.

The question of adverse impact on the local fisheries was not an issue. The release requirements from the dam were jointly developed 10 years previously with representatives of State and Federal governments to enhance the fisheries below Rollins Dam. (No fish passage facilities exist at Rollins because there are no migratory runs within the river.)

TABLE 2-2
SCHEDULE OF APPLICATION AND PERMITS

Permits or Applications	Responsible Agency	Date Filed	Approval Granted	Months Before Approval
1. Water Right Application to Develop Power at the Site	State Water Resources Control Board	1/29/76	9/27/77	20
2. Environmental Impact Negative Declaration in accordance with California Environmental Quality Act	Nevada Irrigation District	9/27/76	7/11/77	9
3. Amendment to License to Develop Power at the Site	Federal Energy Regulatory Commission (F.E.R.C.)	10/1/76	10/14/77	12
4. Water Quality Certificate (Sec 401 F.W.P.C.A 1/)	Regional Water Quality Control Board	2/20/77	5/11/77	3
5. Request to Lower Rollins Reservoir below Minimum Level	State Dept. of Fish and Game and F.E.R.C.	4/15/77	4/21/77	0
6. Application to Alter Permit Application No. 6333 (Sec. 404 F.W.P.C.A. 1/Permit)	Corps of Engineers Sacramento District	6/8/77	6/29/77 2/	1
7. Application to Make Alterations to a Dam	California Divison of Safety of Dams	6/10/77	7/11/77	1
8. Permission to Sell Phase I Bonds 3/	California Districts Security Commission	6/22/77	7/29/77	1
Phase II		1/11/78	9/20/78	8
9. Permission to Work in tunnel	California Division of Industrial Safety 4/		4/28/78	n/a

1/ Federal Water Pollution Control Act
2/ Not required
3/ For permission to sell revenue bonds
4/ Article 8422 D, Title 8, California Administration Code.
Classification of tunnel work required. Work was classified as non-gassy.

SECTION 3
DESIGN

General

The standards and criteria used for the design of the Rollins Power Plant were organized during the preliminary design stage. The contract between the District and PG&E stated "the power plant shall be equal to completeness of features and quality of design and materials in all respects to recent installations as in Pacific (PG&E) Feather River, McCloud River and Pit River projects." Therefore, the Rollins Project was designed as a major hydroelectric plant installation. As such, the costs of the Rollins Power Project are greater than the costs provided in the guide manual since the manual suggests the reduction of the requirements for control and protection on various items from those that would be necessary for major installations. A general description of the major electromechanical and civil features follows. Thereafter, a section is included which points out the specific discrepancies between the manual and the Project as designed.

Electromechanical Equipment

The appropriate turbine parameters were determined by a series of systematic routing operation studies in which the size of the unit and the design head (the head of maximum efficiency) were optimized. The turbine efficiency curve used was similar to the curve shown on Figure 3-8, Volume V. The curve is given in a different form in Figure 3-1 of this Appendix. For Rollins, a 12,700 kW unit with a maximum gross head of 204 feet was determined to be the most cost-effective installation. During the course of the investigation, it was learned that PG&E was decommissioning a power plant with two 13,000 kW Francis turbines and generators installed in 1927 with similar head and flow characteristics. An investigation was made of the desirability of using one of those turbine/generator combinations for the Rollins Project. After a thorough study, it was decided it was feasible to refurbish one of the units for the Project. Not all of the old parts could be reused, however. Manufacture of a new draft tube, spiral case and stay ring was required. The total cost of refurbishing the 13,000 kW unit was found not to be significantly different than the purchase cost of a new 12,700 kW unit. However, the time for procurement was reduced by 9 months by the reuse of the old equipment, providing a significant savings in cost and a one year reduction in the construction schedule. Furthermore, the old unit, being substantially heavier than a new unit, provided increased rotational inertia for better speed regulation and more durability.

Civil/Structural Design

The standards and criteria used in the civil/structural design were generally in accordance with common utility practice. Several features were particularly worth noting. A semi-outdoor design was used for the power plant. This design was selected primarily for economy. The power plant structure was designed to accommodate a portable gantry crane. The crane, however, was not included in the Project since its use would be infrequent and it could be rented when needed. An office building with storage area for spare parts and maintenance equipment was furnished as a separate building. This building, although built with power plant funds, was needed for other Yuba-Bear River Project purposes and would not have been necessary for the power plant alone.

Comparison of Guide Manual and Actual Costs

Construction Costs. A comparison of the power plant cost derived by use of the manual with the actual construction costs bid by the contractor for the Rollins Power Project is provided in Table 3-1. The cost level for manual costs is July 1978. The project was bid and awarded in about the same time-frame. Upon comparison, it can be seen that the actual costs are higher than those estimated by use of the manual. This difference can be attributed to the fact that the actual Rollins construction cost contains several items in addition to the basic power plant cost addressed by the manual. A listing, by account number, of the differences between the manual estimate and actual costs follows.

Account No. 331

1. The Rollins turbine was designed for bottom removal of the runner, a feature which adds to powerhouse depth. Bottom removal is not normally required and was therefore not considered in the manual.
2. At Rollins, the turbine is a refurbished older unit. This unit is considerably larger in physical size than a new unit of the same capacity. Because of the larger turbine, the powerhouse structure is larger than would have been required to house the turbine. Also, the PG&E required that certain equipment be installed in the powerhouse which normally would not be required and was therefore not considered in the manual. The larger turbine and additional equipment resulted in the Rollins powerhouse area being nearly 20 percent greater than the area that would have been calculated by use of the manual.

Account No. 332

1. The owner furnished the upstream shut off valve for the Rollins project. Consequently, the costs determined by use of the manual were higher than the actual Rollins costs.

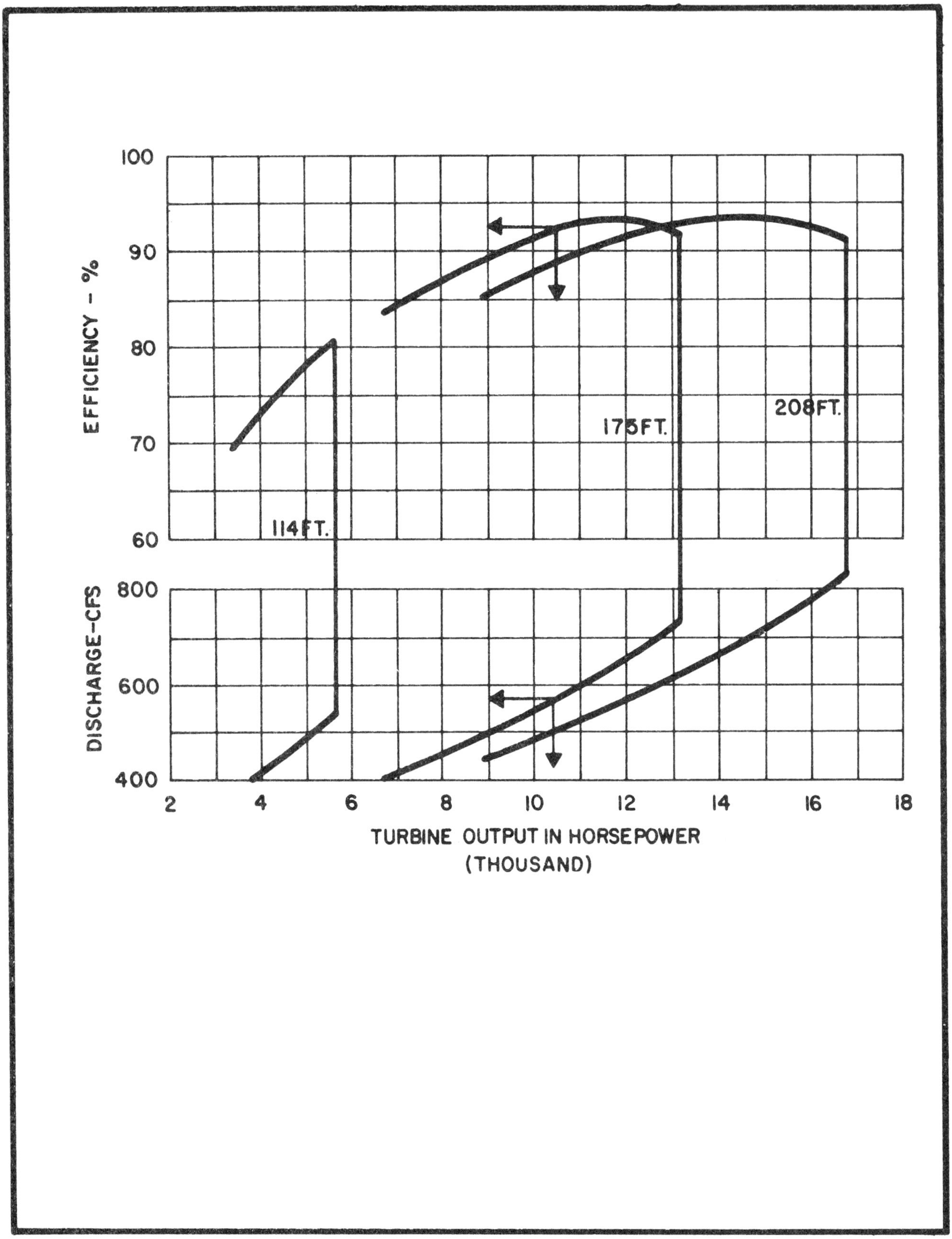

Figure 3-1. Francis Turbine Efficiency Curve

TABLE 3-1
COMPARISON OF CONSTRUCTION COSTS VS. MANUAL COSTS

Account No.		Guide Manual	Actual
331	Structures and Improvements	$ 642,000	$ 876,000 1/
332	Waterways	765,000	531,000
333	Turbine & Generator	2,300,000	2,479,000 2/
334	Electrical	785,000	897,000
335	Mechanical	125,000	292,000
		Total: $4,617,000	Subtotal: $5,075,000
Additional Work Items:			
Road and traffic control			$ 110,000
Toe drainage for dam			37,000
Office and warehouse building 3/			38,000
Channel diversion and afterbay excavation			90,000
Remote control (including equipment in Chicago Park Powerhouse)			150,000
	Total Phase II Contstruction Contract		$5,500,000
	Other Costs:		
	Tunnel plug contract		352,000
	Turbine/Generator purchase		112,000
	Contingencies		223,000
	TOTAL CONSTRUCTION COST		$6,187,000

1/ Concrete for waterways included in structures

2/ Includes governor $206,000
Includes governor housing $80,000

3/ Electro-mechanical included in accont 334 and 335

Account No. 333

1. The Rollins turbine has an automatic grease lubricating system which is not normally required and was not included in the manual.

2. To enable the unit to be motored and operated as a synchronous condensor, provisions were included at Rollins for water lubrication of the wearing ring at an increased cost not considered in the manual.

3. A special requirement of the power purchase agreement at Rollins was that the unit be capable of operating in an isolated system. This requirement mandated the installation of an Electric-hydraulic Speed Regulating Cabinet Type governor. Normally, a gate shaft governor would be adequate.

Account No. 334

1. Due to additional mechanical equipment in the powerhouse which was requested by the power purchaser, it was necessary to install an additional motor starter center and a low voltage distribution system.

Account No. 335

1. A heating system, not normally required and not considered in the manual, was included in the Rollins powerhouse.

2. At the Rollins project, the generator is water-cooled. The manual addresses air cooling only.

3. Rollins has an automatic fire protection system as opposed to the manually operated fire stations addressed by the manual.

4. A station air compressor with outlets at work areas is included in the Rollins project but not considered by the manual.

5. Rollins has hoists and jib cranes which are not normally required for a small hydroelectric project and were not addressed by the manual.

As a general commentary, the design of the Rollins power plant was greatly influenced by the requirements of the power purchaser. The plant operating criteria were based upon recently constructed major hydroelectric projects in the power purchaser's system. There are several major features which could be eliminated or modified, with an attendant reduction in cost, if the design has been consistent with normal small hydroelectric plant design practices.

Total Project Costs. In comparing the total project costs, Table 3-2 is presented. As can be seen, the percentages assigned in the manual to estimate indirect costs are relatively close to the actual percentages experienced at Rollins.

TABLE 3-2
ESTIMATED TOTAL COST OF PROJECT

Costs:			
Construction Contract			5,500,000
Design and Development			640,000
Construction Supervision			395,000
Surveys and Testing			30,000
Tunnel Plug (Construction already completed)			352,000
Equipment Purchases from Pacific			112,000
District Counsel			125,000
Costs of Issuance			110,000
State Treasurer's Review and Certification			28,000
Contingencies			223,000
	Gross Project Costs		$7,515,000
Less:	Investment Income (Estimated at 6.5%) 1/		390,000
	Net Costs		$7,125,000
Add:	Funded Interest (Two years at estimated 9-1/2% 2/)	$1,482,000	
	Less: Accrued Interest (Estimated 1-1/2 months)	93,000 3/	1,389,000
	Total Costs		$8,514,000

	Recap of Total Project Cost by Categories Cost	Actual Percentage of Construction Cost	Guide Manual Percentage of Construction Costs
Construction Costs	$6,187,000		
Indirect Costs	1,065,000	17.2	20
Financing Fees	138,000	2.2	1.7 - 3.3
Interest During Construction	999,000	16.2	15.8
Legal Fees	125,000	2.0	Not estimated
Total Cost	$8,514,000		

1/ Includes investment income from Interest Fund for approximately 16 months and assumes Construction Fund balances available for approximately 9 months.

2/ From July 1, 1978 to and including the July 1, 1980 payment.

3/ Received as part of the proceeds from the sale of the Bonds.

VOLUME II: ECONOMIC AND FINANCIAL ANALYSIS

SECTION 1: OVERVIEW

This volume presents guidelines for preparing the economic and financial portions of an overall feasibility investigation. The other volumes used in conjunction with this one will assist investigators in making a fair and accurate assessment of small hydro project feasibility.

The body of this volume is broken into four major subject areas. It is preceded by an introduction and followed by a summary. The introduction presents general information describing the purpose of the report and the differences between small hydro and large installations. Other discussions on sources of information, ownership characteristics and inflation are also presented.

The market analysis section describes in detail a wide variety of factors that affect the value of a small hydroelectric project. The information in this chapter will be of particular importance to the economic and financial analyst who must prepare the market assessment.

The economic analysis section first discusses the meaning and scope of economic analysis. Recommendations are given on formulating the cost and benefit streams and the appropriate evaluation criteria, and a generalized procedure is developed for applying these techniques.

Several topics that are financial in nature are discussed in the section on project implementation. The institutional requirements, timing of expenditures and sources of feasibility funding are explored.

The financial analysis section presents information pertinent to establishing project financial feasiblity. Funding sources are reviewed as is a method for establishing financial feasibility. A thorough discussion of the important role played by the project's financial advisor is also given.

The report concludes with a summary and cost guidelines for the preparation of the economic and financial portion of the feasibility assessment.

SECTION 2
INTRODUCTION

Scope and Objectives

The primary objective of an economic and financial feasibility investigation is to provide the economic basis for deciding whether to implement a project. An additional objective is to examine the promising development options in sufficient detail to determine which are most attractive.

To achieve these objectives, the scope of the economic and financial portion of the feasibility study must encompass all pertinent engineering, institutional, economic, and financial factors of the project that influence the implementation decision. With the basic project revenue and cost information arrayed in the feasibility study, the project sponsors should then be able to determine if implementing the project is in their best interest. At this point concerns beyond the project, such as capital availability, contractual problems and other factors, are taken into account. These concerns, which relate to the sponsor's overall goals and constraints, are typically not the subject of the feasibility investigation.

Feasibility studies are usually undertaken only when there is a reasonable expectation that the project will be feasible in some form. This may be determined with an inexpensive prefeasibility or reconnaissance study or by expert judgment of a qualified individual. Since all funds spent prior to the decision to implement a project are subject to total loss if the project is not implemented, it is clearly desirable to minimize these expenditures. To do so, intermediary studies that do not yield a definitive answer on feasibility should generally be avoided. Instead, sufficient funds should be expended to determine feasibility, and these results may then be used to either implement the project or reject it and end unnecessary expenditures.

The body of the economic and financial portions of a feasibility investigation are performed in the latter part of the study for the simple reason that they require input from the engineering and other investigations. However, close coordination and exchange of information are maintained with the other investigations. During these investigations, many problem areas may turn up that can render the project infeasible. If the project gets past the engineering and other hurdles, it can then be judged on its economic and financial merits.

The confidence that may be placed in the results of this portion of the feasibility study is a function of the quality of the information used and the analysis performed with this information. The investigation must:

— obtain the best relevant information concerning the value of power production from a small hydro site,

— using this and other information, determine the economic and financial feasibility of the project.

The aim of this report is to describe concepts and provide guidelines for their use in evaluating small hydro developments. A wide variety of situations will occur; therefore, no single procedure will suffice for all projects. For this reason, emphasis has been placed on the proper conceptual framework while providing as much information specific to small hydro projects as possible.

Differences Between Small and Large Hydro Projects

"Small hydroelectric power facilities" are defined in terms of the total nameplate capacity of the generating units and include installations with less than 15 MW of installed capacity. Most projects that fall under this definition would be located at existing impoundments throughout the United States. The U.S. Army Corps of Engineers (Institute of Water Resources, 1977) has estimated the potential at these existing small impoundments to be over 25,000 MW.

These small projects differ from the over 60,000 MW in existing conventional hydroelectric facilities in four significant ways important to the economic and financial feasibility analysis. First, most projects have relatively low heads (less than 100 feet). Because turbine and other powerhouse costs are more closely correlated to flow than head, the per-kilowatt (kW) cost of powerhouse, switchyard and other miscellaneous equipment can be relatively high. Figure 2-1 illustrates the strong dependence of cost per kW on gross head for new large installations. This will lead to a relatively high capital cost component of total cost in most instances.

Second, the analysis of small projects is usually conducted in the context of a single-purpose, non-essential project. The decision to construct or not construct will generally be based solely on the benefits versus the costs of power production. This is in contrast to many major, multi-purpose projects justified on flood control, recreation and other benefits in addition to the value of power.

Third, most small hydro projects will have little working storage dedicated to power production. This will simplify the operational plan of the project and will also result in the nature of the project's power being different than in most major projects. In the typical small project with little or no storage, there is no ability to store water and schedule peak power generation. Consequently, the project is run-of- the-river, with little, if any, dependable capacity.

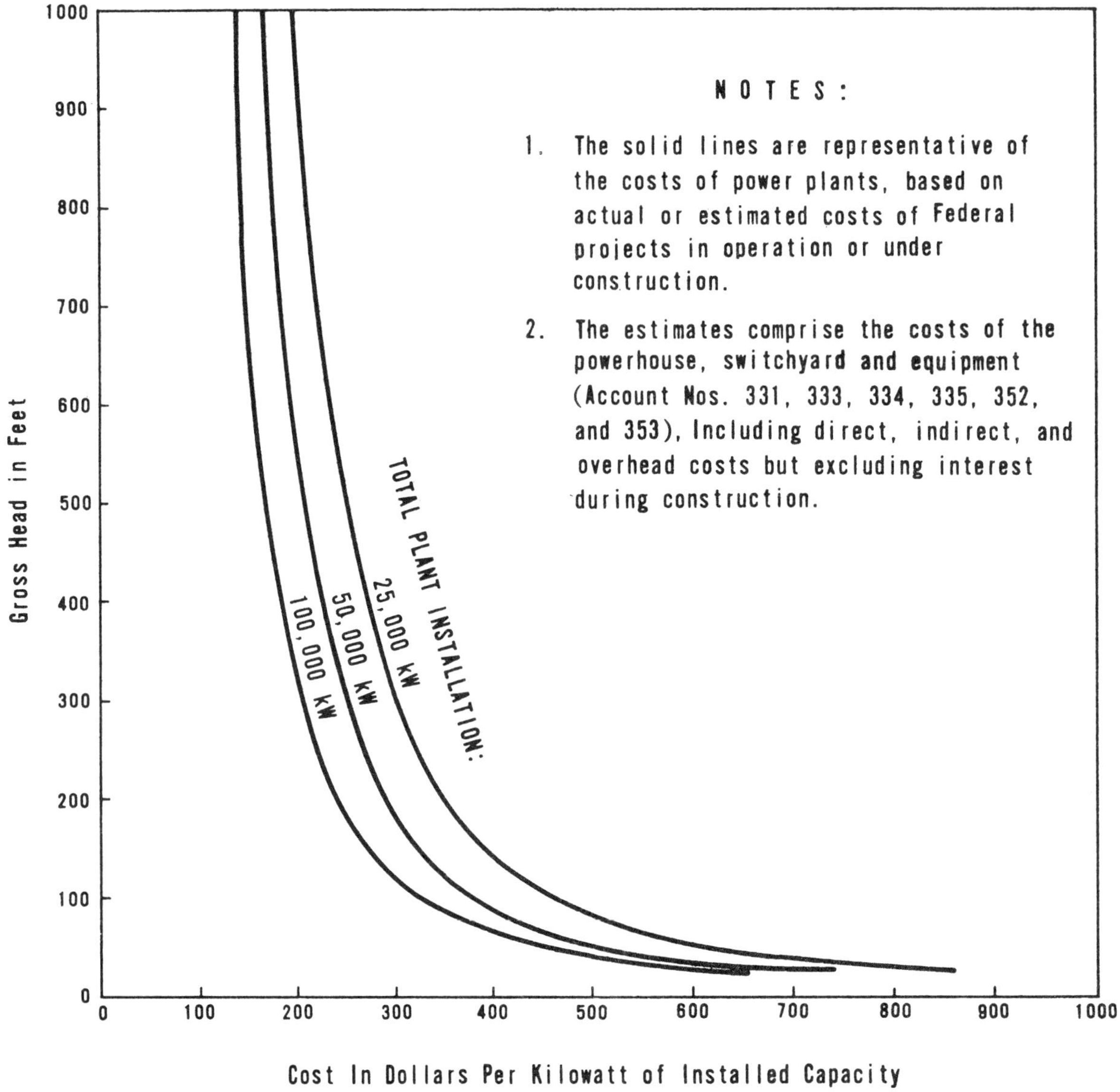

SOURCE: Federal Energy Regulatory Commission,
'HYDROELECTRIC POWER EVALUATION', Draft, Aug. 1978)

Figure 2-1. Plant installation costs for large hydro projects

Fourth, the cost of service of large hydro projects will include the transmission system to a substation capable of handling a large power input. In small hydro projects with much smaller power output, transmission line costs should typically represent a lesser portion of the total project cost because of the availability of substations and transmission lines that can handle up to 15 MW of additional input. Because of this, the treatment of transmission system costs and losses will be easier to evaluate.

Informational Requirements

All cost, marketing, performance and financial information must be assembled in an orderly fashion. The annual costs and capital requirements will be developed in the civil, mechanical and electrical portions of the feasibility study. These estimates will:

1. Be stated in current dollars of the year the study is performed.
2. Provide a capital cost expenditure pattern for each year of construction. (This will typically be expressed as percentages of the lump sum cost estimate per year.)
3. Indicate whether the costs are subject to escalation.
4. Provide funds for repair and replacement of major equipment necessary for project operation through the financing period. Power production information will be developed by the hydrologic analysis in conjunction with the turbine and generating equipment selections. This analysis will establish the dependable capacity and expected energy production for the development options being considered. The power marketing study will establish the value of the project's capacity and energy output.

The sources and description of the information required for the economic and financial analysis are summarized in Table 2-1.

The power marketing information will frequently be developed by the economic and financial analyst. This information must be carefully prepared since it will be used by the project sponsor in negotiations with the ultimate purchaser. Whether the purchaser is the local utility or one of its customers, the bulk of the information required deals with the utility's existing and planned operations. Some major sources of this information, other than the utilities themselves, are listed and described below.

1. Securities and Exchange Commission Form 10-K — If a privately owned corporation publicly offers securities (stock and bonds) and has over $1,000,000 in assets, it is required to file an annual Form 10-K. This form contains management's detailed statement of operations and audited financial statements and is a valuable source of information. Note that municipal or other public utilities are exempt from SEC reporting requirements.
2. Federal Energy Regulatory Commission (FERC) — Formerly the Federal Power Commission, the FERC requires detailed annual information from both publicly and privately owned utilities. Exhibit I lists and describes the forms electric utilities are required to submit and indicates how they may be ordered.
3. National Electric Rate Book (by state) — The rate book, updated periodically, presents summaries of rate schedules under which electric service is sold to general ultimate consumers by all privately and publicly owned electric utilities operating in urban areas throughout the United States. Many libraries will have the Rate Book or

TABLE 2-1
INFORMATION REQUIRED FOR ECONOMIC AND FINANCIAL ANALYSIS

Source	Information Supplied
(1) Facility Integrity Investigation	Capital and recurring costs or other work required to allow power production at an existing impoundment.
(2) Civil Facilities Investigation	Capital and maintenance costs of site, water-ways, powerhouse and other appurtenant civil facilities.
(3) Electromechanical Investigation	Capital, maintenance and operational costs of turbines, generators and other electrical or mechanical equipment. Also required is the timing and cost of future major repairs and replacements necessary for continued operation.
(4) Hydrologic Study	Annual and seasonal energy production, year to year variations, and dependable capacity. Existing water uses and rights and potential costs that might be incurred to assure water availability.
(5) Power Market Analysis	Value of capacity and energy production from the project.
(6) Economic and Financial Analysis	Cost of capital if not specified by sponsor, and general escalation rate.
(7) Project Sponsor	Capital limitations and cost, cost of land or other right-of-ways, other implementation costs (such as financial consultants) not included elsewhere.

it may be ordered from the U.S. Government Printing Office.

4. State Public Utility Commissions and Public Service Commissions — These are the agencies at the state level charged with regulating utilities; as such, they are important sources of information. It is common practice to establish a formal proceeding to review supply planning, and the record of these proceedings will contain much information on the utility.

5. Industry organizations — Two main industry organizations that have useful information are the Edison Electric Institute (90 Park Avenue, New York, NY 10016) and the Electric Power Research Institute (Box 10412, Palo Alto, CA 94303)

6. Moody's Public Utility Manual — Most public libraries will receive this publication, which provides investment-oriented information.

In some cases, the cost of financing available to the project sponsors will be unknown and must be determined. Generally, large project sponsors will supply this information, so it is the small sponsors, usually public entities, that may be uncertain of their cost of capital. Most of these sponsors will issue bonds to finance their project; hence, the current bond yields will approximate the appropriate cost of capital. *Moody's Bond Record* contains this information and is available in most public libraries. The tax status of the bond interest payments will be an important factor and is discussed in the section on financial feasibility.

Public and Private Ownership

There are two important differences between public and private project sponsors: (1) the taxes levied on private project sponsors, and (2) the differing costs of capital for the two types of owners.

Taxes levied on a privately owned project will effectively increase the project cost and reduce its return when compared to public ownership. Property taxes levied on both real and tangible personal property will result in a direct and escalating annual cost to the project. Because of the definition of property, property tax will be levied on virtually all of the capital cost of the project, and it will usually amount to between 1.0 and 3.0 percent of the capital cost. A publicly owned project will not have this cost.

Private ownership entails a higher cost of capital than does public ownership. State and local government obligations (bonds) are unique in that their holder is exempt from paying federal income tax on their interest payments (D. F. Jacobs, 1972) with certain exceptions contingent on the disposition of the power production. These exceptions are discussed in Section 6. Many bond issues are also exempt from income taxation by the states in which they are issued, through they are seldom exempt from the income taxes of other states. This exemption allows governments to borrow through bond issues at a lower rate than corporations whose interest payments are not exempt from taxation. In addition, the return on corporate equity is taxed on two levels, the corporate income tax and individual income tax dividends, pushing the cost of this component of the corporate capital structure even higher.

Because of the difference in the cost of capital, which can be as much as four to six percent, public entities will generally find a capital-intensive project more attractive than a private sponsor would. In small hydro, it is possible that a project infeasible for a private promoter will be attractive to a public entity.

Dichotomy Between Economic and Financial Feasibility

Economic justification deals primarily with the development and application of benefit-cost analysis. Benefit-cost analysis is an analytical procedure used in the economic evaluation of a project to:

1. Indicate the relative merits of different project configurations by identifying, measuring, timing and comparing project economic benefits and economic costs.

2. Determine the size, geographic scope and capacity of projects.

3. Establish the construction priorities and develop time schedules in energy service areas.

The objectives of the economic feasibility are met by relating all project benefits to project economic costs. This relationship provides relevent comparisons of the feasibility of different small hydroelectric configurations at a given site.

Financial feasibility, on the other hand, takes into account the availability of funds and relates financial costs to project revenues. Project financial costs are those incurred in constructing, operating, and maintaining project work and facilities, and they are elements of the total cost considered in the benefit-cost analysis (economic feasibility).

Inflationary Effects

Inflation will affect both the capital cost of a project and the continuing operations of the project. Furthermore, the effects of inflation must be explicitly accounted for if funds set aside for future repairs and replacement are to be sufficient to accomplish their purpose.

In capital-intensive projects with multi-year construction periods, inflation will lead to substantial increases in completed cost over the lump sum cost estimate. This is because prices for components will escalate between the time of the estimate and their actual procurement. Section 4 illustrates this and shows how to incorporate inflation in estimating completed cost. However, once the project is completed, the repayment of capital costs will generally remain fixed through the project life. In contrast, other project annual costs and revenues will be escalating with the result that capital costs become a decreasing proportion of total cost. This tends to enhance the cash flow later in the project but has little effect in the project's early years of operation.

In performing the economic analysis of a project, it is important that the effects of inflation on the cost and benefit streams be handled in a consistent manner. It is common for governmental agencies to use constant price levels in effect at the time of the study. If this is done, all future costs and benefits need to be expressed in constant price level dollars.

While it is possible to adopt this posture for the economic analysis, inflation must be accounted for in the financial analysis to correctly determine cash flow. Inflation can be explicitly incorporated in the cost and revenue streams by escalating future values by the expected inflation rate. It may also be desirable to escalate different portions of the projects at differing rates, depending on the expected escalation rate. This is particularly true of energy values, since there is a general expectation that the value of energy will rise faster than the general inflation rate.

SECTION 3
MARKET ANALYSIS

A variety of complex factors affect the marketability and value of output from a small hydro project. This chapter provides guidance on establishing what the project's power production characteristics are and how these characteristics relate to the value of the project. Also, institutional considerations and potential marketing arrangements are considered.

Institutional Factors

The ability to market power from a small hydro project may be affected by institutional factors at the federal, state and organizational level. This discussion provides background information concerning these factors and is intended to highlight items important to the marketability of small hydro power output.

Purchasing Utility. Under certain circumstances private, or investor-owned utilities (IOUs) may be less inclined than public utilities to purchase output from small hydro projects. This will be particularly true if the plant has significant quantities of dependable capacity and the total development cost is borne by the sponsor. The potential disincentive to IOUs for leasing capacity from another organization has been discussed at length in the economic literature (for instance, Alfred Kahn, 1971), and the explanations for this are briefly put forth below.

Marketing power to investor-owned utilities may be complicated, particularly if the project has significant quantities of dependable capacity. Like any other business enterprise, one of the objectives of an IOU is to make a profit. In contrast to unregulated enterprises, the amount of profit an IOU can make is limited by the size of their rate base (capital assets) and the regulated fair rate of return on this rate base. Consequently, to show an earnings growth requires growth in the rate base, which is primarily accomplished by the addition of companyowned capacity. If the company were to lease all of its capacity additions, there would be no earnings growth; conversely, earnings growth can be maximized by owning all capacity additions. For this reason, an IOU may not be inclined to purchase capacity and the associated energy production. It should be noted that this concept has yet to be empirically proven as a real tendency.

Publicly and cooperatively owned electric utilities encompass federal, state, municipal and cooperatively owned organizations. They are discussed below (excerpted from U.S. Senate Report No. 95-1292):

> In 1975, there were 1,835 municipals, 946 cooperatives, 306 investor owned, 123 State and county, 72 Federal and 22 industrial producers or distributors. The type of ownership tends to vary geographically. For example, in New England only 2 percent of the capacity is publicly owned, whereas in the East South Central Region 63 percent is publicly owned. By and large, public ownership tends to be more common in the Western states. There are five major Federal organizations which market power. The largest by far is the Tennessee Valley Authority (TVA) followed by the Bonneville Power Authority, Southwest Power Authority, Southeast Power Authority, and the Bureau of Reclamation. TVA is the largest electric utility in the United States, and like the other federally owned organizations, is primarily a wholesaler.
>
> The non-Federal public systems include municipals and States. These often purchase their energy from Federal installations, as well as from investor-owned utilities. In some cases, they produce a portion of their energy requirements.
>
> The most common form of non-Federal publicly owned system is the municipal system. Included in this group are several State-owned authorities. The municipals vary from very small to quite large, as in the case of the Los Angeles Department of Water and Power. The State-owned systems tend to be wholesalers operating hydro facilities. Some, such as the Power Authority of the State of New York, have both hydro and thermal power.
>
> Cooperatives tend to be small in terms of number of customers but also tend to have more circuit miles in distribution facilities than do other utilities. These utilities, owned by their consumers, are located primarily in rural areas and are almost always exclusively distributors. Some cooperatives, however, have joined together to create generation and transmission (G. & T.) cooperatives. There are approximately 50 G. & T.'s in the United States which generate approximately 27 percent of the cooperative requirement. Cooperatives obtain the bulk of their financing from a Federal agency — the Rural Electrification Administration — usually at relatively low interest rates.

The primary motivation of these organizations is to deliver the lowest-cost service while meeting reliability and other constraints. Marketing small hydro output to these organizations should be relatively easy if it offers the system a cost savings.

National Energy Act. The Public Utility Regulatory Policies Act of 1978, one of the five sections of the President's National Energy Act legislative package, has a number of provisions affecting small hydroelectric developments. These provisions can be grouped as those concerning power marketing (discussed here) and those providing funding for feasibility investigation and construction (discussed in Section 6). The Public Utility Regulatory Policies Act of 1978 contains provisions on wheeling, which, in specific situations, could result in an order from the FERC to the local utility to wheel power on behalf of a small hydro producer.

Wheeling may be defined as an electric utility providing transmission services for another utility, power producer, or power purchaser. If a small hydro producer could wheel output to end users or other utilities, this wider market might allow the power to be marketed more successfully. Consequently, the possibility of wheeling should be addressed in the economic and financial investigation.

Sections 202 and 203 of the Act give the FERC authority to order interconnection and wheeling of power produced from a "small power production facility" if such an order is in the public interest and would:

a) Conserve a significant amount of energy,

b) Significantly promote the efficient use of facilities and resources, or

c) Improve the reliability of any electric utility system to which the order applies.

Small hydro as defined herein qualifies as a "small power production facility".

There are a number of restrictions on the FERC's authority but the most important one to small hydro is: "No (wheeling) order may be issued...which provides for the transmission of electric energy directly to an ultimate consumer."

The FERC's authority appears to be restricted to wheeling power to organizations reselling the power. State agencies, however, may have broader authority than the FERC.

More important than the wheeling provisions are the rules concerning the sale and purchase of power from cogenerators and small power producers. Section 210 requires the FERC to prescribe rules that require electric utilities to:

1. Sell electric energy to qualifying cogeneration facilities and qualifying small power production facilities, and
2. Purchase electric energy from such facilities.

The rules are prohibited from authorizing a small power producer to make any sale for purposes other than resale.

The rates for purchases by electric utilities are to be set such that they:

1. Shall be just and reasonable to the electric consumers of the electric utility and in the public interest, and
2. Shall not discriminate against qualifying cogenerators or qualifying small power producers.

The purchase rules are required not to exceed the incremental cost of the electric utility for alternate electric energy.

Clearly, these regulations, when promulgated by the FERC, will have an important impact on small hydro power marketing. The small hydro power marketing analysis must examine the regulations governing the rates for purchases and interpret them in the context of the project at hand. The regulations should be available by the end of 1979 at the latest.

Regulatory Commissions. Early in the 1900s, the electric utility business started being regulated at the state level to protect the general public welfare. Regulation in its modern form confers on the IOUs certain advantages such as protection from direct competition in its service area by another private utility, the right to use streets and highways, and the right to condemn property. There are also certain obligations and disadvantages that arise from regulations which include the limitation of earnings, the obligation to serve all who apply for service, and the prohibition against withdrawal of service without regulatory approval.

The state-level organizations that oversee the investor-owned utilities (IOUs) are the Public Utilities Commission or Public Service Commissions (PUC/PSC), depending on the particular state. In some cases these agencies have been placed in an overall state energy agency that has a broader purview. To locate these agencies, see the *Directory of State Government Energy-Related Agencies*, National Energy Information Center, Federal Energy Administration, 1975 or updated versions.

While one of the main concerns of the regulatory commissions is limiting utility earnings to a fair rate, their main objective is protecting the public interest by seeing that the lowest-cost reliable service is provided. In this role, the Commissions frequently examine supply planning, managerial efficiency in general and other pertinent subjects. Because of these responsibilities, a PUC or PSC would likely intervene if an IOU were to refuse to purchase small hydro power output that offered the system a genuine cost saving.

It is recommended that the PUC/PSC in the state involved be contacted early in the power marketing assessment. The staff will be knowledgeable about any applicable laws and other pertinent information on the marketability of small hydro power in the state of the project's location.

Hydroelectric Capacity and Energy

There are essentially three types of hydroelectric developments in the United States:

1. *Run-of-the-river* plants whose generation is solely controlled by available flow as it occurs or is dictated by some controlling concern, such as irrigation needs.
2. *Storage* plants where there is storage available for use with the hydroelectric plant to control its power output over more than a short period.
3. *Pumped storage* plant where reversible turbines are

installed to use low-cost off-peak energy to pump water to an upper reservoir where it is stored for subsequent use to generate high-value peak-load power.

Combined projects are ones with both storage and pumped storage, and they have recently become more common. As a general rule, small hydroelectric developments will be run-of-the-river plants with little, if any dependable capacity and widely varying annual energy production.

The value of hydroelectric development is based on two components — capacity and energy costs of the most likely alternative developments. To establish the value of a hydro project, the amount of alternate capacity that the hydro development can substitute for, or is equivalent to, must be determined, as well as the cost of the energy the project will displace or replace.

Capacity. A large body of literature examines the interrelated power system concepts of system reliability, effective load-carrying capability, loss of load probability and other concepts. After maintenance and the probability of forced outages have been accounted for, the portion of peak demand that a unit will carry at a stated reliability level is termed the "Effective Load Carrying Capability" (ELCC). There has been less discussion concerning the amount of thermal generation capacity a run-of-the-river plant can substitute for. To establish the capacity value of a small hydro project, this substitute capacity is what needs to be determined.

The current FERC definition of "dependable hydro capacity" is explained and presented in Figure 3-1. In essence, dependable capacity is the amount of load a hydroelectric plant can carry under adverse hydrologic conditions during the period of peak system load. The adverse hydrologic conditions are usually based on the most adverse year of record. The period of peak system load depends on the particular utility and may occur during the winter or summer months.

This definition addresses two of the criteria necessary for determining the amount of thermal capacity a small hydro plant can substitute for. These are the annual flow variability in the river and the most critical period for the utility. The measure is conservative because no consideration is given to the low forced-outage and maintenance rates of hydro plants when compared to thermal plants. It is also conservative to base the assessment on the most adverse year of record. Doing so may subject the project to extremely stringent standards if the most adverse year is a rare occurrence with frequency of less than once in 100 years.

While capacity credits could be negotiated based on the FERC definition, a number of adjustments in the capacity credit may be justified. Several possibilities are suggested below.

The FERC recognizes that the low forced-outage rates for hydroelectric equipment, when compared to thermal-based generation, may warrant a capacity credit to the hydro project (FERC, 1978). Average forced-outage rates are published periodically by the Edison Electric Institute.

The FERC recommends that consideration of the particular utility in question should usually justify a capacity credit of 5 to 15 percent due to low forced-outage rates and rapid emergency start-up for hydro facilities. The FERC does not provide any guidance on determining what is justified.

Another technique that might be used to account for both adverse years and forced-outage rates is illustrated in Figure 3-2. The power availability curve for a small hydro plant can be constructed from daily stream flow records during the operation study. The following procedure is applicable in cases where the project is likely to have some dependable capacity.

1. The critical period of utility system load must be established. This will generally include several months on either side of the system peak.

2. The stream flow records during this period of the year must be examined to establish if any of the periods of low flow are extremely rare occurrences during this period. If so, excluding them from the record *may* be justified.

3. With the stream flow records from 2 above, a histogram of the daily power producible from the proposed installation can be calculated.

4. The histogram can then be converted into the power availability curve shown in Figure 3-2. Note that the horizontal axis of the power availability curve is equal to one minus the cumulative probability that the capacity available will be less than or equal to the stated capacity.

5. The forced-outage rate adjustment and its rationale are clearly illustrated in Figure 3-2 by showing the power availability curve for a thermal plant. Note that this two-state on-and-off reliability model of a thermal plant is the simplest and most commonly used. The thermal-equivalent capacity can then serve as the basis for negotiating capacity credits.

A slightly different procedure achieving the same results would be to use the stream-flow records in 2 above to construct a flow-duration curve. This curve can then be converted into the power availability curve.

The amount of dependable capacity arrived at by any of the procedures described will almost always be less than the generator nameplate rating. Depending on the specific circumstances, assigning some value to the non-dependable capacity may be justified.

Energy. Project energy production is the amount of kilowatt-hours (kWh) input into the utility system or delivered to a final user. The power factor of generation can be an important factor in the value of energy, and, hence, it should always be stated.

Because project revenues will ultimately be based on the energy delivered to the ultimate purchaser, care

SAMPLE--The sample is precented to avoid a lengthy explanation of the manner of preparation of Schedule 2

Schedule 2

SYSTEM HYDROELECTRIC DATA

A. AGGREGATE DEPENDABLE HYDROELECTRIC CAPACITY AND POTENTIAL ENERGY.

This schedule need not be completed if there have been no changes affecting the data previously reported. In such case the following notation should be made at the bottom of the page: "Data reported on FPC Form 12 for the year 19. . . , correct as of December 31 of the year herein reported." *Furnish data indicated below in accordance with the instructions in paragraphs 1-5, page 7.*

ADVERSE FLOW CONDITIONS*

Month	PLANNED USE OF STREAM FLOW AND STORAGE Energy (Megawatt-hours)					MACHINE CAPABILITY (Megawatts)		Dependable Capacity (Megawatts)
	Storage Plants: Natural flow	Storage Plants: Storage[1]	Run-of-River Plants	Total Available (Col. 2 plus col. 3 plus col. 4)	In Storage End of Month[2]	Run-of-River Plants	Storage Plants	
(1)	(2)	(3)	(4)	(5)	(6)	(7)	(8)	(9)
Dec.	xxxxxx	xxxxxx	xxxxxx	xxxxxx	2,800	xxxxxx	xxxxxx	xxxxxx
Jan.	33,200	(2,000)	12,500	43,700	4,800	40.0	126.3	148.0
Feb.	32,000	(3,100)	11,900	40,800	7,900	40.0	127.5	149.0
Mar.	48,900	(14,200)	18,900	53,600	22,100	40.0	133.0	165.0
April	52,700	(17,700)	21,700	56,700	39,800	39.5	138.0	176.0
May	47,100	(11,700)	18,200	53,600	51,500	40.0	140.0	171.0
June	39,700	(3,500)	15,400	51,600	55,000	40.0	140.0	166.0
July	22,800	0	8,400	31,200	55,000	40.0	140.0	149.0
Aug.	11,000	11,600	4,200	26,800	43,400	40.0	139.0	142.0
Sept.	13,200	9,800	4,900	27,900	33,600	40.0	136.6	143.5
Oct.	14,300	15,600	5,600	35,500	18,000	40.0	131.5	141.0
Nov.	19,900	11,100	7,700	38,700	6,900	40.0	127.2	141.0
Dec.	27,900	5,400	10,500	43,800	1,500	40.0	125.0	143.0
Year	362,700	1,300	139,900	503,900	xxxxxx	xxxxxx	xxxxxx	xxxxxx

AVERAGE OR MEDIAN FLOW CONDITIONS*

Month	PLANNED USE OF STREAM FLOW AND STORAGE Energy (Megawatt-hours)					MACHINE CAPABILITY (Megawatts)		Dependable Capacity (Megawatts)
	Storage Plants: Natural flow	Storage Plants: Storage[1]	Run-of-River Plants	Total Available (Col. 2 plus col. 3 plus col. 4)	In Storage End of Month[2]	Run-of-River Plants	Storage Plants	
(1)	(2)	(3)	(4)	(5)	(6)	(7)	(8)	(9)
Dec.	xxxxxx	xxxxxx	xxxxxx	xxxxxx	1,500	xxxxxx	xxxxxx	xxxxxx
Jan.	47,300	(7,100)	19,400	59,600	8,600	40.0	128.0	161.0
Feb.	43,400	(6,800)	18,200	54,800	15,400	40.0	130.5	164.5
Mar.	58,200	(13,600)	24,600	69,200	29,000	36.5	135.5	172.0
April	62,700	(17,400)	25,500	70,800	46,400	36.0	139.7	175.7
May	58,200	(6,300)	24,000	75,900	52,700	37.0	140.0	177.0
June	51,600	(2,300)	21,600	70,900	55,000	39.5	140.0	177.5
July	42,000	0	18,200	61,100	55,000	40.0	140.0	171.0
Aug.	36,300	2,300	14,800	53,400	52,700	40.0	140.0	165.0
Sept.	33,500	6,600	13,700	53,800	46,100	40.0	139.5	163.5
Oct.	35,200	15,200	14,700	65,100	30,900	40.0	136.0	161.5
Nov.	39,000	13,100	15,900	68,000	17,800	40.0	132.0	155.0
Dec.	41,200	15,000	17,100	73,300	2,800	40.0	125.5	150.5
Year	549,500	(1,300)	227,700	775,900	xxxxxx	xxxxxx	xxxxxx	xxxxxx

[1] When energy is drawn from storage, show as a positive quantity. When energy is stored, show as a negative quantity in parentheses.

[2] Change in storage based on entry in column 3.

*NOTE.—The method or basis used in determining the above data for adverse flow and average or median flow conditions should be explained in accordance with instructions 2 and 3 of this schedule.

SAMPLE EXPLANATION

Notes:

Data reported under "Adverse Flow Conditions" are based on stream flows in the calendar year (19.......), which is the most adverse year of record. The critical flow period normally occurs during the last 6 months of the calendar year.

Data reported under "Average or Median Flow Conditions" are based upon the average of monthly stream flows during the period of record (19.......19.......).

(6-a) Rev. (12-75)

Figure 3-1. Source: FERC Form 12, "Power System Statement," for the year ended December 31, 1977.

Power System Statement of. . . . . for the Year Ended December 31, 1978

Schedule 2—Continued

SYSTEM HYDROELECTRIC DATA—Continued

1. The data to be reported in Part A of Schedule 2 are intended to present a realistic picture of the potential energy and capacity of system hydroelectric plants under the specified flow conditions. The data to be reported should be based upon an assumed schedule of system operation that would permit serving the maximum possible annual system load with existing facilities and arrangements for purchase or sale of firm power, assuming a continuance of the relative seasonal and hourly variations of load that occurred during the year of this report. Contracts for purchase or interchange of off-peak energy also may be taken into account. In determining the magnitude of the seasonal load that could be carried by the system and the necessary scheduling of system operations, provisions for necessary maintenance scheduling and reserve capacity to be supplied by own system should be taken into account. Explanatory notes relative to Schedule 16 should be referred to in connection with this schedule. If the seasonal and hourly variations in load are expected to change materially, the information given may be based on the expected load shape, explaining in a footnote.

2. The information to be reported under adverse flow conditions should, in general, be based on stream flows equivalent to the year giving the most adverse flow conditions of record during the critical period of system operation. Where stream-flow records indicate that the most adverse flows are not likely to occur except at long intervals of time and are likely to be of a very short duration, the figures used in determining the capacity and energy available from hydro plants may be modified, treating such abnormal limitations as emergency conditions to be covered by the reserve capacity; such modifications, however, should be fully explained. Any system which maintains comparable data based on flows during a year which would give the minimum potential annual output, or based on minimum flow or output for each month, may report on whichever basis it believes will present the most realistic condition for its system. The basis of reporting should be fully explained in the space provided for notes with addenda sheets if needed.

3. Information to be reported under average or median flow conditions may be made on the assumption of the recurrence of flows equivalent to a year which would give the average annual potential output or may be based on median flow or output for each month, or average flow or output for each month, whichever it is believed will present the most realistic condition for its system. The basis of reporting should be fully explained in footnotes or addenda sheets.

4. "Run-of-river" refers to those plants whose operation cannot be regulated over a period of more than a few hours, either from storage at site or above, but whose operation is, in general, controlled by the volume of flow which must be utilized as it occurs or be wasted.

"Storage" refers to those plants whose operations can be varied as desired because of storage at site or above. This regulation may be weekly, monthly, or seasonal.

"Total available energy" refers to the maximum potential output of the existing hydro-generating facilities on the basis of the regulated stream flow, regardless of whether such output can be fully utilized in serving system load or by transfer to other systems. The monthly distribution of storage energy should be such as to permit the serving of the maximum annual peak load under the conditions outlined in instruction 1. However, where required releases for irrigation, navigation, flood control, and other water-use are controlling, the monthly distribution of available energy should reflect the effect of such requirements and full explanation should be given in footnotes.

"Capability" in any month is the machine capability under the most adverse conditions to be expected in that month under the assumed flow conditions without respect to the energy available or the characteristics of the load to be served other than the power factor conditions normally to be expected.

"Dependable capacity" in any month is that capacity that can be relied upon for serving system load and firm power commitments on the basis of the energy available in that month and its use as limited by the characteristics of the load to be served.

5. Dependable hydroelectric capacity as used in this power system statement is intended to be the capacity value of the system hydroelectric plants in serving, together with the other available system capacity, the maximum annual system peak load under the conditions given in instruction 1. For any specified period it represents, on the basis of complete utilization of available storage energy over the critical flow periods, the difference between the peak load for that period and the maximum other capacity required. Where a portion of storage energy is scheduled to be held as a reserve for emergency use only, the dependable capacity should also include the reserve capacity value of such energy reserve. The dependable hydroelectric capacity shown in column 9 under adverse flow conditions for the month of annual peak demand may not necessarily be the same as the annual dependable hydroelectric capacity to be reported in schedule 16, as the annual peak demand may not occur in the month requiring the maximum capacity from other than system hydroelectric plants. This is illustrated by the following graph:

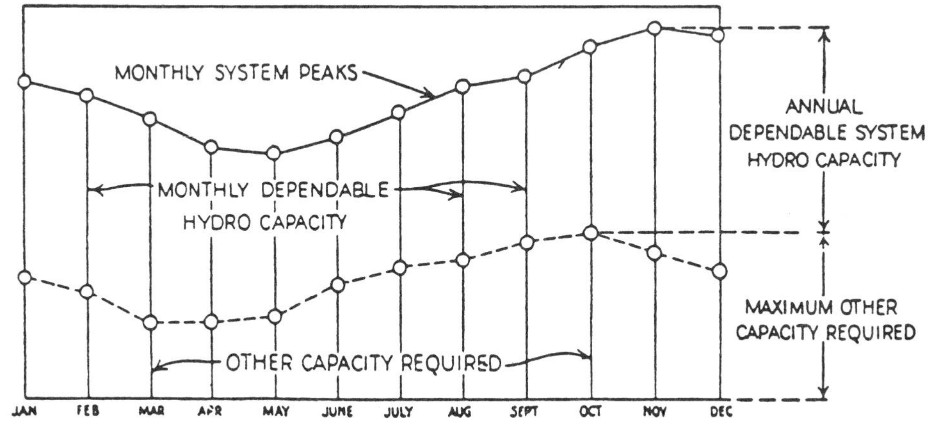

(7) Rev. (12-75)

Figure 3-1. (continued)

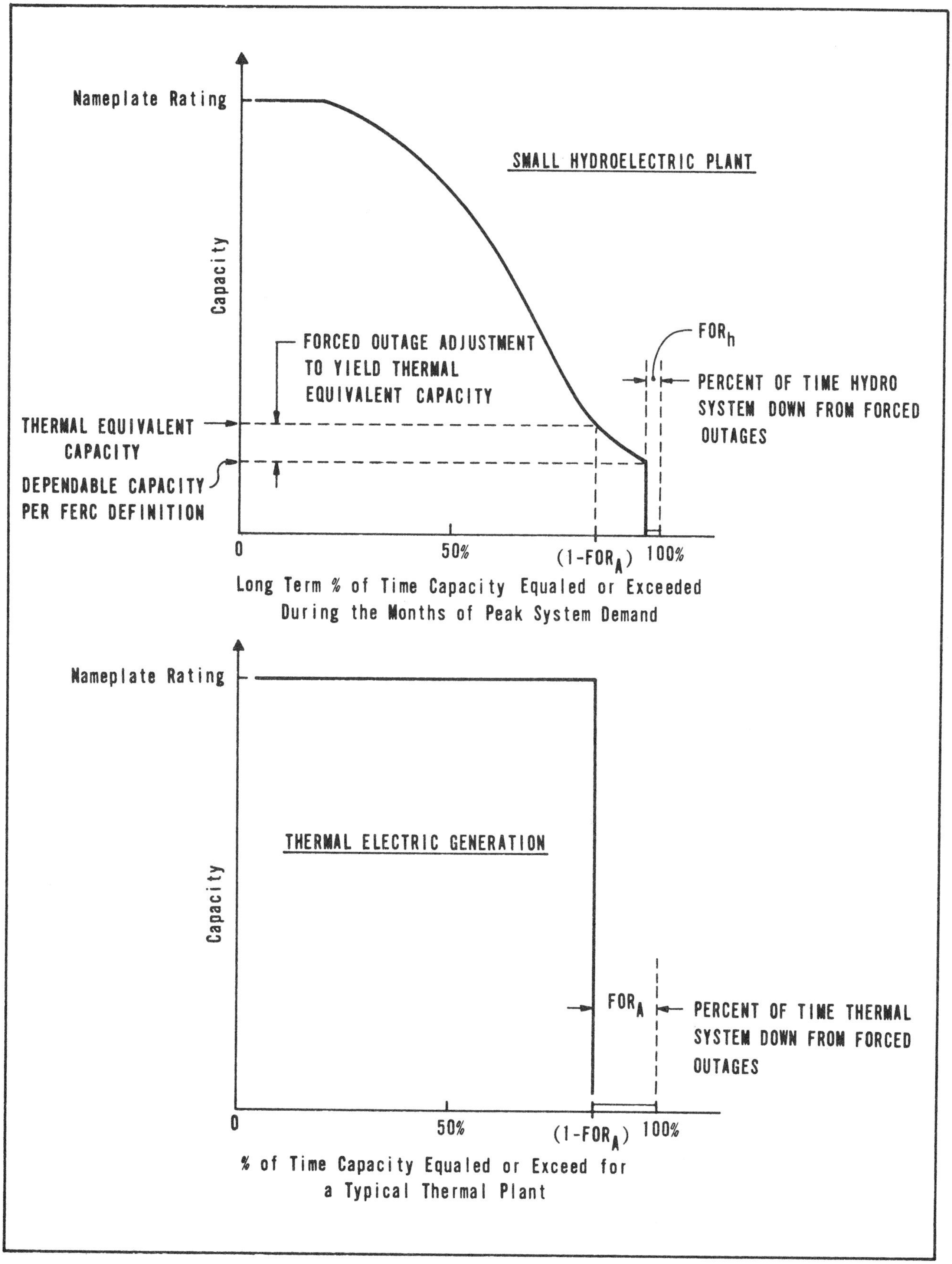

Figure 3-2. Capacity availability curves for small hydro and thermal plants.

should be taken to account for all losses up to the point of ownership transfer. If extensive transmission is required, these losses must be included as well as step-up transformer losses, generator and speed increaser losses, and station service use. Also, a loss due to forced outage should be included to avoid overstating the average annual energy output.

Energy production will vary on a yearly, monthly, and daily basis. The effects of daily fluctuations and the impacts on dependable capacity have already been discussed. Annual and monthly variability can be portrayed in a number of ways. One desirable method is to consider the annual energy production as a random variable and construct annual production histograms and cumulative probability distributions as in Figure 3-3. This curve can be useful in assessing project risk and will be discussed in Section 6 on financial feasibility.

The seasonality of power production can be portrayed as in Figure 3-4. This curve is useful for assessing in broad terms how the project output would fit into a utility system and the effects of adding capacity. For example, if the project of Figure 3-4 were located in a summer peaking utility, it is apparent that adding to installed generation capacity will do little to increase the project's ability to serve system peak-load.

At a minimum, the average annual energy production and its annual variability must be established. Additional information on the seasonality of energy production can be helpful both in project design and in establishing whether dependable capacity is present. To establish that the project has dependable capacity, very detailed energy production estimates will be required, possibly on a daily basis.

Peaking Capability. For a small hydro plant to serve as a peaking unit, it must incorporate storage. Operationally, water is accumulated for release through the turbines during the hours of peak demand. The storage capability allows the energy available to be scheduled at the time of maximum value.

When the small hydro project does have working storage available for power operations, a peaking operation may be explored as a way of increasing project value. The dependable capacity from a storage reservoir which is to be operated as a peaking unit can be established using the FERC definition (Figure 3-1). Note that this is not an easy task. Even if no dependable capacity is present, operating the storage reservoir and powerhouse as a peaking unit will generally increase its value to the local utility over what it would be in run-of-the-river operation. Storage capacity, turbine capacity and the flow regime must be integrated into a model by the hydrologic study to determine the amount of energy that may be shifted to peak periods. The value can then be calculated as indicated later in the discussion of the value of energy to a utility purchaser.

Value of Capacity and Energy

The value of small hydroelectric capacity and energy output is based on the costs of equivalent alternatives available to the prospective power purchaser. Consequently, the value of a small hydro project can vary widely, based on the potential purchaser. This discussion first considers in broad terms how the value of a project is established and then presents detailed examples of how the value of power can be calculated for an industrial and utility purchaser.

Opportunity Cost as a Basis for Establishing Small Hydro Project Value. The value of a small hydro project is determined by the power purchaser's opportunity to reduce existing costs while maintaining the same level of service. To do so, equivalent situations with and without the small hydro project are determined. The difference in total cost between the two cases, *without assigning any cost to the small hydro project*, will be the project's maximum value to the purchaser. The difference in total cost, after including the actual cost of the small hydro project, is the *net* value of the project and represents the opportunity cost of foregoing the project.

The proviso of maintaining the same level of service is important. While small hydro may allow a purchaser to reduce some costs, such as power purchases or fuel expenditure, maintaining the same level of service required without the small hydro project may entail additional costs such as standby service or generation capacity. The project information developed on dependable capacity and annual energy production allows the equivalent situations to be determined.

Since the project's value is established by looking at the power purchasers and the costs of their alternatives, a particular purchaser can significantly alter a project's value. Some general observations in this regard follow.

Industrial or Other End User Power Purchasers. Generally, industrial electric users require electric service more reliable than that afforded by the typical run-of-the-river small hydro project. Consequently, they will have to maintain some sort of a standby service arrangement with the local utility. This type of service may increase or decrease the electricity displacement benefits of the small hydro project, thereby altering the incremental cost savings attributable to small hydro.

Utility Systems. In general, utilities with higher-cost fuels will find small hydro projects have higher value to them because of the cost of the fuels displaced by small hydro. This is particularly true of utilities using oil to fire base load units. Some Eastern and Western utilities, by necessity, will be generating baseload energy with oil for a number of years.

Publicly owned utilities will place less value on capacity than IOUs. This is because their lower cost of capital and exemptions from property and income taxes significantly lower their fixed costs when compared to IOUs.

User as Power Purchaser. The gross value of small hydro output to an end user, such as an industrial plant,

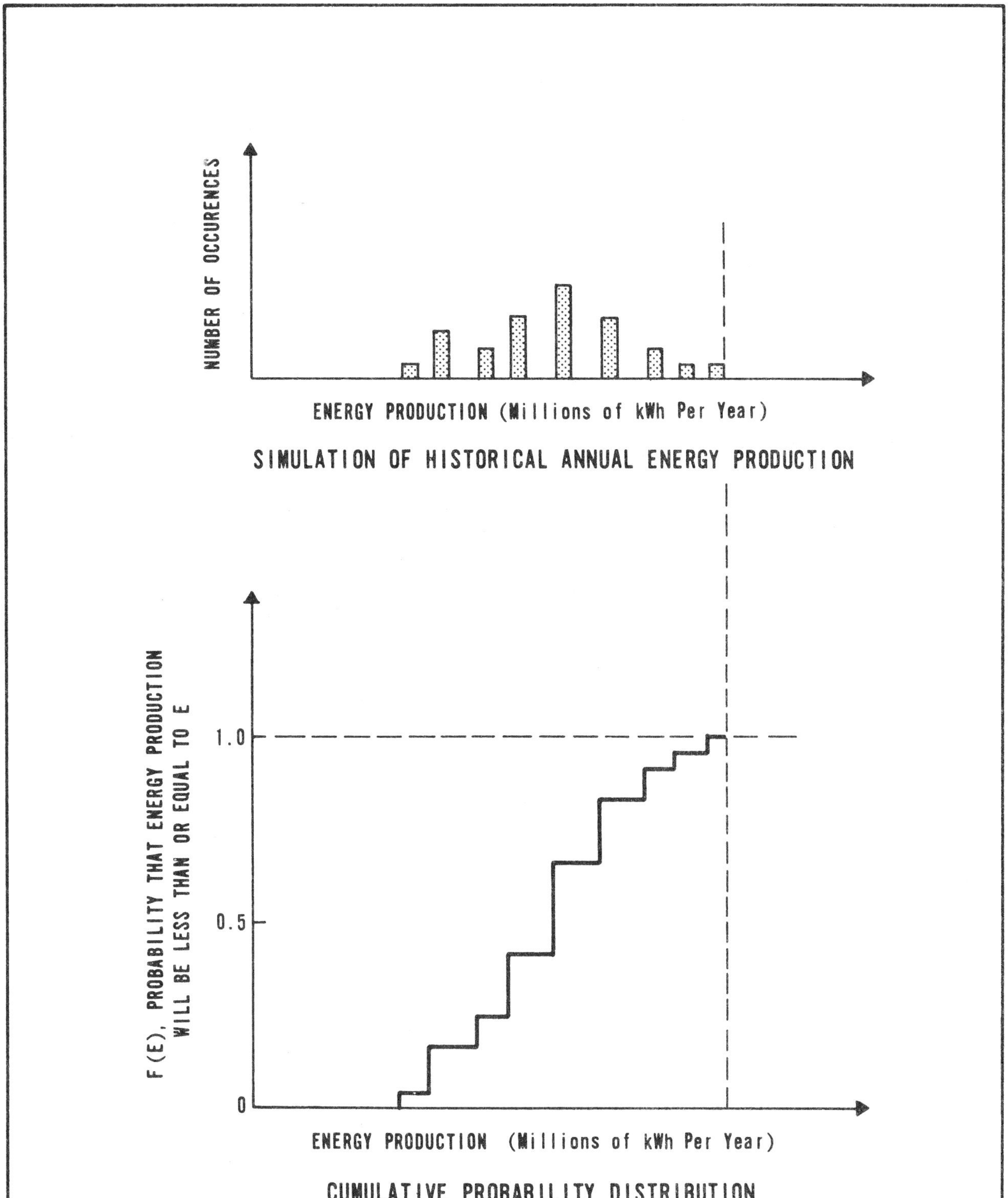

Figure 3-3. Annual energy production histogram and cumulative probability distribution.

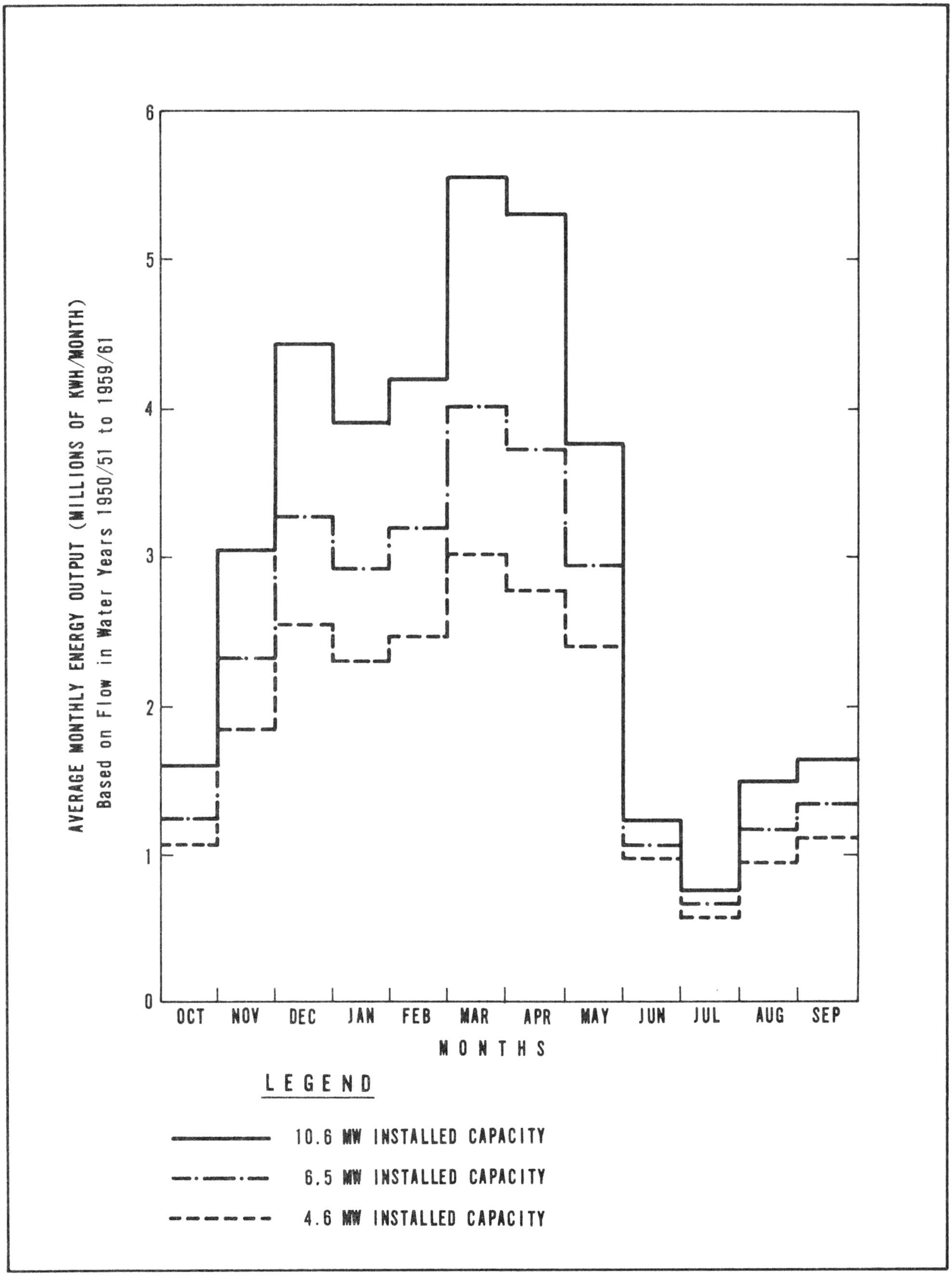

Figure 3-4. Average monthly energy output.

municipality or irrigation district, is the maximum cost reduction the purchaser can achieve without assigning any cost to the small hydro plant. In this calculation make certain the user is receiving the same level of service before and after the addition of small hydro output. If not, the cost comparison will be between different situations and will not truly reflect the value of small hydro. The purchaser will find the small hydro output attractive if the actual hydro costs are less than the maximum cost reduction. If so, a net cost reduction will be achieved.

Information about the small hydro project's output, the user's load characteristic and the applicable utility tariffs is necessary to establish the value of small hydro to the user. In many cases, the average monthly electric production from the small hydro plant will be sufficient for the analysis. Using average data will lead to "expected" benefits, but yearly variations in these benefits must be expected. The user load should be readily typified either through utility or user-metering records. The utility tariffs are also accessible from a number of sources. The *National Electric Rate Book* gives summaries by state of utility rates nationwide. The state level regulatory commissions will have detailed rates and the local utility will also supply any necessary rate information.

The following example demonstrates the calculation of the maximum value of a small hydro project to an industrial purchaser. The example is simplified but contains all the essential elements that need to be accounted for. Figure 3-5 specifies the load characteristics of the industrial purchaser and the average monthly power production of the small hydro project. Also shown is the minimum monthly power production from the hydro plant. This value will determine the billing demand. The industrial plant is assumed to have a continuous demand of 5,000 kW. The small hydro project has maximum production in the winter months and drops to zero during the summer. No dependable capacity is present. Figure 3-5(c) shows the industrial purchaser's demand on the utility system after including the small hydro power.

A simplified utility tariff for general and standby service is given in Table 3-1. A common type of rate, the Hopkinson demand rate, with flat demand and energy charge has been assumed (for more information on rates, Caywood, 1972). Typically, a flat monthly customer charge is present, but has been left out for simplicity.

Two other common rate provisions are provided. Minimum charges are frequently levied and may be calculated in a number of ways. In this case, the minimum bill is based on the maximum amount of demands. A billing demand ratchet has also been included. This clause associates the billing demand to the highest demand in the last X months where X may be between 2 and 12, or on the average demand over some time period or on a percentage of these two. The effect of a billing demand ratchet is to increase demand charges to a customer.

Table 3-2 and 3-3 calculate the annual utility-supplied electricity cost to the industrial purchaser with and without the small hydroproject. With all other things equal, the difference in total annual costs, $587,300, is the *maximum* value of the small hydro output to the industrial user. Note that on a per kWh basis, this value is 3.83¢ per kWh, which is greater than just the cost of

---

TABLE 3-1
SIMPLIFIED RATE SCHEDULE

GENERAL SERVICE

Rate:
- Demand Charge:
  - $6.00 per kW demand per month
- Energy Charge:
  - 3.5¢ per kWh

Minimum Bill: The demand charge on 10 percent of maximum demand.
Billing Demand: The maximum 15-minute measured demand during the month, but not less than 90 percent of the highest demand in the preceding three months. (Note: This type of clause is known as a billing demand ratchet clause.)

STANDBY OR AUXILIARY SERVICE

Contract Demand: The maximum demand the customer will place on the utility system. The utility will not meet a demand higher than the contract demand.
Rate: Same as general service.
Minimum Bill: $3.00 per kW of contract demand.

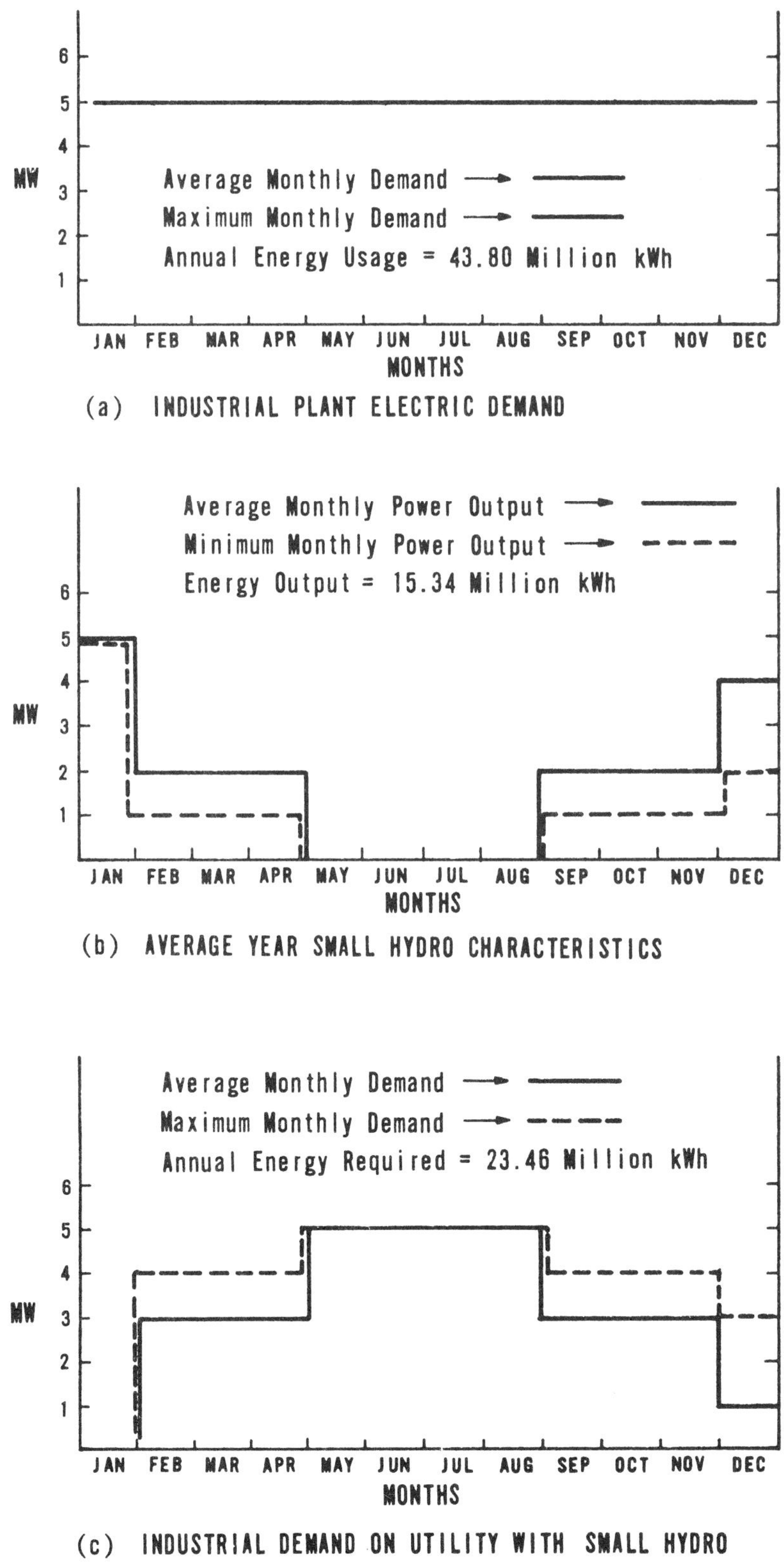

Figure 3-5. Demand pattern and small hydro output for example calculation of small hydro benefit to an industrial purchaser.

the energy displaced. This will not always be the case, and only the facts of the individual situation will determine the results.

Utility as Purchaser. The correct way of determining the value of a small hydro project to a utility is to determine the reduction in total system cost that would result from adding the small hydro plant to the utility system, without assigning any cost to the small hydro project. To be valid, the comparison must be between like systems before and after the small hydro addition.

A small hydro project will displace fuels, and if it has dependable capacity, it will reduce the need for new utility investment. Some operational cost savings may also be possible. These cost reductions can be reasonably approximated by considering a simple production cost model of utility generating units. In the following material, the production cost model will be explained and the connection between the type of hydro development and the appropriate production cost will be discussed. The value of the small hydro plant will then be calculated as the cost savings indicated by the production cost model.

The basic production cost model of thermal electric generation is composed of three components: capital costs, fuel costs, and operation and maintenance costs. (For more information, see Sullivan, 1977, or the draft *Hydroelectric Power Evaluation*, Federal Energy Regulatory Commission, August 1978.) This is:

$$TC_i = CC_i + FC_i + OM_i$$

where:

TC_i = Total cost of generation type i
CC_i = Capital associated costs
FC_i = Fuel costs
OM_i = Operation and maintenance

Capital Associated Costs. It is common practice to calculate the annual fixed costs per unit of generating capacity by specifying a fixed charge rate as a percentage of capital cost. The annual capital cost per unit of generator type i is then:

$$CC_i = FCR \times I_i$$

where:

FCR = Fixed charge rate
I_i = Investment per unit of capacity i, \$/kw

The fixed charge rate is composed of five components:

1. The weighted average cost of new capital.
2. Depreciation or amortization.
3. Insurance.
4. Ad valorem or property taxes.

TABLE 3-2
EXAMPLE INDUSTRIAL GENERAL SERVICE ANNUAL CHARGES

Month	Maximum Actual Demand (kw)	Billing Demand (kw)	Energy Used (10^6kwh)	Demand Charge[1] (\$)	Energy Charge[2] (\$)	Total Charge[3] (\$)
(1)	(2)	(3)	(4)	(5)	(6)	(7)
January	5000	5000	3.72	\$ 30,000	\$ 130,200	\$ 160,200
February	5000	5000	3.36	30,000	117,600	147,600
March	5000	5000	3.72	30,000	130,200	160,200
April	5000	5000	3.60	30,000	126,000	156,000
May	5000	5000	3.72	30,000	130,200	160,200
June	5000	5000	3.60	30,000	126,000	156,000
July	5000	5000	3.72	30,000	130,200	160,200
August	5000	5000	3.72	30,000	130,200	160,200
September	5000	5000	3.60	30,000	126,000	156,000
October	5000	5000	3.72	30,000	130,200	160,200
November	5000	5000	3.60	30,000	126,000	156,000
December	5000	5000	3.72	30,000	130,200	160,200
	TOTALS		43.80	\$360,000	\$1,533,000	\$1,893,000

1/ Calculated as Billing Demand, column (3), times General Service Demand Charge, \$6/hr.
2/ Calculated as energy used, column (4), times energy charge, 3.5¢/kWh.
3/ Sum of (5) + (6), *or* the minimum bill.

Assumptions

1. Demand as in Figure 3-5(a)
2. Rate schedule in Table 3-1
3. Minimum bill = .1 × 5000 kW × \$4/kW = \$2,000 per month

5. Income taxes (federal, state or local).

As a general rule, the cost of capital and depreciation will be the largest components of the fixed charge rate.

Accurately calculating the fixed charge rate for a utility from basic financial data is difficult. For this reason, it is recommended that the appropriate fixed charge rate be obtained by contacting the local utility or state regulatory commission.

Fuel Costs. The annual fuel cost of operating unit i for t_i hours in the year is given by the linear approximation:

$$FC_i = HR_i \times EC_i \times t_i$$

where:

HR_i = the heat rate of unit i defined as the number of Btu's of energy input required to produce one kWh.

EC_i = the energy cost of the fuel used in unit i expressed in \$/Btu

t_i = hours of operation of unit i in the year.

The heat rate used can be based on generic heat rates or the actual values for the utility in question. Generic values by type of plant and fuel are available from numerous sources. A few examples are the FERC, Edison Electric Institute, Electric Power Research Institute and trade journals. The actual values of a specific utility's existing plants are also available in the annual FERC Form 12 filed by all utilities and in SEC Form 10-K filed by publicly traded investor-owned utilities (see Exhibit I).

Operation and Maintenance Costs. O&M costs, exclusive of fuel use, are usually broken into fixed and variable components.

Many factors, such as kind of plant, location, size, plant factor, operational plan, and age affect the O&M costs. These costs are generally much less than fuel and capital costs.

Because small hydro plants have a capacity of 15 MW or less and will usually be run-of-the-river, utility system O&M cost reductions will be small if they exist. As a general rule, in calculating the value of a small

TABLE 3-3
EXAMPLE STANDBY SERVICE ANNUAL CHARGES FOR INDUSTRIAL USER PURCHASING SMALL HYDRO OUTPUT

Month	Maximum Actual Demand (kw)	Billing Demand (kw)	Energy Used (10^6kwh)	Demand[1] Charge (\$)	Energy[2] Charge (\$)	Total[3] Charge (\$)
(1)	(2)	(3)	(4)	(5)	(6)	(7)
January	-0-	-0-	-0-	-0-	-0-	\$15,000 [4]
February	4000	4000	2.02	\$ 24,000	\$ 70,700	94,700
March	4000	4000	2.23	24,000	78,050	102,050
April	4000	4000	2.16	24,000	75,600	99,600
May	5000	5000	3.72	30,000	130,200	160,200
June	5000	5000	3.60	30,000	126,000	156,000
July	5000	5000	3.72	30,000	130,200	160,200
August	5000	5000	3.72	30,000	130,200	160,200
September	4000	4500 [5]	2.16	27,000	75,600	102,600
October	4000	4500 [5]	2.23	27,000	78,050	105,050
November	4000	4500 [5]	2.16	27,000	75,600	99,600
December	3000	3600 [5]	0.74	21,600	25,900	47,500
	TOTALS		28.46	\$294,600	\$966,100	\$1,305,700

1/ Calculated as Billing Demand, column (3), times General Service, Demand Charge, \$6/kW.
2/ Calculated as energy used, column (4), times energy charge, 3.5¢/kWh
3/ Sum of (5) + (6), *or* the minimum bill.
4/ Minimum bill effective.
5/ Billing demand ratchet clause effective.

Assumptions

1. Demand as in Figure 3-5(c)
2. Rate schedule in Table 3-1
3. Minimum bill = \$3/kW × 5000 kW = \$15,000 per month

hydro project to a utility system no cost saving for O&M should be assigned.

This type of production cost model is used by utilities in power system planning at both simple and extremely sophisticated levels (Knight, 1972). A basic application is to generate linear cost curves for electric production from different generating technologies. These cost curves, and their application for determining target amounts of generating capacity of each type, are shown in Figure 3-6.

Levelized Cost. Utilities will frequently make comparisons between generation technologies based on levelized annual costs. The technique is used to account for differences in the rate of escalation of total costs for the different production alternatives. While it will generally be unnecessary to use levelizing procedures to establish the value of small hydro plants, the technique will be outlined here for completeness.

Levelized annual cost of an alternative is calculated by first projecting the total annual costs for the life of the alternative using the best estimates of escalation in energy and other costs. This escalating cost stream is converted into a constant annual cost by finding the pesent value of the cost stream in the first year of operation and then calculating the constant annual cost over the project life that is equivalent to the present value of the cost stream. The appropriate interest rate to use is the company's weighted average cost of capital. This constant annual cost equivalent to the escalating actual cost stream is known as the levelized cost.

Time of Day. Figure 3-7 shows how the type of generating units are used to meet daily demands. The unit-cost lines in Figure 3-6 shown that the greatest decrease in cost for an hourly reduction in operation is for peaking units, then intermediate sources, and finally, baseload units. This is because the fuel-cost component of generation is arrayed in this order. Mathematically, this is shown by noting that for each generation type:

$$\partial TC_i/\partial t = HR_i \times EC_i$$

From Figure 3-7, it is also apparent that the value of replacing a unit of energy is a function of the time of day. This is why hydro is used whenever possible as a peaking unit to replace the highest-cost energy.

The characteristics of the hydro project will determine the type of thermal unit it can substitute for or the value of energy it can displace. If the project has dependable capacity, then the project will have both capacity and energy value to the utility. If no dependable capacity is available, only energy displacement value will be possible.

Run-of-the-River Projects. In the typical run-of-the-river project with no dependable capacity, the time or source of energy a small hydro project will be replacing will generally be unknown. The *minimum* value of energy displaced will be the energy cost of the most costly baseload source. The example following this section will illustrate this calculation.

Rather than using this minimum value for the energy displaced, an alternative and more accurate method is applicable if the energy production from the hydro plant is fairly randomly distributed throughout the year. The method is to determine the amount of time that each major generation type is the marginal (most expensive) energy source. These times can then be used to calculate a weighted average fuel displacement value for the system. With information as shown in Figure 3-8, this is a feasible technique in small hydro analysis.

Projects with Peak Power. If peaking power is present, the amount of energy produced on peak must be determined. The value of the energy will then be based on the energy displacement of the thermal peaking unit.

The balance of the project's energy production can be valued in the same manner as in run-of-the-river projects.

To summarize, the maximum value of small hydro to an electric utility is the reduction achievable in total system costs without assigning any cost to the small hydro project. This value is determined by the production characteristics of the small hydro project and the production costs of the utility.

Example of Utility Power Value Calculation. The following example illustrates how the value of power from a small hydro plant is calculated. To establish the value of power, information about both the small hydro project and the utility must be specified.

Small Hydroelectric Project. A typical run-of-the-river plant has been assumed for this example, with the following characteristics:

Installed capacity	7.5MW
Plant factor	49%
Average annual energy	32.2 million kWh
Peak production	February to August
Dependable capacity	None

Electric Utility. The electric utility is assumed to be a major utility with a 6000 MW summer peak and a lesser winter peak. Figure 3-8 is assumed to be the company's load-duration curve. Tables 3-4 and 3-5 summarize pertinent information typical for such a utility. This information would be available for an actual utility in the FERC publication *Steam Electric Plant Construction and Annual Production Expenses* and SEC Form 10-K.

Value of the Small Hydro Project. Since the small hydro project has no dependable capacity, its value is based on the cost of the fuels it can displace. The energy costs for each type of fossil-fired generation are calculated below using the information in Tables 3-4 and 3-5. These costs are the plant heat rate times the cost of fuels expressed in the correct units. This is:

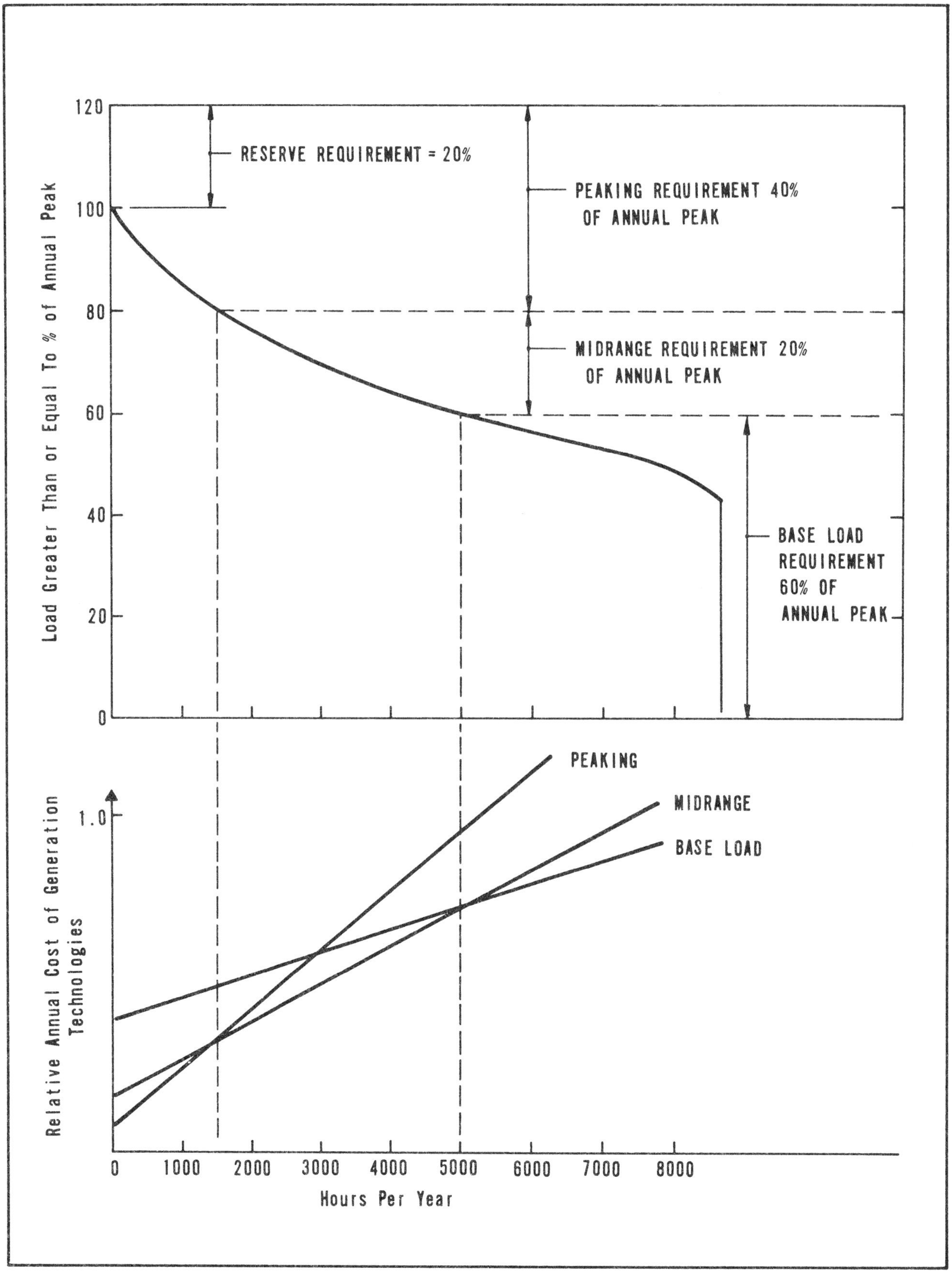

Figure 3-6. Basic application of production cost curves to power system planning.

Plant Type	Average Heat Rate (Btu/kWh)	Fuel Cost (¢/million Btu)	Energy Cost of Electricity (¢/kWh)
Coal-fired steam	9,409	143.4	1.35
Combined cycle	9,044	276.5	2.50
Gas turbines	13,777	276.5	3.81

From the load duration curve, Figure 3-8, at a minimum the small hydro plant would displace energy from baseload coal-fired units. Therefore, the minimum value of the small hydro energy is 1.35¢/kWh.

However, the value of this small hydro project is probably higher than this because it will frequently be displacing higher-cost electricity than that from the coal-fired units. Making the assumption that the small hydro output occurs randomly with respect to the load-duration curve, the small hydro plant will be displacing energy from the three sources in proportion to the time these sources are the marginal energy source. From Figure 3-8, it is seen that gas turbines are the marginal source 16 percent of the time, combined cycle units 44

TABLE 3-4
EXAMPLE UTILITY POWER PLANT HEAT RATES

Baseload generation — coal-fired steam plants

Plant Name	Capacity (MW)	Heat Rate (Btu/kWh)
Coal - 1	600	9700
Coal - 2	1100	9200
Coal - 3	600	9500
	2300	

Weighted average heat rate = 9409 Btu/kWh

Intermediate generation — distillate-fired combined cycle

Plant Name	Capacity (MW)	Heat Rate (Btu/kWh)
CC-1	625	9200
CC-2	675	8900
	1300	

Weighted average heat rate = 9044 Btu/kWh

Peaking Units — distillate-fired gas turbines

Plant Name	Capacity (MW)	Heat Rate (Btu/kWh)
GT - 1	450	16,100
GT - 2	550	13,750
GT - 3	450	14,250
GT - 4	700	12,000
	2150	

Weighted average heat rate = 13,777 Btu/kWh

TABLE 3-5
EXAMPLE FUEL COSTS

Year	Coal $/ton	Coal ¢/million Btu[1]	Distillate $/bbl	Distillate ¢/million Btu[1]
1972	10.70	48.6	5.25	90.8
1973	11.06	50.2	5.38	93.1
1974	14.72	66.9	9.35	161.5
1975	19.50	88.6	11.86	208.8
1976	23.79	108.1	13.04	229.8
1977	27.23	123.8	15.09	266.7
1978	31.55	143.4	15.98	276.5

[1] Assuming coal with 22.0 million Btu/bbl
[2] Assuming distillate with 5.78 million Btu/bbl

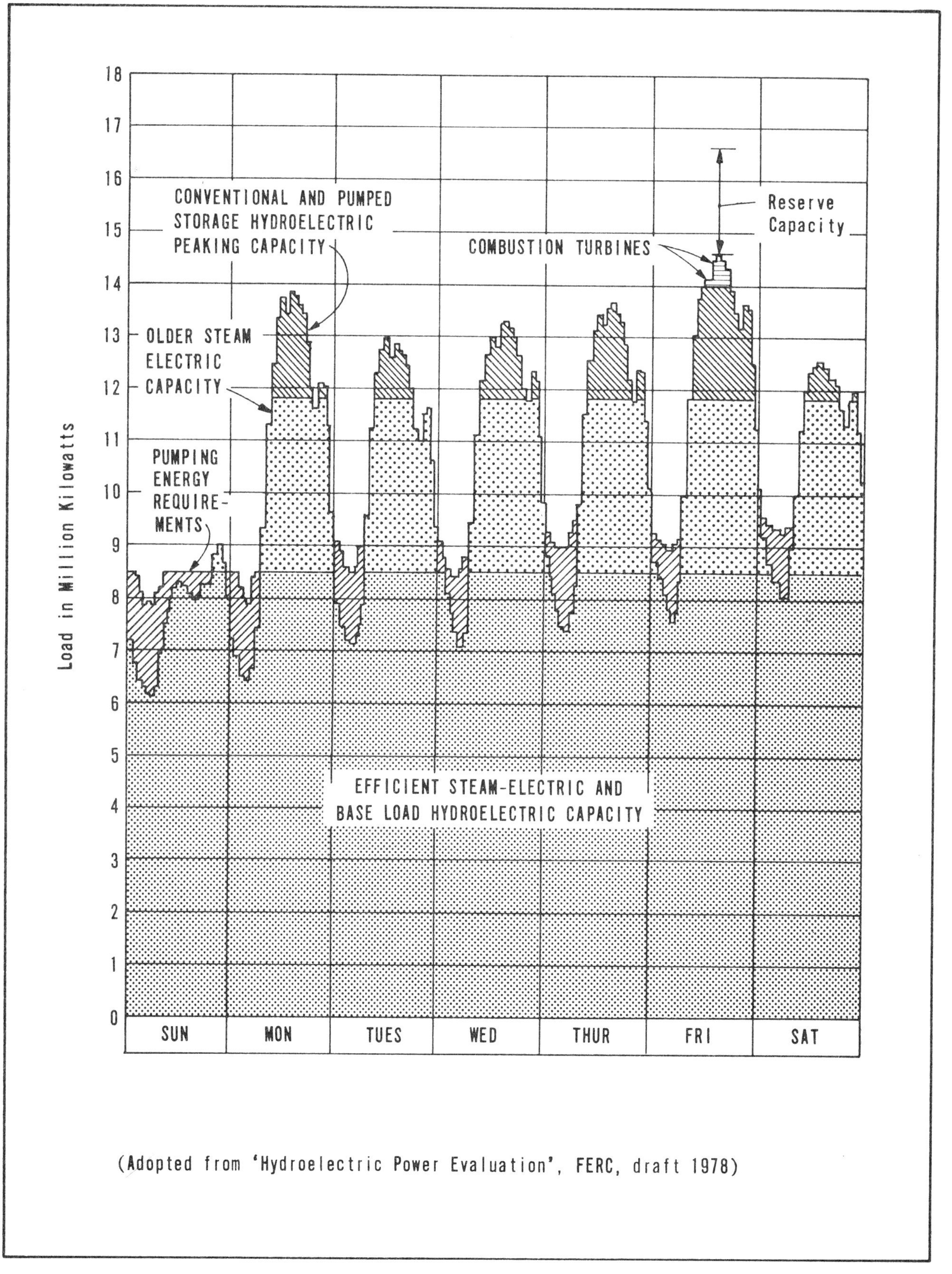

Figure 3-7. Weekly load curve of a large electric utility system.

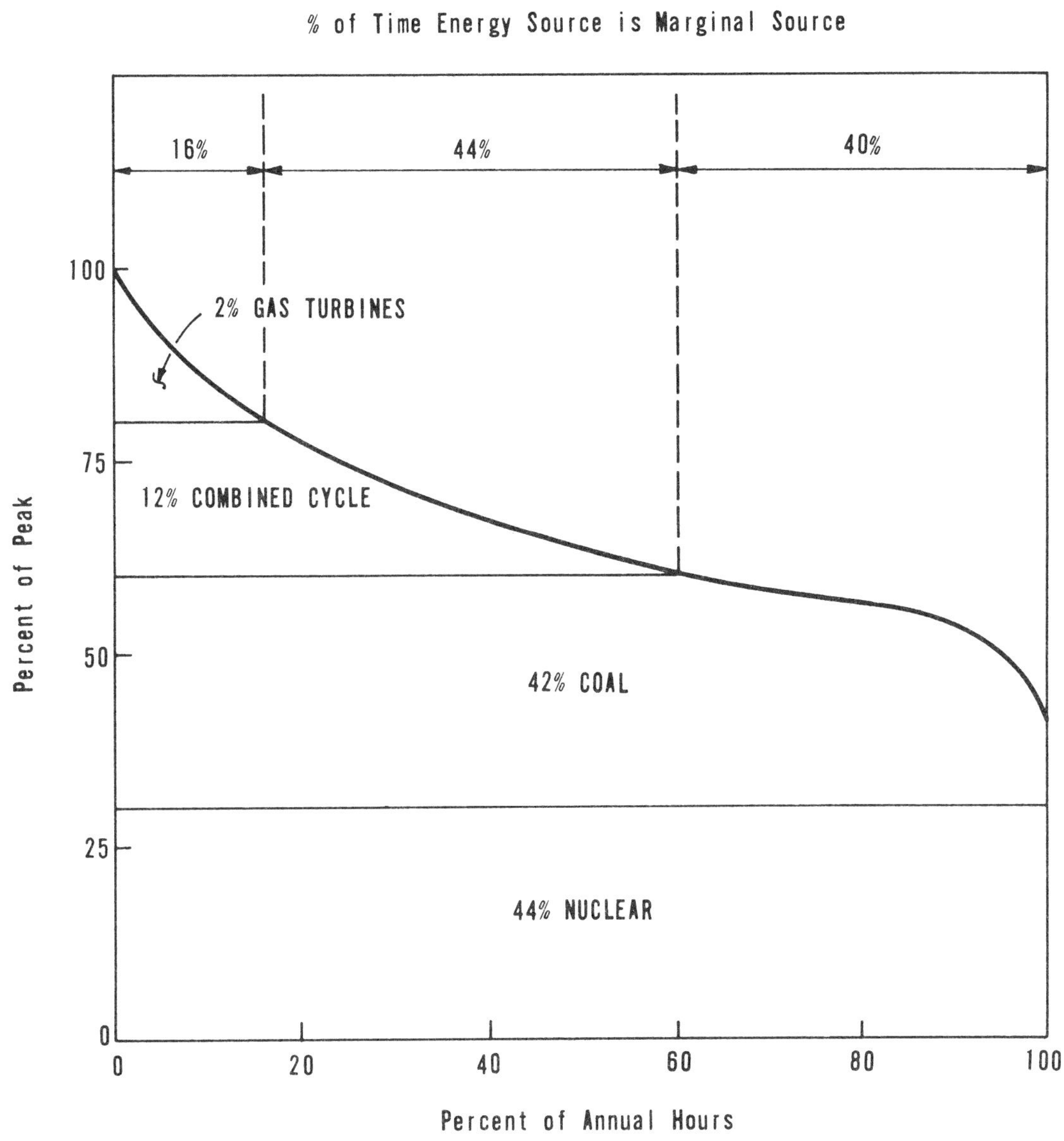

Figure 3-8. Annual load duration curve with energy by source superimposed (used in example utility power value calculation).

percent of the time, and coal fired steam units 40 percent of the time. These percentages can be used to calculate the weighted average value of small hydro output as:

$$\text{Value} = (.16 \times 3.81) + (.44 \times 2.50) + (.40 \times 1.35) = 2.25¢/\text{kWh}$$

As is seen, this procedure substantially and justifiably increases the small hydro value.

Regional Power Values

The Federal Energy Regulatory Commission (FERC) has prepared generalized estimates of the value of electrical power on a regional basis. These estimates can be used for preliminary analysis of the value of energy and capacity from small hydro installations.

The regional power values can be obtained from the FERC's five regional offices whose locations and service areas are listed in Exhibit I.

Market Arrangement

The manner in which small hydro output is marketed is an important factor in determining if financing will be available and at what price. It is imperative that adequate financial and legal consultation be obtained prior to entering into the actual power marketing agreement. Section 6 discusses the role of the financial advisor in more detail.

The capital investment in small hydro projects will be committed for a long period at a fixed price. Consequently, the "investors" will be unwilling to assume any business or technical risk associated with the project. As used here, "investors" may be a bank, insurance company or other long-term lenders, in addition to bond purchasers. This means the principal and interest obligation associated with project financing must be assured with a high degree of certainty. This assurance can be obtained in four ways: (1) Occasionally the project will have sufficient financial strength on its own so that the risk to investors is acceptable without any guarantees; (2) Guarantees can be given by a credit-worthy sponsor; (3) A credit-worthy power purchaser can "guarantee" the debt service through the marketing agreement; or (4) A third party, such as a state government, can guarantee the debt service. These guarantees will generally be required for the duration of the project's financing.

Time-of-Day Considerations. The section titled "Utility as Purchaser" discussed utility production economics in basic terms and the incremental cost of electric energy as a function of the time of day. If the small hydro project being analyzed has significant quantities of storage available for peaking power generation, then the marketing agreement should account for the higher value of energy displaced. This could be done in a simple fashion by adjusting a flat rate per kWh charge. At the other end, a complex rate, fully reflecting time-of-day factors, could be negotiated for use with a time-of-day meter to record energy production.

Whatever method is used, the value, if any, associated with project peaking capability should be established and set forth in the market analysis.

This discussion will examine four potential types of marketing agreements and examine the security effects of these arrangements on project financing.

Cost Plus a Percentage of Debt Service. This is a potential marketing arrangement which has been used to secure financing for hydroelectric development in the United States. An example of this arrangement is the June 1978 issue of $10,000,000 of revenue bonds by the Public Utility District No. 1 of Chelan County, Washington, to expand the Columbia River-Rock Island Hydroelectric System (Public Utility District No. 1 of Chelan County, Washington, 1978). A portion of the output was sold under this type of power contract to Puget Sound Power and Light Company.

The power purchaser and the project sponsors enter into a "power contract" for sale of all or a portion of the electric output. The essential elements of this contract are that the project sponsor agrees to deliver all or a part of the output, and, in return, the purchaser agrees to pay, "in all events", a pro rata share of "all costs" of the plant, plus an additional fixed percentage of the pro rata share of debt service. "All costs" include operating costs; taxes and other payments to governmental agencies; debt service, including principal and interest; amounts required for repairs and replacements not provided for otherwise; and any other costs associated with ownership, operation and maintenance allocated along the percentage of output sold.

The security of debt service repayment is obtained through the operation of the in-all-events clause. Such a clause will contain language similar to this:

> Payment to be made whether or not the operation of said facilities is interrupted, suspended, or interfered with, in whole or in part, for any cause whatsoever during the term of the power contract.

With this type of clause included in the contract and a credit-worthy power purchaser, the holders of the project debt will have sufficient security to place their funds in the project and allow implementation.

The major drawback to this type of agreement is that the compensation to the project sponsor is fixed at a constant amount for the duration of the power contract, which may be for 30 to 40 years. With the consensus expectation that the real value of electric and other forms of energy will be increasing, the fixed percentage of debt service may become a lesser percentage of the true value of the electricity. While the arrangement may be fair at the start of the power contract, as time passes the power purchaser may receive a disproportionate share of the benefits. The next type of arrangement discussed can rectify this problem.

Cost Plus a Royalty Subject to Escalation. This type of power contract has been used to secure financing for hydroelectric development in the U.S. An example is the July 1978 issuance of $7,800,000 of revenue bonds by Nevada Irrigation District (NID, California) for construction of a powerhouse at Rollins Reservoir, which is a part of NID's Yuba-Bear River Development. The security for the bonds was obtained through a power contract with Pacific Gas and Electric Company (Nevada Irrigation District, CA., 1978).

The contract is very similar to the one just discussed. Once again the project sponsor agrees to deliver power and the power purchaser agrees to pay all costs. The difference is that in lieu of having the project sponsor receive a fixed percentage of debt service as compensation, the sponsor receives a minimum per kWh payment, *which is subject to escalation.*

With this type of agreement, the power purchaser's payment to the project sponsor above "all costs" will fluctuate based on actual energy production. However, the per kWh rate of payment has a floor and is subject to escalation. The escalation clause will generally work as follows: At periodic intervals of one to five years, the per kWh factor will be adjusted for use in the following period by the same percentage some index of energy costs has changed in the same period. A logical index to use is the fuel-cost component of the utility's thermal electric generation.

This type of contract provides the debt service security needed to obtain funds and also recognizes that the future value of the project's output is likely to rise. This combination leads to a desirable marketing plan for the project sponsors to pursue.

Sales per Kilowatt-Hour. Project output could be sold on a per kWh basis, with the price being subject to adjustment based on an index. In this case, the power purchaser would not guarantee to pay "all costs", but would simply pay for energy actually produced. This arrangement could lead to wide variations in yearly revenues as annual power production varied.

Without purchaser guarantees to cover debt service in all events, some other method of assurance is needed before financing is possible. Usually, either sponsor or third-party guarantees will be necessary; however, occasionally the project will be strong enough on its own to lower the risk of revenue deficits to acceptable levels. In section 6 the method for calculating the probability of a revenue deficit under a per kWh sales agreement is discussed.

The difficulties in assuring debt service payments with this type of sale will usually preclude the possibility of obtaining project financing. Consequently, except with unusually attractive projects, one of the other forms of marketing the power output should be attempted.

Sales per Kilowatt-Hour with Cost Guarantee and Balancing Account. This type of arrangement values the plant output on a per kWh basis but also provides the revenue security necessary to obtain financing. Once again, the project sponsor agrees to supply electricity that the power purchaser agrees to purchase at a per kWh rate that is indexed. In addition, to provide security for debt service, the power purchaser agrees to pay "all costs"; the excess is used to reduce the balancing account balance, if any, with the remainder going to the project sponsor.

With this arrangement, the power purchaser is, in effect, providing short-term financing to assure the project's debt service. If the project is economically sound, at the end of the financing period the balancing account balance will be zero.

This contract has the two desirable characteristics of providing sufficient security to obtain financing and recognizing that the future value of electricity will rise. This arrangement will also lead to greater sponsor revenues than in the cost plus escalating royalty contract described earlier. This is because a larger value will be subject to escalation.

Market Information Used in Project Sizing

Market information is necessary for project sizing since it provides data on expected project value versus installed capacity. Using cost-versus-capacity information generated by the project engineers, the appropriate project size can be chosen. Clearly, at the feasibility level where market and cost information are both estimates, only an approximate "best" project size may be selected. The actual installed capacity will generally be chosen after equipment bids are received.

In general, project sponsors will want to maximize "profits" from the project. A well-established body of economic theory deals with the conditions for profit maximization. In a non-inflationary and competitive business environment, the conditions for maximum profit are satisfied if total revenues are equal to marginal costs. Inflation complicates the picture, since both total revenues and costs are escalating, but at different rates. However, a useful approximation to the best project size comes from maximizing the profits in the first year of operation. This will enhance the ability to obtain financing and market power by reducing project risk.

SECTION 4
ECONOMIC ANALYSIS

Traditionally, economic analysis for projects has meant development of benefit-cost ratios. This section shows how economic analysis of small hydro projects can be performed by public and private organizations. Guidelines are provided for formulating the benefit and cost streams and several commonly used procedures for comparing the benefits and costs are explained.

Definition of Economic Analysis

Economic analysis deals primarily with the development and applications of benefit-cost analysis which is the most frequently used procedure for project economic evaluation. The objective of this type of analysis is to relate all project economic benefits to all project economic costs accruing to the project sponsor. The appropriate scope of the analysis (the benefits and costs that should be included in the analysis) depends largely on the nature of the sponsoring organization.

Important components of the economic analysis are the project's initial and recurring annual costs and annual revenues which are the primary concern in the financial analysis. However, other costs and benefits not included in the project financial analysis may properly be included in the economic analysis. An example would be recreational benefits accruing to a county's population from reestablishing an impoundment for small hydro purposes. Such benefits would accrue to the area, but probably would not influence the finances of the project.

Analytical Scope and Framework

Framework of Economic Evaluation. The most efficient use of resources is the objective of economic analysis as measured by economic evaluation criteria such as the B/C ratio. This objective will generally be met if the project sponsor maximizes their net benefits and the scope of the analysis is properly formulated.

Within this framework, many small hydroelectric projects can be analyzed as single-purpose, stand-alone ventures if they are additions to, or replacements of, already existing facilities and their purpose is strictly power production. Such things as irrigation and urban water supply, flood control, navigation, recreation, and fish and wildlife might not be considered in the benefit-cost analysis because rehabilitation or add-on projects frequently have little or no effect on these items. If this is the case, the benefits are those associated with selling power, and the costs are those associated with supplying the power including rehabilitation.

If other objectives are of importance to the project's sponsors, such as environmental quality or employment, the analysis may be structured to include these additional objectives. Multi-objective analysis is used to analyze this type of project. In multi-objective analysis, each separate objective served by the project is considered independent but not necessarily of equal rank or priority. Each objective generates its own benefit stream, and carries its own costs and its fair share of any joint project costs. The multi-objective project is economically justified if, at a minimum, total economic benefits exceed costs and if each project purpose provides benefits at least equal to its separable costs.

Price level escalation, or inflation, may or may not be included in the economic analysis. The present federal government practice is to not escalate prices. Many private and other governmental analysts do escalate prices. This manual will explicitly include inflation in the analyses. The equivalent analysis without escalation can be obtained by using zero percent inflation and adjusting the discount rate.

Scope of Economic Analysis. A properly formulated small hydro project proposal attempts to maximize the net benefits of the project as determined by the scope of the analysis. The scope of the analysis, or the objectives, benefits, and costs to be included, depends on the nature of the sponsoring organization. The appropriate scope of analysis is to include costs and benefits which accrue to the sponsoring organization. If the sponsor is a private organization then the analysis would include items directly affecting profitability (revenues and expenses). Local governments might have a broader scope and include flood control, recreation or other local benefits. The federal government, whose purpose is broadest, would include all costs and benefits on the local, regional, and national level.

Cost and Benefit Streams

Benefits and costs are broadly categorized as monetary and non-monetary. Most nonmonetary benefits and costs can be quantified into dollar values if certain assumptions are made during the evaluation procedure. For example, in a local government sponsored project, recreation could be quantified into the user-days of recreational facilities and a dollar value determined for a user-day.

Components of Economic Costs and Benefits. In all small hydroelectric projects, the largest components of economic costs and benefits will be the present value of future cash inflows on the benefit side and the present value of the original and any future cash outlays on the cost side. Many of the elements from which these costs and benefits are calculated are contained in the *Uniform System of Accounts* prescribed for public utilities and licensees and published by the Federal Energy Regulatory Commission (FERC). Excerpts of these accounts are contained in Exhibit II. The accounts established by the FERC include balance sheet, electric plant, income, retained earnings, operating revenue, and operation and maintenance expense accounts. The various elements of these accounts, when properly quantified into present value, become the components of the economic costs and benefits.

As previously noted, other costs and benefits will properly be included in the analysis depending on the sponsoring organization. The individual situation determines which benefits and costs should be included. Examples of the types of considerations of interest are water supply, flood control, recreation, fish and wildlife, permanent employment, land use, and historical preservation.

Inflation. Escalation in the market value of power and project costs will occur over the project life. This escalation in price levels is composed of two components: inflation, or generalized price level increases, and real price increases due to shifts in supply-demand relationships for commodities.

Real price increases cause some items to escalate more rapidly than others. For instance, construction costs have increased at a substantially greater pace than inflation in recent years. This is also true of energy values. In some cases it may be desirable to escalate various cash inflows and outflows at different rates. This decision must be based on judgment about the project at hand and anticipated changes in the general economy and the future real price increases in the value of energy.

If inflation is explicitly included in the economic analysis, the future benefit and cost streams must be escalated by the expected inflation rate. This is done by using the factor for the future value of a present sum with the inflation rate in the place of interest. This is

$$P_t = P_o \times (1 = e)^t$$

where:

P_t = price t years in the future
P_o = current price
t = years in future
e = inflation rate.

This factor is multiplied times the future unescalated estimates of costs and benefits in the appropriate year to obtain the escalated amount.

Table 4-1 illustrates how escalation during construction is calculated for a four-year project. Also shown is the calculation of completed capital cost. First the lump sum cost estimate is broken into the amount to be spent in each year of construction. This unescalated cost estimate is then escalated to the expected future cost by using the factor to calculate the future amount of a present sum with the appropriate escalation rate. The contribution to complete cost includes the interest to finance the expenditure until the construction is complete.

This technique can be used for each separate portion of a construction project. In this manner, variation in escalation rates for different project components, such as the civil works or the mechanical equipment, can be incorporated in the completed cost estimate. (Table 4-2 is a complete example showing how inflation is incorporated in the benefit and cost streams.)

Formulating Benefit and Cost Streams. The period over which the benefit and cost streams must be calculated is the economic, or useful, life of the project. In the case of small hydro, this will frequently be the length of the financing period since periodic major replacements are usually required for continued operation and the financing plan will typically provide these funds only through the financing period.

If escalation is going to be included in the analysis, all the costs and benefits must be escalated in a consistent manner. Depending on the given project, different escalation rates for different portions of the project may be desirable. In particular, the general expectation that energy values will escalate more rapidly than general inflation should be considered.

The cost stream is composed of the capital costs, operation and maintenance costs, future replacements, quantified nonmonetary costs and any other cost associated with the project affecting the project sponsor. The benefit stream will include the value of power generation, quantified nonmonetary benefits accruing to the sponsor, and other benefits. The timing of these streams is important and must be accurately established.

Note that the receipts and outlays associated with the actual financing of a project, together with any effects on income taxes that follow, are excluded from the benefit and cost streams. Payments made into sinking funds to provide for future replacements are also excluded.

For more detail, a private sponsor should consult a basic text on managerial finance (Bolten, 1976) and a public sponsor a text on benefit/cost analysis (Mishan, 1976).

Economic Evaluation Criteria

A number of frequently used decision criteria are available for evaluating the economic feasibility of small hydro projects. All of the theoretically correct criteria are based on the time-value of the project's benefit and cost streams formulated according to generally accepted practices.

Discount Rate. A discount rate is used in calculating the economic evaluation criteria which reflects the time-value of money. For private project sponsors and local governments, this is properly the cost of capital. The private sector will use their weighted average cost of capital and the public sector their cost of borrowing in the bond market or from other sources. The federal government and some state governments have their discount rates, and economic practices, set by law. For example, federal projects use a constant dollar analysis and a discount rate set at 6-7/8 percent as of October 1, 1978, which is adjusted annually.

TABLE 4-1
IMPACT OF ESCALATION DURING CONSTRUCTION AND CALCULATION OF COMPLETED CAPITOL COST

CONSTRUCTION WITHOUT ESCALATION

PROJECT DATA

ITEM	AMOUNT
Cost of Financing	10.0% Per Year
Financing Period	30 Years
Construction Period	4 Years
Construction Costs Per Year	Shown Below
Construction Cost Escalation	0.0% Per Year
Lump Sum Project Cost Estimate	$1,000,000

CAPITAL COST CALCULATIONS

YEAR	UNESCALATED COST ESTIMATE	ESCALATED TO YEAR OF PAYMENT	CONTRIBUTION TO COMPLETED COST INCLUDING INTEREST DURING CONSTRUCTION
1	$ 300,000	$ 300,000	$ 399,300
2	300,000	300,000	363,000
3	200,000	200,000	220,000
4	200,000	200,000	200,000
TOTALS	$1,000,000	$1,000,000	$1,182,300

Completed Cost = $1,182,300

Fully amortized over 30 years at 10% interest,
Annual Debt Service = $125,417 Per Year

CONSTRUCTION WITH ESCALATION

PROJECT DATA

ITEM	AMOUNT
Cost of Financing	10.0% Per Year
Financing Period	30 Years
Construction Period	4 Years
Construction Costs Per Year	Shown Below
Construction Cost Escalation	10.0% Per Year
Lump Sum Project Cost Estimate	$1,000,000

CAPITAL COST CALCULATIONS

YEAR	UNESCALATED COST ESTIMATE	ESCALATED TO YEAR OF PAYMENT	CONTRIBUTION TO COMPLETED COST INCLUDING INTEREST DURING CONSTRUCTION
1	$ 300,000	$ 300,000	$ 399,300
2	300,000	330,000	399,300
3	200,000	242,000	266,200
4	200,000	266,200	266,200
TOTALS	$1,000,000	$1,138,200	$1,331,000

Completed Cost = $1,331,000

Fully amortized over 30 years at 10% interest,
Annual Debt Service = $141,191 Per Year

TABLE 4-2
EXAMPLE CALCULATION OF
NET PRESENT VALUE

*** NET PRESENT VALUE CALCULATION ***

(0.0% PRICE ESCALATION , 10.0% INTEREST)

YEAR	CAPITAL COSTS	OTHER COSTS	BENEFITS	NET ANNUAL BENEFITS (4-2-3)	PRESENT VALUE FACTOR	PRESENT VALUE (5)*(6)
(1)	(2)	(3)	(4)	(5)	(6)	(7)
0	$600,000			$ -600,000	1.000	$-600,000
1	900,000			- 900,000	0.909	- 818,181
2		45,000	245,000	200,000	0.826	165,289
3		45,000	245,000	200,000	0.751	150,262
4		45,000	245,000	200,000	0.683	136,602
5		45,000	245,000	200,000	0.620	124,184
6		45,000	245,000	200,000	0.564	112,894
7		45,000	245,000	200,000	0.513	102,631
8		45,000	245,000	200,000	0.466	93,301
9		45,000	245,000	200,000	0.424	84,819
10		45,000	245,000	200,000	0.385	77,108
11		45,000	245,000	200,000	0.350	70,098
12		45,000	245,000	200,000	0.318	63,726
13		45,000	245,000	200,000	0.289	57,932
14		45,000	245,000	200,000	0.263	52,666
				NET PRESENT VALUE OF PROJECT =		$- 126,662

(7.0% PRICE ESCALATION , 10.0% INTEREST)

YEAR	CAPITAL COSTS	OTHER COSTS	BENEFITS	NET ANNUAL BENEFITS (4-2-3)	PRESENT VALUE FACTOR	PRESENT VALUE (5)*(6)
(1)	(2)	(3)	(4)	(5)	(6)	(7)
0	$600,000			$ -600,000	1.000	$-600,000
1	963,000			- 963,000	0.909	- 875,454
2		51,520	280,500	228,980	0.826	189,239
3		55,126	300,135	245,008	0.751	184,078
4		58,985	321,145	262,159	0.683	179,058
5		63,114	343,625	280,510	0.620	174,174
6		67,532	367,678	300,146	0.564	169,424
7		72,260	393,416	321,156	0.513	164,803
8		77,318	420,955	343,637	0.466	160,309
9		82,730	450,422	367,691	0.424	155,937
10		88,521	481,952	393,430	0.385	151,684
11		94,718	515,688	420,970	0.350	147,547
12		101,348	551,786	450,438	0.318	143,523
13		108,443	590,412	481,969	0.289	139,609
14		116,034	631,740	515,706	0.263	135,801
				NET PRESENT VALUE OF PROJECT =		$619,738

Screening and Ranking. Economic decision criteria can be grouped into two classes: those most suitable for screening and those most suitable for ranking.

Screening refers to determining if a project has an acceptable economic return. In a small hydro development, a number of potential development plans must be considered. Screening the various plans will yield those that have acceptable results; all others will be rejected as uneconomic developments.

Ranking refers to determining the order of economic preference among projects. In a small hydro situation, the screening process may yield two or more installed capacities or turbine types that are viable development alternatives. The ranking process helps choose which is the most economically desirable project among the group of acceptable plans.

The example presented below will be useful to illustrate the discussion of the various criteria. The example project parameters are:

The example presented below will be useful to illustrate the discussion of the various criteria. The example project parameters are:

1.	Installed capacity	2 MW
2.	Annual energy production	9.8 million kWh/year
3.	Plant factor	56 percent
4.	Lump sum cost per kw	$750
5.	Annual O&M	$45,000
6.	Expected financing cost	10 percent
7.	Construction period	2 years
8.	Financing period	12 years
9.	Escalation	0.0 and 10.0 percent
10.	Value of energy	2.5¢/kWh

Net Present Value (NPV). The net present value criterion incorporates all of the pertinent economic data into a consistent one-figure decision rule that allows projects to be both screened and ranked. The criterion requires that a discount rate be specified for use in present value calculations.

The general procedure is to determine the present value (at the time of the first expenditure) of the future stream of net benefit flows. The screening decision criterion is to reject the project if the NPV is less than or equal to zero. Without constraints on the amount of capital available for the project, the project with the highest NPV is ranked highest. If capital is constrained, as may very possibly be the case, the project with the highest NPV within the budget constraint is ranked highest.

Explicitly, NPV is calculated as

$$NPV = \sum_{i=0}^{n} (CF_i/(1+k)^i) + (S_n/(1+k)^n)$$

where:

Σ = summation

CF_i = *net* cash flow in period i, starting with the initial outlay.

n = last period of cash flow

S_n = salvage value if any

k = discount rate

The example presented in Table 4-2 illustrates the calculation. Without escalating the benefit and cost streams the project has a negative NPV while including escalation indicates an economically feasible project.

Benefit-Cost Ratio (B/C). The B/C ratio, the most commonly used decision rule, reduces the analysis to a single consistent figure like the NPV. The rule incorporates all the essential elements of a valid economic comparison. The ratio compares the present value of future cash inflows to the present value of the original and all subsequent outflows by dividing the inflows by outflows. The decision rule is to reject projects that have B/C ratios less than one. For the example in Table 4-2, the present value of the escalating stream of benefits is $2.567 million and of the escalating stream of costs is $1.947 million. The B/C ratio is then 1.32 indicating an economically feasible project.

Internal Rate of Return (IRR). The IRR, which is primarily a screening criterion, is the discount rate that results in the project's NPV being zero. Like the NPV, internal rate of return incorporates all the pertinent economic data. IRR is calculated through an iterative process.

The decision criterion is to reject projects whose IRR is less than the expected cost of financing used to implement the project. This criterion has the appeal of being expressed as a percentage that is readily comparable with the expected cost of financing. The criterion does not, however, reflect any information on project scale, and, consequently, it cannot be used as the sole ranking criterion.

The IRR for the example project in Table 4-2 was calculated and is presented below for a range of initial energy values.

Energy Value (¢/kWh)	IRR (percentage)
2.3	14.1
2.4	15.0
2.5	15.9
2.6	16.8
2.7	17.7

Note that for energy at 2.5¢/kWh, the project's IRR is 15.9 percent. Consequently, for financing at less than 15.9 percent, the NPV of the project must be greater than zero, as is the case.

Other Criteria. Several other decision criteria are available for evaluating investment alternatives, but these are considered less competent at providing adequate evaluation information. These include the average rate of return (ARR) and the payback method (PB), among others. The ARR method is similar to the IRR, but does not discount future cash inflows and outflows; thus it does not take into account the time value of money. The payback method is one of the most commonly used methods in the United States, but it also fails to take into account the time value of money. PB is actually a measure of how quickly the original investment is returned in absolute dollars, and it ignores potentially great future gains.

Uncertainty

Uncertainty is the lack of sureness about an outcome or quantity. In small hydro projects, uncertainty surrounds capital cost estimates, future annual costs, escalation rates, and the future value of energy. Because these quantities are not known with certainty, an outcome unfavorable to the project sponsor is possible. This risk should be analyzed and minimized to the extent feasible. The discussions on sensitivity and risk analysis address the analysis of risks.

Analytical Procedure

Sensitivity analysis and risk analysis are two of the techniques used in analyzing investment decisions. The purpose of these techniques is to explore more fully the ramifications of uncertainty on the economic and financial decision criteria. Following a discussion of these techniques, a general procedure for the economic analysis of small hydroelectric projects is put forth.

Sensitivity Analysis. Sensitivity analysis, when applied to investment decision criteria, may be defined as the investigation of the impact on the decision criteria of variations in the important project parameters taken one at a time. The analysis is very useful for examining the degree to which the overall project desirability could be affected by changes in parameters whose values may vary.

The procedure is to determine the range over which the parameter being investigated might vary. The value of the decision criteria is then calculated over the range of the parameter. The results are then usually presented graphically as in Figure 4-1, which shows an example of the sensitivity of IRR with respect to the initial value of the project's energy production.

Some of the variables whose effect on the project might be investigated are complete cost, operation and maintenance costs, interest rates, and the initial value of the project's energy.

Risk Analysis. The risk associated with a small hydro project may need to be evaluated. Risk may be defined as the probability of the occurrence of an unacceptable outcome. Several methods of evaluation account for risk. Two of these are discussed here: the discount rate approach, and the Monte Carlo simulation approach.

The discount rate approach accounts for risk by increasing the discount rate associated with a project. An increase in the discount rate lowers future net benefits, thereby decreasing the NPV, IRR, or B/C ratio. In this way, a project with more risk, identified by a higher discount rate, would have to meet higher requirements in order to be judged economically feasible.

A more advanced technique for evaluating risk is the Monte Carlo simulation analysis. Monte Carlo simulation allows uncertainty in a number of the project's parameters to be simultaneously accounted for and the impacts on the decision criteria to be quantified. A brief description of the method is given below.

The procedure entails first deciding which of the project's economic parameters are uncertain either initally or year by year. Next, a probability distribution associated with each uncertain parameter is specified to embody the uncertainty in the parameter's value. A typical method for doing so is to use the triangular probability distribution as shown in Figure 4-2.

The evaluation criterion is calculated many times (as many as 400 times in some cases) each time using the probability densities for the uncertain parameters to choose values for the parameters. The resulting set of values for the evaluation criterion forms a histogram of possible outcomes, such as shown in Figure 4-3. In the figure, if A represents the minimum acceptable outcome, then the shaded area represents the probability of an unacceptable outcome and the risk associated with implementing the project.

The use of this simulation technique is becoming more widespread, and financial simulation packages are available from a number of computer software vendors. Occasionally, this level of analysis may be justified for small hydro projects.

Economic Evaluation Procedure

Table 4-3 summarizes the steps in the economic evaluation procedure for a small hydro development option.

TABLE 4-3
ECONOMIC EVALUATION PROCEDURE

Step	Description
1	Determine if inflationary or constant dollar analysis will be used. In an inflationary analysis, establish the general escalation rate. If items such as energy values or construction costs will be escalated at a rate different than the general inflation rate, determine the appropriate rate(s).
2	Establish the project economic life.
3	Assemble the unescalated cost stream (by year) for the economic life of the project. This includes the capital costs by year, operation and maintenance, replacements, quantified nonmonetary costs and other costs.
4	Assemble the unescalated benefit stream (by year) for the life of the project. This includes the value of power generation, quantified nonmonetary benefits, and other benefits.
5	Escalate costs and benefits as determined in Step 1.
6	Establish the appropriate discount rate.
7	Calculate the economic evaluation criterion chosen for use.

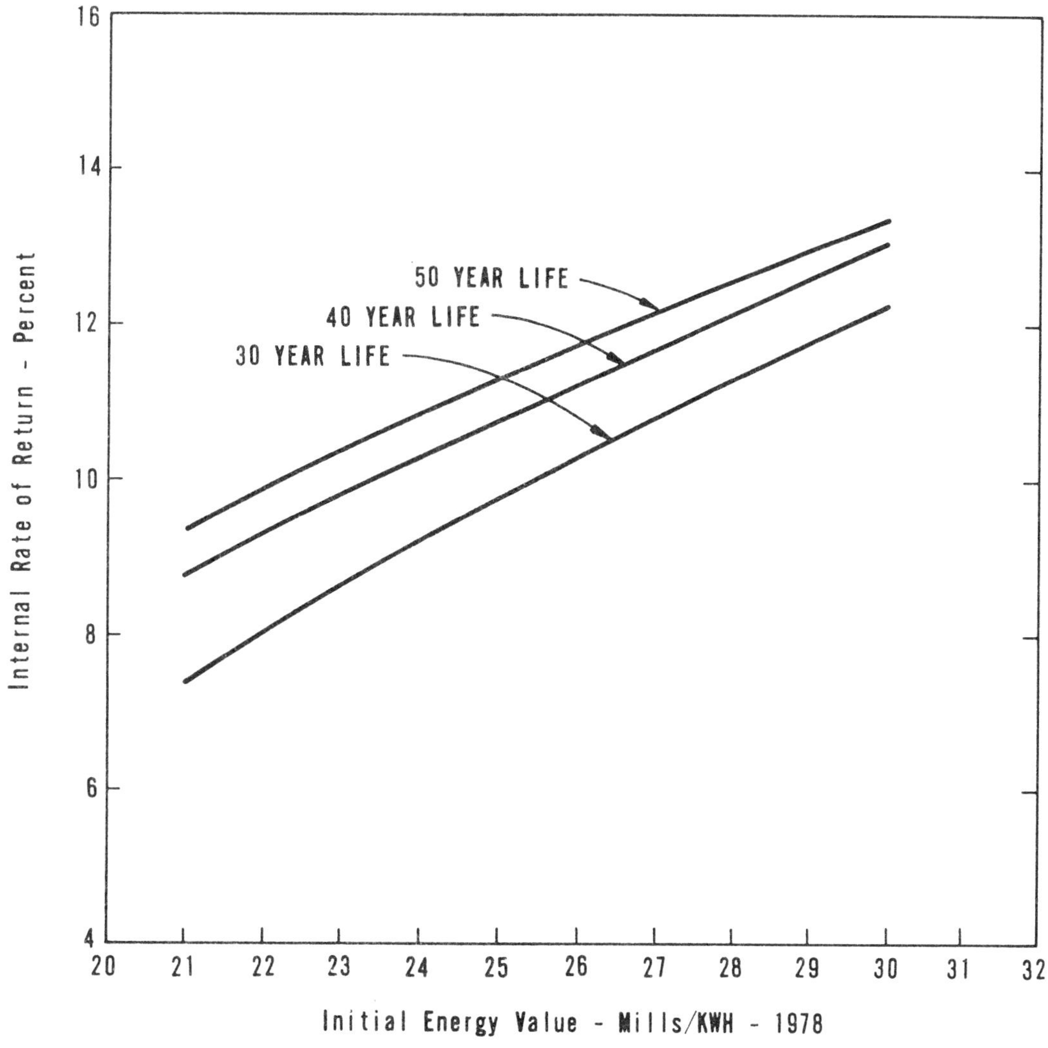

Figure 4-1. Example of sensitivity analysis.

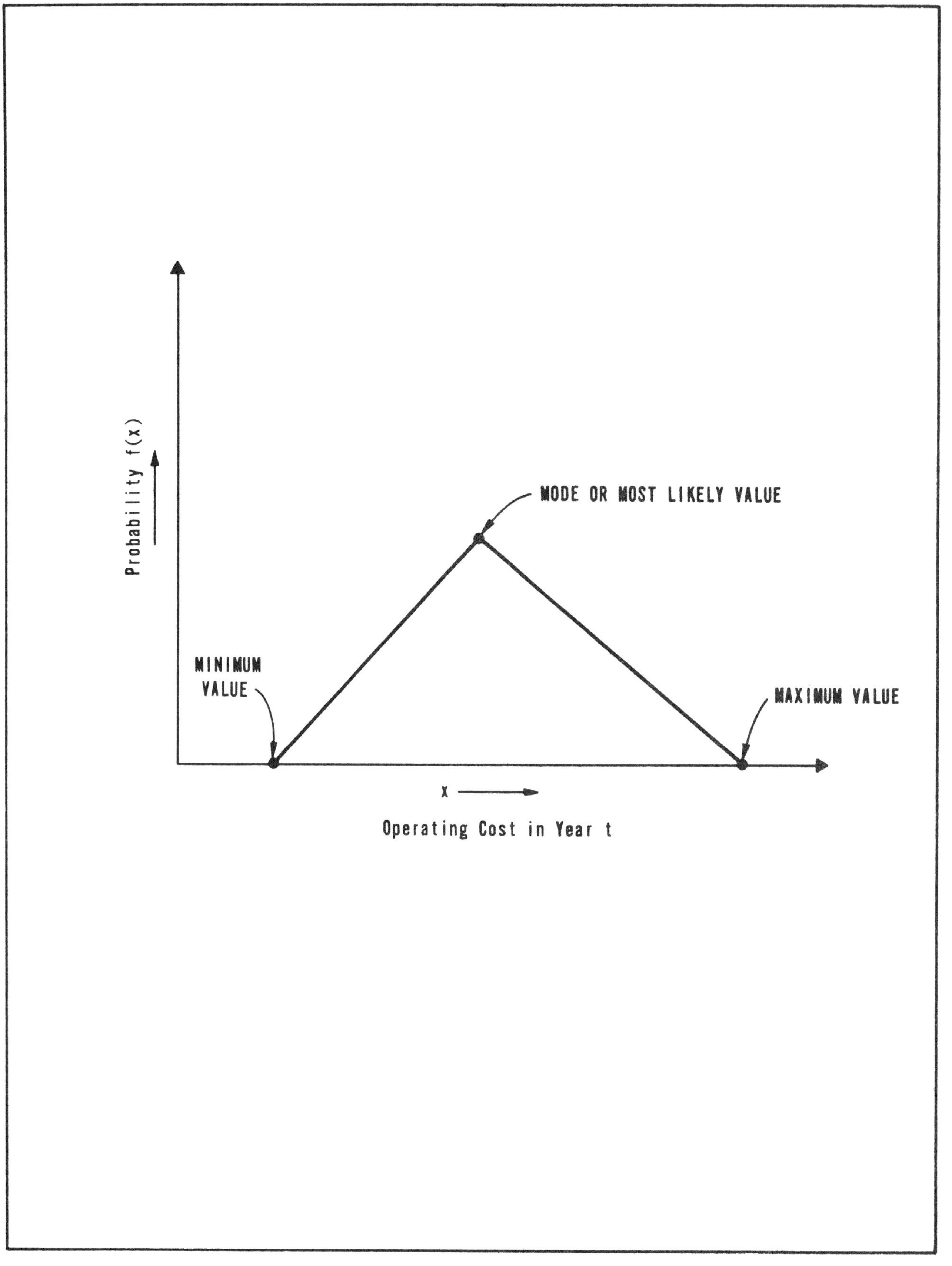

Figure 4-2. Example of triangular probability distribution of a project parameter.

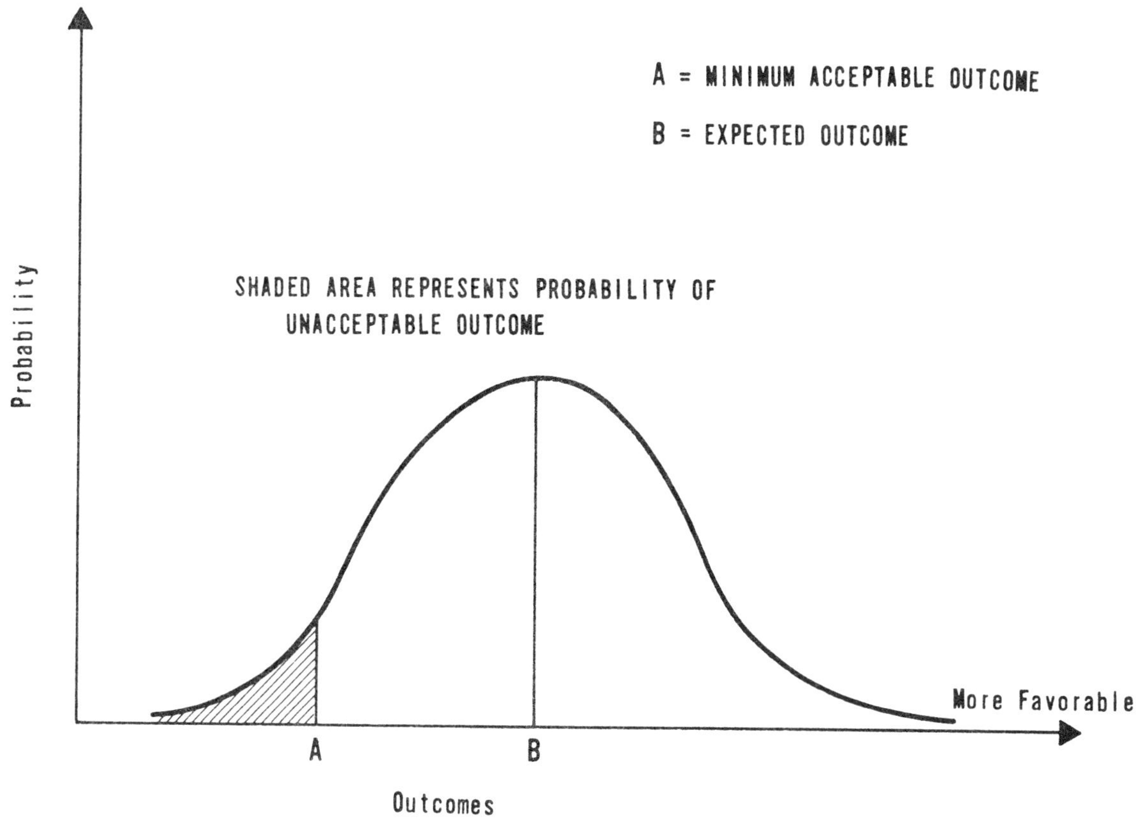

Figure 4-3. Probability of possible outcomes from Monte Carlo simulation.

SECTION 5
PROJECT IMPLEMENTATION

A number of economic and financial factors affect the implementation of small hydro projects. Financial reporting requirements and environmental considerations are discussed first under the broad heading of institutional considerations. The other major area of concern is the timing and usage of funding, particularly relating to funding the feasibility determination. These topics are discussed in this section.

Institutional Considerations

A large number of permits will be required for the operation of hydroelectric developments. Some of these permits relate to financial information and others can have substantial cost impacts.

Federal Energy Regulatory Commission (FERC). The FERC licensing requirements contained in Title 18 of the Federal Code of Regulations include the disclosure of substantial financial information for all projects within their jurisdiction.

Major Projects. The license application for "major" projects, defined as installations of 2000 horsepower (1500 kW) or greater, must include "Exhibit N," an estimate of the project development costs including: land and land rights; power plant structures and improvements; reservoirs, dams and waterways; waterwheels, turbines and generators; accessory electric equipment; miscellaneous power plant equipment; roads, railroads and bridges; and transmission facilities.

These cost categories are the same as those used in the FERC Uniform System of Accounts excerpts of which are in Exhibit II. The Commission may require that quantities, unit costs and total costs be shown for each of the above items. The Commission may also require the inclusion of indirect construction costs, such as construction equipment and Workmen's Compensation if the work is not to be done by contract. If work is to be done by contract, estimates of indirect cost would include engineering and administrative overhead, construction supervision, legal expenses, taxes, interest on construction funds, and contingencies.

Annual cost estimates may also be required by the FERC. These estimates would include: rate of return or interest; local, state and federal taxes; depreciation; insurance; and operation and maintenance, and administration

In addition, it may be necessary to furnish to the Commission the costs of obtaining an equivalent amount of power from an alternate source of power in terms of dollars per kilowatt-year of capacity and mils per kilowatt-hour of average energy.

Minor Projects. FERC license applications for "minor" projects defined as projects with less than 2000 horsepower (1500kW) of installed capacity are not required to include any of the Exhibit N cost information discussed above.

Completed Cost Statements. For all projects constructed under a FERC license, the licensee must, within one year after the project is ready for service, file a statement of actual project costs with the Commission. This statement would include construction costs, cost of water rights, right-of-way costs and land costs. Similar statements are required annually for any project additions or improvements. Annual operating expenses and revenues shall also be reported to the Commission in accordance with their Uniform System of Accounts. All reports will be evaluated by the Commission and all records are subject to audit.

Securities and Exchange Commission (SEC). The SEC requires issuers of securities making public offerings in interstate commerce or by mail to file registration statements containing financial and other pertinent data about the issuer and the securities being offered. Unless a registration statement is in effect for such securities, it is unlawful to sell the securities in interstate commerce or through the mails. There are certain limited exemptions, such as government securities, nonpublic offerings, and intrastate offerings.

The effectiveness of a registration statement may be rescinded or suspended after a public hearing if the statement contains material mis-statements or omissions, thus barring sale of the securities until the statement is appropriately amended. Registration of securities does not imply approval of the issue by the SEC or that the SEC has found the registration disclosures to be accurate. Persons connected with the public offering may be liable for damages to purchasers of the securities if the disclosures in the registration statement and prospectus are materially defective. Also, antifraud provisions apply generally to the sale of securities, whether registered or not.

The SEC also requires the filing of registration applications, annual reports and other reports prepared for national securities exchanges by the following: companies whose securities are listed upon the exchanges, companies that have assets of $1 million or more and 500 or more shareholders of record, and companies that distributed securities pursuant to a registration statement declared effective by the SEC under the Securities Act of 1933. Such applications and reports must contain financial and other data prescribed by the SEC as necessary or appropriate for the protection of investors and to insure fair dealings. Special provisions provide for regulation by the SEC of the purchase and sale of securities and assets by companies in electric utility holding company systems.

State-Level Requirements. Most states have given themselves the power to regulate the activities of investor-owned utilities (IOUs). This authority is usually delegated to a public utility commission (PUC) or public service commission (PSC) or board.

A PUC typically is both a court and administrative agency. Some of its powers may be set forth in the state's constitution. It may issue decisions and orders, cite for contempt and subpoena records, and hold hearings on any of the regulated utilities.

Generally, a PUC does not have regulatory power over cities and other public entities, although applicable laws in each state should be ascertained.

The power of a PUC to approve rates also may apply to approval of contracts for the purchase of power. In addition, a state may have another agency to regulate all corporate securities. Such an agency typically would provide control over the marketing of securities to the residents of the state, require disclosure of relevant financial and legal information consider essential in the public offering, maintain safeguards against unscrupulous promotional schemes, and take suitable enforcement action. State laws should be checked for any regulatory powers additional to those prescribed by a PUC. Such additional regulations will probably be minimal.

Environmental Considerations. Federal, state and local governmental environmental and other regulatory agencies require varying degrees of environmental assessment that could result in significant costs and affect the project schedule. The FERC license application requirements for "major" projects (1500 kW or more) include Exhibits W and S, comprehensive environmental and fish and wildlife assessments. "Minor" projects require brief environmental assessments such as a description of the existing environmental setting, impacts due to project construction or operations, mitigation measures, and alternative means of obtaining the power to be produced by the project.

After review of the license application, the FERC may also require that a complete Environmental Impact Statement be prepared in accordance with the National Environmental Policy Act. Such an EIS may also be required where the project involves the use of federal lands or funds.

State and local public agencies may also require the preparation of environmental impact reports for any projects within their jurisdiction. The environmental assessment process will be of particular importance in areas with significant aesthetic, recreational, fish and wildlife, and historical values. All the environmental assessment processes include requirements for public involvement and provisions for legal challenge.

Project Facilities to Mitigate Impacts. Significant costs can result from facilities required to mitigate potential environmental impacts, including:

1. Fish facilities — such as ladders, elevators, screens, bypasses — and collection, handling and storage facilities.
2. Architectural treatment of the powerplant with respect to historical, recreational or aesthetic values of the site.
3. Modification of locations and alignments and possible undergrounding of transmission lines due to aesthetic or other environmental considerations.
4. Special recreational facilities require to compensate for the loss of existing values.

Project Operations. Effects on project operations can include:

1. Minimum-flow requirements for water quality, fisheries, aesthetic and recreational purposes.
2. Restrictions on peaking operations to limit reservoir fluctuations and rapid variations in streamflow due to fisheries and recreational considerations.

Among the federal laws that must be considered in the implementation of any project are:

- National Environmental Policy Act (P.L. 91-190)
- Fish and Wildlife Coordination Act (P.L. 85-624)
- Endangered Species Act (P.L. 93-205)
- Historic Preservation Act (P.L. 89-665)
- Water Pollution Control Act (P.L. 92-500)
- Water Quality Improvement Act (P.L. 91-241)
- Wilderness Act (P.L. 88-577)
- Wild and Scenic Rivers Act (P.L. 90-542)
- Costal Zone Management Act (P.L. 93-612)
- Federal Land Policy and Management Act of 1976 (P.L. 94-579)

In addition, numerous state and local statutes, ordinances, and administrative regulations could impact the economics of project developments.

Timing and Usage of Funds

To avoid significant delays in the implementation of a small hydroelectric project, it is important that the funding required for each task in the process be anticipated in advance and procured in a timely manner. Figure 5-1 illustrates the implementation process for a hypothetical typical project and the cumulative funding requirements.

The time requirements and costs of the various implementation tasks will vary widely as determined by project magnitude, site conditions, and institutional factors. The implementation process includes the following significant tasks.

Prefeasibility of Reconnaissance Study. A brief prefeasibility study should be conducted to determine if the project appears sufficiently attractive to justify further, more detailed assessment. The guidelines and cost curves contained in this manuel will greatly facilitate this early study. Such a study will normally require two to four weeks to complete and should cost, at most, no more than one-half to one percent of the total project implementation cost. Any political or environmental ramifications that could stop the project should be iden-

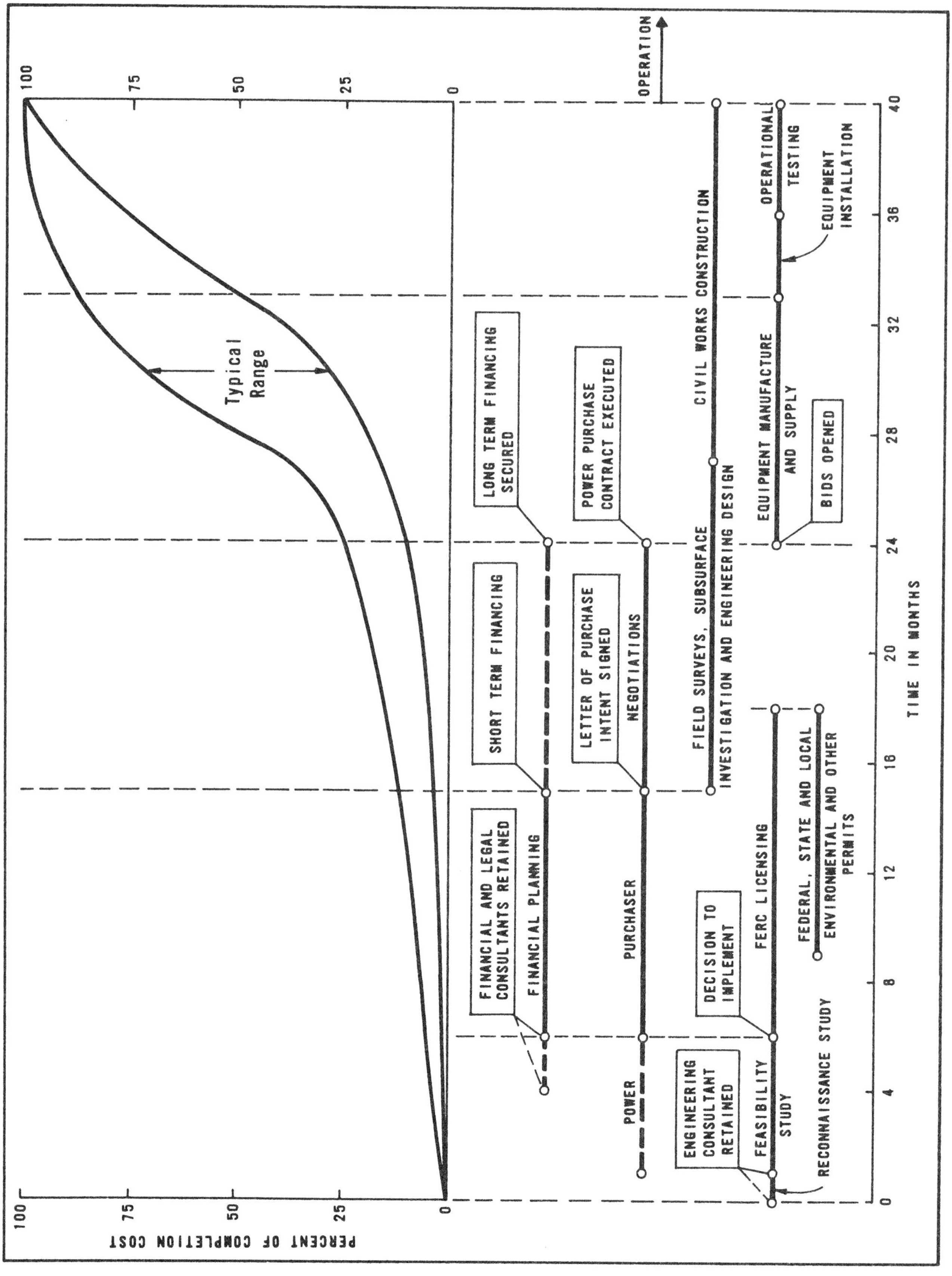

Figure 5-1. Typical project implementation schedule and expenditure pattern.

tified at this time.

Feasibility Studies. Detailed feasibility studies normally require three to six months to complete and should cost no more than two to five percent of the total costs.

Licensing and Permits. Preparation of FERC license application for minor projects of less than 2,000 horsepower of capacity takes two to four weeks, while major project applications will require three to six months for projects of less than 15 MW capacity. The time required to receive a license is in a state of flux. The FERC (Washington, D.C.) should be contacted for particulars.

Time requirements for other federal, state and local permit applications and environmental reports will vary widely depending on the institutional factors involved. These activities could take from four months to one year before all approvals are received. Much of this work can be concurrent with the FERC licensing process.

The cost of the licensing and permit processes should normally be between two and five percent of the project cost; however, for complex or controversial projects it could approach ten percent.

Engineering Design, Construction Supervision and Administration. Project design including site surveys and subsurface investigation will normally require six to 12 months for a typical project, with a possible overlap of two to four months with construction. The total exploration, design, construction supervision and administration can range from six to 12 percent depending on project size and complexity. (Guidelines for such costs are included in the American Society of Civil Engineers' Manual No. 45)

Manufacture, Construction, and Installation. Approximately 12 to 30 months will be required to manufacture and install equipment and construct civil works, depending on the size of the project.

Normally, all contracts include provisions for an advance payment to the contractor of about ten percent of the contract amount. An equipment supply contract normally calls for a payment of 80 percent upon delivery; construction and installation contract costs are normally paid monthly based on the actual work performed, excluding a ten percent retainer. The final ten percent is paid after all work is deemed complete and acceptable by the Owner or Engineer.

Funding Feasibility Studies

Prior to issuance of the FERC license it is unlikely the project sponsor will be able to obtain funds for any purpose from the long-term financing source based on the strength of the project. As indicated in Figure 5-1, the cumulative funds required to this point can amount to a significant fraction of total project cost. In addition, funds spent to demonstrate feasibility are generally subject to total loss if the project is not shown to be viable.

Many project sponsors, particularly public entities, may find it difficult to fund feasibility investigations. Consequently, obtaining these funds can represent a significant barrier to implementing a small hydro project. The next two sections discuss some possibilities in regard to feasibility funding.

Power Purchasers. An interested power purchaser may be induced to advance funds for feasibility studies and other investigations. Because the financial strength of the purchaser will typically be much greater than that of the project sponsor/site owner, such an arrangement may make sense. In the event the project is established to be feasible and is implemented, the advance will be recovered.

The project sponsor may wish to consider this type of arrangement if other funding sources are not available. It is likely that advancing funds for feasibility will enhance the negotiating strength of the potential purchaser. This, in addition to potentially being required to limit the ultimate sale of power to the advancing organization early in the feasibility stage, is a factor the project sponsor must consider prior to approaching a potential purchaser for feasibility funding.

Other Feasibility Funding Sources. A number of other sources of feasibility funding can be explored by the project sponsor/site owner.

Title IV of the Public Utility Regulatory Policies Act (Exhibit III) contains important provisions that provide funding for feasibility studies and loans for project costs. The act requires the Secretary of Energy to establish a program to provide loans of up to 90 percent of the cost of feasibility studies and license applications. In the event the project studies is not feasible, the Secretary may forgive loan repayment. Ten million dollars per year has been authorized for this purpose through September 30, 1980. These funds will be an important source of feasibility funding for project sponsors.

Several other potential sources of funding are regional development commissions, state energy agencies, and equipment manufacturers and engineering consultants, contingent upon the use of their equipment or services.

SECTION 6
FINANCIAL FEASIBILITY

General

Financial feasibility may be defined as a project's ability to obtain funds for implementation and repay these funds on a self-liquidating basis. Whether a project is feasible depends on the project's characteristics, the sponsor and purchaser, the requirements of those providing funds, and the overall credit market as it affects the cost of borrowing.

Generally, a project will be financially feasible if it can be shown it is self-liquidating with acceptable risk at realistic interest rates. An "acceptable" level of risk is generally very low. If these capabilities can be demonstrated, funds can usually be obtained.

An important part of establishing financial feasibility is the expected borrowing cost. The cost of capital for debt is the return potential investors require of the debt securities, such as bonds. This cost is generally considered to be the sum of the real interest rate that compensates the lender for surrendering the use of funds, the purchasing-power risk premium that compensates for expected inflation, the business and financial risk, and the marketability risk associated with low-liquidity of a debt security.

All of these factors must be considered in determining financial feasibility since the projects will usually be sensitive to the costs of financing.

Inflation. Inflation has two important effects on the financial feasibility of capital-intensive projects such as small hydroelectric developments.

First, inflation contributes to the cost of capital, since one component of the cost of capital is the long-run expectation of the inflation rate. Therefore, high inflation rates lead to higher costs of borrowing and annual debt service requirements. Note, however, that most financing plans will fully amortize project debt, which means the combination of principal and interest payments will be constant for the financing period.

Second, once a project is financed, inflation will generally enhance the project's net cash receipts as time passes. In capital-intensive projects, debt service will usually be a large portion of annual cash disbursements in the early portion of project life. Since the financing plan generally fixes debt service payments, only a portion of annual costs (operation, maintenance, replacements, etc.) is subject to escalation. However, the total market value of the product will be escalating, thereby increasing the difference between market value and project cost as time passes. This increase is comprised of two components — inflationary price increases and real price increases due to shifts in the supply-demand relationship for energy. The small hydro power marketing agreement should reflect these increases.

The consequences of inflation are that the first few years of operation will be the most difficult financially. It is therefore usually sufficient to show that the project is self-liquidating in its early years to be assured of overall financial feasibility. Financial feasibility is also usually assured if the project can be shown to be feasible assuming no inflation.

Security of the Minimum Revenue Requirement. The project's annual minimum revenue requirement (MRR) is the amount of funds required to pay all costs incurred by the project. The debt service portion will not be escalating, while other costs will. Consequently, the project's MRR can be expected to increase with time. For the project to be feasible, the MRR must be met with a high degree of assurance. Doing so will be a prime consideration when project financing and the power market agreement are arranged.

Funding Sources and Arrangements

A variety of long-term funding sources may be used to finance small hydroelectric developments. Several federal programs may provide funding, in particular the loan program being administered by the Department of Energy described below. In addition, the traditional methods of public entity financing will be important. These sources of financing, along with the methods available to privately owned businesses, will be reviewed here.

Federal Programs. The Department of Energy loan program is the most important federal source of long-term financing available for small hydroelectric development. This program, along with two other potential sources of federal funds, may provide financing for small hydro developments.

Department of Energy Loan Program. Title IV of the Public Utility and Regulatory Policies Act requires the Secretary of Energy to establish a loan program to provide long-term financing for small hydroelectric development. The pertinent sections of the Act (Exhibit III) should be consulted for the complete details of the program.

The loan program will provide funds for up to 75 percent of project costs to be paid off in up to 30 years. The interest rate charged will be the rate used for water resources planning projects at the time the loan is made. One hundred million dollars per year has been authorized through September 30, 1980.

The project sponsor should consider submitting a loan application under this program. While the authorized funding may not satisfy the demand for loans, the program will make an important contribution to small hydro financing.

Economic Development Administration (EDA). The EDA is concerned with communities burdened with too few jobs and too little income. Such areas typically suffer from high unemployment or low family income, and lagging or even declining population growth. They often are too poor to provide public facilities to attract new businesses and new jobs. The EDA has several programs to mitigate these kinds of problems. These include grants to help provide public works and development facilities, loans up to 100 percent to assist in financing public works, loans up to 65 percent for industrial and commercial expansion, guarantees of up to 90 percent of working capital and fixed asset loans, and technical assistance grants for planning. EDA assistance is also provided to redevelopment areas, economic development districts, and economic development regions.

EDA's Public Works and Economic Development Act program provides financial assistance for a variety of public works facilities. However, no financial assistance may be provided for projects involving the generation, transmission, or distribution of electric energy or for projects that would compete with an existing privately owned public utility. This program would appear to specifically exclude any power-related project. However, if the lack of an adequate power supply is a deterrent to community growth or aggravates unemployment, EDA financial assistance might be obtained for enterprises that are either under-employing people or would employ more people. With this assistance, these enterprises would then help to implement a small hydro power project through contracts for electrical service.

Small Reclamation Projects Act. The U.S. Bureau of Reclamation administers this program, which provides loans and grants to state and local entities for water-related projects that otherwise could be constructed under reclamation law, including multipurpose projects. Loan proceeds allocatd to irrigation and drainage are repayable without interest, while in a multipurpose project, the portions allocated to municipal and industrial water and to hydroelectric power are repaid with interest. Grants are made toward costs allocated to flood control and to fish and wildlife and recreation benefits. The maximum loan or grant is $18 million for a project whose costs cannot exceed $27 million (as of 1978). Loans must be repaid within 40 years. This financing program is available only to areas located west of the Mississippi River.

If a small hydro project can be incorporated into an otherwise federally acceptable irrigation project, resulting in a multipurpose project, substantial federal financing assistance may be possible through this Act.

Financing by Public Entities. Most public entities operate on a cost-recovey, non-profit basis. Revenues derived from taxes or commodity sales (e.g. water) or services (e.g. electricity or garbage pickup) are set annually at a level that will cover only debt amortization costs, O&M and replacement costs. Typically, little or no cash reserve is available to finance construction of capital programs even on a modest basis. Consequently, when a sizable capital expenditure program is to be undertaken, the entity is forced to borrow funds to finance it.

The two most common methods of borrowing are (1) issuance of general obligation bonds and (2) issuance of revenue bonds. Within these two types are numerous variations. Therefore, when an issue of bonds is contemplated by the entity, financial and legal bond consultants usually are retained to provide counsel to aid in the sale of such securities. These services are discussed under the heading of the "Financial Consultants in Public Sector Financing."

A successful marketing of bonds requires, among other things, that the proposed issue can be legally marketed by the entity. Sometimes legislation at the state level may be required to permit it to engage in a particular activity, such as the generation and sale of power, and to incur debt in connection with the activity. The financial markets (e.g. Wall Street) must also be in a state of receptiveness towards purchase of the bonds.

General Obligation (G.O.) Bonds. G.O. bonds are unique to the public sector in that their repayment ultimately is secured by the taxing power of the issuing entity. If revenue from the sale of electricity at any time during the payout period of the bonds becomes inadequate to cover the debt amortization, O&M and replacement costs, then the bond-issuing entity is required to impose taxes, increase taxes, or take all other measures necessary to cover such costs.

Fundamentally, the taxing power of the bond issuing entity undergirds the security of G.O. bonds. If, however, the entity has a mediocre record of financial management of its affairs or already is heavily in debt from prior issues of bonds that have priority to income over subsequent issues, the importance of the taxing power is diminished. In such cases, usually either one of two things occurs, the interest rate on the bonds is increased as a tradeoff to the increased risk inherent, or revenue bonds are issued.

Two principal types of G.O. bonds are issued: self-liquidating and non-self-liquidating. As the name implies, self-liquidating bonds are secured by revenues from the sale of a commodity or service without resorting to taxes to aid in bond payment. However, in cases of emergencies or other unforeseen events, tax revenues may be used. Non-self-liquidating bonds usually are secured largely, if not solely, by revenues from taxes. The credit rating of the entity is enhanced as the ratio of self-liquidating bonds to non-self-liquidating bonds increases, and the resulting interest rate on its borrowed funds tends to decrease.

Inasmuch as G.O. bonds become a legal obligation of all property owners within the entity, approval of the voting electorate must first be obtained. Usually a two-thirds majority vote is required for approval.

Revenue Bonds. Revenue bonds, of which there are several types, are secured only by revenues obtained from the marketing of commodities or services. Such bonds may be of a general revenue type in which first claim is made on all revenues or be more restrictive in that bond payout and security are limited to a single source (e.g. a project). Authorization for issuance of revenue bonds usually is not required by a two-thirds vote, but by a majority vote of the electorate.

Revenue bonds are not secured by the taxing power of the issuing entity. Consequently, a project to be financed by such bonds has to be financially sound and demonstrate in the financial feasibility report supporting the proposed bond issue that the required annual revenues will be forthcoming. If the project revenues are the sole security for the debt service, annual revenues, less operating costs, are usually required to exceed debt service by 25 to 30 percent. Such a margin of safety is required by the bond buyer to provide a cushion, so to speak, against unforeseen adversities that may befall the project and yet assure coverage of annual debt amortization costs.

When revenue bonds are used to finance a small hydro project, the reliability of the rvenues becomes most important. Close scrutiny needs to be given to contracts for the purchase of power from the project. The contract should cover the payout period of the bonds, the credibility of the power purchaser needs to be examined, and any loopholes adverse to the security of the revenue flow need to be dealt with.

Other Forms of Indebtedness. A public entity may find it desirable to issue notes or warrants based upon the advice of financial counsel, the size of the proposed capital expenditure and indebtedness to be incurred, or other factors. These are general obligations of the district, with maturity periods of up to ten years; oftentimes they are purchased by one buyer, such as a bank or insurance company.

Tax Status. Virtually all local public entities' bonds and other forms of debt are tax free. That is, the interest accruing to the bondholder is exempt from federal income taxes and state income taxes in the state in which the bonds are issued. It is customary to obtain a legal opinion from bond counsel prior to the issuance of the debt form as to the tax-exempt status. Such income tax exemption results in very favorable interest rates being obtained on the borrowed money. Assuming they have a good credit rating, non-federal public entities are able to borrow funds at a much lower interest rate than the federal government or private enterprise.

Public entity revenue bonds may become Industrial Development Bonds (IDBs) under certain conditions and will generally lose their taxexempt status. For small hydro developments, the bonds are IDBs if over 25 percent of the output is used by an investor-owned utility. However, the interest paid on IDBs may be exempt from taxation in small hydro developments — for example, when they are used to finance a facility to furnish local electrical energy solely within an area consisting of a city and one contiguous county, or when the facility furnishes water to members of the general public, including an electric utility. Thus, any dam built or modified to provide generation of hydro power to be used by the general public through an electric utility would be eligible for financing by tax-exempt bonds. Certain small issues are also tax exempt. Small issues are issues of $1 million or less, the proceeds of which are used for the aquisition or construction of depreciable property or land, such as a small hydro facility. At the election of the issuer, the $1 million size limit can be raised to $10 million due to amendments to the Internal Revenue Code in the Revenue Act of 1978. This small issue exemption should benefit many small hydro facilities.

If non-public entity participation is involved, it is strongly recommended that the tax status of a proposed revenue bond issue be determined in the early stages of the project proposal.

Some Repayment Provisions. Bond repayment provisions may vary depending on local circumstances and the money market situation. Serial bonds are bonds that mature annually according to the serial number. For example, if 1000 bonds are issued, bonds numbered 1 through 40 would mature and be redeemed the first year, bonds numbered 41 to 80 the second year, and so on.

Term bonds mature and are redeemed at the end of a term or period of years, with only the interest on the bonds paid during the interim. Usually, a sinking fund is built up to pay off the bonds at the end of the term. Consequently, the cash needs of the issuer are similar to what would be required if the project were financed with a fully amortized loan.

Generally, the larger the obligation is, in relation to the financial size of the issuing entity, the longer is the maturity period. The maximum period ranges from 30 to 50 years depending on the statutes that govern the entity. These statutes will vary by state and also by the type of public entity within a state.

Costs. When a bond issue is to be sold, usually it is put out to bid, and other things being equal, the bidder (underwriter) offering the lowest average interest rate will be awarded the bid. If the maturity period of the bonds extends over a considerable period of time, then often the bonds that mature early will bear a different interest rate than later-maturing bonds. The rate may be either higher or lower depending on the supply-demand situation in the financial markets at the time of issue. Usually, however, longer maturing bonds require a higher interest rate.

All or part of a bond issue may be callable before maturity. That is, the issuing entity may wish to call in the bonds ahead of their maturity date and pay off the bondholders. Usually, though, a small bonus must be paid by the issuer to the bondholder. Refunding bonds are similar to callable bonds in that the bonds are called in and replaced by another bond issue with differne terms (usually lower interest rates) and conditions.

When a bond issue is sold, a covenant or contract between the seller and buyer is executed in which all of the terms and conditions are set forth governing such things as the coupon (interest) rate for each bond during its life and callable and refunding features, if any. A bond may be subsequently resold many times before it matures and is redeemed by the issuer.

Investor-Owned Project Financing. Corporations in the utility business have choices broader even than public entities in funding a capital improvement. Many types of bonds, notes, warrants, preferred stock and common stock may be issued, subject, of course, to state and federal regulatory approval. Original issues may be sold only in states where they comply with the securities laws.

Investor-owned utilities (IOU) may issue various types of bonds that, like a public entity's bonds, are simply promises to pay back to the lender the principal and interest thereon over a specified period of time. The bondholder is the creditor. He has no voting power, but has first claim on the assets of the firm in case of liquidation. However, prior issues of bonds still outstanding have a higher priority claim. Such bonds also may have callable or refunding features and other terms and conditions as set forth in the bond covenant.

IOU bonds are not general obligation bonds because the utility does not have the power to tax. They are more similar to revenue bonds in that the project revenues or other revenues are used to pay off the indebtedness. Also, interest paid on the bonds (or notes and warrants) is subject to federal and state income taxes. Consequently, because of the higher risk and income tax law provisions, the cost of borrowed money is much higher (about 50 percent more) for IOUs than for public entities in the utility business.

An IOU also sells common stock that is evidence of ownership or equity in the firm as contrasted with that of a creditor position. The stockholder has a voting right, and therefore controls corporate policy, and a residual claim on profits after all prior claims have been satisfied. Unlike bonds, dividends are paid if profits have been made and, again, unlike bond interest, these dividends may vary from time to time or even not be paid at all if the financial condition of the firm is poor.

Preferred stock may be issued. Such stocks are in an intermediate position between bonds and common stock. They have a lower priority on corporate assets than the bondholder but higher than the common stockholder. The dividend rate is fixed, as in the bond interest rate. The priority on assets and earnings is below that of the bondholder, but higher than the common stockholder. And usually there is no voting privilege.

Dividends paid on common and preferred stock are taxable income (unless the dividend paid is a return of capital due to poor earnings).

IOUs try to maintain a balance between bonds and stock in their financing. Inasmuch as IOUs are regulated as to their rates and require approval of public regulatory agencies to issue bonds and stocks in return for being given a franchise or monopoly position for a given area, their securities do not fluctuate much in price, and their dividends are relatively secure.

Establishing Financial Feasibility

Cost of service is the term commonly used for the cost of producing electrical energy at the point of ownership transfer. In the case of small hydroelectric development, this will typically be the annual costs of delivering power to the high voltage side of the step-up transformer divided by the annual energy production.

If the cost of service is less than the value of the energy produced, it should be possible to negotiate a marketing agreement that allows the project to be implemented while providing the needed security in debt service payments. This is because both parties can financially benefit from the project, which is the essential requisite for entering into relationships. Because inflation will generally increase the value of energy faster than the cost of services, it will usually be sufficient to show that the cost of service is favorable within the first few years of project operation.

Occasionally it may be desirable to calculate the levelized cost of service from small hydro for comparison with alternate utility production costs. This technique is outlined.

Cost of Service Calculations. This section briefly describes how cost of service is calculated and presents an example of the cost of service for average energy production throughout the life of a project.

The lump-sum capital-cost estimate is used to establish the completed project cost. The method illustrated in Table 4-1 is applicable. The cost of capital used in calculating annual debt service is also used in construction financing unless some circumstances particular to the project indicate otherwise. With the completed cost estimate and the cost of financing specified, the annual debt service can be calculated.

The debt service payments plus other escalating and constant annual costs are then summed to estimate total annual cost through the project financing period. Total annual cost divided by average annual energy production yields the expected cost of service.

A brief example of a municipal utility project is presented to illustrate the method. Assume:

(1) completed cost equals \$6,000,000

(2) annual O&M in the first year of operation equals \$135,000

(3) cost of financing from Figure 6-1 is approximately 6 percent

(4) 30-year financing period

Then the capital recovery factor is CRF = .07265 and

annual debt service is approximately $435,900.

Table 6-1 shows the results of the cost-of-service calculations.

Over the 30-year financing period used in this example, the cost of service approximately doubles. Over this same period, the value of energy — which was close to the original cost of services and escalated at the same rate as O&M — increased by a factor of about *five*. This example illustrates how inflation will generally enhance the project's long-run annual value.

Occasionally, it may be desirable to convert the escalating cost of service into a levelized cost. This can be accomplished by discounting and summing the cost of service stream to the first year of operation and then calculating the constant annual cost, which is equivalent to the summed costs. Since the procedure would only be used to compare the cost of the hydro project to an alternative available to the power purchaser, the appropriate interest rate to use in these calculations is the weighted average cost of capital to the power purchaser. If the levelized cost of the hydro plant is less than the cost of the power purchaser's alternatives, it should be possible to negotiate a marketing agreement that allows project implementation.

Sensitivity Analysis. Frequently, some form of sensitivity analysis should be performed to provide additional information for the decision makers. This is particularly true when some parameter is not known with certainty or will be fixed at some time in the future.

A good example of an uncertain parameter that might be the subject of a sensitivity analysis is the cost of financing. At the completion of the feasibility study, the actual financing may not be obtained for one or even two years even if the sponsor decides to implement the project in a timely manner. As an examination of Figure 6-1 shows, over this period the cost of financing can range a full two and one-half percentage points. For this reason, the project sponsor may need a sensitivity analysis of the effect the financing cost has on the cost of service.

The results of this analysis will allow the implementation decision to be made with more complete knowledge.

To illustrate, the example from the preceding section was used to perform this sensitivity analysis. Completed cost was assumed to be constant at the $6,000,000 figure, and the interest rate was varied over the five to eight percent range after examining Figure 6-1. The impact on the cost of service in the first year of operation is shown in Figure 6-2. If the value of power from this project is 2.5¢/kwh, the sensitivity analysis shows a definite risk in going ahead with the project, even though the current interest rate yields reasonably favorable results. The utility of the analysis is evident.

Other project parameters that may be desirable to investigate include initial value of the project's energy, completed cost, operation and maintenance costs, and escalation rates.

Coverage of Revenue Requirements. The project's minimum revenue requirement must be assured with a high degree of certainty for the project to be able to attract funds for implementation. Discussed below are several ways the necessary level of security can be obtained.

Marketing Arrangements. Most small hydro projects are expected to obtain revenue security through a power contract executed with the ultimate power purchaser. For the power contract to be an effective device to secure debt service on the long-term project financing, several conditions must be met.

1. The contract must require payments sufficient to cover debt service in *all events*. This requirement is necessary to transfer *force majeure* and other risks to the power purchaser and away from the holders of project debt.

2. The capability of the power purchaser to give this assurance must be proven. In the case of large IOU's where the state-level PUC approves of the power contract, the assurance will generally be present.

3. The power contract should generally be in force for the length of the financing period.

Sponsor Guarantees. If security for the project debt service is not present in the marketing agreement, the financial integrity of the project sponsor may be used as security. If the sponsor is a public entity, issuing general obligation bonds effectively secures the debt service with the overall integrity of the project sponsor.

In a similar manner, a private sponsor can guarantee debt service. One method is pledging specific real assets, in addition to the project itself, as security. Large corporations are frequently able to issue bonds or otherwise borrow funds not specifically secured by real assets but relying on the general credit worthiness of the borrower.

In the typical small hydro project, sponsor guarantees are not expected to be the source of security.

Power Production as Security. Conventional projects financed with revenue bonds are sometimes secured by the projected revenues generated by selling the project's output. The rule of thumb often used to determine if the expected revenue from the project is adequate is to calculate the excess of revenues over operating expenses and debt service. If this excess exceeds 25 to 30 percent of annual debt service, as a general rule the project can be financed.

If small hydro output is sold on a per kWh basis, the situation is similar to the conventional project. However, because power production will vary based on the flow conditions, the rule of thumb applied to average production may be inadequate to determine if the project is sound.

The risk of a revenue deficit from low-flow conditions

TABLE 6-1. Cost of Service Calculation

*** COST OF SERVICE CALCULATIONS ***

(ANNUAL ESCALATION OF 6.0 % PER YEAR)

YEAR OF OPERATION	BOND AMORTIZATION	OPERATION & MAINTANENCE	TOTAL ANNUAL COST	AVERAGE ANNUAL ENERGY PRODUCTION (MILLIONS OF KWH)	COST OF SERVICE (CENTS/KWH)	VALUE OF ENERGY (CENTS/KWH)
1	$435,900	$135,000	$570,900	22.500	2.537	2.500
2	435,900	143,100	579,000	22.500	2.573	2.650
3	435,900	151,686	587,586	22.500	2.611	2.809
4	435,900	160,787	596,687	22.500	2.651	2.977
5	435,900	170,434	606,334	22.500	2.694	3.156
6	435,900	180,660	616,560	22.500	2.740	3.345
7	435,900	191,500	627,400	22.500	2.788	3.546
8	435,900	202,990	638,890	22.500	2.839	3.759
9	435,900	215,169	651,069	22.500	2.893	3.984
10	435,900	228,079	663,979	22.500	2.951	4.223
11	435,900	241,764	677,664	22.500	3.011	4.477
12	435,900	256,270	692,170	22.500	3.076	4.745
13	435,900	271,646	707,546	22.500	3.144	5.030
14	435,900	287,945	723,845	22.500	3.217	5.332
15	435,900	305,222	741,122	22.500	3.293	5.652
16	435,900	323,535	759,435	22.500	3.375	5.991
17	435,900	342,947	778,847	22.500	3.461	6.350
18	435,900	363,524	799,424	22.500	3.552	6.731
19	435,900	385,335	821,235	22.500	3.649	7.135
20	435,900	408,455	844,355	22.500	3.752	7.563
21	435,900	432,963	868,863	22.500	3.861	8.017
22	435,900	458,941	894,841	22.500	3.977	8.498
23	435,900	486,477	922,377	22.500	4.099	9.008
24	435,900	515,666	951,566	22.500	4.229	9.549
25	435,900	546,606	982,506	22.500	4.366	10.122
26	435,900	579,402	1,015,302	22.500	4.512	10.729
27	435,900	614,166	1,050,066	22.500	4.666	11.373
28	435,900	651,016	1,086,916	22.500	4.830	12.055
29	435,900	690,077	1,125,977	22.500	5.004	12.779
30	435,900	731,482	1,167,382	22.500	5.188	13.545

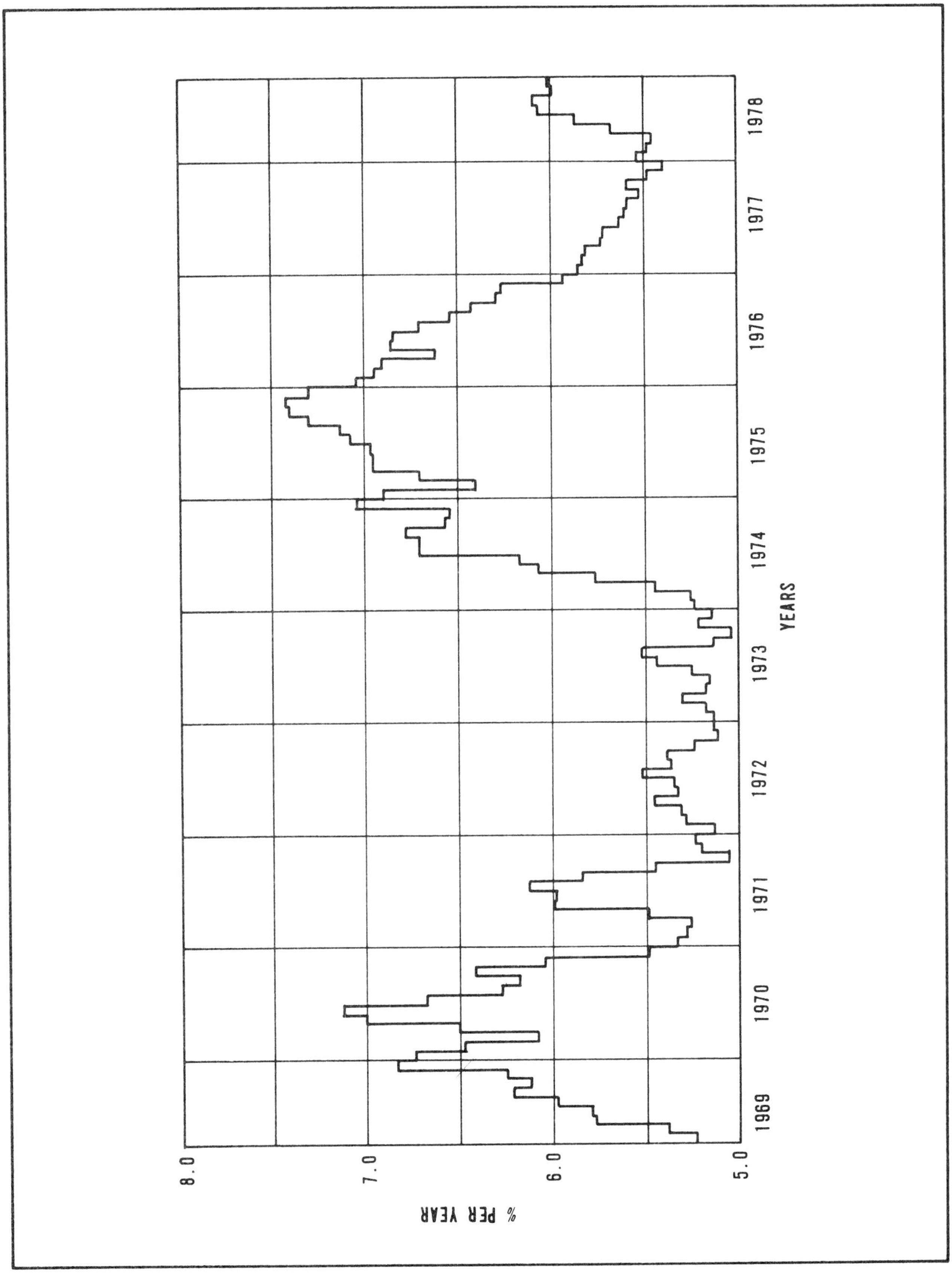

Figure 6-1. Municipal bond yield averages for general obligation bonds. Source: Moody's Bond Record.

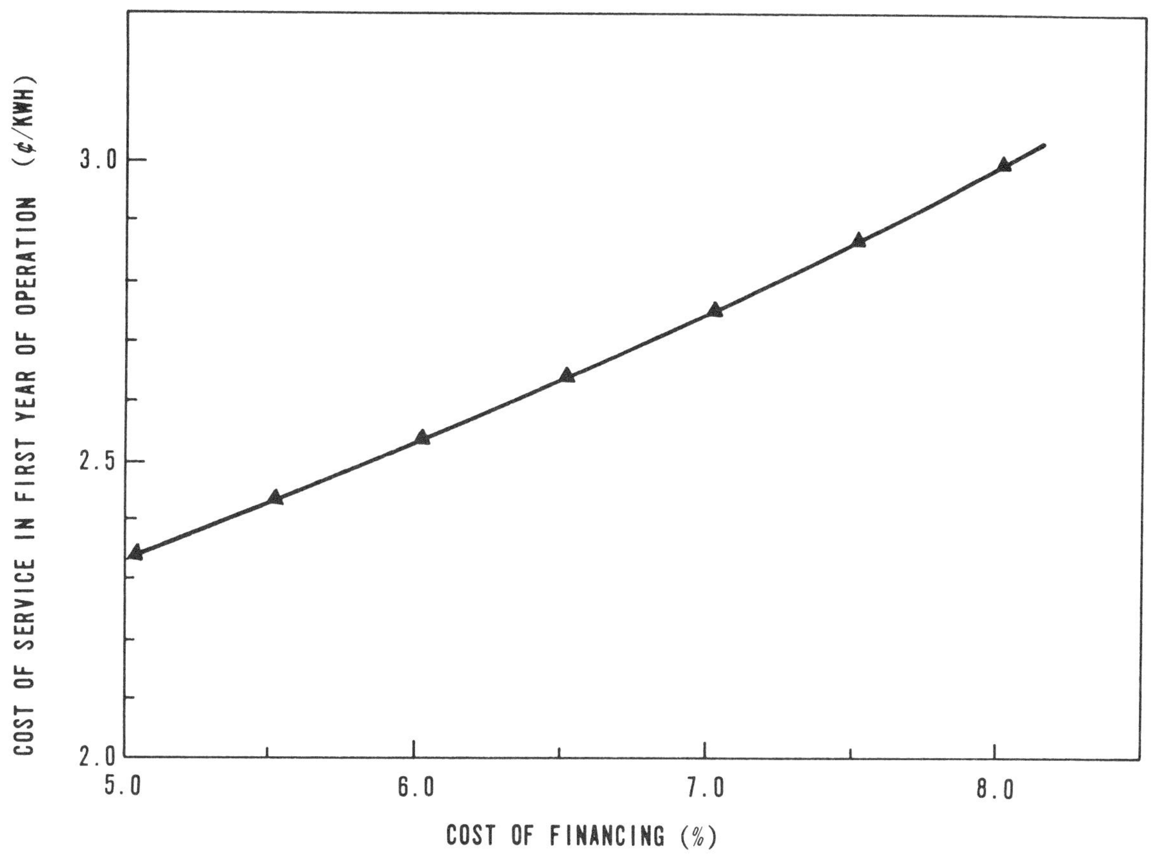

Figure 6-2. Example of financial sensitivity analysis.

can be calculated in a reasonable manner. The procedure is to first calculate the minimum revenue requirement to meet all costs. Then, using the expected per kWh price, the annual production necessary to meet these requirements can be calculated. Finally, an annual energy production histogram (as shown in Figure 3-3) can be used to calculate the risk of a revenue deficit. The risk would be measured by the proportion of the time annual energy production is less than the minimum requirement.

As a general rule, there can be no risk of a revenue deficit from low flow if the project is to be acceptable to the bond purchasers.

Financial Consultants in Public Sector Financing

Bond financing of a capital improvement project requires the services of the finance, legal, accounting and engineering professions. Each contributes to a process that requires development of a financing plan and its subsequent implementation through the preparation of bond sale documents, creation of a marketing program and finally the bond sale. While the issuer's staff, the auditors and consulting engineers all provide essential information for the documents required to market bonds, the financial advisor, drawing on the legal expertise of bond counsel, is responsible for creating and preparing the market for the securities being issued to raise the required capital.

Financial advisory or bond counsulting services are provided by several different types of organizations, including investment banking firms, commercial banks, and independent consulting firms. These differ from one another in their activities in addition to bond consulting. Investment banking firms underwrite (buy for subsequent resale) and distribute all types of securities, while commercial banks underwrite and distribute only U.S. government and general obligation bonds, as they are prohibited by federal law from underwriting most types of revenue bonds. Independent consultants do not underwrite or distribute securities. Consequently, an investment banking firm may have more experience in hydroelectric revenue bond financing than the other two, and it also could act as investment banker in the event the bonds are sold by negotiation rather than competitive bid.

Tax Status. When the small hydro project is constituted by a municipal or other publicly owned utility, the common practice is to finance the project with tax-exempt electric revenue bonds.

The tax-exempt nature of these securities is of major importance and is the reason their interest rates are substantially lower than corporate securities.

When a municipality or other public entity is building a generating facility that will be used by an investor-owned utility, the bonds may be deemed to be industrial development bonds, and as a result there may be a loss of the federal income tax exemption. The terms of the power sales contract must be drafted to avoid creating an industrial development bond as defined by federal law and Internal Revenue Service regulations and rulings. In simple terms, the exemption is preserved if less than 25 percent of the capacity is used by an investor-owned utility or if the utility's system serves no more than two counties.

Recent changes to the Internal Revenue Code contained in the Revenue Act of 1978 may increase the ability of a project sponsor to retain tax-exempt status even if more than 25 percent of the capacity is used by an IOU. Funds used to finance portions of the water-related facilities may be tax exempt due to these changes. In addition, increases in size limits of certain exempt small issues may allow the powerhouse to be financed with tax-exempt securities.

The financial advisor and legal counsel consider these matters in recommending terms for the power sales contract. This is generally done during the planning stage.

Financial Advisor's Role. The role, in detail, that a financial advisor plays in financing a project is described below for each stage in the process: financial planning, document preparation, market development and bond marketing.

Development of a Financing Plan. The development of the financing plan should be based, among other things, upon (1) the engineering studies on the construction program of the proposed project for which financing is required, including the estimates of construction and acquisition costs and the schedule of drawdown of construction funds, (2) studies on the economic and financial feasibility of such a program, (3) studies on the future revenue base of the client to support its existing indebtedness and proposed future indebtedness to be incurred in connection with the construction program for the project, and (4) the existing corporate, statutory, financial and legal structure of the client as it pertains to the project. This financing plan, which must be drafted in complete concert with the appropriate members of the client's staff and its legal and engineering consultants, should cover, among other things, the following areas:

1. The results of a complete review of the client's existing financial and legal structure as it pertains to the project, and more particularly the provisions incorporated into the bond resolutions of the client.
2. The contemplated amount of bonds or other forms of indebtedness necessary to be issued to finance the immediate as well as anticipated future capital requirements of the project. The amount should include the costs of construction, land acquisition, funded interest during the construction period, appropriate amounts for reserves, contingencies and fees, financing costs, etc. This area should also encompass suggestions and recommendations on a shorter medium-term financing program to be implemented prior to or in conjunction with the long-term financing program. This evaluation of alternative financing concepts should include a

review of legal constraints, market conditions, timing of capital requirements, comparative cost of money, and impact on the client's credit standing.

3. The proposed financial structure and the suggested security provisions covering the proposed revenue bonds of the client to be issued to finance the project.

4. The establishment of maturity schedules for the revenue bond issue and subsequent issues or shorter medium-term notes to be issued to finance the project, and the establishment of appropriate redemption provisions for the initial issue of the bonds. This should also encompass a complete review of the benefits or detractions of term bonds versus serial bonds, or a combination of serial and term bonds, for the purpose of developing, among other things, the most favorable cost of money.

5. The provisions of a financial nature to be incorporated, where applicable, into any participation agreement, power sales agreements, and any other agreement necessary to implement the improvement and construction program of the client in regard to the project. Such contracts are very important to the successful sale of the revenue bond issue for small hydro projects due to the fact that the security and quality of the issue is frequently based on not only the strength of the client but also on the strength of the other participants involved.

6. The financial provisions, in depth, to be incorporated into the bond resolution under which the revenue bonds to finance the project will be issued and will be secured. These provisions must be carefully developed in order to provide the project sponsor with the maximum degree of flexibility and an acceptable financial and legal structure to sophisticated institutional investors throughout the United States. These provisions should address:

—The establishment of a specific construction fund or funds, the methods of disposition and investment of the moneys in said fund or funds, and the disposition of any surplus money therein.

—The establishment, if deemed appropriate and in concert with the existing resolutions of the client, of specific funds within such bond resolution to cover (1) operation and maintenance expenditures, including necessary provision for working funds, (2) the payment of interest on and principal of bonds when due, and reserves therefore, (3) necessary reserves for extraordinary renewals and replacements, depreciation, public liability claims, etc., and (4) purchase of new or replacement equipment.

—The proposed covenants or revisions thereto governing the issuance of additional revenue bonds.

—The establishment of such additional covenants regarding rates, consulting engineers, audits, annual reports, etc., as may be deemed appropriate or necessary.

7. The timing of issuance of the bonds or the drawdown of note or loan funds that should be based in part upon the anticipated drawdown of construction funds, as well as the anticipated construction or acquisition contractual obligations of the project.

8. In collaboration with the bond counsel or general counsel to the client, the provisions of a financial nature to comply with such rules and regulations of the Securities and Exchange Commission, the Internal Revenue Service and any other federal agency that may have a bearing on the financing and construction program of the project.

9. To the extent deemed necessary for the development of the financial plan, financial analysis of pertinent data furnished by the client, its engineers, or other consultants on sources and estimated amounts of revenues, and other funds that might reasonably be expected to be available to the client to aid in financing the construction, acquisition, operation and maintenance of the project or the payment of principal and interest on its prospective future revenue bonds.

10. In conjunction with the financial plan, the financial advisor would perform any additional financial analysis and attend any hearings, to the extent necessary and proper, in matters required by administrative, judicial, legislative and other government bodies that would be necessary to the successful completion of the revenue bond issue.

Development of All Necessary Documentation. Upon completion and acceptance by the client of the principles incorporated within the plan for financing the project, the duties and responsibilities of the financial advisors should encompass the coordination of work with the attorneys of the client, including bond counsel, regarding the financial and security provisions to be contained in the instruments authorizing and securing the bonds.

In addition, in collaboration with the client's financial staff, its legal counsel and its engineering consultants, the financial advisors will prepare all necessary underwriting documents and the "Official Statement." The Official Statement includes the:

1. Amount and title of the bond issue, with maturities, interest rates, call feature, paying agents, registrability features, approving attorneys, etc.
2. History and description of the client and the source of its authorization to issue bonds.
3. Full disclosure of the purpose of the bond issue and description of the project to be financed.
4. Feasibility studies.
5. Detailed disclosure of historical operating records of the client.
6. Description of the revenues or other moneys, if any, pledged to the payment of the bonds.
7. Full disclosure as to use and application of the bond proceeds.
8. Summaries of the authorizing bond resolution, trust indenture, power sales agreement, and any lease and related proceedings.

Market Development for Revenue Bonds. One of the major functions of the financial advisors is the development of a market for the revenue bonds to finance the project prior to the actual sale. This is probably one of the more difficult functions to successfully accomplish in view of, among other things, the constant competitive pressures within the national money market from bond issues of federal agencies, corporations and municipalities, and municipal agencies. Some of the specific tasks the financial advisor would perform for the client in preparing and developing a market for its proposed bonds would include:

1. Develop an extremely broad and comprehensive nationwide mailing list of all institutional investors who have or could have an interest in the financing programs of the client. Through this mailing list, all pertinent documents and additional information that is deemed useful and appropriate would be disseminated.
2. Assist the client in developing a presentation and making the presentations concerning its financial and legal structure and the security aspects of the bonds to finance the project to the appropriate rating and credit agencies.
3. Assist the client in arranging and conducting such tours by representatives of institutional investors of the physical properties and operations of the client as are deemed appropriate or advisable.
4. Assist the client in conducting information or due diligence meetings in major financial centers as is deemed appropriate or necessary. This function is very important for the success of the financing of the project at hand as well as the future financing programs of the client through the medium of his revenue bonds. Since 5,000 commercial and savings banks and fire and casualty insurance companies constitute the institutional market for municipal securities, this broad market must be effectively developed on behalf of the client and the project in order to insure to the greatest extent practicable the lowest cost of money.
5. Arrange on behalf of the client special meetings with major institutional investors throughout the United States to fully inform such institutions on all aspects of the client and his construction and financing program.

Marketing the Bonds. The fourth major function of the financial advisor is the formal marketing or sale of the client's bond issue. This should be accomplished when the market has been developed and conditions in general are opportune. In implementing the sale of bonds, some of the more important steps would include the following:

1. Determine the most appropriate method of sale of the bonds, whether private placement with institutional investors or public offering on a competitive or negotiated basis. The major factors affecting this discussion are the statutory rights and power of the client, the condition of the market, the availability of funds with institutional buyers and an evaluation and comparison of the possible interest costs under either financing method.
2. Within the financial requirements, recommend the most appropriate issue structure to insure the broadest possible market acceptance. This would contemplate utilizing serial bonds, term bonds or a combination of both.
3. Assist in the preparation of the official notice of sale, if appropriate, or any public announcement regarding the sale of notes or bonds.
4. Determine, again within financial requirements, the most appropriate time to market the bond issue. While it is admittedly impossible to precisely predict bond markets, the advisor should follow and analyze money market trends, the future supply of new debt issues, secondary market activity and buying patterns of investors — all important considerations in scheduling a sale date.
5. Attend any sales of notes or bonds and assist the client in the analysis of the bids. The purpose is to determine the accuracy and appropriateness of all proposals that may be submitted.
6. Advise and assist the client in arranging for printing, execution and signing and delivery of the bonds after the bond sale.

Cost of Issuing Bonds. The table below shows a range for the costs of issuing bonds. The financial advisor's fees are established by con- tract following discussions with the issuer. The remaining expenses are normally provided for out of the gross spread, which is the difference between what the issuer is paid for the bonds and the price at which they are sold to the public.

	Percent of Issue
Financial advisor	0.3% – 0.5%
Expenses	0.2 – 0.4
Underwriter's fee	0.2 – 0.4
Salesman's takedown	1.0 – 2.0
Total Financial	1.7% – 3.3%

Additional to these financial costs are other financially related fees paid to legal counsel, bond counsel and auditors.

SECTION 7
SUMMARY AND COST GUIDELINES

Summary

Small hydroelectric development is an important renewable electric resource in the United States. The Public Utility Regulatory Policies Act of 1978, part of the national energy plan, contains specific provisions that may enhance small hydro development. The Act has required the FERC to prescribe rules for wheeling and purchasing small hydro output by electric utilities. It also contains important provisions that authorize funding for both construction and feasibility studies.

The feasibility stage is a period of major risk for the project sponsor, since all funds spent during this time are subject to total loss if the project is not viable. In many cases the sponsor will be unwilling or unable to take this risk, and the project will not be able to proceed. The feasibility funding provided in the Act should help, in many cases, to reduce the sponsor's risk to acceptable levels. The project sponsor can also minimize the potential financial loss by avoiding intermediary feasibility studies that do not allow the implementation decision to be reached.

Small hydro projects are capital-intensive, and consequently debt service comprises a major portion of cash disbursements, particularly in the project's early years. Because of this, project feasibility is sensitive to the cost of financing. The public sector, with its low-cost capital, will find small hydroelectric development more attractive than private promoters. Continuing escalation in the value of energy may reduce the importance of low-cost capital.

However, under current law, debt securities issued by a public sponsor may lose their tax-exempt status, depending on the disposition of the project's power output. If a public project sponsor intends to sell the power output to investor-owned utilities, the tax status of any debt securities used to finance construction must be determined at an early stage.

With many of the potential small hydro sites controlled by public entities, congressional legislation on the tax status of revenue bond financing for small hydroelectric developments may be desirable. Suitable legislation could decrease the current uncertainty in regard to the financing costs of many projects. Also, if small hydro developments were added to the list of categorically tax-exempt activities (Internal Revenue Code SEC. 103 (b) (4)), publicly developed small hydro developments would be assured of low-cost of capital. This would make more projects financially feasible, thereby accelerating small hydro development and furthering the nation's energy plan.

The marketing of small hydro output will play a central role in achieving feasible projects. The marketing agreement will, in many cases, provide the security necessary to obtain project financing. Consequently, adequate financial and legal counsel must be obtained to ensure that the ultimate power contract contains all the essential elements for financing to be arranged.

Summary Procedure. Figure 7-1 summarizes a procedure applicable to the economic and financial analysis of small hydroelectric projects. There are three major stages in the analysis. First, all of the cost, power production and marketing information must be assembled and organized in an understandable manner. Second, the economic and financial analysis is done using the best estimates of all the project parameters. The economic analysis screens and ranks the development options. If none of the options is viable, the analysis can be terminated. The financial feasibility of options that appear economically viable is then investigated. Once again, if no viable financing plan can be formulated the project may be infeasible. Third, if one or a number of the development options are viable, the impact of changes in the major project parameters may need to be investigated. In the case of small hydro, sensitivity analysis will usually suffice. However, in some cases a risk analysis may be necessary. This would most likely occur when the project alone is the security for the lenders.

Cost Guidelines for the Study

Level-of-effort guidelines that can be used to determine the costs to perform the power market and economic and financial tasks of a feasibility study have been based on experience in developing cost proposals for small hydroelectric and other related projects in addition to having documentation as to the actual expenditures incurred for completed projects. Fifteen to 25 percent of the total feasibility costs is generally required to complete the market and economic and financial tasks.

Power Market Analysis. This task generally consists of peforming the market analysis as discussed in Section 3 and preparing the narrative portion, which would include tables and figures as appropriate.

Completing the power market task will take anywhere from 15 to 25 man-days and approximately 10 to 15 percent of the total feasibility study cost.

Economic and Financial Analysis. Preparation of the economic and financial analysis has been discussed extensively in this volume of the Feasibility Investigation Manual. While this task is central to project feasibility and integrates all of the information into a measure of economic desirability, the level of effort involved is relatively modest.

The economic and financial analysis task will take anywhere from six to ten man-days and approximately five to 10 percent of the total feasibility effort. In addition, the level of effort estimate takes into account the preparation of the narrative portion, which would include supporting tables and charts as appropriate.

The level of effort that can be expected in completing the power market and economic and financial tasks of a small hydroelectric project feasibility study is summarized below.

Task	% of Total Cost	Total Man-Days	Level of Effort by Professional Classification (Man-Days) Draftsman/Tech	Jr. Associate	Sr. Associate
Power Market	10-15	15-25	2-3	10-15	3-7
Economic and Financial	5-10	6-10	1	3-5	2-4

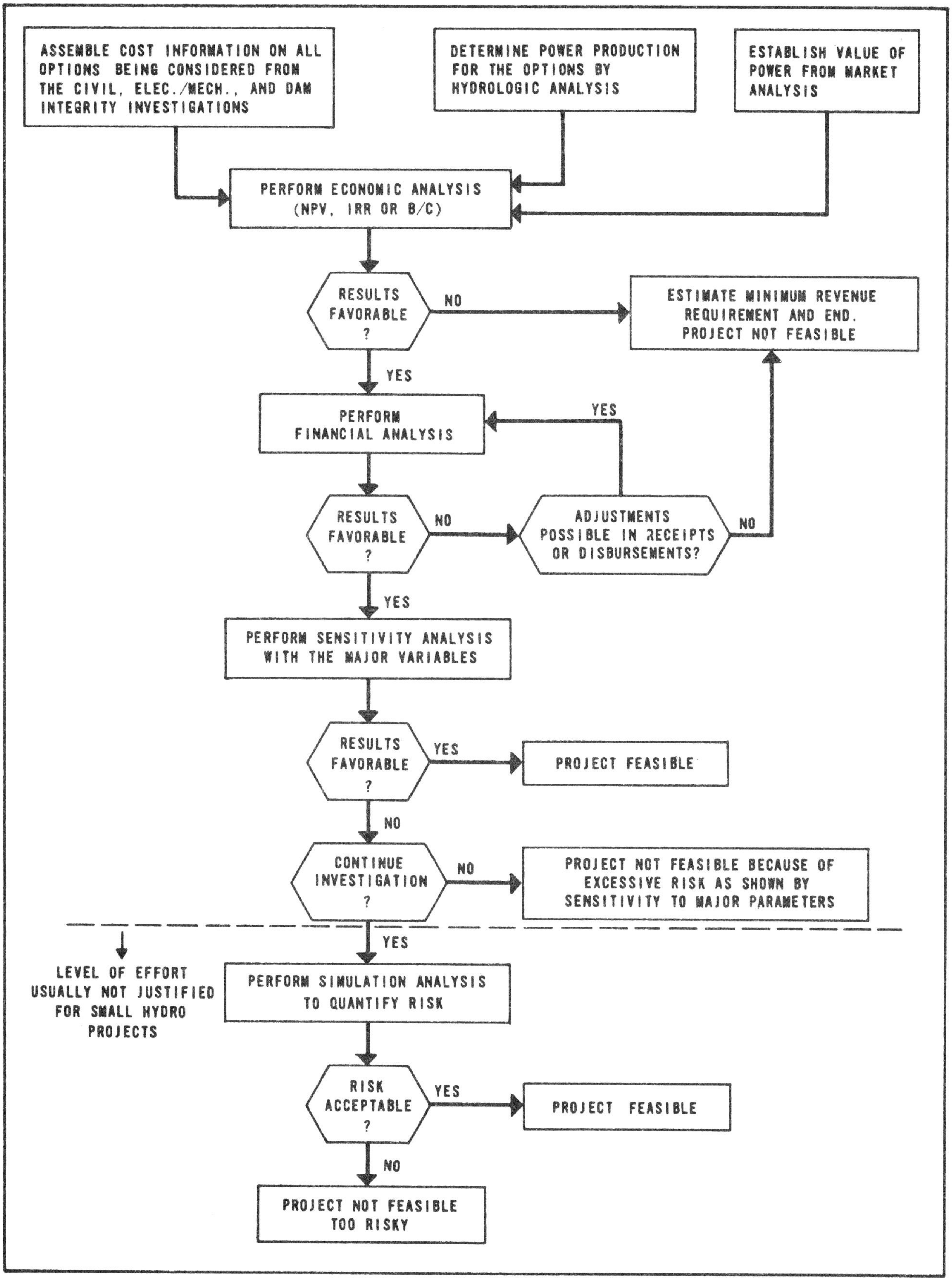

Figure 7-1. Summary procedure.

REFERENCES

American Society of Civil Engineers, Manuel No. 45, *Consulting Engineering: A Guide for the Engagement of Engineering Service.*

Bolten, Steven, E., *Managerial Finance - Principles and Practice,* Houghton Mifflin Co., 1976.

Caywood, Russel E., *Electric Utility Rate Economics,* McGraw Hill, Inc., 1972.

Edison Electric Institute, *Report on Equpment Availability,* New York, issued at intervals.

Energy Information Administration, *Statistics of Privately Owned Utilities in the United States, Class A&B,* yearly, U.S. Department of Energy.

Federal Energy Regulatory Commission (FERC), *Annual Summary of Cost and Quality of Electric Utility Plant Fuels,* 1976 or later, U.S. Department of Energy.

Federal Energy Regulatory Commission, *Hydroelectric Power Evaluation,* U.S. Department of Energy, draft, 1978.

Federal Energy Regulatory Commission, *National Electric Rate Book,* U.S. Department of Energy, by state and regularly updated.

Federal Energy Regulatory Commission, *Steam Electric Plant Construction Cost and Annual Production Expenses,* yearly, U.S. Department of Energy.

Federal Power Commission, *Hydroelectric Power Evaluation,* 1968, and *Supplement No. 1,* 1969.

Institute of Water Resources, *Estimate of National Hydroelectric Power Potential at Existing Dams,* U.S. Army Corps of Engineers, July 1977.

Jacobs, D. F., Farwell, L.C., and Neave, E., *Financial Institutions,* Fifth Edition, Richard D. Irwin, Inc., 1972.

Kahn, Alfred E., *The Economics of Regulation: Principles and Institutions,* Vols I and II, John Wiley and Sons, Inc., 1971.

Misham, Edward J., *Cost-Benefit Analysis,* Praeger, 1976.

Moody's Investors Service, Inc., *Moody's Bond Record,* Monthly.

Moody's Investor Service, Inc., *Moody's Public Utility Manual,* annual with semi-weekly supplements.

National Energy Information Center, *Directory of State Government Energy-Related Agencies,* Federal Energy Administration, 1975 or update.

Nevada Irrigation District, (Nevada, Placer, Sierra and Yuba Counties), California, *Official Statement, $7,800,000 Yuba-Bear River Development Revenue Bonds,* Fifth Issue, Second Division, 1978.

Public Utility District No. 1 of Chelan County, Washington, *Official Statement, $10,000,000 Columbia River - Rock Island Hydro-Electric System Revenue Bonds,* Series of 1978.

Public Utility Regulatory Policy Act of 1978. See U.S. Senate Report No. 95-1292 for the full text and the Joint Explanatory Statement of the Committee Conference.

Securities and Exchange Commission, *Form 10-K,* annually submitted by all registered companies.

U.S. Department of Energy, *Monthly Energy Review,* monthly.

EXHIBIT I
FEDERAL ENERGY REGULATORY COMMISSION (FERC)
ELECTRIC REPORTS

ANNUAL REPORTS OF FINANCIAL ACCOUNTS, PLANTS, OPERATIONS

Form 1: Annual Report (Classes A and B)

Detailed financial and operating information, filed by federally and privately owned electric utilities with electric operating revenues of $1 million or more. Due March 31

Form 1-F: Annual Report (Classes C and D)

Financial and operating information, filed by privately owned electric utilities with annual electric operating revenues of between $25,000 and $1 million. Due March 31.

Form 1-M: Annual Report, Municipal Electric Utilities

Similar information from municipal electric utilities with annual revenues of $250,000 or more. Due March 31.

Form 9: Licensed Project Annual Report

From owners of major privately-owned FERC-licensed hydroelectric projects, covering all projects with installed capacity of more than 2,000 horsepower owned by the licensee. Due April 30.

ANNUAL POWER SYSTEM OPERATIONS

Form 12: Power System Statement

Filed by all investor or publicly owned systems which generate all or part of their requirements and whose net energy for the year was over 20 million kilowatt-hours. Contains annual information on electric power generation, energy exchanges, and sales to ultimate consumers. Due May 1. Full information on generating equipment is filed on a five-year cycle only.

Form 12-A: Power System Statement

Same annual system generation and power exchange information, filed by systems with from 5 to 20 million kilowatt-hours net energy, systems engaged primarily in sale for resale or sales to industrial users, and systems which obtain their entire energy requirements from other systems. Due May 1.

Form 12-C: Industrial Electric Generating Capacity

From all industrial establishments which own or operate generating capacity, other than motor generators, under 5,000 kilowatts. Due May 1.

Form 12-D: Power System Statement

Filed annually by each utility with energy requirements under 5 million kilowatt-hours, containing information on generation, energy exchanged, deliveries to ultimate consumers by type of use, and projected changes in system generating facilities. Due May 1.

Form 12-F: Power Line Data

Power line data, filed by electric utility systems with power lines operating at 69 kilovolts and above. Data submitted is as of June 30 each year and is due at the Commission on July 31.

MONTHLY POWER GENERATION AND OPERATION

Form 12-E-2: Monthly Supplement to Power System Statement

Filed by approximately 270 major electric utility systems. Several systems operated under some form of power pool or common dispatching submit only a summary report. This monthly report provides energy generation and monthly peak load data. A semi-annual supplement provides near and long term load projections as well as generator and transmission line construction schedules. Due 15 days after end of month reported.

Form 4: Monthly Power Plant Report

Filed by all electric utilities with generating capacity, monthly information on generation of electricity and consumption and stocks of fuel (Form 4-white), and from industrial establishments with installed generating capacity of 10,000 kilowatts or more (Form 4-pink). Due 10 days after close of month reported.

RETAIL POWER RATES

Form 3: Typical Net Monthly Bills for Residential Service

Filed annually by selected power suppliers in each state for specified communities, typical net monthly bills for power at retail for residential service for communities of 2,500 or more population; and commercial and industrial service for communities of 50,000 or more, or if there are no cities that size, the three largest. Due about Feb. 15.

Form 3-A: All-Electric Homes Data Sheet

Filed annually be power suppliers in all cities having population of 50,000 or more or supplying the three largest cities, net annual retail bills for all-electric homes computed under rates applicable January 1. Also, latest information on number of all-electric customers and average electric consumption. Due April 1.

Form 3-P: Monthly Electric Bill Data

Residential, commercial, and industrial electric bill data for the U.S. Bureau of Labor Statistics, Consumer and Wholesale Price Indexes.

Form 13: Summary for National Electric Rate Book

Selected retail rate schedules of electric utilities, both public and private, for inclusion in the FERC National Electric Rate Book. Filed periodically as requested by

FERC.

Form 82: Retail Rate Level Change

All changes in retail rates, filed within 60 days of date of change, from all electric utilities serving at least one community of 2,500 or more population.

OTHER ELECTRIC REPORTS

Form 5: Monthly Statement of Electric Operating Revenues & Income

Monthly information on operating revenues and income, filed by all privately owned electric utilities with annual electric operating revenues $2.5 million and over, and certain publicly owned utilities. Due about 40 days after end of month.

Form 6: Initial Statement of Actual Legitimate Original Cost

An initial statement of actual legitimate original cost of FERC licensed hydroelectric projects, filed by all licensees of projects over 2,000 horsepower installed capacity and above.

Form 7: Statement of Actual Legitimate Original Cost

A statement filed after determination by the FERC of actual legitimate original cost of construction of an FERC-licensed hydroelectric project.

Form 67: Steam-Electric Plant Air and Water Quality Control Data

Annual information on steam-electric plant air and water quality control, for each generating plant 25 megawatts or more. Due May 1.

Form 80: Licensed Projects Recreation Report

From all licensees, a bi-annual report showing recreational use and development at FERC-licensed hydroelectric power projects. Due June 30 in odd numbered years only.

Form 237-A: (Yellow) Fuel Emergency Report, Coal as Principal Fuel

To be filed within two days after end of reported week when any generating electric utility faces a fuel emergency.

Form 237-B: (Blue) Fuel Emergency Report, Oil as Principal Fuel

To be filed within two days after end of reported week when any generating electric utility faces a fuel emergency.

Form 423: Monthly Report of Cost and Quality of Fuels for Electric Plants

Filed by each electric power producer for each plant (steam, internal combustion, gas turbine or any mix) 25 megawatts capacity or greater, monthly data on cost, quality and source of fuels delivered. Due days after end of month reported.

Regional Reliability Council Annual Reports

The geographical area of the 48 contiguous states is divided into nine electric reliability councils. Each of these nine regional electric reliability councils submits annually a non-formatted report detailing the regional coordinated bulk power supply plans. Listings of projected loads, existing and projected generation, and proposed transmission lines (over 230 kilovolts) are included. Information is also provided on the communications systems and other coordinated operating practices in each council area. Detailed information is given for the upcoming 10-year period with more general data for the 11-20 year upcoming period. Due April 1.

Index of Electric Rate Schedules

Issued quarterly.

HOW TO OBTAIN COPIES

Files and records of the Federal Energy Regulatory Commission are available for public inspection in the FERC's Office of Public Information. Photocopies of public records may be obtained through a private firm under FERC contract, requiring payment of a fee directly to the firm. Written requests for photocopies must be directed to the Commission's Office of Public Information. The reproduction company cannot accept orders submitted directly by members of the public. Completed orders are mailed C.O.D. by the Company. Orders may also be picked up at the office of the company or may be delivered by messenger within the District of Columbia.

Most FERC publications are sold in the Office of Public Information, on a cash, over-the-counter basis only, in addition to availability from the Superintendent of Documents, U.S. Government Printing Office.

INFORMATION AVAILABLE OUTSIDE WASHINGTON, D.C.

The Federal Energy Regulatory Commission requires companies under its regulatory jurisdiction to keep the following types of information available for inspection by members of the public in a convenient form during each company's business hours at the places specified.

At an Electric Utility's Principal and District or Division Offices in the Territory Served:

> Complete rate schedules clearly setting forth all rates and charges for any transmission of electric energy at wholesale for resale in interstate commerce subject to FERC jurisdiction, and the classifications, practices, rules and regulations affecting such rates and changes, and all contracts which in any manner affect or relate to such rates, charges, classifications, services, rule and regulations, or practices.

LIST OF PROPOSED REGULATORY INFORMATION SYSTEM FORMS

Form No.	Title	Would Replace Part or All of Existing Form Nos.
156	Report of Generating Plant, Technical, Environmental, and Operating Data	1, 1F, 1M, 4 4A, 12, 12A, 12D, 67, 423
157	Annual Report of Power System Transmission and Distribution Technical Data	1, 1F, 1M, 12, 12F
158	Annual Report of Power Systems Energy Accounting, Peak Demands, and Intersystem Purchases and Sales	1, 1F, 1M, 12, 12, 12D
159	Report on Retail Electric Bills and Rate Changes	3, 3A, 82
160	Report of Industrial Generation of Electric Energy and Electric Generating Capacity	4B, 12C
161	Licensed Project Information	9, 80
162A	Annual Corporate and Financial Report for Class A and B Electric Utilities	1
162M	Monthly Corporate and Financial Report for Class A Electric Utilities	5
163	Corporate and Financial Report for Classes C and D Electric Utilities and Licensees	1F
164	Financial Report for Municipal Electric Utilities and Federal Projects	1, 1M

FEDERAL ENERGY REGULATORY COMMISSION OFFICES

Office	Serves
WASHINGTON, D.C.: Director Office of Electric Power Regulation Federal Energy Regulatory Commission 825 North Capitol Street, N.E. Washington, D.C. 20426 Tel: (202) 275-4006	
ATLANTA: Regional Engineer Federal Energy Regulatory Commission 730 Peachtree Building Atlanta, Georgia 30308 Tel: (404) 526-5134	Alabama, Florida, Georgia, North and South Carolina, Tennessee and parts of Kentucky, Mississippi, Virginia, and West Virginia
CHICAGO: Regional Engineer Federal Energy Regulatory Commission 31st Floor, Federal Building 230 S. Dearborn St. Chicago, Illinois 60604 Tel: (312) 353-6171	Illinois, Indiana, Iowa, Michigan, Minnesota, North Dakota, Wisconsin, and parts of Kansas, Missouri, Montana, Nebraska, Ohio, South Dakota, and Wyoming
FORT WORTH: Regional Engineer Federal Energy Regulatory Commission 819 Taylor Street Fort Worth, Texas 76102 Tel: (917) 334-2631	Arkansas, Colorado, Louisiana, Oklahoma, Texas, and parts of Kansas, Mississippi, Missouri, Nebraska, New Mexico, South Dakota, and Wyoming
NEW YORK: Regional Engineer Federal Energy Regulatory Commission 26 Federal Plaza (22nd Floor) New York, New York 10007 Tel: (212) 264-3687	Connecticut, Delaware, District of Columbia, Maine, Maryland, Massachusetts, New Hampshire, New Jersey, New York, Pennsylvania, Rhode Island, Vermont, and parts of Kentucky, Ohio, Virginia, and West Virginia
SAN FRANCISCO: Regional Engineer Federal Energy Regulatory Commission US Customs House 555 Battery Street San Francisco, California 94111 Tel: (415) 556-3581	Alaska, Arizona, California, Hawaii, Idaho, Nevada, Oregon, Utah, Washington, and parts of Montana, New Mexico, and Wyoming west of the Continental Divide

EXHIBIT II
EXTRACTS FROM THE UNIFORM SYSTEMS OF ACCOUNTS APRIL 1973 FEDERAL ENERGY REGULATORY COMMISSION

APPLICABILITY OF SYSTEMS OF ACCOUNTS

These systems of accounts are applicable in principle to all Classes A, B, C and D licensees subject to the Commission's accounting requirements under the Federal Power Act, and to all Classes A, B, C and D public utilities subject to the provisions of the Federal Power Act. The Commission reserves the right, however, under the provisions of section 301(a) of the Federal Power Act, to classify such licensees and public utilities and to prescribe a system or classification of accounts to be kept by and which will be convenient for and meet the requirements of each class.

These systems of accounts are applicable to public utilities, as herein defined, and to licensees engaged in the generation and sale of electric energy for ultimate distribution to the public.

The systems of accounts shall also apply to agencies of the United States engaged in the generation and sale of electric energy for ultimate distribution to the public, so far as may be practicable, in accordance with applicable statutes.

In accordance with the requirements of section 3 of the act, the "classification of invstment in road and equipment of steam roads, issue of 1914, Interstate Commerce Commission," is published and promulgated as a part of the accounting rules and regulations of the Commission, and a copy thereof is located at Part 103. Irrespective of any rules and regulations contained in these systems of accounts, the cost of original and betterments thereof, shall be determined under the rules and principles as defined and interpreted in said classification of the Interstate Commerce Commission so far as applicable.

CLASSIFICATION OF UTILITIES

For the purpose of applying systems of accounts prescribed by the Commission, electric utilities and licensees are divided into four classes, as follows:

Class A. Utilities having annual electric operating revenues of $2,500,000 or more.

Class B. Utilities having annual electric operating revenues of $1,000,000 or more but less than $2,500,-000.

Class C. Utilities having annual electric operating revenues of $150,000 or more but less than $1,000,000.

Class D. Utilities having annual electric operating revenues of $25,000 or more but less than $150,000.

ACCOUNTS FOR CLASS A, B, C, AND D PUBLIC UTILITIES AND LICENSEES
ELECTRIC PLANT ACCOUNTS

PLANT PRODUCTION

Hydraulic Production

330 Land and land rights.
331 Structures and improvements.
332 Reservoirs, dams and waterways.
333 Water wheels, turbines and generators.
334 Accessory electric equipment.
335 Miscellaneous power plant equipment.
336 Roads, railroads and bridges.

Transmission Plant

350 Land and land rights.
351 (Reserved)
352 Structures and improvements.
353 Station equipment.
354 Towers and fitures.
355 Poles and fixtures.
356 Overhead conductors and devices.
357 Underground conduit.
358 Underground conductors and devices.
359 Roads and trails.

ACCOUNTS FOR CLASS A AND CLASS B OPERATION AND MAINTENANCE EXPENSE ACCOUNTS

POWER PRODUCTION EXPENSES

Hydraulic Power Generation

Operation

535 Operation supervision and engineering.
536 Water for power.
537 Hydraulic expenses.
538 Electric expenses.
539 Miscellaneous hydraulic power generation expenses.
540 Rents.

Maintenance

541 Maintenance supervision and engineering.
542 Maintenance of structures.
543 Maintenance of reservoirs, dams and waterways.
545 Maintenance of miscellaneous hydraulic plant.

TRANSMISSION EXPENSES

Operation

560 Operation supervision and engineering.
561 Load dispatching.
562 Station expenses.
563 Overhead line expenses.
564 Underground line expenses.
565 Transmission of electricity by others.
566 Miscellaneous transmission expenses
567 Rents.

Maintenance

568 Maintenance supervision and engineering.
569 Maintenance of structures.
570 Maintenance of station equipment.
571 Maintenance of overhead line.
572 Maintenance of underground lines.
573 Maintenance of miscellaneous transmission plant.

SALES EXPENSES

Operation

911 Supervision.
912 Demonstrating and selling expenses.
913 Advertising expenses.
916 Miscellaneous sales expenses.

ADMINISTRATIVE AND GENERAL EXPENSES

Operation

920 Administrative and general salaries.
921 Office supplies and expenses.
922 Administrative expenses transferred - Cr.
923 Outside services employed.
924 Property insurance.
925 Injuries and damages.
926 Employee pensions and benefits.
927 Franchise requirements.
928 Regulatory commission expenses.
929 Duplicate charges - Cr.
930 Miscellaneous general expenses.
931 Rents.

Maintenance

932 Maintenance of general plant.

ACCOUNTS FOR CLASS C AND CLASS D OPERATION AND MAINTENANCE EXPENSE ACCOUNTS

POWER PRODUCTION EXPENSES

Hydraulic Power Generation

Operation

530 Operation supervision and labor.
531 Water for power.
532 Operation supplies and expenses.
533 Rents.

Maintenance

535 Maintenance of hydraulic production plant.

TRANSMISSION EXPENSES

Operation

550 Operation supervision and labor.
551 Operation supplies and expenses.
552 Rents.

Maintenance

553 Maintenance of transmission plant.

SALES EXPENSE

Operation

910 Sales expenses.

ADMINISTRATIVE AND GENERAL EXPENSES

Operation

920 Administrative and general salaries.
921 Office supplies and expenses.
922 Administrative expenses transferred - Cr.
923 Outside services employed.
924 Property insurance.
925 Injuries and damages.
926 Employee pensions and benefits.
927 Franchise requirements.
928 Regulatory commission expenses.
929 Duplicate charges - Cr.
930 Miscellaneous general expenses.
931 Rents.
933 Transportation expenses.

Maintenance

935 Maintenance of general plant.

EXHIBIT III
PUBLIC UTILITY REGULATORY POLICIES ACT
PUBLIC LAW 95-617—Nov. 9, 1978
TITLE IV — SMALL HYDROELECTRIC POWER PROJECTS

SEC. 401. ESTABLISHMENT OF PROGRAM.

The Secretary shall establish a program in accordance with this title to encourage municipalities, electric cooperatives, industrial development agencies, nonprofit organizations, and other persons to undertake the development of small hydroelectric power projects in connection with existing dams which are not being used to generate electric power.

SEC.402. LOANS FOR FEASIBILITY STUDIES.

(a) LOAN AUTHORITY.—The Secretary, after consultation with the Commission, is authorized to make a loan to any municipality, electric cooperative, industrial development agency, nonprofit organization, or other person to assist such person in defraying up to 90 percent of the costs of—

(1) studies to determine the feasibility of undertaking a small hydroelectric power project at an existing dam or dams and

(2) preparing any application for a necessary license or other Federal, State, and local approval respecting such a project at an existing dam or dams and of participating in any administrative proceeding regarding any such application.

(b) CANCELLATION.—The Secretary may cancel the unpaid balance and any accrued interest on any loan granted pursuant to this section if he determines on the basis of the study that the small hydroelectric power project would not be technically or economically feasible.

SEC. 403. LOANS FOR PROJECT COSTS.

(a) AUTHORITY.—The Secretary is authorized to make loans to any municipality, electric cooperative, industrial development agency, nonprofit organization, or other person of up to 75 percent of the project costs of a small hydroelectric power project. No such loan may be made unless the Secretary finds that—

(1) the project will be constructed in connection with an existing dam or dams,

(2) all licenses and other required Federal, State, and local approvals necessary for construction of the project have been issued.

(3) the project will have no significant adverse environmental effects, including significant adverse effects on fish and wildlife, on recreational use of water, and on stream flow, and

(4) the project will not have a significant adverse effect on any other use of the water used by such project.

The Secretary may make a commitment to make a loan under this subsection to an applicant who has not met the requirements of paragraph (2), pending compliance by such applicant with such requirements. Such commitment shall be for period of not to exceed 3 years unless the Secretary, in consultation with the Commission, extends such period for good cause shown. Notwithstanding any such commitment, no such loan shall be made before such person has complied with such requirements.

(b) PREFERENCE.—The Secretary shall give preference to applicants under this section who do not have available alternative financing which the Secretary deems appropriate to carry out the project and whose projects will provide useful information as to the technical and economic feasibility of—

(1) the generation of electric energy by such projects, and

(2) the use of energy produced by such projects.

(c) INFORMATION.—Every applicant for a license for a small hydroelectric power project receiving loans pursuant to this section shall furnish the Secretary with such information as the Secretary may require regarding equipment and services proposed to be used in the design, construction, and operation of such project. The Secretary shall have the right to forbid the use in such project of any equipment or services he finds inappropriate for such project by reason of cost, performance, or failure to carry out the purposes of this section. The Secretary shall make information which he obtains under this subsection available to the public, other than information described as entitled to confidentiality under section 11(d) of the Energy Supply and Environmental Coordination Act of 1974.

(d) JOINT PARTICIPATION.—In making loans for small hydroelectric power projects under this section, the Secretary shall encourage joint participation, to the extent permitted by law, by applicants eligible to receive loans under this section with respect to the same project.

SEC. 404. LOAN RATES AND REPAYMENT.

(a) INTEREST.—Each loan made pursuant to this title shall bear interest at the discount or interest rate used at the time the loan is made for water resources planning projects under section 80 of the Water Resources Development Act of 1974 (42 U.S.C. 1962-17(a)). Each such loan shall be for such term, as the Secretary deems appropriate, but not in excess of—

(1) 10 years (in the case of a loan under section 402) or

(2) 30 years (in the case of a loan under section 403).

(b) REPAYMENTS.—Amounts repaid on loans made pursuant to this title shall be deposited into the United States Treasury as miscellaneous receipts.

SEC. 405. SIMPLIFIED AND EXPEDITIOUS LICENSING PROCEDURES.

(a) ESTABLISHMENT OF PROGRAM.—The Commission shall establish, in such manner as the Commission deems appropriate, consistent with the applicable provisions of law, a program to use simple and expeditious licensing procedures under the Federal Power Act for small hydroelectric power projects in connection with existing dams.

(b) PREREQUISITES.—Before issuing any license under the Federal Power Act for the construction or operation of any small hydroelectric power project the Commission—

(1) shall assess the safety of existing structures in any proposed project (including possible consequences associated with failure of such structures), and

(2) shall provide an opportunity for consultation with the Council on Environmental Quality and the Environmental Protection Agency with respect to the environmental effects of such project.

Nothing in this subsection exempts any such project from any requirement applicable to any such project under the National Environmental Policy Act of 1969, the Fish and Wildlife Coordination Act, the Endangered Species Act, or any other provision of Federal law.

(c) FISH AND WILDLIFE FACILITIES.—The Commission shall encourage applicants for licenses for small hydroelectric power projects to make use of public funds and other assistance for the design and construction of fish and wildlife facilities which may be required in connection with any development of such project.

SEC. 406. NEW IMPOUNDMENTS.

Nothing in this title authorizes (1) the loan of funds for construction of any new dam or other impoundment, or (2) the simple and expeditious licensing of any such new dam or other impoundment.

SEC. 407. AUTHORIZATIONS.

There are hereby authorized to be appropriated for each of the fiscal years ending September 30, 1978, September 30, 1979, and September 30, 1980, no to exceed $10,000,000 for loans to be made pursuant to section 402, such funds to remain available until expended. There are hereby authorized to be appropriated for each of the fiscal years ending September 30, 1978. September 30, 1979, September 30, 1980, not to exceed $100,000,000 for loans to be made pursuant to section 403, such funds to remain available until expended.

SEC. 408. DEFINITIONS.

For purposes of this title, the term—

(1) "small hydroelectric power project" means any hydroelectric power project which is located at the site of any existing dam, which uses the water power potential of such dam, and which has not more than 15,000 kilowatts of installed capacity:

(2) "electric cooperative" means any cooperative association eligible to receive loans under section 4 of the Rural Electrification Act of 1936 (7 U.S.C. 904);

(3) "industrial development agency" means any agency which is permitted to issue obligations the interest on which is excludable from gross income under section 103 of the Internal Revenue Code of 1954;

(4) "project costs" means the cost of acquisition or construction of all facilities and services and the cost of acquisition of all land and interests in land used in the design and construction and operation of a small hydroelectric power project;

(5) "nonprofit organization" means any organization described in section 501(c)(3) or 501(c)(4) of the Internal Revenue Code of 1954 and exempt from tax under section 501(a) of such Code (but only with respect to a trade or business carried on by such organization which is not an unrelated trade or business, determined by applying section 513(a) to such organization);

(6) "existing dam" means any dam, the construction of which was completed on or before April 20, 1977, and which does not require any construction or enlargement of impoundment structures (other than repairs or reconstruction) in connection with the installation of any small hydroelectric power project;

(7) "municipality" has the meaning provided in section 3 of the Federal Power Act; and

(8) "person" has the meaning provided in section 3 of the Federal Power Act.

VOLUME III: HYDROLOGIC STUDIES

SECTION 1: INTRODUCTION AND OVERVIEW

Scope and Objectives

This volume presents techniques and examples of procedures and references on investigations leading to investments in small hydroelectric power additions to existing facilities. Many of the procedures discussed are equally applicable to larger power installations but generally this volume is restricted to those structures which presently make use of present reservoir release patterns and authorized project purposes. Small hydropower additions are intended to make a nonconsumptive use of water presently flowing past the site or released from the impoundment for other purposes, generally consumptive in nature. Even if storage is not available at the damsite for other purposes, the hydraulic head created by the structure can often be economically utilized to generate electrical energy.

The definition of "small" as adopted in this guide manual, refers to installed capacities less than 15 MW. References are made to various publications containing detailed procedures beyond the intent of this volume. More comprehensive discussions can be found in these references on the concepts addressed herein.

Two levels of study are assumed when discussing techniques of investigation procedures. A reconnaissance level of study is discussed first. More detailed studies are then covered, which are intended to serve as the basis for investment decisions and licensing application requirements.

This volume presents procedures for developing data concerning stream flow, evaporation, capacity vs. average annual energy, spillway design, dam safety from overtopping and statistical data concerning generation patterns and power availability.

Overall Strategy for Hydrologic Study

The general procedure is to establish how much water is available to divert through a turbine and the hydraulic head associated with this flow. Information is needed on the variability of the flow presently passing or released from the structure. These data may be readily available from the project owner-operator or may require estimation from such records as are available at nearby points. Estimates should first be made with reconnaissance level of detail and later, if a feasibility level of study is warranted, they can be refined and prepared in greater detail. Net power head can be estimated based on pool level and tailwater elevations which prevail at least 50 percent of the time. Estimates of hydraulic losses can be based on engineering judgment. If average annual energy estimates appear to have a value exceeding the cost of adding the power plant to the existing facility, the next step is to evaluate the spillway for structural and hydraulic adequency. This entails the estimation of a spillway design discharge and an evaluation of the hydraulic characteristics of the existing spillway. Any structural rehabilitation or improvement costs are included in a second economic evaluation while still in a reconnaissance level of study. All costs for the power plant, including rehabilitation and improvements, should be compared with the expected value of average annual energy. If the project revenue from power exceeds power costs by a wide margin, a more detailed analysis should be made of all of the same basic items but to a greater level of accuracy. Figure 1-1 presents a diagram outlining the various tasks necessary to reach a meaningful conclusion to hydrologic aspects associated with the investment decision process.

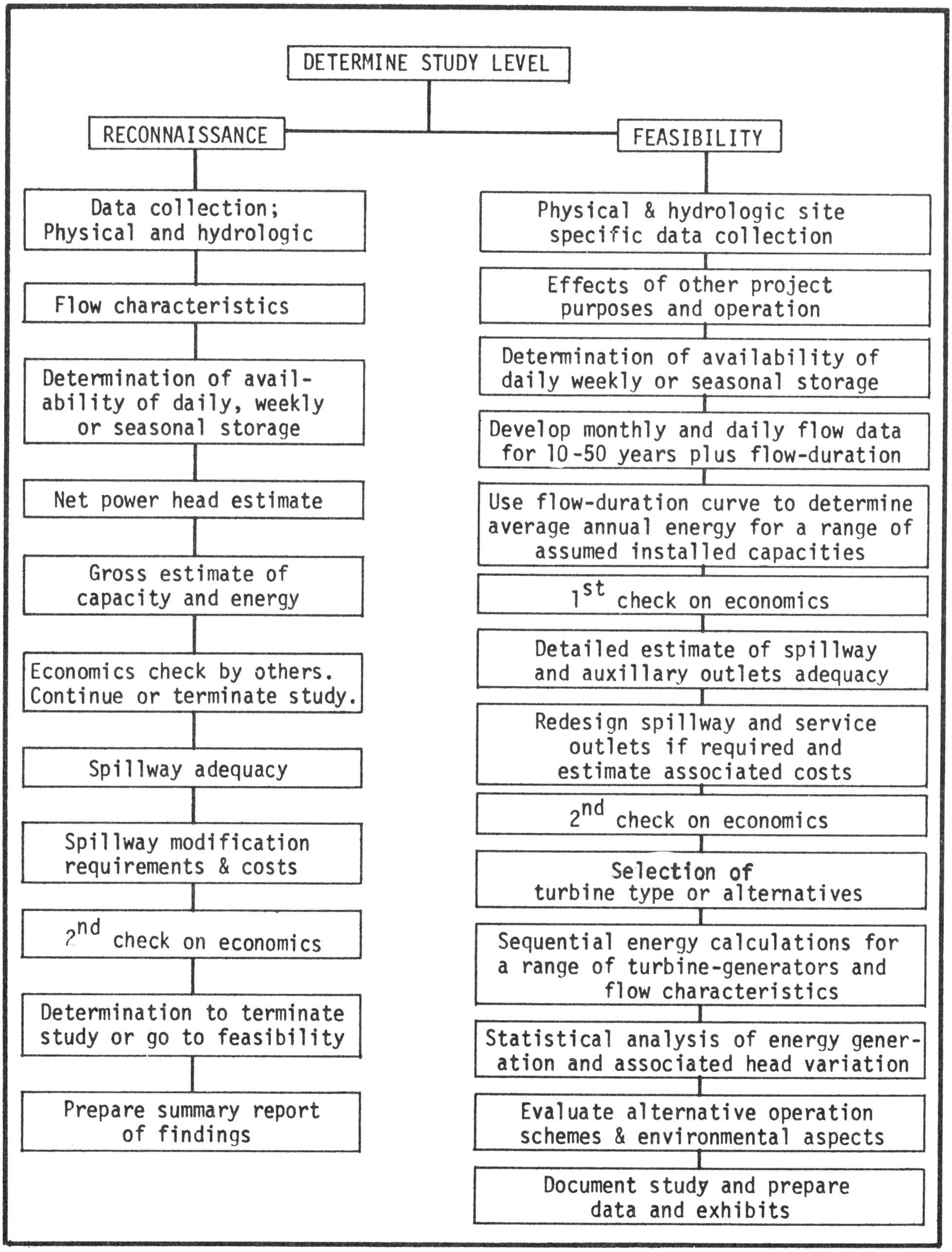

Figure 1-1. Hydrology and hydraulics study task outline for small hydropower additions to an existing facility.

SECTION 2
DATA NEEDS AND SOURCES

Level of Study

Although the basic data needs are not much different between the reconnaissance level of study and the feasibility level of study, the detail and accuracy of these estimates and the manpower expended to obtain them usually will be significantly different. For instance, all that may be needed in a reconnaissance level of investigation is an estimate of average annual flow, and average net power head. Some idea of the flow availability during low flow seasons and years is needed in order to estimate the likelihood of credit for dependable capacity. However, power benefits will typically be based on average annual energy generation since capacity will usually not meet the standard definition of "dependable".

Physical and Operational Data

Physical and operational data concerning the existing structure are fundamental to even a gross reconnaissance estimate of power potential and energy estimate. The following list indicates those items needed in the feasibility level of data collection with those minima data required for estimates at the reconnaissance level shown with an asterisk (*).

1. Maximum hydraulic height of dam.
2. Emergency spillway elevation, type and dimensions.
3. Maximum elevation at which water can be stored.
4. *Normal water surface elevation.
5. Maximum allowable drawdown or inactive pool elevation.
6. Outlet size, location and rating curve.
7. *Tailwater elevation at normal flow.
8. Surface area and storage versus elevation relationships.
9. Storage purposes, if applicable, and operation rules.

Terminology frequently applied to a dam and storage facility are shown schematically on Figure 2-1.

Hydrologic Data

Basic information and data are needed about the drainage area and run off characteristics of the watershed and any major water usage or diversions upstream of the dam. Usually these data are available in the files of the owner-agent or reports by State or Federal water resources agencies. Recorded pool elevations and releases should be compiled and adjusted to flow at the site under expected future conditions in order to make reliable estimates of hydopower potential. If no records have been kept, a search must be made for stream gages in the surrounding region for which comparisons and adjustments can be made to develop long term (10-50 years) daily and/or monthly flow data.

If daily flow data are readily available flow-duration data can be constructed from which to make average annual energy estimates. The accuracy of the capacity and energy estimates is dependent on the combined accuracy of estimating flow characteristics and corresponding head variability. The following list of hydrologic data required in feasibility level energy calculations shows those items needed for reconnaissance level studies marked with an asterisk (*).

1. Drainage area.
2. Daily and/or monthly flow data for an extensive period of time (10-50 years).
3. *Flow-duration curves.
4. *Tailwater elevation versus flow relationship.
5. Spillway and outlet rating curves.
6. Spillway design flood hydrograph.
7. *Project purposes, operation rules and storage available.
8. Evaporation rates.
9. *Seepage losses, fish ladder water requirements, diversions direct from storage.
10. Pool elevation-duration data.
11. *Annual peak discharge data may be needed to assess the adequacy of the spillway capacity at some projects.
12. Minimum flow requirements downstream of the site.

Data Sources

The most logical source for both the physical and hydrological data is the operator-owner of the existing facility. The U.S. Corps of Engineers have been given the responsibility to prepare Phase I safety inspection and evaluation reports on high hazard non-Federal dams. These reports are a primary source of both reconnaissance and feasibility level data. State Division of Dams permit and inspection agencies files are a primary data source in many states.

The majority of continuous flow data are published by the U.S. Geological Survey (USGS). Mean daily flow data are published annually by state and five year summary reports are published by major river basin grouping. Data published by States and by the USGS are usually available in the State libraries, University libraries or libraries of Federal agencies such as the U.S. Army Corps of Engineers, Bureau of Reclamation, or Soil Conservation Service. District and Sub-District offices of the Geological Survey can obtain computer

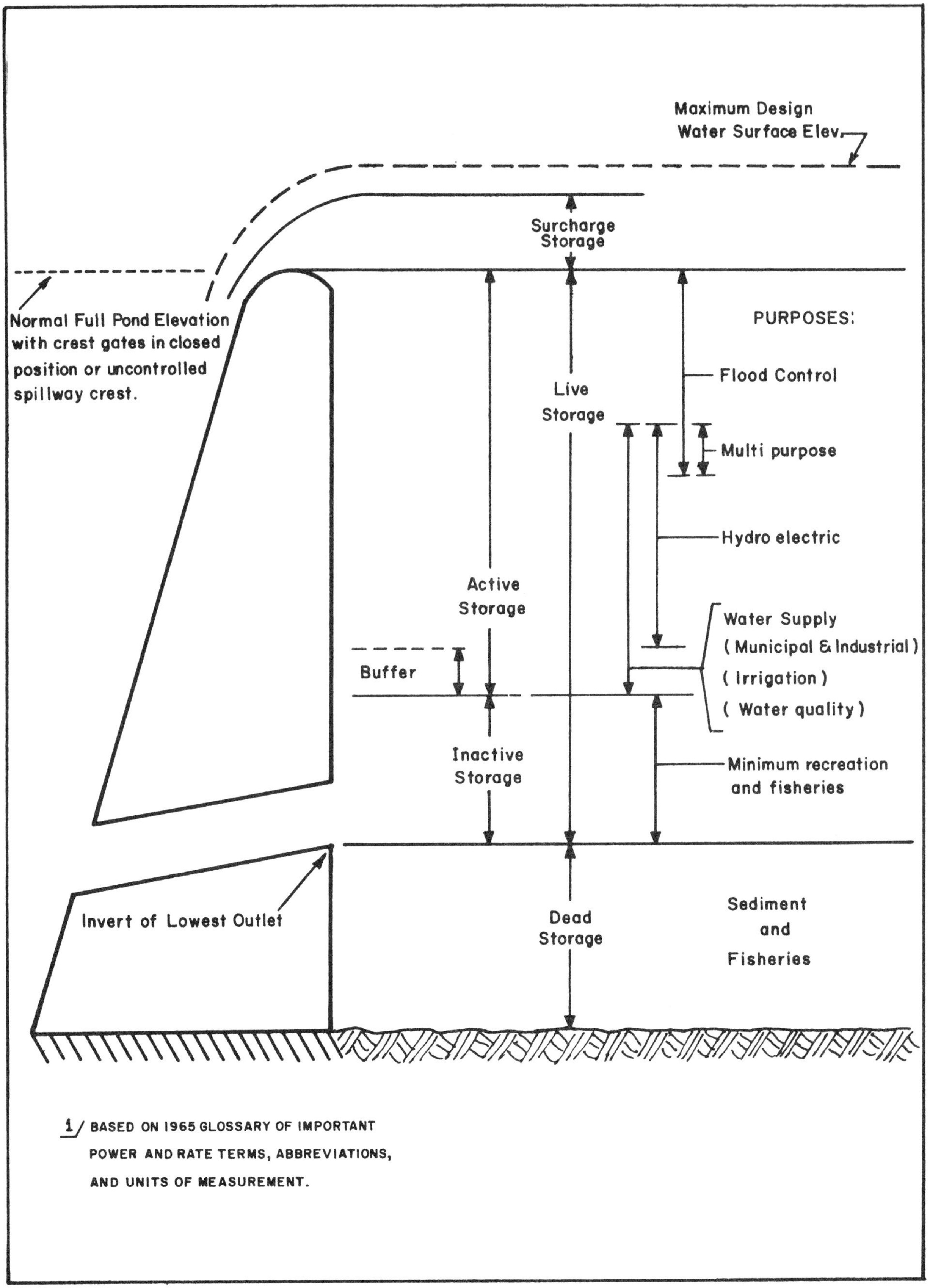

Figure 2-1. Illustration of reservoir terms.

listings from their National Water Data Storage and Retrieval System (WATSORE). Both daily values and annual peak discharges are available along with several statistical analysis capabilities. Frequently, utility companies, irrigation districts, water companies, and other water using organizations collect similar surface runoff data which may be published separately from the Geological Survey publications or may be unpublished but available if one is willing to spend the necessary effort to compile the data in a usable form.

Streamflow Correlation Studies

If streamflow data are not available at the project site, the nearest site of similar size and hydrologic characteristics should be evaluated as a source of data that can be proportioned by drainage area ratio. It would be preferable to have observed data as near as possible downstream of the project site in order to require a minimum of adjustment for runoff between the project site and the gage. This situation can also circumvent the necessity of adjusting for evaporation and diversion from the project. If comparison must be made strictly by site similarities or from a nearby upstream gage, adjustments must be made for any significant evaporation losses, diversions, seepage losses and fish ladder flow requirements. Sophisticated regional studies and correlation procedures are generally not warranted during reconnaissance studies and probably only infrequently even during feasibility studies. In a situation where a large investment cost and where installed capacities approach the upper boundary of this manual may be involved, it may be worthwhile to utilize a stochastic procedure for estimating long term flow sequences to evaluate extreme droughts. This would be particularly applicable if dependable capacity were an issue. Detailed discussion of correlation procedures and examples are contained in Hydrologic Data Management, Vol. 2, Corps of Engineers IHD 1972 and in most textbooks on hydrology and statistics.

Introduction. Stochastic procedures are only justified at the feasibility level of investigation and only then in those cases where dependable capacity is a significant issue and where project benefits warrant the extra study expenditure. The term "simulation" is applied to the mathematical or physical modeling of a phenomenon or process. In this section, it is used to denote only the mathematical modeling of a stochastic process. A stochastic process is one in which there is a chance component in each successive event and ordinarily some degree of correlation between successive events. Modeling of a stochastic process involves the use of the "Monte Carlo" method of adding a random (chance) component to a correlated component in order to construct each new event. The correlated component can be related, not only to preceding events of the series, but also to concurrent and preceding events of a series of related phenomena. Work in stochastic hydrology has related primarily to annual and monthly streamflows, but the results often apply to other hydrologic quantities such as precipitation and temperatures. A computer program, HEC-4 Monthly Streamflow Simulation, number 723-X6-L2340, that can be used for this purpose is available from The Hydrologic Engineering Center, Corps of Engineers, Davis, California.

Data Fill-In. Ordinarily, periods of recorded data at different locations do not cover the same time span, and therefore, it is necessary to estimate missing values in order to obtain a complete set of data for analysis. In estimating the missing values, it is important to preserve all statistical characteristics of the data, including frequency and correlation characteristics. To preserve these characteristics, it is necessary to estimate each individual value on the basis of multiple correlation with the preceding value at that location and with the concurrent or preceding values in all other locations. There are many mathematical problems involved in this process, and the details involved are discussed in the computer program description for HEC-4, 1971.

Reliability. While the simulation of stochastic processes may be able to add some dependability in hydrologic design, the techniques have not yet developed to the stage that they are completely dependable. All mathematical models involve some simplification of the physical phenomena represented. In most applications, simplifying assumptions do not cause serious discrepancies. It is important at this state of the art, however, to examine carefully the results of hydrologic simulation to assure that they are reasonable in each case.

Flow-Duration Curve

After monthly flow estimates have been completed, these can be analyzed to find critically low flow periods where several months or perhaps several years of daily flow data should be estimated. These data will be used to make more precise evaluations of electrical generation during average years and critical drought periods. If daily flow data are available, or can be developed with a reasonable degree of reliability, this should be done in order to compute a flow-duration curve for the complete period of record.

A duration curve of the observed, or estimated, flow characteristics at the site should be based on daily data. Adjustments for errors in estimates based on monthly curves can be made but results would likely be less reliable than those obtained from daily data. A duration curve is developed by grouping all the daily flow values into groups or classes within set ranges of discharge. Enough classes should be specified to reasonably define the curve (usually 10 to 30 classes). Starting at the highest discharge class, the number of days when the lowest range limit was exceeded is accumulated for successive classes and expressed as a percent of the total number of recorded days. An example of this procedure is illustrated in Table 2-1. A curve is then plotted with the lower range limit of each class as the ordinate and the precent of total events as the abcissa as shown in Figure 2-2.

Flow-duration curves developed from monthly data generally become increasingly less reliable if power storage is relatively small or nonexistent. Average discharge estimates made from flow-duration curves developed from monthly data will overestimate the average flow through a given turbine capacity by as much as 15 to 50 percent, depending on the day-to-day variability of flow. Figure 2-2 illustrates this possible source of error. Use of flow-duration curve will be discussed in Section 3.

Evaporation Data

Loss of water by evaporation can be a significant quantity in the arid western United States if there is a large surface area associated with the project storage. Generally this refinement is ignored at the reconnaissance stage of investigation. Gross evaporation for the reservoir area may be obtained from "Class A" pan records in the locality. These data are published by the Environmental Data Services of NOAA by States each month. Evaporation data obtained from Class A pans are too high and a coefficient averaging about 0.70 is commonly used to reduce them to equivalent evaporation values from a reservoir surface. Estimates can also be made by theoretical formulae but the availability of wind velocities and vapor pressure data required for the formula are less likely to be available than evaporation data. A good source of evaporation data or estimates is Federal, State, municipal, and private water agencies which collect these data at their existing projects.

Often the same monthly evaporation is used for each year of analysis, but if added refinement appears warranted, a greater evaporation rate can be used during drought years. Estimates of *net*evaporation at about 130 locations throughout the United States are contained in Exhibit I taken from EM 1110-2-1701 (U.S. Army Corps of Engineers, 1952). Average annual values in the sited reference range from 96 inches at Yuma, Arizona, to a minus 20 inches at Mobile, Alabama.

If energy calculations are based on flow data representing observed reservoir releases, canal flow or similar type data, no adjustment need be made to lake evaporation since it is already imbedded in the data.

Losses and Efficiencies

Losses. There are several reasons why all of the energy of flowing or stored water cannot be converted to useable electrical energy. Besides evaporation losses, there are seepage losses to groundwater, through the dam, and around gate seats, leakage losses through idle turbines, station use for sanitary and drinking purposes, cooling water use for generator bearings, and water use by navigation locks and fish ladders.

For existing structures, many of the possible sources of loss can be evaluated by observation or measurements. Large earth dams may exhibit losses as great as 5 to 10 cfs. Leakage losses through power plants vary, depending on the number, type, and size of turbine units and percent of time not operating. Estimates can be obtained from similar operating plants or from turbine manufacturers.

Efficiencies. Efficiencies of generators are dependent on design peculiarities but generally they can be expected to average about 97 percent within the operating range of the connected turbine. Turbine efficiencies depend on blade angle and design as well as draft tube design and placement. Best efficiencies generally occur at about 0.8 gate opening, at design head. Turbine efficiencies drop off as the net head falls below the rated head. Eighty-nine percent is frequently assumed for an average turbine efficiency in preliminary studies. If a speed increasing gear set is used to increase the rotational speed of the generator over that of the turbine, another 2 percent in efficiency is usually lost. The various turbine designs and efficiency characteristics are discussed in Volume V "Electromechanical Equipment".

TABLE 2-1
FLOW DURATION CURVE COMPUTATION

07144200 LITTLE ARKANSAS RIVER AT VALLEY CENTER, KS

LOCATION.--Lat 37°49'56", long 97°23'16", river gage is in NE¼NW¼SW¼ sec.36, T.25 S., R.1 W., Sedgwick County, Hydrologic Unit 11030012, at downstream side of highway bridge, 0.5 mi west of Valley Center, and 17.5 mi upstream from mouth. Little Arkansas River Floodway gage is in NE¼NE¼NE¼ sec.34, T.25 S., R.1 W., at downstream side of highway bridge, 1.2 mi northwest of river gage.

DRAINAGE AREA.--1,327 mi^2, of which about 77 mi^2 is probably noncontributing.

PERIOD OF RECORD.--June 1922 to September 1976.

REMARKS.--Natural flow of stream affected by diversions and ground-water withdrawal for irrigation and municipal supply. Since May 1957, part of high-water flow bypasses river gage through floodway channel for which separate records are computed; figures representing combined discharge are given herein.

AVERAGE DISCHARGE.--54 years, 273 ft^3/s, 197,800 acre-ft/yr.

Source: U.S. Geological Survey, Reston, VA.

DURATION TABLE OF DAILY VALUES FOR YEAR ENDING SEPTEMBER 30

DISCHARGE, IN CUBIC FEET PER SECOND
MEAN
L ARKANSAS R AT VALLEY CENTER, KS

CLASS	0	1	2	3	4	5	6	7	8	9	10	11	12	13	14	15	16	17	18	19	20	21	22	23	24	25	26	27	28	29	30	31	32	33	34
YEAR													NUMBER OF DAYS IN CLASS																						
1923										1	13	68	124	49	26	10	20	9	5	4	3	6	4	3	3	1	4	1	6	1	3	1			
1924												5	34	33	55	114	60	20	8	10	8	3	6	2	1	2		2	2	1					
1925											79	54	105	103	12	6	4	1		1															
1926										49	95	156	32	7	3	6	5	6	1	3	1	1													
1927													28	151	49	26	25	17	8	6	7	5	10	10	7	3	5	3		4	1				
1928													9	62	156	28	32	20	11	7	8	10	5	3	6	6	1	1	1						
1929													21	67	118	46	33	20	10	7	7	5	4	5	3	6	2	4	4	2	1				
1930											2	24	45	102	100	49	17	4	6	3	5	2	1		1	3	1								
1931									3	29	36	45	175	60	8	5		1	1	1	1														
1932										6	38	41	51	117	38	25	12	8	4	3	4	5	6	2	4	1	1								
1933								27	40	21	4	98	111	16	6	9	6	3	4	3	1	6	1	1	5	3									
1934								2	24	99	65	122	27	10	4	3	5	3	1																
1935									7	6	64	153	45	18	5	5	11	9	6	7	3	4	2	2	1	4	3	6	2	2					
1936								18	21	41	31	171	54	14	5	3	3	1	2				2												
1937											66	105	57	39	19	16	8	10	4	5	7	7	6	4	6	4	2								
1938											2	169	38	26	29	19	14	10	4	8	6	6	3	11	6	6	4	2	1		1				
1939												16	152	96	26	15	15	13	5	4	5	1	3	3	3	3	3	2							
1940											12	131	96	38	21	17	16	14	3	2	4	3	2	2	3	1	1								
1941											14	37	99	100	26	23	17	8	9	4	7	3	1	4	3	2	2	3	1	1	1				
1962															62	77	69	35	27	16	9	15	7	13	9	7	7	10	2						
1943													24	24	63	76	76	26	14	17	11	10	7	4	5		4	1	3						
1944													30	80	59	34	29	31	18	10	7	11	8	5	10	2	8	4	6	4	7	2		1	
1945														26	16	63	96	52	22	15	11	11	12	10	7	6	6	2	2	2	1	3	1		1
1946												41	41	39	99	73	34	13	5	9	4	1		2	2		1		1						
1947												24	72	78	47	26	27	17	12	14	8	4	8	1	6	10	3	5	1	1		1			
1948												16	61	61	46	36	28	14	12	18	9	10	9	7	7	3	3	9	8	5	4				
1949														7	65	61	54	31	21	18	15	19	15	10	9	11	10	17	2						
1950													9	179	67	21	15	10	2	6	7	8	6	9	3	9	5	3		4	2				
1951													4	18	137	23	14	13	21	17	16	15	14	12	10	11	6	7	7	8	3	4	4	1	
1952												6	54	39	9	23	109	47	27	17	11	6	9	1	3	3	2								
1953										6	29	47	116	113	20	11	10	2	4	4	1	2													
1954									28	58	42	132	57	19	9	5	7	2	1	1			3		1										
1955						10	30	31	8	99	98	41	21	3	6	3	5	1	1	1	1	1	1			1	1	2							
1956				1	30	23	10	18	26	42	120	57	12	4	5	3	3	3	1	1	1		1	2		1	2								
1957		3		5	7	6	3	35	73	38	1	1	38	20	8	20	17	14	10	10	13	6	6	5	7	2	1	2	2	6	4	1		1	
1958												10	123	28	11	19	55	28	13	17	8	7	5	3	4	8	9	8	6	2		1			
1959											6	34	12	77	110	42	29	16	14	4	3	4	4	1	3	2	1	3							
1960													1	64	48	43	39	47	23	17	13	15	8	6	11	9	4	5	3	6	2	2			
1961														12	110	85	48	30	20	15	10	10	3	4	4	6	1	1	2	2	1	1			
1962													6	25	56	89	61	32	21	18	10	3	8	7	8	3	6	1	4	1	2	2		2	
1963										2	9	26	48	68	130	33	16	7	4	4	6	3	1	5	1	1	1								
1964										8	10	39	138	72	25	17	14	10	8	5	4	4	2	2	2	5	1								
1965											23	12	10	107	52	29	28	15	8	10	16	9	4	5	4	7	5	6	8	4	1	1	1		
1966											29	30	40	66	122	36	17	8	3	6	2	2	1		2	1									
1967										14	84	121	14	12	11	11	8	10	9	9	10	10	6	4	10	3	9	5	2	3					
1968										4	17	25	51	107	76	28	15	13	7	4	3	2	1	1	6	2		2	1	1					
1969											4	4	23	30	76	41	46	26	23	14	14	12	6	9	10	6	7	4	5	3	1	1			
1970										7	27	18	30	133	49	27	25	13	5	4	5	2	4	1	5	1	5	2	1	1					

SUMMARY FOR 1923-1970

CLASS	VALUE	TOTAL	ACCUM	PERCT
0	0.00	0	19724	100.0
1	1.00	3	19724	100.0
2	1.50	0	19721	100.0
3	2.00	6	19721	100.0
4	2.70	37	19715	100.0
5	3.70	39	19678	99.8
6	5.00	43	19639	99.6
7	6.70	131	19596	99.4
8	9.10	230	19465	98.7
9	12.00	530	19235	97.5
10	17.00	1037	18705	94.8
11	22.00	2178	17668	89.6
12	30.0	2552	15490	78.5
13	41.0	2986	12938	65.6
14	56.0	2582	9952	50.5
15	75.0	1710	7370	37.4
16	100.0	1537	5660	28.7
17	140.0	913	4123	20.9
18	190.0	556	3210	16.3
19	250.0	458	2654	13.5
20	340.0	372	2196	11.1
21	460.0	319	1824	9.2
22	620.0	258	1505	7.6
23	840.0	214	1247	6.3
24	1100	240	1033	5.2
25	1500	197	793	4.0
26	2100	170	596	3.0
27	2800	156	426	2.1
28	3800	98	270	1.3
29	5100	78	172	.8
30	6900	46	94	.4
31	9400	30	48	.2
32	13000	11	18	
33	17000	5	7	
34	23000	2	2	

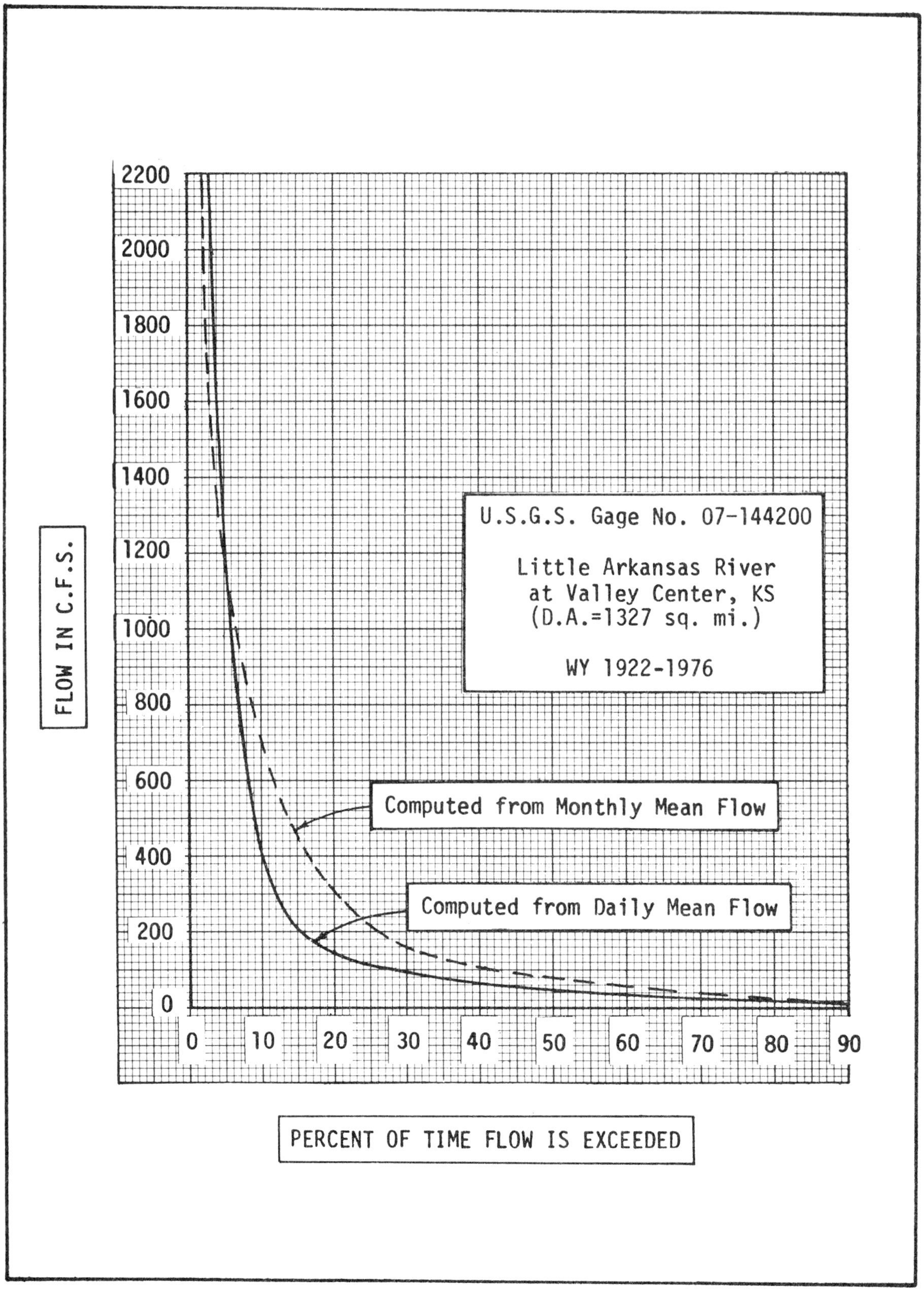

Figure 2-2. Flow duration curves.

SECTION 3
CAPACITY AND ENERGY CALCULATIONS

Energy-Flow-Head Relationship

The fundamental procedure for generating electrical energy from flowing water between different elevations is to convert kinetic energy to electrical energy by means of a prime mover (turbine, et al.) connected to a generator which is in-turn connected to an electrical load. The energy, in foot pounds, is measured by the weight of the water in pounds (equal to 62.4 lbs/ft^3) times the quantity of water (Q) in cubic feet (ft^3) multiplied by the elevation difference (head) in feet (H) through which the water drops. Mechanical power is the rate of this energy transformation or work done in a specified time. The usual unit of power is horsepower (550 foot-pounds per second) and is equal to 62.4 times Q in c.f.s. times H divided by 550.

$$\text{Mechanical power (hp)} = (62.4 \times Q \times H) / 550 = (Q \times H) / 8.81 \qquad (3\text{-}1)$$

This is the theoretical power at 100 percent efficiency. The actual power developed on the turbine shaft is adjusted by multiplying by the turbine efficiency (E_t). The kilowatt output of the generator is determined by multiplying by the conversion factor from horsepower to kilowatts (.746 hp/kW) and by the generator efficiency E_g thus

$$\text{Electrical power (kW)} = (Q \times H \times E_t \times E_g) / 11.8$$

A major effort of the hydrologic investigations deals with estimating the long term values and sequential variability of the flow and developing operational criteria which will lead to a determination of the corresponding change in head (H). Existing project purposes must generally be met while providing the additional hydro—power benefits.

Reconnaissance Sizing Procedures

Reconnaissance Estimates. Simplified methods using estimates for the variables in the power equation presented are typically used to make estimates of capacity and energy at potential power sites in order to determine the desirability of expending more time and funds to refine these preliminary estimates. Also, these approximate methods are used to "screen" large numbers of potential sites to a more select group of most likely candidates for development. Screening based on factors other than capacity and energy is also a necessary study step, but this section is limited to capacity and energy aspects.

Duration Curve Analysis. A duration curve of the observed, or estimated, flow characteristics at the site should be based on daily data. A typical curve for a stream with low base flow is shown in Figure 3-1. The area under the curve represents the average flow. The average daily observed runoff at this site for the period June 1922 to September 1976 was 273 cfs. If a run-of-river site evaluation were to be made for a dam with an estimated net power head of 30 feet at an assumed plant efficiency of 86 percent we could use the power formula to estimate the site capability:

$$\begin{aligned}\text{Site capability} &= (Q \times H \times E) / 11.8 \\ &= (273 \times 30 \times .86) / 11.8 \\ &= 597 \text{ kW}\end{aligned}$$

If the plant could generate continuously at this rate it would produce 5.2×10^6 kWh of energy in a years time. However, it is apparent from inspection of Figure 3-1 that a flow rate of 273 cfs is available about 13 percent of the time and with no storage available to capture water during these periods of above average flow, 87 percent of the time the generator would be operating at less than name plate capacity.

Assume that regional studies have developed a guidance rule that turbines should be designed for a flow that will be exceeded at least 15 percent of the time. From the flow-duration curve, a flow of 200 cfs is shown to be exceeded 15 percent of the time. This would establish a preliminary turbine-generator selection of

$$(200 \times 30 \times .86) / 11.8 = 437 \text{ kW}.$$

The allowable operating range of the turbine is determined by the type of turbine and its characteristics as discussed in Volume V. If the selected turbine can only be operated within a flow range of 30 to 110 percent of the design flow, the lower limit of operation would be about 60 cfs (.30 × 200). The flow duration curve indicates the flow of the river is less than 60 cfs about 58 percent of the time. Also, it is likely that at extremely high flows the tailwater will rise so high that the net power head will become too small for the powerplant to function. If this should occur when discharges exceed 3,000 cfs, an additional two percent of the time or about seven days a year on the average would be unsuitable for power production. Therefore, about 60 percent (58 + 2) of the time the plant would be inoperable, unless there is available storage to regulate flows to more favorable discharge rates. The energy potential from the site would now be restricted to the area shown cross hatched on the flow-duration curve (Figure 3-1). The cross-hatched area under the curve is equivalent to 54.5 cfs flowing 100 percent of the time. Converting this flow

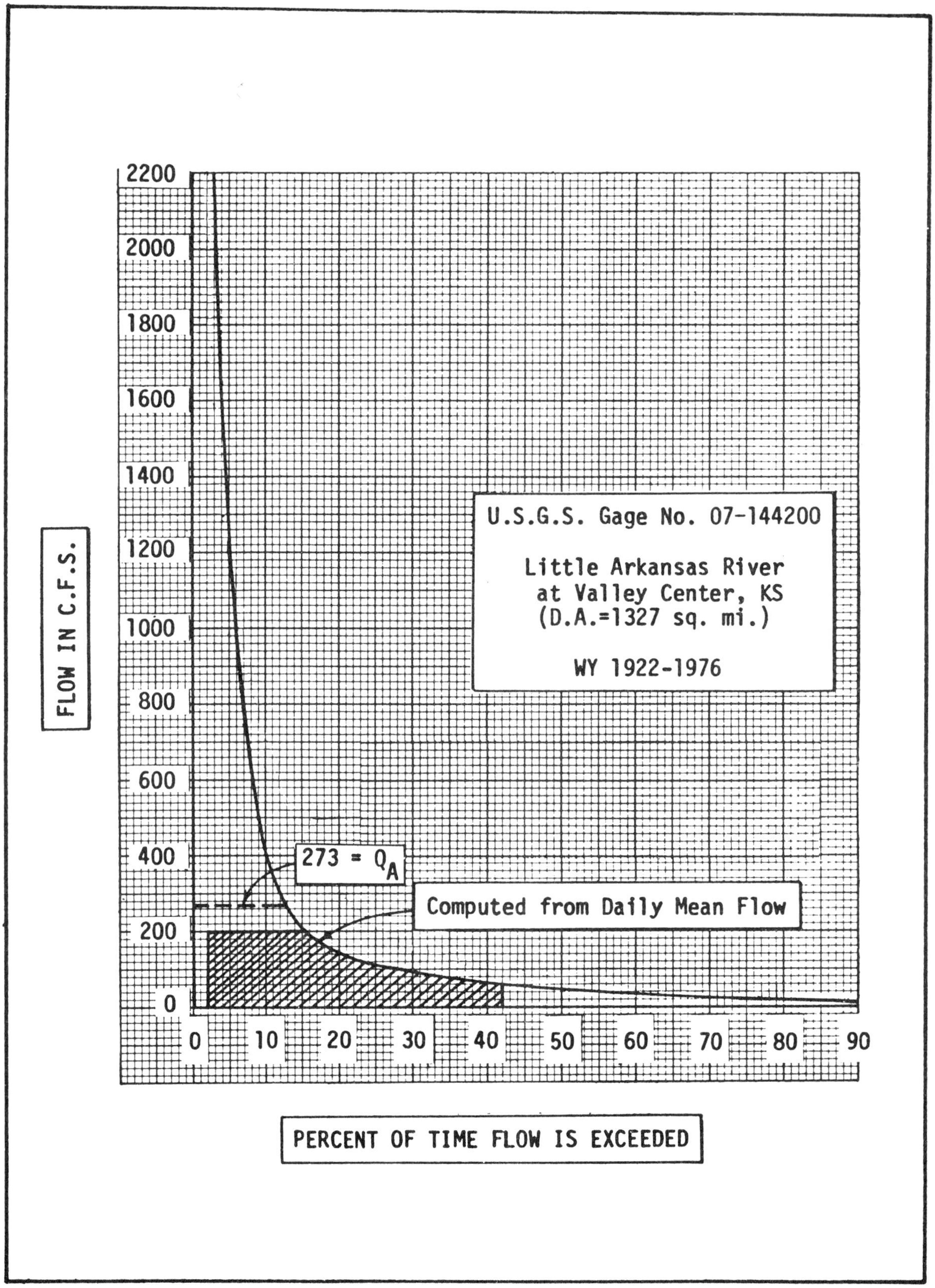

Figure. 3-1. Flow duration curve.

to average annual energy we get

Average annual energy
$= (54.5 \times 30 \times .86 \times 8760) / 11.8$
$= 1.044 \times 10^6$ kWh.

An average annual capacity factor or plant factor of 27 percent ($1.044 \times 10^6 / (437 \times 8760)$). Installation of two units of 218 kW each would allow generation until the flow fell below 30 cfs and would result in approximately 200,000 kWh per year more energy and 23 percent more time when at least one unit of the plant could operate. However, the value of this additional energy may not justify the added expense of 2 units, instead of one unit twice the size.

A similar procedure could be used to work through reconnaissance estimates of several assumed plant sizes. With appropriate cost and energy value curves, rough economic analysis could be completed. If some pondage (storage) were available to store low flow and release it during a shorter period each day, electrical energy could be generated from the stored flow. For example, a continuous flow of 60 cfs accumulates to about 85 acre feet in 17 hours time. So, with that amount of pondage, water could be stored for 17 hours and used to generate at capacity during the other 7 hours each day. As the inflow dropped to 30 cfs storage would be required for a longer period of time or generation would be at less than nameplate capacity, or some combination of the two. There could be water quality, environmental, recreational, and other reasons why a store-release pattern of flow would be undesirable. If greater amounts of storage were available in this hypothetical problem, surplus flow could be stored during times when flow exceeds 200 cfs and released during periods of flow deficiencies, depending on water rights, project purposes, and other operating constraints.

The above analysis is based on a run-of-river situation where net power head is likely to be nearly constant. If existing project purposes are such that this is not true, a reconnaissance estimate would use an estimate of average net power head. If the project were evaluated to be economically favorable at this point, more detailed energy evaluations would be conducted using a sequential monthly or daily analysis.

Sequential Period of Record Routing

The most reliable estimates of energy yield from a given set of inflow and storage data can be obtained from sequential analysis. The time interval chosen for sequential analysis should be consistent with the accuracy desired. In the case of power estimates during feasibility studies the maximum time interval used should not exceed one month. Feasibility estimates of firm energy should be based on daily or weekly time intervals during critical periods using all available information on project purposes, diversions, seasonal storage levels, losses, tailwater rating, and plant efficiency data. If "dependable capacity" is not a consideration, a monthly analysis for the entire period of record will usually suffice.

Importance of Load Pattern. If dependable capacity is a serious consideration, the seasonal load pattern is an important variable in determining firm power and firm energy estimates. This is true because the project must be capable of delivering its credited firm power during the most critical drought period and coincident load pattern. The importance of whether the load pattern (curve) is synchronized with the seasonal flow pattern can be seen in Figure 3-2. This example is taken from a water supply demand but is illustrative of the increased storage or decreased yield which comes from flow versus demand patterns. A project that has either the water demand or energy demand schedule "out-of-sync" with the inflow pattern will require a greater amount of storage from which to draw the needed demand. Generally, increasing storage is not an alternative in small hydropower additions. If existing project purposes require release patterns which are near enough to the energy demand, or useable on the load, some dependable capacity can be credited to the project.

Typically load patterns fluctuate throughout the day and are lower on Saturdays and Sundays. Figure 3-3 shows an hourly load curve for a typical week of a large electric utility system. The peak demands on a system vary from week to week and from month to month throughout the year. The system related to Figure 3-3 has its highest demand in August and its average annual demand is about 60 percent of its annual peak (annual load factor = 60%) and monthly load factors range from 65 to 75 percent.

Figure 3-3 shows the role played by hydroelectric energy sources in meeting peak power demands each day. Run-of-river plants could be used to assist in meeting base load requirements. It is apparent that if a hydroplant only generates during the hours of high demand each day, reservoir storage (or pondage) must be available to store water during the remainder of the day or water will pass thru the project without producing power. Energy generated to meet peak demands has greater value as it would replace more expensive fuel consuming sources as discussed in Volume II of this guide manual. However, when used to replace non-renewable energy sources, hydropower has considerable value, regardless of its position in the load curve.

Seasonality of Storage Allocation. Multipurpose projects usually allocate the total available storage to the various purposes proportional to some cost and benefit relationship or to achieve prescribed objectives. Often these objectives have conflicting demands on storage, such as when flood control storage must be evacuated as soon as possible after an occurrence of surplus inflow, whereas a power purpose would prefer to hold it until it could be evacuated through the turbines. If the season when major floods occur is a different season than when the highest demand for energy occurs, some of the flood control space can be seasonally assigned to power and

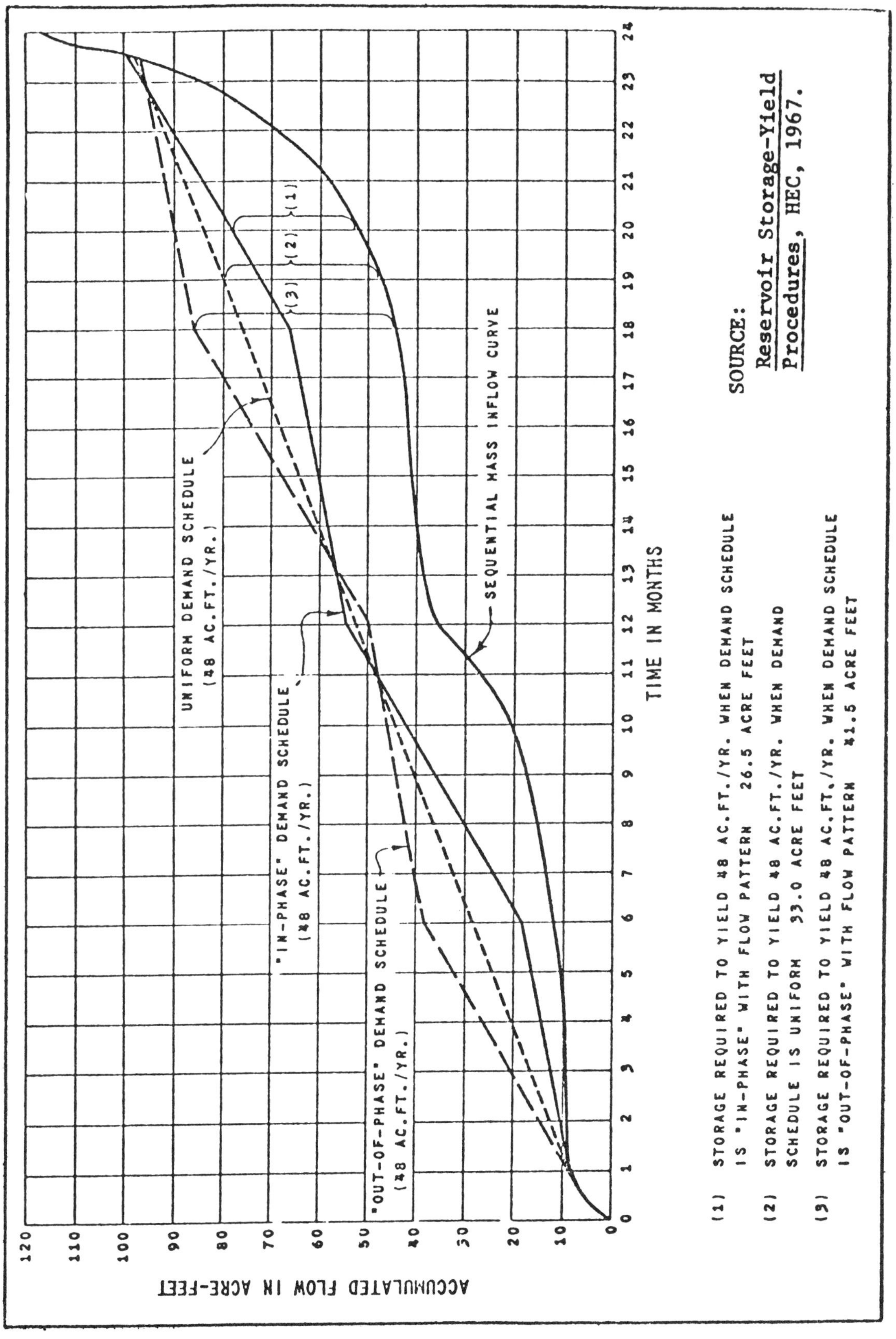

Figure 3-2. Effect of demand sequence on storage requirement.

thus obtain a multiple-use of common storage space. The depth of such studies is generally beyond the scope of the small hydropower investigation.

Head Limitations. Each turbine type and design has its own efficiency characteristics as discussed and illustrated in detail in Volume V. Even reconnaissance estimates of power potential at a site should account for efficiency characteristics and an operating range limitation consistent with the turbine type likely to be installed. Operating head ranges of 60 to 120 percent of the design or rated head are typical of the limitations which must be kept in mind when determining the amount of active storage which can be used for energy generation. When performing sequential routings during feasibility studies it is common practice to incorporate the efficiency characteristics of the turbine-generator system into the computations rather than using a uniform efficiency at all head values.

Computational Aids. It is almost a practice of the past to do sequential routing by hand computations and "spread-sheet" accounting, but there are several computational aids that provide valuable tools for checking computer output and assisting in making better estimates than can be otherwise made. These include curves or tables of storage-elevation-area, tailwater rating, storage-efficiency curves and storage-evaporation-month of year tables. A typical format of an elevation-area-storage table is illustrated in Table 3-1 and several formats for hydropower sequential analysis are shown in Table 3-2. Several of these can be combined to develop the kW/cfs nomograph shown in Table 3-3 and Figure 3-4 which is almost a necessity for sequential routing by desk top calculator.

Computer Programs

With the increasing availability of computer service firms and reasonably priced but powerful mini com-

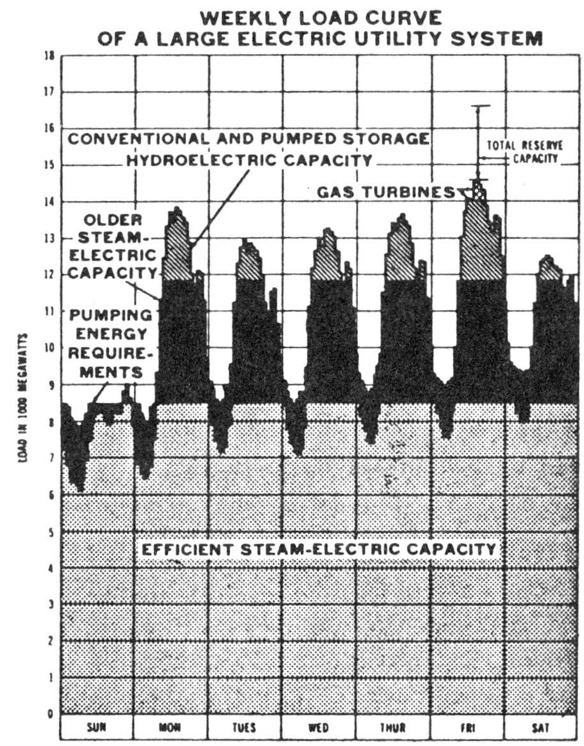

Source: FCC P-35 Hydroelectric Power Evaluations, Federal Power Commission (FERC) 1968

Figure 3-3. Weekly load curve of a large electric utility system.

TABLE 3-1.
EXAMPLE OF ELEVATION-AREA-STORAGE TABLE

ROLLINS RESERVOIR NR COLFAX, CA (11421800) AREA AND CAPACITY TABLE (CONIC)

ELEV FEET	CAP AREA .0	CAP AREA .1	CAP AREA .2	CAP AREA .3	CAP AREA .4	CAP AREA .5	CAP AREA .6	CAP AREA .7	CAP AREA .8	CAP AREA .9
2060.0	11027 242	11051 242	11075 243	11100 243	11124 243	11148 244	11173 244	11197 245	11222 245	11246 245
2061.0	11271 246	11295 246	11320 246	11345 247	11369 247	11394 248	11419 248	11444 248	11468 249	11493 249
2062.0	11518 249	11543 250	11568 250	11593 250	11618 251	11643 251	11669 252	11694 252	11719 252	11744 253
2063.0	11769 253	11795 253	11820 254	11846 254	11871 255	11897 255	11922 255	11948 256	11973 256	11999 256
2064.0	12024 257	12050 257	12076 258	12102 258	12128 258	12153 259	12179 259	12205 260	12231 260	12257 260
2065.0	12283 261	12309 261	12335 261	12362 262	12388 262	12414 263	12440 263	12467 263	12493 264	12519 264
2066.0	12546 264	12572 265	12599 265	12625 266	12652 266	12678 266	12705 267	12732 267	12759 268	12785 268
2067.0	12812 268	12839 269	12866 269	12893 269	12920 270	12947 270	12974 271	13001 271	13028 271	13055 272
2068.0	13082 272	13110 273	13137 273	13164 273	13192 274	13219 274	13246 275	13274 275	13301 275	13329 276
2069.0	13357 276	13384 276	13412 277	13440 277	13467 278	13495 278	13523 278	13551 279	13579 279	13607 280

CAP = STORAGE CAPACITY IN ACRE-FEET

TABLE 3-2.
TYPICAL FORMAT FOR HYDROPOWER SEQUENTIAL ANALYSIS

PROJECT NAME ____________ Computed by ____________

Date ____________

FIXED DATA:

Installed Capacity ____________ Representative Gage ____________

Overload Factor ____________ Data Sources ____________

Effeciency ____________ Spillway (Elev/Stor) ____________

Penstock Loss ____________ Top of Power Pool (Elev/Stor) ____________

Tailwater Elev. ____________ Bottom of Power Pool (Elev/Stor) ____________

Yr.	Mo.	Inflow (cfs)	End of Mo. Storage (Ac. Ft.)	End of Mo. Elevation (Ft.)	Net Power Head. (Ft.)	Released Water Irr. (cfs)	W.Q. (cfs)	Fish (cfs)	Spill (cfs)	Flow Through Turbine	Unit Power (KW/ cfs)	Power (KW)	Energy (KWh)	Energy Potential Wasted (KWh)
														(Spill, Fish ladder, leakage, etc)
		*	*			*	*	*						*

On a simple analysis, columns marked (*) may be unnecessary.

A more complex accounting of variables might require adding the following column headings:

Area (Ac.)	Evap. (Ac. Ft.)	Pool Index Level	Tail-Water Elev. (Ft.)	Plant Effic. (%)	Diversion From Pool (cfs)	Total Flow Past Project (cfs)	Required Power (KW)	Power Shortage (KW)	Release Case (Index No.)	Remarks

Legend:

Irr. = Irrigation requirement

W.Q. = Water quality

Fish = Fish ladder requirements

Eff. = Efficiency of generator and turbine

puters, it is almost easier to make a monthly sequential analysis than to plot a duration curve and make reconnaissance estimates. The results are more accurate and costs are comparable. Basic input requirements of well documented computer programs can be expanded and upgraded to the level of precision required in later feasibility estimates.

Utility computer programs, which can develop detailed tabular data of elevation-storage-area relations and tailwater and spillway rating curves, are readily available from State and Federal water resources agencies at minimal handling charges. One such source available to both public and private sectors is the U.S. Army Corps of Engineers, Hydrologic Engineering Center. Abstracts of several such applicable programs are contained in Exhibit II. A comparison of several computer models developed by the Corps of Engineers is contained in Table 3-4. An example of user specified output format using HEC-5C for a run-of-river project, where outflow is dependent on criteria other than power demand, is illustrated in Table 3-5.

TABLE 3-3
SAMPLE KW/CFS NOMOGRAPH COMPUTATION PROCEDURE

Pool Elevation (ft, m.s.l.)	Storage(1) (wsf/1000)	Net Head (2) (ft)	Efficiency(3) (%)	kW/cfs(4)
1131	145.2	203.5	83.2	14.34
1128	136.0	199.5	84.0	14.19
1124	127.3	195.5	84.6	14.01
1120	119.0	191.5	85.1	13.80
1116	111.0	187.5	85.5	13.58
1112	103.5	183.5	85.9	13.35
1108	96.3	179.5	86.1	13.09
1104	89.5	175.5	86.3	12.83
1100	83.0	171.5	86.1	12.51
1096	76.9	167.5	85.9	12.19

Based on constant average tailwater at elevation 927.8 ft, m.s.l. with assumed constant penstock losses of 0.7 ft.

(1) The use of storage in week-second-feet (wsf) for this example is based upon the selection of a week as the routing interval and week-second-feet as the flow units.

(2) Net head = pool elevation - penstock losses - average tailwater (Both penstock loss and average tailwater may be varied with pool elevation if relationship known).

(3) Overall station efficiency (may be assumed constant at all pool elevations).

(4) kW/cfs = Head × Eff × .08474

(Source: U.S. Army Corps of Engineers, HEC, 1967, **Methods Systemization Manual, Reservoir Storage-Yield Procedures**.)

TABLE 3-4
COMPARISONS OF HYDROLOGIC MODEL CAPABILITIES IN HYDROPOWER STUDIES

		HEC-5C	HEC-3	SWD SUPER	HYSSR	NPD HYSIS(1)	HLDPA(2)
a.	Routing Intervals	Any	Monthly	Daily	Monthly or 2 weeks	1-4	Hourly
b.	Routing Methods	6	No	Puls	No	SSARR	SSAAR
c.	System Power Operation	Yes	Yes	Yes	Yes	Yes	Yes (also thermal)
d.	Yield Maximization	Yes	Yes	No	No	No	No
e.	Peaking Capability	Yes	Yes	Yes	Yes	Yes	Yes
f.	Evaporation	Yes	Yes	Yes	Yes	SSARR	SSARR
g.	Power Benefits	Yes	Yes	Yes	No	No	No
h.	Flood Control	Yes	No	Yes	No	Limited	No
i.	Pumped Storage	Yes	No	No	No	No	Yes

(1) Basically a model used in operation
(2) Model used primarily for planning
SWD Southwestern Division, Corps of Engineers
NPD North Pacific Division, Corps of Engineers
SSARR The NPD Stream Simulation and Reservoir Routing Model (storage routing and loss procedures)

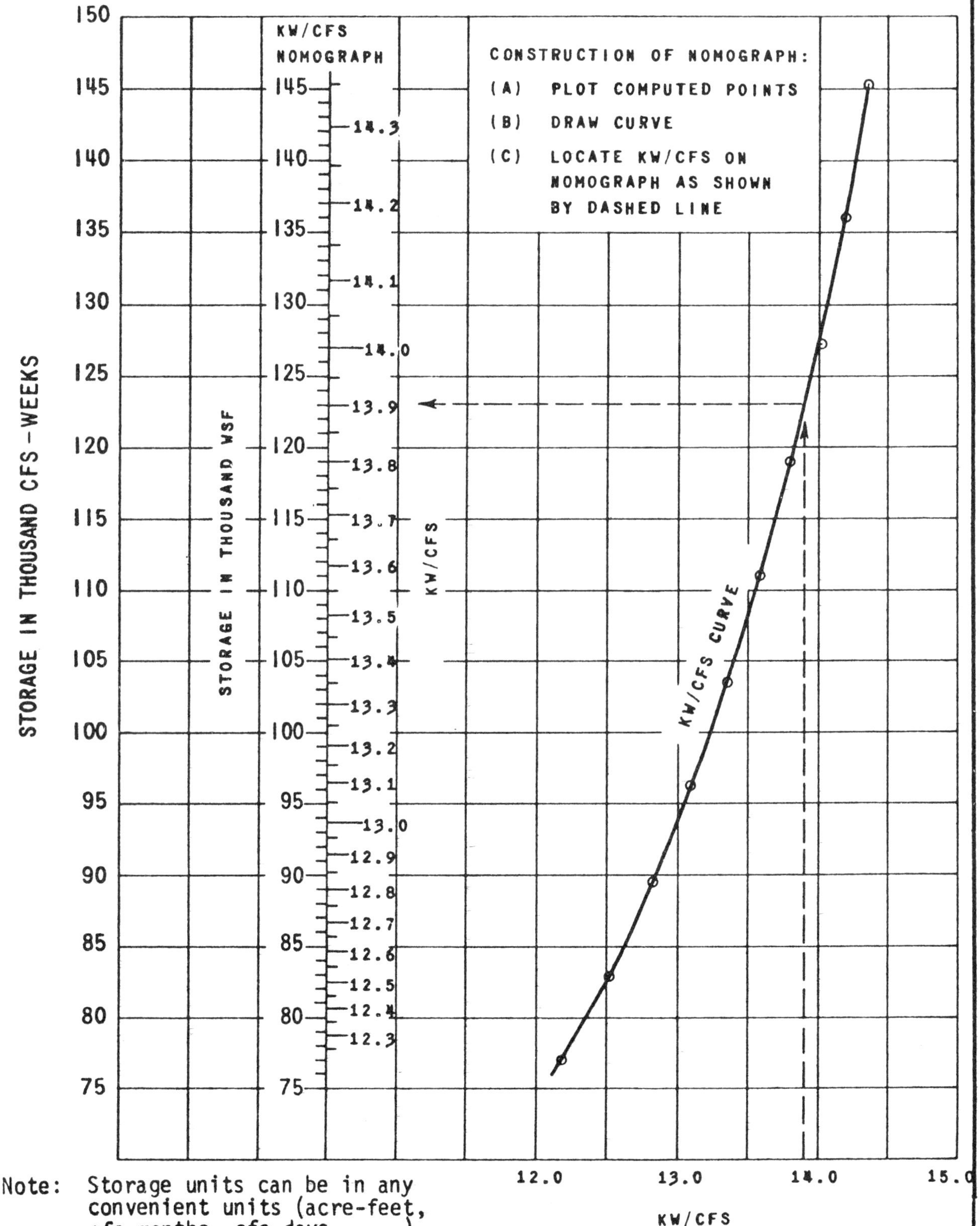

Note: Storage units can be in any convenient units (acre-feet, cfs-months, cfs-days . . .) depending on flow units and time interval.

Source: Reservoir Storage-Yield Procedures, HEC, 1967.

Figure 3-4. KW/CFS nomograph.

TABLE 3-5.
EXAMPLE OF USER SPECIFIED FORMAT USING HEC-5C

ROLLINS RES. BEAR RIVER CA   EXISTING PROJECT   APR 1979
SEQUENTIAL ROUTING MONTHLY   TAILWATER=1958.3   INST CAP=13000KW
USING GENERATOR-TURBINE DATA FROM TUDOR ENGR INC.   SP. CREST=2171.

PER	DY	MO	YR	DW	ROLLINS INFLOW	ROLLINS OUTFLOW	ROLLINS EOP STOR	ROLLINS LEVEL	ROLLINS EL/STAGE	ROLLINS ENERGY G	ROLLINS PEAK CAP
1	0	10	64	1	253.20	253.20	66000.00	3.00	2171.00	2950.52	13433.77
2	0	11	64	1	371.50	371.50	66000.00	3.00	2171.00	4189.42	13433.77
3	0	12	64	1	1079.70	1161.00	61000.98	2.92	2167.90	9646.08	12965.16
4	0	1	65	1	1803.90	1825.00	59703.58	2.89	2163.94	9354.12	12572.75
5	0	2	65	1	836.50	1292.00	34406.02	2.47	2145.00	7335.05	10915.26
6	0	3	65	1	767.50	356.00	59708.54	2.89	2145.01	3767.06	10915.57
7	0	4	65	1	1114.00	993.00	66908.64	3.12	2167.66	9317.69	12941.24
8	0	5	65	1	689.00	698.00	66355.24	3.05	2171.75	8144.67	13557.24
9	0	6	65	1	644.30	651.00	65956.56	3.00	2171.19	7343.64	13464.23
10	0	7	65	1	600.00	600.00	65956.56	3.00	2170.95	6990.71	13425.26
11	0	8	65	1	88.00	307.60	52453.69	2.77	2162.43	3499.93	12429.76
12	0	9	65	1	74.50	282.90	40052.84	2.56	2143.76	2877.55	10815.35
13	0	10	65	1	75.40	216.90	31352.22	2.41	2125.59	2056.43	9244.86
14	0	11	65	1	602.60	400.90	43354.38	2.62	2128.72	3747.26	9555.77
15	0	12	65	1	855.70	481.60	66357.23	3.05	2156.48	5370.32	11887.63
16	0	1	66	1	816.00	816.00	66357.23	3.05	2171.43	9516.16	13503.57
17	0	2	66	1	536.80	544.00	65957.35	3.00	2171.19	5727.68	13464.50
18	0	3	66	1	804.10	796.00	66455.41	3.06	2171.25	9279.69	13474.09
19	0	4	66	1	970.00	970.00	66455.41	3.06	2171.54	9736.38	13522.75
20	0	5	66	1	574.10	592.00	65354.77	2.99	2170.88	6896.28	13415.22
21	0	6	66	1	269.10	354.80	60255.19	2.90	2167.04	3956.38	12878.47
22	0	7	66	1	336.00	336.00	60255.19	2.90	2163.81	3836.89	12560.64
23	0	8	66	1	470.50	460.70	60857.78	2.91	2164.21	5266.28	12598.20
24	0	9	66	1	377.80	443.30	56960.20	2.85	2162.05	4876.22	12392.84
25	0	10	66	1	559.60	548.20	57661.17	2.86	2159.94	6195.71	12193.62
26	0	11	66	1	672.20	526.00	66360.80	3.05	2166.05	5848.70	12779.50

ANNUAL SUMMARY

YEAR	ROLLINS INFLOW AVG	ROLLINS OUTFLOW AVG	ROLLINS EOP STOR AVG	ROLLINS EOP STOR MAX	ROLLINS EOP STOR MIN	ROLLINS ENERGY G SUM
65	693.51	732.60	58708.55	66908.64	34406.02	75416.85
66	557.34	534.35	59164.36	66455.41	31352.22	70266.18
67	952.45	939.68	65587.58	67461.44	57661.17	107102.27
68	593.60	615.04	59030.10	66473.86	43972.04	79389.27
69	1061.10	1049.38	65549.70	66974.89	59681.16	101431.36
70	914.84	927.67	63076.32	66883.65	50789.33	97081.72
71	943.17	938.28	65398.70	67090.33	54790.25	110582.39
72	662.85	658.03	61944.92	66795.38	50494.58	86002.91
73	895.17	909.00	64870.94	66886.23	49089.27	104235.52
74	1151.38	1164.67	62902.29	67393.01	39597.83	109467.31
75	752.41	742.28	56770.26	66696.01	37495.52	68739.43
76	406.62	437.47	46595.30	66695.22	23092.53	52871.07
77	111.98	135.91	8939.48	13385.28	4959.57	7735.30
SUM =	9696.41	9784.35	738538.51	816099.55	537381.51	1090321.50
MAX =	1151.38	1164.67	65587.58	67461.44	59681.16	110582.39
MIN =	111.98	135.91	8939.48	13385.28	4959.57	7735.30
MPER=	74.00	74.00	67.00	67.00	69.00	71.00
AVG =	745.88	752.64	56810.65	62776.89	41337.04	83870.89

SECTION 4
SPILLWAY ADEQUACY

Introduction

The determination of spillway adequacy is an essential aspect of studies involving the addition of hydropower units to an existing structure. A spillway acts as a safety valve to protect a dam from being overtopped, an occurrence which, especially in the case of an earthfill or rockfill dam, can have catastrophic consequences. Only a cursory inspection and evaluation are warranted until a reconnaissance estimate of hydro-electric energy resources indicates favorable economics and further study. A long history of trouble-free operation is not a certain indication of a safe spillway design. Spillway adequacy is a function of spillway capacity, storage capacity of the reservoir, physical condition and reservoir operation procedures. If operation for hydropower requires that a reservoir be kept at higher levels than intended by the original design, a presently adequate spillway may become inadequate. Also, the addition of power facilities at the toe of a dam will, in some cases, require that a spillway be relocated so that it does not discharge in the vicinity of the powerhouse.

Two main topics are addressed in this Section. The first deals with hydraulic characteristics of outlet structures, and the second pertains to criteria for spillway adequacy and techniques for calculating the Spillway Design Flood.

The overall strategy for evaluating spillway adequacy is outlined below:

Reconnaissance

- Determine if a State or Federal safety report is available.
- Determine spillway crest type.
- Obtain physical dimensions, number of piers, and type of abutments; and relative crest elevation and top of dam elevation.
- Estimate discharge coefficients from experience or appropriate references.
- Determine evaluation criteria for size and hazard classification (Table 4-3).
- Estimate spillway design discharge from regional envelope curves, if data are available.
- Estimate maximum water surface and remaining freeboard or depth of probable overtopping.

Feasibility

- Same as reconnaissance but with greater accuracy.
- Determine spillway design flood hydrograph from probable maximum precipitation and watershed runoff model.
- Route hydrograph through reservoir surcharge storage and spillway.
- Determine freeboard adequacy or required modifications.
- Evaluate location adequacy of stilling basin relative to power plant site.

Hydraulic Characteristics of Spillways

A spillway is a hydraulic passageway designed to conduct flood flows safely past a dam. Some dams are designed with two spillways - a service spillway to discharge floods likely to occur fairly frequently, and an auxiliary or an emergency spillway to handle larger, infrequent flows. The latter type of spillway is frequently constructed with unpaved channels; hence, maintenance costs associated with erosion of the structure, and possibly with downstream deposition, may be incurred following periods of operation.

The configuration of a spillway is tailored to a particular dam site and is dependent on the type of dam and intended operation and on the economic tradeoff between spillway capacity and dam height. Types of spillway are overflow, chute, shaft, side-channel, and siphon.

Overflow. An overflow spillway is a portion of a dam designed for water to pass over. Many overflow spillways are designed with a shape that closely approximates the shape of the lower nappe of flow over a sharp-crested weir, because a profile of this shape produces near-maximum discharge efficiency at the design head. The curved shape of the nappe is found to be a function of the head on the weir, the slope of the front side of the weir, and the velocity of approach. Consequently these three characteristics affect the magnitude of discharge coefficients for various weir shapes and heads.

Chute. A chute spillway is basically an open channel designed to convey water from a control section to the downstream river channel. Chute spillways are commonly used with earthfill dams. Flow down steep chutes is rapid and unstable, and a chute must be carefully designed for safe and proper operation. Chutes may be constructed along the abutment of a dam, down the face of a dam, or down a saddle at some distance from the dam.

Shaft. A shaft spillway typically consists of a vertical shaft which makes a 90-degree bend into a horizontal tunnel that passes through or under a dam or abutment. This type of spillway is often used where space or site conditions preclude the use of other types of spillway. Because the inlet for a shaft spillway is often a funnel-shaped overflow crest, the name "morning glory" is commonly applied to a shaft spillway. Under low heads, the overflow crest will act as a control. However, as the head increases, control will shift to the spillway "throat", and finally full pipe flow will occur. Under full pipe flow conditions, discharge will vary in proportion to the square-root of the head, in which case there is little increase in capacity with increasing head. Vortex formation at the intake, surging in the vertical shaft, and erosion of concrete in the vicinity of the vertical elbow are problems associated with shaft spillways.

Side-Channel. A side-channel spillway is one in which water enters a channel by passing over an overflow crest that parallels the channel. Side channel spillways are sometimes used in narrow canyons where there is insufficient width for a suitable crest length for an overflow or chute spillway.

Siphon. A siphon spillway is sometimes used where it is desirable to develop full discharge capacity quickly, for example in the event of a turbine shut-down. Siphon spillways are capable of providing automatic regulation of reservoir levels within fairly narrow limits. However, the siphon spillway, like the shaft spillway, cannot handle flows much greater than the design flow, because under high flow conditions discharge is proportional to the square-root of the head. Inability to pass ice and debris, and potential for surging are also disadvantages of the siphon spillway.

Although from the operational viewpoint an uncontrolled spillway is often desirable, there are many situations where the advantages of having a gated spillway outweigh the disadvantages. Gates are used when sufficient crest length for an uncontrolled spillway cannot be developed or when sufficient head cannot be developed for the required discharge capacity. Gates are also used where it is necessary to initiate spillway releases when the reservoir is below the normal full pool elevation. Numerous gate types are in use, including rectangular lift gates, roller gates, radial gates, and drum gates. In some instances flashboards or stoplogs may be utilized. For information on the hydraulics of gated spillways, the reader is referred to *Hydraulic Design of Spillways* (U.S. Army Corps of Engineers, 1965), *Design of Small Dams* (USBR, 1977), and *Handbook of Applied Hydraulics* (Davis and Sorenson, 1969).

Discharge Over an Ogee Spillway Crest

Overflow spillways behave as weirs and, therefore, if the spillway is not submerged the flow will pass through critical depth over the crest. One of the most common crest shapes is the "ogee". The discharge over an uncontrolled ogee crest is given by the following equation:

$$Q = C \times L \times H_e^{3/2} \qquad (4\text{-}1)$$

where
Q = discharge
C = variable discharge coefficient
L = effective crest length
H_e = total head on the crest, including velocity of approach head, h_a

The discharge coefficient, C, is influenced by the following factors: (1) the depth of approach, (2) relation of crest shape to "ideal" nappe shape, (3) slope of upstream weir face, (4) downstream apron interference, and (5) submergence if it occurs downstream. Where the design of the approach channel results in appreciable additional losses, they must be added to He to determine reservoir elevations that correspond to discharges determined with equation (4-1).

As the approach depth of the flow to a weir decreases the approach velocity increases, thus affecting the discharge coefficient. The discharge coefficient for a vertical-faced ogee crest ranges between 3.8 and 3.9 for values of P/H_0 ranging from .5 to 3.0, where P is the weir height and H_0 is the design head for the spillway (USBR, 1977). The discharge coefficient will also vary for heads other than the design head. The ratio of discharge coefficient to design-head discharge coefficient varies from 0.85 to 1.07 as the ratio of head to design head varies from 0.2 to 1.6 (USBR, 1977). The reader is referred to *Design of Small Dams* (USBR, 1977), *Hydraulic Design of Spillways* (U.S. Army Corps of Engineers, 1965), and *Hydraulic Design Criteria* Chart 111-3/3 WES 2-72 (U.S. Army Corps of Engineers, 1968), for relationships for discharge coefficients, including adjustments for downstream apron interference and downstream submergence.

Piers and abutments cause side contraction of the overflow and therefore decrease the effective crest length of a spillway. These effects may account for a reduction of 1 to 5 percent, depending on pier and abutment types and spacing. The references sited above should be consulted for details.

Sources of Criteria for Determining Spillway Discharge Characteristics

The previous paragraphs briefly reference criteria that are applicable for estimating discharge characteristics for an ogee spillway crest. Detailed criteria for ogee crests and for the other spillway types mentioned previously may be found in a number of technical publications. Some of these are *Design of Small Dams* (USBR, 1977), *Handbook of Applied Hydraulics* (Davis and Sorenson, 1969), *Handbook of Hydraulics* (King and Brater, 1963), *Hydraulic Design of Spillways* (U.S. Army Corps of Engineers, 1965), and *Hydraulic Design Criteria* (U.S. Army Corps of Engineers, 1968). Computer programs are available as an aid to determining spillway ratings. Exhibit II, program 7 is one such program available through U.S. Army Corps of Engineers, Hydrologic Engineering Center, 609 Second Street, Davis, California 95616.

Although many past physical model studies provide a good insight to the range of values for coefficients and transition losses, large investment costs in major spillway and stilling basin modifications can justify consideration of site specific physical model studies.

Hydraulic Characteristics of Conduits and Outlet Works

Outlet works are a means of controlling the release of water from a reservoir. They are used to control downstream flows for a variety of purposes, such as irrigation, water supply, recreation, fisheries, and water quality control. Outlet works may be used during flood controi regulation to augment spillway discharges or to evacute storage in anticipation of a flood event. They are also used to empty an impoundment for inspection and repair.

Definition of the discharge characteristics for outlet works typically involves both open channel and pressure flow computations. For the situation where free flow occurs over the crest at the outlet works, the weir flow equation, equation 4-1, is applicable. When open channel outlet flows are controlled by partly opened surface gates or radial gates, or sluice flows are controlled by partly opened surface gates or radial gates, sluice flow will result. Discharge for sluice flow may be calculated with the equation:

$$Q = 2/3 \times \sqrt{2 \times g} \times C \times L \times (H_1^{3/2} - H_2^{3/2}) \qquad (4\text{-}2)$$

where
- Q = discharge
- g = gravitational acceleration
- C = discharge coefficient
- L = width of gate
- H_1 = total head to overflow crest
- H_2 = total head to bottom of gate

The discharge coefficient in equation 4-2 will vary with gate type and as a function of flow conditions upstream and downstream of the gate. Typically "C" ranges from .65 to .85.

For the case where the control opening is either partly or entirely submerged, discharges are calculated with submerged orifice or tube flow relations such as:

$$Q = C \times A \times \sqrt{2 \times g \times H}$$

where
- Q = discharge
- A = area of opening
- g = gravitational acceleration
- H = difference between the upstream and downstream water levels
- C = discharge coefficient for submerged orifice or tube flow

Coefficients for various conditions and orifice configurations are found in *Design of Small Dams* (USBR, 1977), and *Hydraulic Design of Reservoir Outlet Structures* (U.S. Army Corps of Engineers, 1963).

Discharge through an outlet conduit that is flowing full may be determined by application of the Bernonilli equation and estimation of loss coefficients for the various components of loss, which may include trashrack, entrance, bend, expansion, contraction, gate, and exit losses in addition to friction losses. Friction losses are commonly estimated with the Darcy-Weisbach formula. Loss coefficients and friction factors are provided in numerous textbooks and publications which deal with pipe flow. See, for example, *Hydraulic Design of Reservoir Outlet Structures* (U.S. Army Corps of Engineers, 1963), or *Handbook of Hydraulics* (King and Brater, 1963). The former publication contains a detailed example computation of the discharge rating curve for outlet works.

Spillway Design Floods

The determination of a standard against which to base judgment on the spillway adequacy is not a clear cut decision upon which the engineering profession has fully agreed. State and Federal agencies have varying standards and an owner of any dam of sufficient height and storage to come within the approval requirements of the State within which it is located, must obtain approval from the appropriate State agency. In addition, Federal licensing agency (Federal Energy Regulatory Commission) review and approval are required for those structures associated with hydropower installations. The mere fact that the dam has operated "safe and sound" for a significant period of time does not in itself assure an adequate hydrologic design.

Inspection Standards. The occurrence of dam failures during the past ten years resulted in the passage of Public Law 92-367, 92nd Congress, House Resolution 15951, 8 August 1972, wherein the Secretary of the Army, acting through the Chief of Engineers, Corps of Engineers, was directed to carry out a national program of inspection of dams. Appendix D of the Secretary's report to the Congress, "National Program of Inspection of Dams", 1975, contains recommended guidelines for inspection of existing dams. The guidelines are not intended as appropriate standards for the design of new facilities. However, the guidelines provide a satisfactory basis for evaluating existing projects for a reasonable degree of safety under existing conditions. The following three tables have been copied from the above reference.

TABLE 4-1
SIZE CLASSIFICATION

Category	Impoundment Storage (Ac-Ft)	Height (Ft)
Small	≥ 50 and < 1000	≥ 25 and < 40
Intermediate	≥ 1000 and < 50,000	≥ 40 and < 100
Large	≥ 50,000	≥ 100

TABLE 4-2
HAZARD POTENTIAL CLASSIFICATION

Category	Loss of Life (Extent of Development)	Economic Loss (Extent of Development)
Low	Nonexpected (no permanent structures for human habitation)	Minimal (Undeveloped to occasional structures or agriculture)
Significant	Few (no urban developments and no more than a small number of inhabitable structures)	Appreciable (Notable agriculture, industry or structures)
High	More than few	Excessive (Extensive community, industry or agriculture)

TABLE 4-3
HYDROLOGIC EVALUATION GUIDELINES
RECOMMENDED SPILLWAY DESIGN FLOODS

Hazard	Size	Spillway Design Flood (SDF)
Low	Small	50 to 100-yr freq
	Intermediate	100-yr to 1/2 PMF[1]
	Large	1/2 PMF to PMF
Significant	Small	100-yr to 1/2 PMF
	Intermediate	1/2 PMF to PMF
	Large	PMF
High	Small	1/2 PMP to PMF
	Intermediate	PMF
	Large	PMF

1/ PMF = (probable maximum flood); this represents the flood that may be expected from the most severe combination of meteorologic and hydrologic conditions that are considered to be *reasonably possible* in the geographical region encompassing the basin under study.

Design Standards. Whenever overtopping and failure of a dam could cause significant increases in downstream flow and stage, there may be hazards to life and property. Therefore, reservoir projects modified to include hydropower facilities must also provide reasonable security against flood flows. If inspection and evaluation of the project reveals that the spillway capacity is deficient, according to the Corps of Engineers "Recommended Guidelines", plans for improvements to the project spillway capacity must be included with the proposed powerplant design. Because of the potential for future development downstream from hydropower projects, all projects that could significantly increase downstream flooding hazards should be designed to safely pass at least one half of the probable maximum flood (PMF) hydrograph. When potential hazard to life is a major consideration, FERC may require the project to safely pass the full PMF hydrograph.

Indirect PMF Estimates. Many small dams which will be the likely sites for small hydro development will have had recent State or Federal safety inspections as a result of PL 92-367. These reports may have developed PMF estimates which are available. An approximation of the magnitude of PMF inflow can be obtained by means of an envelope curve of drainage area versus PMF discharges at other Federal and FERC licensed public and private sites in the same general region.

A source of PMF data at existing dams is contained in the Nuclear Regulatory Commission's (NRC) Regulatory Guide 1.59, August 1977. Envelope curves for the eastern region of the U.S. were developed by Nunn, Snyder, and Associates for the NRC and are presented in R.G. 1.59 referenced above. These figures are reproduced herein in Figure 4-1 through 4-6 for easy reference. These estimates are intentionally high, but, if a spillway at a site in the region of the map coverage is able to safely pass this discharge, further detailed estimates can probably be delayed until final design studies are warranted.

Probable Maximum Precipitation (PMP)

Well defined procedures have been developed for obtaining PMP estimates for the eastern part of the United States. Various Hydrometeorological Reports have been prepared by the Department of Commerce, National Oceanic and Atmospheric Administration, Office of Hydrology. These reports, listed in the reference section of this volume, address specific large river basins and regions of the country. The most recent version of PMP derivation for basins east of 105th meridian is contained in (HR 51, 1978). These reports are available from the Superintendent of Documents, Washington, D.C. HR 51 is applicable to river basins 10 to 20,000 square miles in size and for storm durations 6 to 72 hours. There are 30 figures which are used to obtain storm depths for a range of durations and sizes. An example of the use of these PMP charts from HR 51 follows:

1. Determine the geographic location and size of the drainage areas under study.
2. From the PMP maps, tabulate the average PMP depths for the basin location. Generally 3 or 4 areal sizes bracketing the size of the study basin are adequate. Tabulate depths for each duration 6, 12, 24, 48, and 72 hours.
3. Plot the PMP depths on semilog paper (depth versus area) and draw smooth curves through the plotted points.
4. From the depth-area-duration graph of step 3, determine the PMP depths at the basin size for each duration.
5. Plot these values on cartesian grid or semilog grid and interpolate for intermediate durations, if required.

Example PMP. Compute PMP values for a 1,200 square mile basin located in central Arkansas, Latitude 35° N, Longitude 93° W. From maps like Figure 4-7, we obtain the following values.

	(A)	(B)	(C)	(D)	(E)
Duration (hr)	6	12	24	48	72
200 sq. mi. Depth (in)	22.0	26.8	30.5	34.0	36.8
1,000 sq. mi Depth	16.3	20.8	25.0	28.5	30.3
5,000 sq. mi. Depth	9.5	13.5	17.0	21.1	23.7

These values are plotted in Figure 4-8, and the adopted 1,200 square mile depth-duration curve is constructed as curve F extrapolated to 3 hours. Values at 3 or 6-hour intervals can be taken from curve F.

PMP West of 105th Meridian. Estimates for basins west of 105 degrees longitude are complicated by the strong influence of the high mountain ranges. The loss of moisture on the windward sides of the ranges and the desiccating effect of the subsidence of the air mass on the leeward side of the range are examples of this influence.

Both general and local type storm genesis are characteristic of the west. The general storms have two components that cause air mass lifting and consequent precipitation. These are orographic and convergence. These two components are considered separately and then combined to develop the storm total precipitation.

Orographic precipitation is defined as that which results from the lifting effect of a topographic feature on a flow of air passing over it. The induced vertical motions in the flow are primarily due to the ground slope, but may also be related to the narrowing of the terrain, such as a constricted valley (e.g., northern Sacramento Valley). Orographic precipitation includes

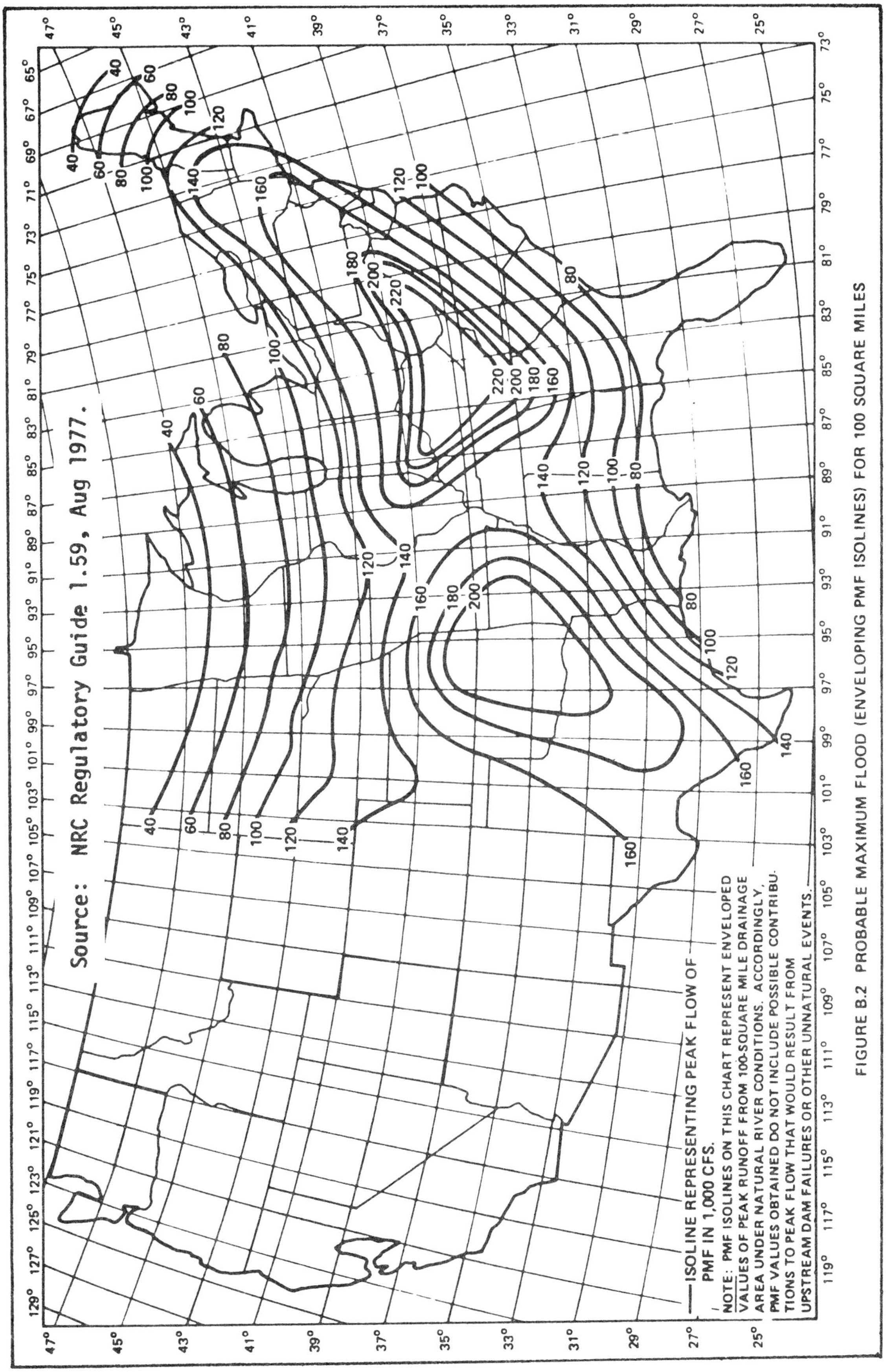

Figure 4-1. Probable Maximum Flood (100 sq. mi.)

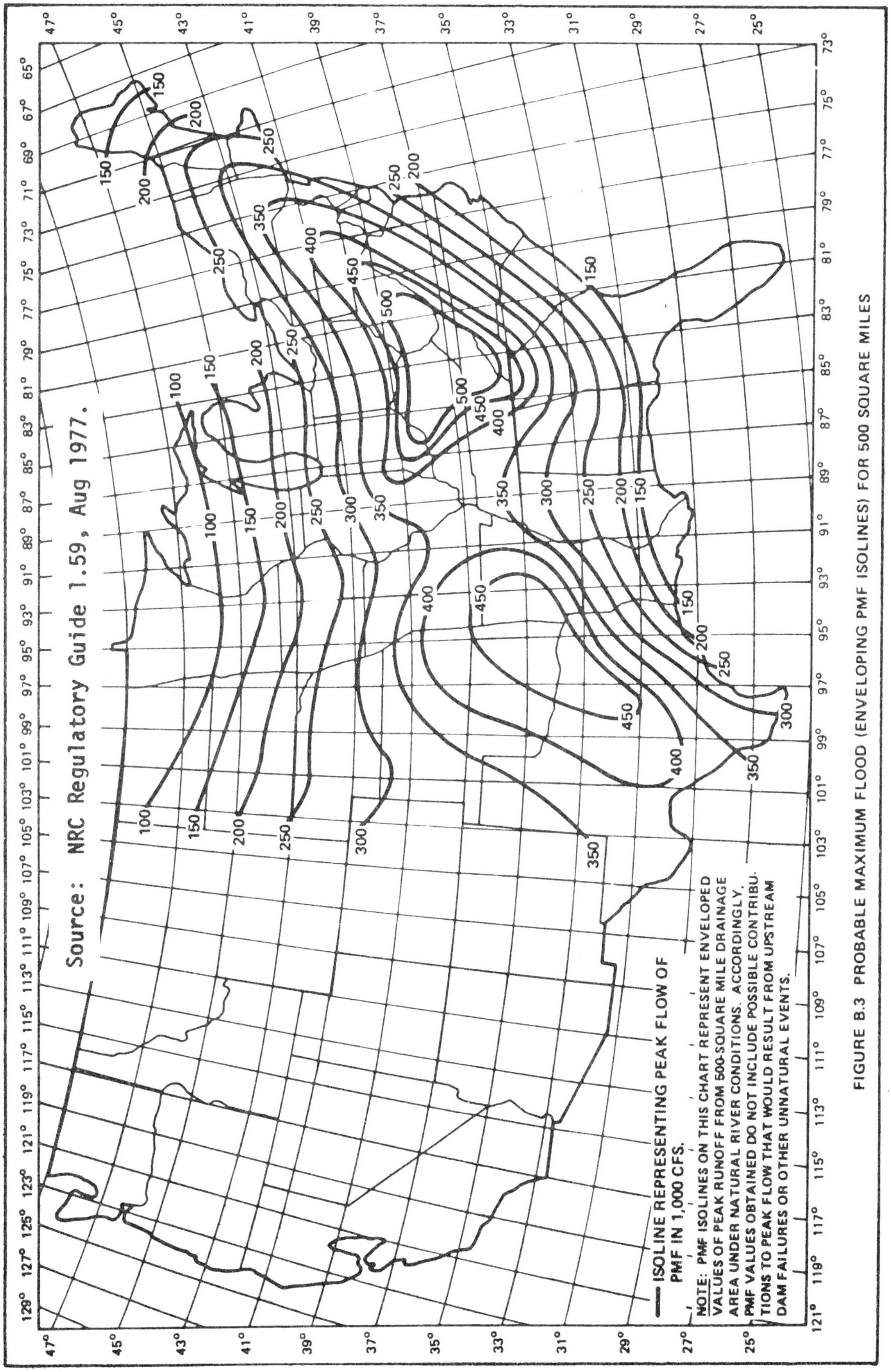

Figure 4-2. Probable Maximum Flood (500 sq. mi.)

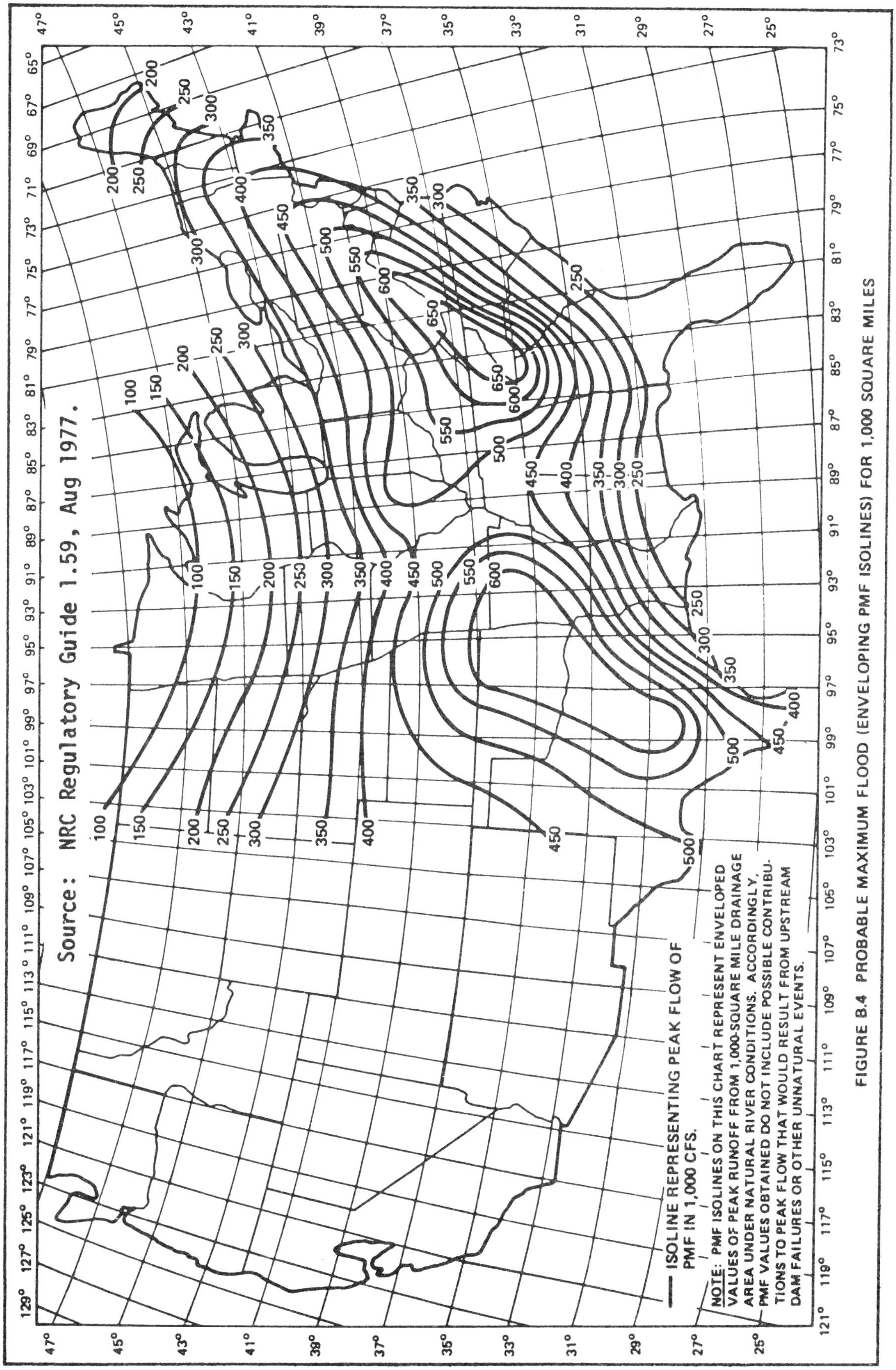

Figure 4-3. Probable Maximum Flood (1,000 sq. mi.)

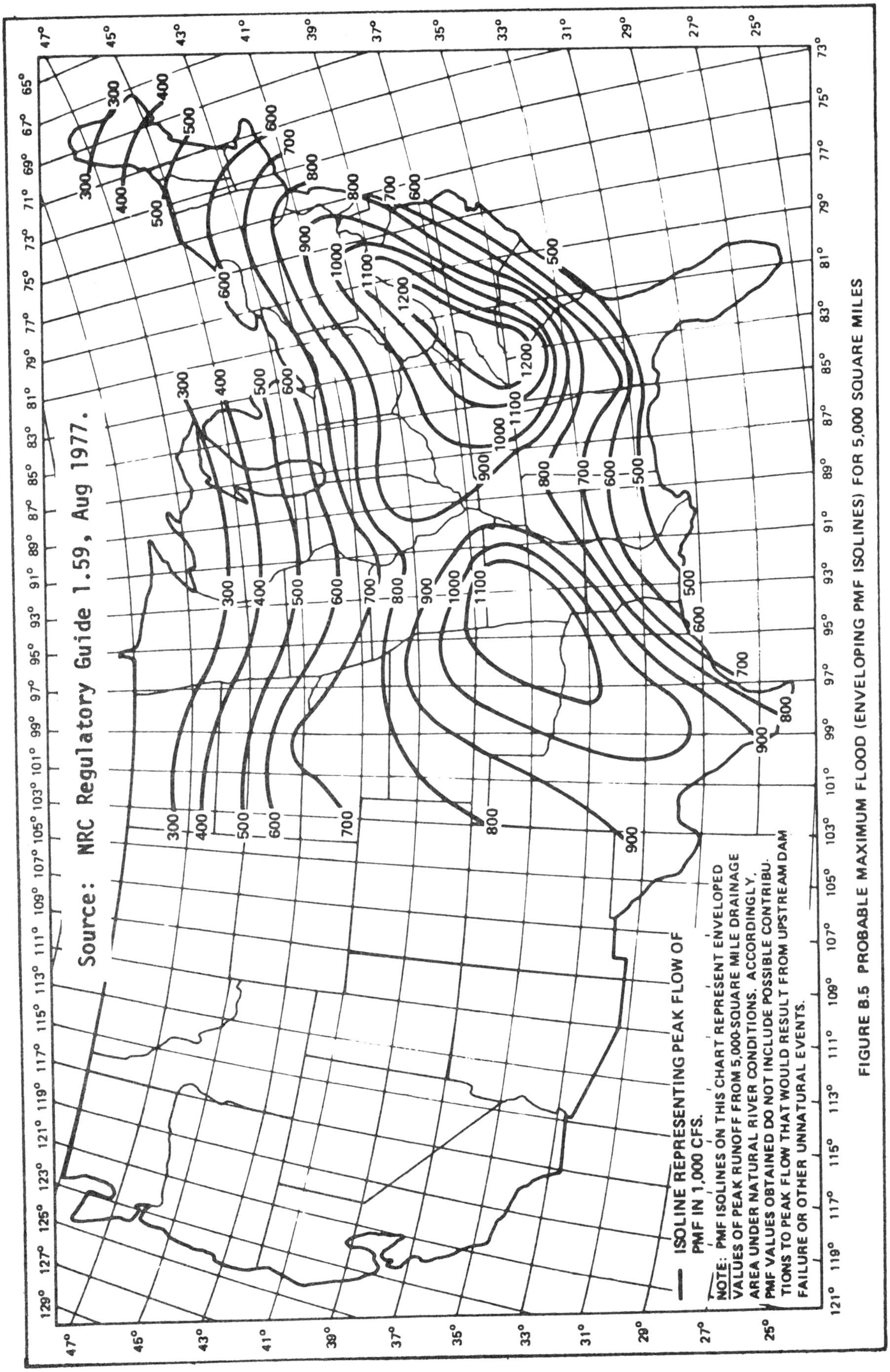

Figure 4-4. Probable Maximum Flood (5,000 sq. mi.)

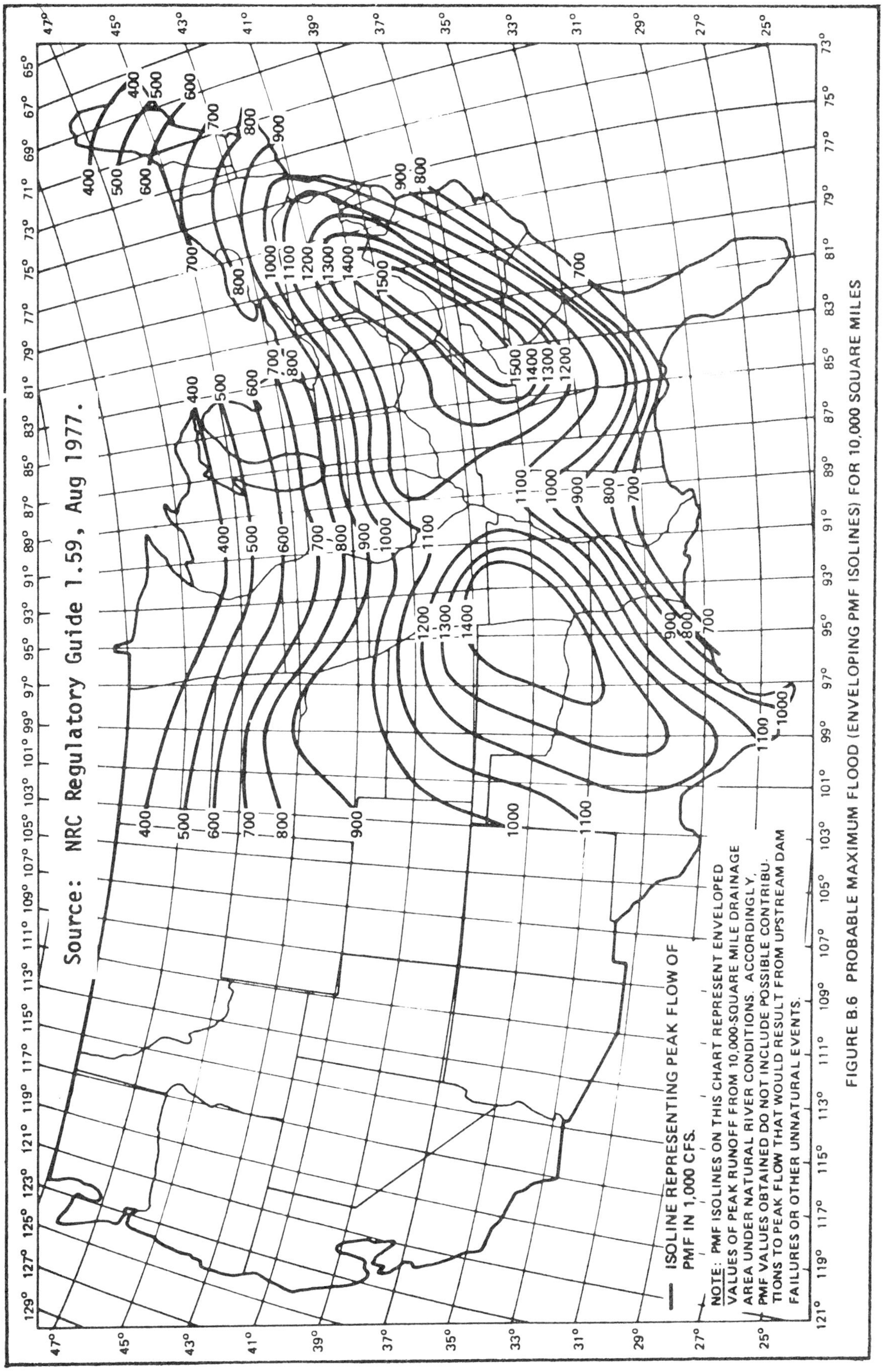

Figure 4-5. Probable Maximum Flood (10,000 sq. mi.)

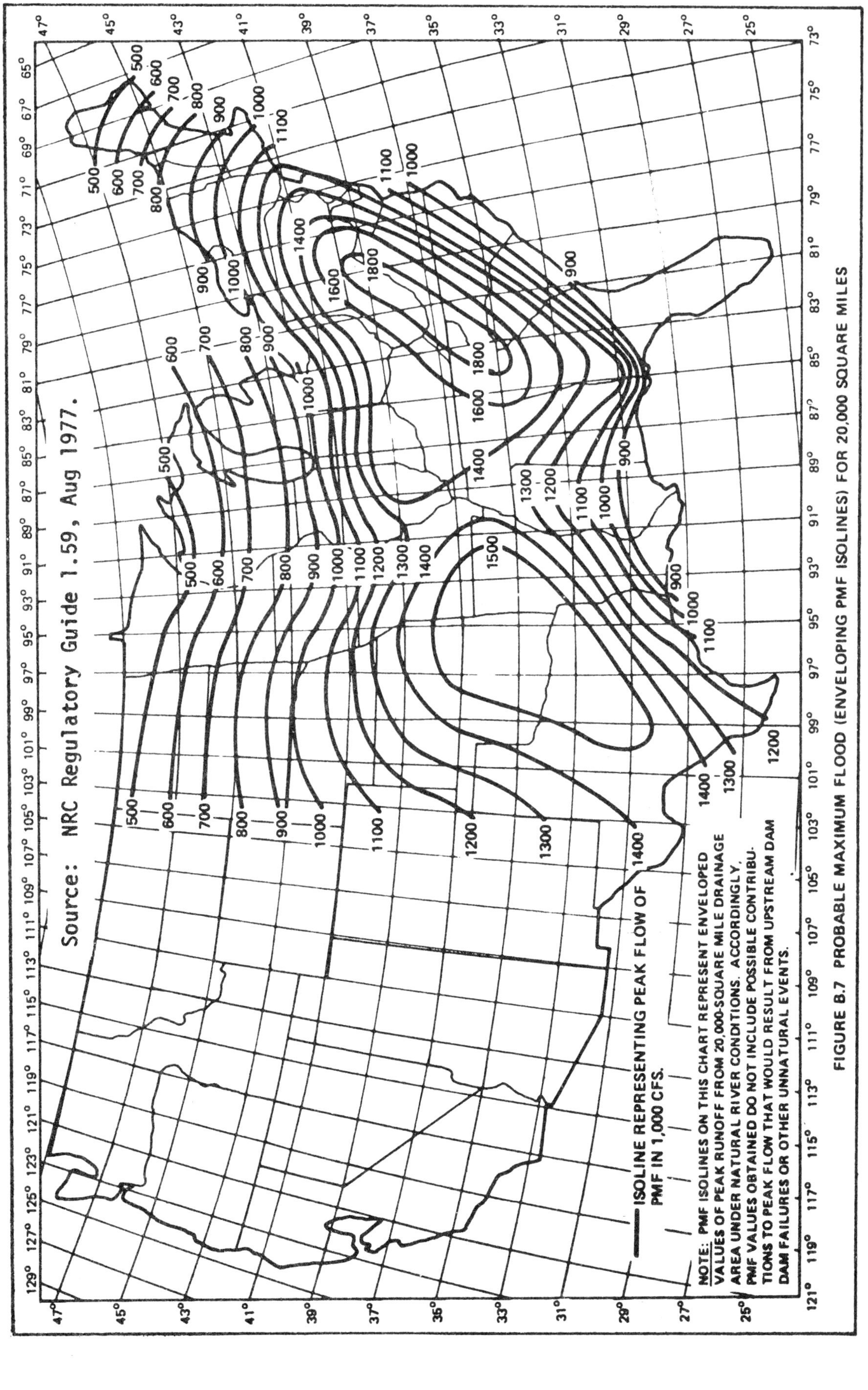

Figure 4-6. Probable Maximum Flood (20,000 sq. mi.)

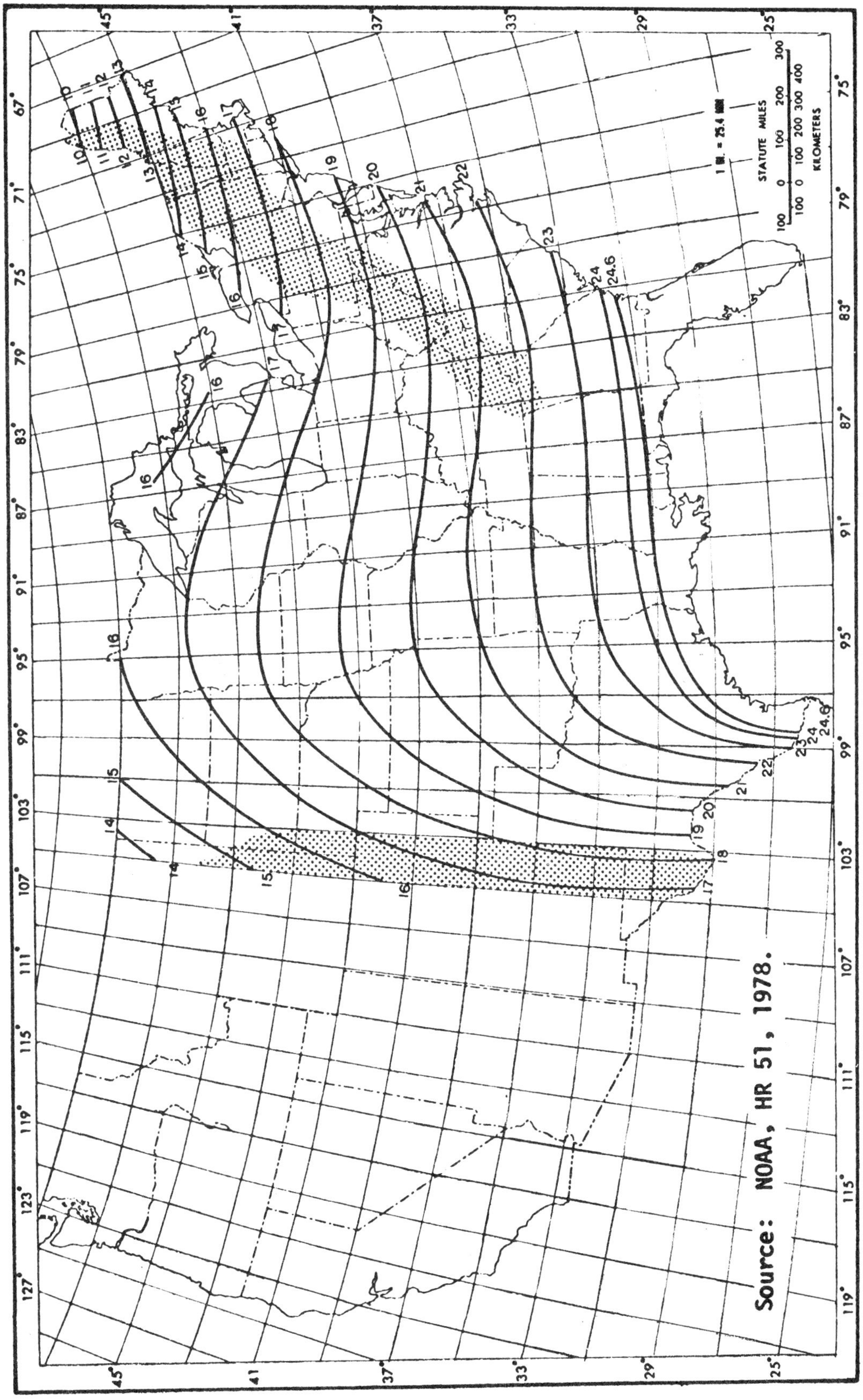

Figure 4-7. All season PMP for 6 hrs, 200 sq. mi.

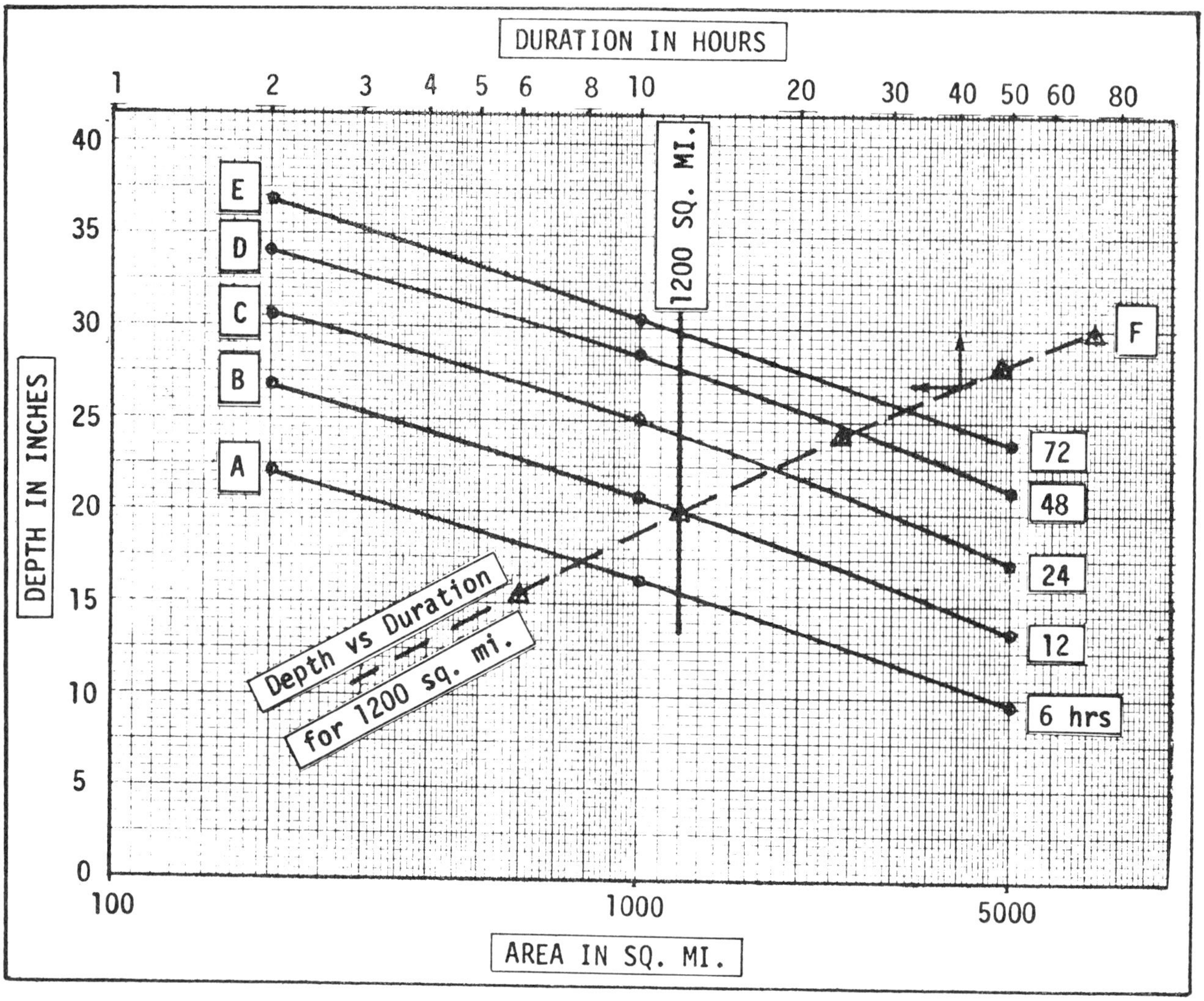

Figure 4-8. Example PMP estimate using HR 51, NOAA, 1978.

---

both that falling on the windward slope and that blown over the barrier (called spillover).

Convergence precipitation includes precipitation resulting from lifting induced by atmospheric processes other than orographic. These are mainly horizontal convergence, frontal lifting, and instability. In mountainous areas these occur simultaneously.

Details for the steps used to estimate the PMP for general storms in the west are contained in various Hydrometeorological Reports (HR) and Technical Papers (TP), as follows:

1. *California*

(1) HR 36, "Interim Report, Probable Maximum Precipitation in California," presents basic steps and charts for making general storm estimates. The report is currently approved for general application in the computation of probable maximum flood hydrographs for drainage basins less than about 5,000 square miles within the area covered by generalized charts, subject to such additional special analyses as may be warranted in special cases. The procedures used include approximate allowances for basin shape, elevations, and orientation, so that no additional adjustments in rainfall quantities are called for. However, in view of the unusually complex problems involved in preparing generalized estimates that are applicable to the region indicated, flood estimates based on the criteria contained in HR 36 should be utilized with utmost care and judgment by experienced hydrologic engineers.

(3) Technical Paper 38, "Generalized Estimates of Probable Maximum Precipitation for the U.S. West of the 105th Meridian for Areas Less Than 400 Square Miles and Durations to 24 Hours," was prepared by the Cooperative Studies Section of the National Weather Service for the Soil Conservation Service. A different, but reasonable, approach to PMP was used in this

report. TP 38 includes small-area intense local warm-season storms while HR 36 is restricted to cool season storms. Also, a single 6/24-hour duration ratio was adopted in TP 38 since local storms are most likely to control for small areas.

2. *Hawaii.* HR 39, "PMP in the Hawaiian Islands," gives generalized rainfall criteria for computation of probable maximum floods for all sized basins encountered in the Hawaiian Islands. Effects of topography, wind exposures, and other pertinent factors are reflected in the estimates. Guide criteria for estimating hyetographs are also included.

3. *Northwest.* HR 43, "PMP, Northwest States," contains generalized PMP rainfall estimates 1 to 72 hours for basins of 10 to 5,000 square miles west of the Cascade Divide, and 10 to 1,000 square miles east of the Divide, in the Columbia River Basin. Seasonal variations are given by months, October through June. Various optional storm patterns are included (see Figure 5 of this material). HR 42, "Meteorological Conditions for the PMF on the Yukon River above Rampart, Alaska," is typical of a number of basin specific reports prepared by the HMB, NWS.

4. *Southwest.* HR 49, "PMP Estimates, Colorado and Great Basin Drainages", gives general-storm PMP estimates for durations between 6 and 72 hours and for area sizes between 10 and 5,000 square miles for the Colorado River and Great Basin. Probable Maximum Thunderstorm Precipitation Estimates, Southwest States, West of Continental Divide, are also included in this report. It covers the areas of California, Nevada, Utah, Arizona, Wyoming, Colorado, and New Mexico, and durations 15 minutes to 6 hours and areas 1 to 500 square miles. Isohyetal patterns are included.

Hyetographs for Use in Computing PMF Hydrographs

The chronological sequence of probable maximum precipitation and snowmelt, associated infiltration losses, and rainfall excess quantities by incremental time intervals of 6 hours or less should be presented in tabular form for each basin sub-area selected for the hydrograph computation. One or more representative hyetographs should be shown graphically by plotting of associated PMF hydrographs, in proper time relationship, with sufficient explanatory notes to show clearly the correlations between PMP increments and runoff.

The sequential arrangements of 24-hour incremental values of PMP, and sub-divisions of these into 6-hour increments (or less) for computation of PMF hydrographs must be compatible with meteorological characteristics affecting the specific basin. Guidance in determining the time sequences may be obtained from observed storms. The following guidelines are acceptable for the development of hyetographs representing PMP sequences derived from generalized enveloping depth-area-duration curves east of 105° longitude.

(1) Group the four heaviest 6-hour increments of PMP in a 24-hour sequence, the next highest four increments in a 24-hour sequence, etc.

(2) For the maximum 24-hour sequence, arrange the four 6-hour increments ranked 1, 2, 3, and 4 (maximum to minimum) in the order 4, 2, 1, 3

(3) Arrange the 24-hour sequences so that the highest period is near the end of the storm, and the second, third, etc., are distributed in a manner similar to (2) above.

(4) Subdivide the most intense 6-hour quantity during the maximum 24-hour PMP series into 1, 2, or 3-hour intervals, if trial computations show such subdivisions would produce significantly higher PMF hydrograph peaks at locations of interest in the studies involved. One-hour increments should equal 10, 12, 15, 38, 14, and 11 percent, respectively, of the maximum 6-hour PMP quantity. These one-hour percentages may be combined into successive two-hour or three-hour increments, if studies show these subdivisions give satisfactory results in hydrograph computations. Rainfall intensities may be assumed as uniform during all other 6-hour increments of the PMP series.

The Southwestern Division of the Corps of Engineers, Dallas, Texas, has studied a large number of storms and concluded that storms in that region have significantly greater intensity than is reflected by the above hourly percentages of 6-hour tota. Distributions made for basins in Texas, Oklahoma, New Mexico, and Southern Kansas may be obtained from hydrologic engineers at the Dallas office, Corps of Engineers. Typical distribution patterns suggested in HR 43 for the northwestern states are presented in Figure 4-9.

Other Aspects of PMP Derivation. The computation of probable maximum flood hydrographs for fairly large drainage basins usually requires determination of runoff from a number of basin subareas. The subareas selected should conform with requirements for unit hydrograph determinations, flood routing computations, and/or other hydrologic-hydraulic evaluations. They should be numbered for convenient reference, depicted on basin maps, and tabulated with areas in square miles and other pertinent data. Various combinations of the following reasons may influence selections of subareas for large basin studies:

1. Physical features of the drainage basin, such as topography, drainage patterns, exposure to air mass movements that affect PMP, and large variations in infiltration characteristics, natural hydraulic efficiencies of principal stream channels and floodways, valley storage capacities during major floods, and other conditions;

2. Major lakes and reservoirs that are capable of impounding or releasing large quantities of water during major floods;

3. Major channel improvements, lock and dam projects, levee systems, flow diversion structures, and other man-made works that influence movements of flows through principal drainage channels; and

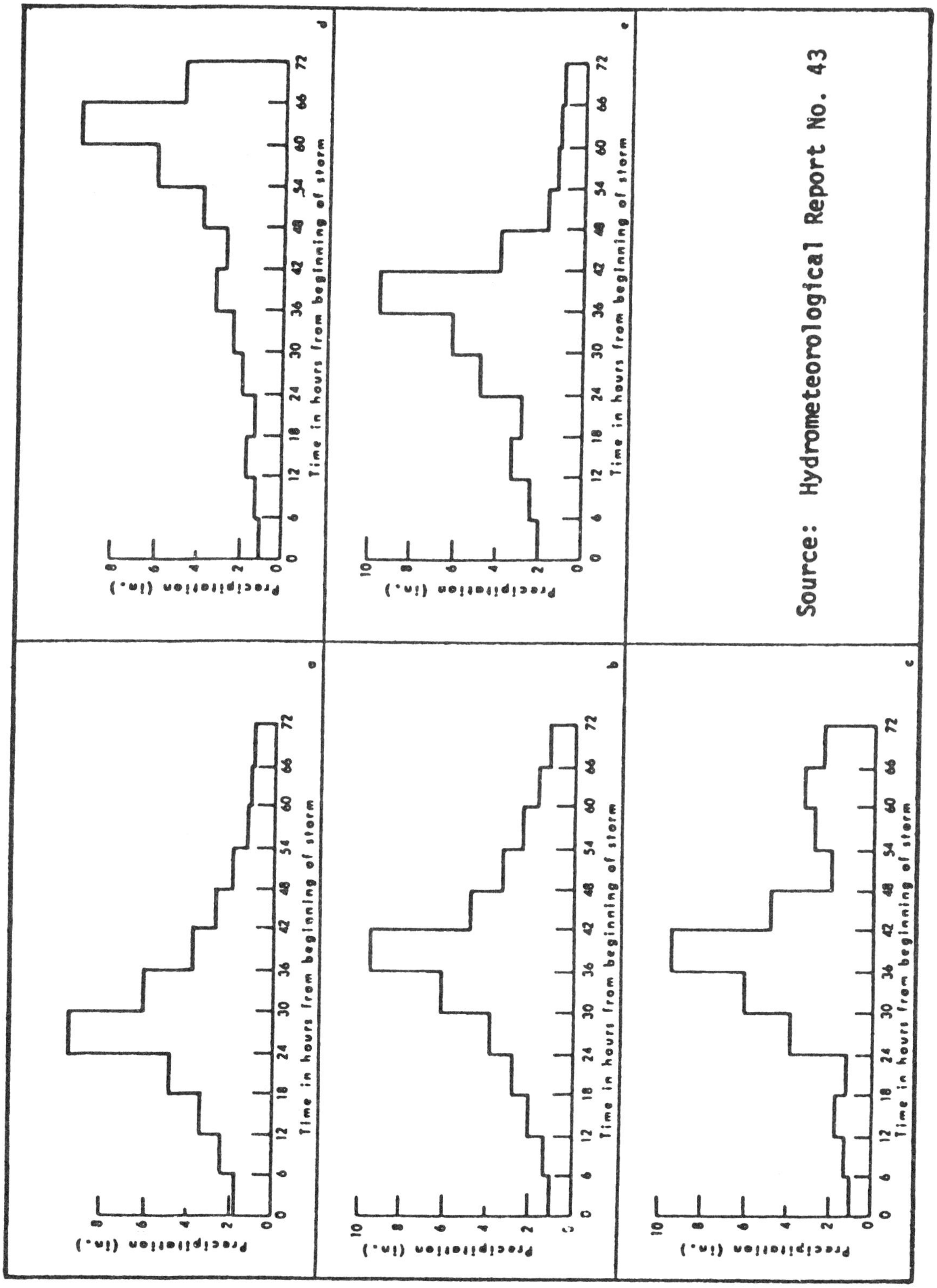

Figure 4-9. Typical PMP Time Distribution.

4. Locations along principal stream channels that have special significance in specific studies, such as cities, industrial centers, and proposed projects.

Areal Distribution of PMP in Small Basins. In utilizing generalized PMP estimates for drainage areas less than approximtely 1,000 square miles, it may be appropriate to assume a uniform depth over the entire drainage basin. However, if a breakdown by major basin subdivision is desirable in specific cases, higher concentrations of PMP intensities may be estimated by meteorologically sound procedures. For example, if there are valid reasons for assuming higher concentrations of PMP over, say, a 400 square mile subdivision of a 1,000 square mile area, computations could be as follows:

1. Determine total-storm PMP for the 1,000-square mile area from Report 51, as explained earlier, expressed in inches depth. In the same manner, determine the value for a 400-square mile area.

2. Convert the values obtained in step (1) to "inch-square miles," subtract to obtain the difference, and divide this by 600 to obtain the average total-storm PMP in inches depth over the 600-square mile subbasin.

Areal Distribution of PMP in Large Drainage Basins. Determinations of critical distributions of PMP quantities over large drainage areas (e.g., exceeding approximately 1,000 square miles) are more difficult to establish, and vastly more significant in estimating PMF hydrographs than in small basins. In large drainage basins located where infiltration capacities of the ground are high, variations in the areal distribution of precipitation during successive 6-hour periods of a total storm may reduce or increase estimates of total net runoff volumes by more than twofold, simply by changing opportunities for infiltration. The critical location of successive 6-hour PMP increments can also change the concentration of runoff at a particular location significantly. The existence of major reservoirs and other runoff controls in some drainage basins may have major influences on characteristics of PMF hydrographs associated with various areal distributions of PMP.

Accordingly, generalized estimates of PMP depth-area-duration relations for specific regions or drainage basins must be supplemented by appropriate analyses to establish critical areal distribution patterns of PMP from the beginning to the end of the overall storm. Normally a breakdown of areal distribution patterns by 6-hour intervals is suitable, in consideration of accuracy limitations in other facets of the computations. Shorter intervals may be needed in special cases.

Some PMP estimates for large drainage basins are based on "transportation" of major storms of record in the region, with certain adjustment in rainfall amounts to represent PMP quantities. In such cases, it is usual practice to retain generally the same isohyetal pattern in the transposed position, and the same chronological sequence of rainfall amounts at individual rainfall stations. These data provide a basis for estimating PMP amounts for any drainage basin subdivisions of interest in the specific studies, with a disaggregation by chronological increments of time needed for PMF hydrograph computations. This "storm transportation" technique provides the most convenient method of accounting for areal distributions of PMP, and the chronological sequence of occurrence.

Rainfall Excess Estimates. For computation of hydrographs of runoff from maximum precipitation, estimates of rainfall amounts exceeding infiltration and other losses are required for successive increments of time. These rainfall excess estimates should be related to basin subareas, in order to account for significant variations in areal distribution of PMP and any differences in infiltration characteristics of the areas. Loss factors tend to vary during successive periods of a rainfall sequence. Ground conditions that affect losses during the probable maximum storm should be the most severe that can reasonably exist in conjunction with maximum probable precipitation. Lowest loss rates that have been observed might be used if there is reasonable assurance that the entire range of possible losses has been experienced. However, loss estimates are subject to major uncertainties, and there are cases where negative loss rates are computed simply because of inadequate precipitation data. Accordingly, some allowance must be made for this uncertainty, and loss rates that are conservatively low should be selected for probable maximum flood computation. Where it is possible for the ground to be frozen at the start of a rainflood or snowmelt flood, it can be concluded that zero or near-zero loss rates should be used for probable maximum flood computation. There may exist a seasonal variation in minimum loss rates, in which case rates selected should be those representative of the most extreme conditions for the season for which probable maximum runoff is being computed. Typical values throughout the United States are in the range of .10 to 1.0 inch initial loss followed by a uniform rate of .05 to .15 inch per hour.

Probable Maximum Snowpack and Snowmelt. Guidance on probable maximum snow conditions is contained in U.S. Army Corps of Engineers Manual 1110-2-1406, "Runoff from Snowmelt", available from OCE Publications Depot, 890 South Pickett Street, Alexandria, VA 22304. Computations of probable maximum snowpack accumulation from winter precipitation, temperatures, and snowpack losses should be estimated from observed snowpack data and should exceed maximum observed accumulations. Adjustments to historical maximum snowpack to obtain probable maximum snowpack should generally be obtained from the National Weather Service.

In the case of rainfloods that have some snowmelt contribution, snowpack used for probable maximum rainflood computation should be the maximum that can contribute toward the peak flow and runoff volume of the flood without inhibiting the direct runoff from rainfall.

The critical snowpack in mountainous regions will ordinarily be located at elevations where most of the rainflood runoff originates. Snowpack is ordinarily greater at higher elevations and less at lower elevations, and hence critical snowpack will not exist at all elevations. Factors to be considered in selecting melt factors for the probable maximum snowmelt are discussed in the referenced EM 1110-2-1406 and in HR 43.

Probable Maximum Base Flow. Base flow for the probable maximum flood is not a critical item because the peak flow, which is not greatly affected by base flow, is the primary characteristic of interest in the probable maximum flood. Nevertheless, it is prudent to adopt a base flow value that is more severe than that which would be used for lesser floods.

Probable Maximum Flood Computation. Rainfall-runoff factors should be selected as the most severe that are reasonably consistent with the storm and flood conditions, and should be more severe than those that have been historically observed. Channel routing coefficients should likewise be modified toward greater translation speed and less storage effects because of the more efficient hydraulic flow conditions during larger floods.

In application of the probable maximum flood for spillway design associated with a large lake surface area, allowance should be made for the accelerating effect of a reservoir in relation to the stream reaches that are inundated, and the reservoir level at the start of the flood should be the highest level reasonable, consistent with probable maximum flood conditions.

Antecedent Conditions. In many spillway design applications, flood conditions that precede the probable maximum flood may have substantial influence on the regulatory effect that the reservoir has on the probable maximum flood. In such cases, it is appropriate to precede the probable maximum flood with a flood of major magnitude at a minimum time interval that is consistent with the causative meteorological conditions. While a special meteorological study is desireable where possible for this purpose, it is often considered that the start of a probable maximum flood reasonable be preceded by the start of a flood of 30 to 60 percent of the PMF by a period of 4 or 5 days.

Computer Programs. Many private and governmental engineering organizations have developed precipitation-runoff models which can be used to convert complex rainfall-snowmelt into discharge hydrographs. Some of the more well known models are:

Stanford Watershed Model IV (Stanford University)
Kentucky Watershed Model (University of Kentucky)
ILLUDAS (Illinois State Water Survey)
HSP (Hydrocomp International)
MITCAT (Resource Analysis Inc.)
SSAAR (Corps of Engineers)
HEC-1 (Corps of Engineers)
TR-20 (Soil Conservation Service)
NWSRFS (National Weather Service)

Users of any of these models are cautioned to obtain proper guidance from experienced hydrologic engineers.

Discharge Exceedence Frequency Estimates

The standard for the "small" size, "low to significant" hazard category of dams allows a minimum design flood having an average return period of 50 to 100 years. More properly stated, this flood would have a 0.02 to 0.01 probability of exceedence in a given year. For reasons previously discussed, design standards will usually exceed this. Estimates of the magnitude of these events are determined by an analysis of the annual observed maximum events during the period of stream flow record at or near the damsite or from regional analysis of recorded events in hydrologically similar basins.

The United States Water Resources Council conducted an investigation of procedures and techniques for analyzing flood events in estimating flood flow frequencies at gaged, essentially unregulated, sites (Bulletin 17A Rev. June 1977). There are numerous other publications available on the subject as well as standard college text books in hydrology (IHD vol. 3, 1975), (Chow, 1964). The recommended distribution to use in the analylsis is the log Pearson Type III with a regional skew. The method of moments is used to determine the statistical parameters of the distribution from the observed annual series. Techniques for identifying and handling high and low outliers are also addressed in Bul. 17A. Computer programs designed to follow procedures recommended in Bul. 17A are available from HEC (Exhibit II), or from the U.S. Geological Survey (USGS), Reston, VA. Data on annual observed maxima at most stream gaging locations throughout the United States are published annually by the State Water agencies and in 5 year summary reports available at USGS offices, Corps of Engineer District libraries, and most State and large university libraries. The WATSTORE data storage system of the USGS is also a readily available source of annual peak discharge data.

Generalized Regional Procedures. The U.S. Geological Survey (USGS) in cooperation with State and other Federal water resources agencies have prepared regional procedures for estimating discharge frequency curves for ungaged areas. These may be in the form of an "index flood" method, regression equations, or other similar methods. These methods can be used for reconnaissance estimates but should be used with caution and considerable hydrologic judgment.

At run-of-river sites, these peak discharge estimates are used directly for spillway hydraulic design evaluation. If reservoir storage is significant, a total design hydrograph must be developed or a reduction of peak inflow to peak outflow can be based on engineering judgement to account for storage effects.

Reservoir Routing

After developing a reservoir inflow hydrograph representing the spillway design flood, the hydrograph must be routed through reservoir storage and spillway. Usually the "storage-indication", also called "modified Puls" method is used to accomplish this. This involves the basic continuity equation: Inflow - Outflow = Change in Storage, and a relationship between outflow and storage. Discussions of this procedure may be found in most hydrology textbooks and in a previously cited reference (USBR, 1977).

All established operating criteria are utilized in this routing, making use of all available outlets. The starting elevation for this flood routing should be normal pool level or, if there is authorized flood control space, additional studies should be made to evaluate the likelihood of a higher starting elevation.

Freeboard Allowance

Freeboard as used herein refers to the vertical height between some reference lake levels and the elevation of some specified feature of the dam (normally the top of the main non-overflow section).

Guidance on minimum freeboard allowance is not within the scope of this volume. Since the primary emphasis of this manual is on existing facilities, it is obvious that some criteria were applied at the time of design and construction. Federal and State agencies have generally not adopted a uniform criterion for "freeboard" requirements. However, parameters that should be considered include: (1) duration of sustained high water levels in the reservoir, (2) effective wind fetch and velocity, (3) reservoir depth, (4) embankment slope and roughness, (5) ability of the dam to resist erosion from overtopping, (6) hazard classification, and (7) reference level. Additional discussion and guidance on this aspect of dam safety can be found in *Design of Small Dams*, USBR, 1977.

SECTION 5
TIME AND MANPOWER REQUIREMENTS

Reconnaissance Level

It is difficult to estimate a generalized time and manpower requirement for the hydrologic and hydraulic aspects of study tasks schematically displayed in Figure 1-1 of this volume. Often a great deal is known about a project or is learned during discussions about managements interest in having the engineering staff conduct the study. A reasonably accurate estimate covering items on the reconnaissance limb of the Figure 1-1 diagram could be accomplished with 3 to 5 man-days of office time by an experienced hydrologic engineer. Time and cost of a field trip to the site should be added to the office time.

Feasibility Level

There is even a greater range of time requirements in the feasibility level of hydrological investigations. This conclusion is based on the great variability in availability of detailed site data, and in the wide variation in spillway and stilling basin evaluation and redesign requirements. Also, the number of alternatives relating to the size, type, number of generator-turbine units, and placement has a great impact on time and cost estimates of conducting these studies. A reasonable range of time required to accomplish hydrologic-hydraulic items on the feasibility limb of Figure 1-1 would be 4 to 8 man-weeks by experienced engineers. The greatest efficiencies can be accomplished by this work being done by no more than one or two engineers.

Documentation

Data sources, assumptions, and study procedures must be well documented in order to be of any lasting value. Some review requirements are typically required, either by the project owner or his representative and by the licensing authority.

Statistical displays and certain minimum basic data should be included in reports or appendices to reports. These would include:

- Pertinent site data.
- Monthly flow data.
- Flow duration curve.
- Average annual energy versus installed capacity table or curve.
- Maximum and minimum annual energy generation.
- Duration frequency data on monthly energy generation including zero generation.
- Spillway design flood.
- Tailwater rating curve.
- Plant efficiency versus head curve.

REFERENCES

Davis, C.V., and Sorenson, I.E., 1969, *Handbook of Applied Hydraulics,* 3rd Ed.; McGraw-Hill Book Company.

King, H.W., and Brater, E.F., 1963, *Handbook of Applied Hydraulics*, 5th Ed.; McGraw-Hill Book Company

U.S. Army Corps of Engineers, 1963, *Hydraulic Design and Reservoir Outlet* Structures, Engineer Manual 1110-2-1602.

U.S. Army Corps of Engineers, 1965, *Hydraulic Design of Spillways*, Engineer Manual 1110-2-1603.

U.S. Army Corps of Engineers, 1952, *Hydro Power*, Engineer Manual 1110-2-1701.

U.S. Army Corps of Engineers, 1960, *Runoff From Snowmelt*, Engineer Manual 1110-2-1406.

U.S. Army Corps of Engineers, 1968, *Hydraulic Design Criteria*, prepared by U.S. Army Engineer Waterways Experiment Station, Vicksburg, Mississippi.

U.S. Army Corps of Engineers, OCE, 1975, *Recommended Guidelines for Safety Inspection of Dams*, Appendix D, National Program of Inspection of Dams, Washington, D.C. 20314.

U.S. Army Corps of Engineers, HEC, *Hydrologic Engineering Methods for Water Resources Development*, IHD,

Vol. 1, *Requirements and General Procedures*, 1971.

Vol. 2, *Hydrologic Data Management*, 1972.

Vol. 3, *Hydrologic Frequency Analysis*, 1975

Vol. 8, *Reservoir Yield*, 1975.

Vol. 9, *Reservoir System Analysis for Conservation*, 1977.

U.S. Army Corps of Engineers, HEC, 1967, Methods Systemization Manual, *Reservoir Storage-Yield Procedures*.

U.S. Bureau of Reclamation, 1977, *Design of Small Dams*, United States Government Printing Office, 816 p.

U.S. Department of Commerce, NOAA, Hydrometeorological Report:

HR 36 - *Interim Report on Probable Maximum Precipitation in California*, 1961.

HR 37 - *Meteorology of Hydrologically Critical Storms in California*, 1962.

HR 39 - *PMP in the Hawaiian Islands*, 1963.

HR 40 - *PMP for the Susquehana River Drainage Above Harrisburg, PA.*, 1965.

HR 42 - *Meteorological Conditions for the Probable Maximum Flood on the Yukon River above Rampart, Alaska*, 1966.

HR 43 - *PMP, Northwest States*, 1965.

HR 48 - *PMP and Snowmelt Criteria for Red River of the North Above Pembina, and Soure's River Above Mifnet, N.D.*, 1973.

HR 49 - *PMP Estimates, Colorado and Great Basin Drainages*, 1977.

HR 51 - *All Season PMP, U.S. East of 105th Meridian, for Areas from 1,000 to 20,000 Square Miles and Durations from 6 to 72 Hours*, 1978.

Technical Paper No. 38 - *Generalized Estimates of PMP for the U.S. West of the 105th Meridian for Areas to 400 Square Miles and Durations to 24 Hours*, 1950

U.S. Geological Survey, 1963, *Surface Water Supply of the United States,* Part 7 Lower Mississippi River Basin, Vol. 2, Arkansas River Basin WSP 1711.

U.S. Geological Survey, 1975, *WATSTORE Users Guide*, National Water Data Storage and Retrieval System, Open-file Report 75-426, Reston, VA22092.

U.S. Nuclear Regulatory Commission, 1977, *Design Basin Floods for Nuclear Power Plants,* Regulatory Guide 1.59.

U.S. Water Resources Council, 1977, *Guidelines for Determining Flood Flow Frequency,* Bulletin 17A of the Hydrology Committee, 2120 L Street, N.W., Washington, D.C. 20037

EXHIBIT I
NET LAKE EVAPORATION ESTIMATES

TABLE NO. 1
PRECIPITATION MINUS EVAPORATION

Station	Dec Jan Feb	Mar Apr Nov	May Oct	Jun Sep	Jul Aug	Average Annual	Critical Year Corrections (1) (Deduct from Avg Annual)
	Rate of Gain or Loss in c.f.s. per 1000 Acres of Designated Period						
Alabama							
Birmingham	+4.7	+1.8	-1.7	-3.0	-1.4	+0.6	-3.1
Mobile	+5.3	+2.9	-0.2	-0.6	+1.8	+2.3	-3.3
Montgomery	+4.8	+1.9	-2.6	-3.8	-2.3	+0.2	-3.3
Arizona							
Phoenix	-3.3	-7.4	-12.2	-17.1	-17.9	-10.6	-3.2
Yuma	-5.5	-9.2	-11.9	-15.2	-17.3	-11.1	-1.7
Arkansas							
Fort Smith	+1.2	-0.6	-1.4	-4.6	-7.2	-2.1	-5.0
Little Rock	+3.7	+1.2	-2.0	-4.7	-5.5	-0.8	-3.7
California							
Eureka	+7.4	+4.6	+0.8	-0.9	-1.6	+2.7	-2.0
Fresno	+0.6	-2.5	-6.7	-12.0	-16.0	-6.3	-1.4
Los Angeles	-0.8	-3.3	-5.0	-5.9	-6.4	-3.9	-2.3
Sacramento	-2.6	-1.7	-5.7	-9.6	-11.9	-4.3	-3.2
San Diego	-0.6	-2.4	-3.3	-4.0	-3.9	-2.6	-1.7
San Francisco	+3.5	-0.4	-2.7	-3.7	-3.5	-0.9	-2.0
Colorado							
Denver	-1.7	-2.5	-4.4	-8.1	-9.6	-4.7	-2.8
Grand Junction	-0.4	-3.5	-7.0	-12.8	-15.5	-6.9	-3.1
Pueblo	1.8	-3.7	-6.0	-10.0	-10.5	-5.8	-3.0
Connecticut							
Hartford	+2.8	+2.1	+0.7	+0.6	+0.2	+1.5	-1.4
New Haven	+3.9	+2.8	+0.7	+0.2	-1.0	+1.7	-1.5
Dist of Columbia							
Washington	+2.8	+0.7	-0.2	-0.7	-1.0	+0.6	-4.0
Florida							
Jacksonville	+0.8	-1.7	-0.3	+1.4	+0.7	+0.8	-2.3
Miami	-1.4	-2.0	+4.7	+5.0	+0.9	+0.9	-3.1
Pensacola	+4.1	+2.1	+0.3	+0.1	+3.0	+2.1	-2.8
Tampa	-0.4	-2.8	-2.8	+1.8	+3.1	-0.5	-2.0
Georgia							
Atlanta	+4.6	+1.5	-1.5	-3.0	-1.7	+0.5	-2.9
Augusta	+2.6	-0.2	-2.4	-2.7	-1.9	-0.6	-3.0
Macon	+2.9	+0.2	-2.6	-3.2	-1.9	-0.5	-2.7
Savannah	+0.8	-1.4	-1.9	+0.6	+1.5	-0.1	-2.8
Idaho							
Boise	+0.6	-1.8	-4.4	-8.5	-13.7	-4.7	-1.6
Pocatello	+0.5	-1.6	-4.2	-8.7	-13.0	-4.6	-2.5

TABLE NO. 1 (Continued)
PRECIPITATION MINUS EVAPORATION

Station	Dec Jan Feb	Mar Apr Nov	May Oct	Jun Sep	Jul Aug	Average Annual	Critical Year Corrections (1) (Deduct from Avg Annual)
	Rate of Gain or Loss in c.f.s. per 1000 Acres of Designated Period						
Illinois							
Cairo	+3.4	+1.7	-0.9	-2.2	-3.4	+0.2	-3.0
Chicago	+1.5	+1.3	-0.2	-1.1	-2.8	0	-2.2
Peoria	+1.7	+1.5	+0.2	-0.2	-2.7	+0.3	-3.0
Springfield	+1.6	+1.2	-0.5	-1.8	-4.6	-0.5	-3.1
Indiana							
Evansville	+3.3	+1.4	-0.9	-2.6	-4.5	-0.2	-3.1
Ft Wayne	+2.3	+2.0	+0.3	-1.1	-3.0	+0.4	-2.4
Indianapolis	+2.7	+2.2	-0.5	-1.7	-3.4	+0.3	-2.6
Terre Haute	+2.8	+2.0	-0.1	-1.8	-4.1	+0.2	-2.4
Iowa							
Charles City	+1.1	+0.8	+0.3	-0.4	2.7	0	-2.4
Davenport	+1.2	+1.1	-0.8	-1.3	-4.1	-0.5	-2.6
Des Moines	+0.5	+0.1	-1.0	-2.0	-5.0	-1.2	-3.5
Dubuque	+1.1	+1.0	0.1	-0.5	-2.9	0	-2.9
Keokuk	+1.0	+0.9	-0.9	-1.2	-4.4	-0.6	-3.6
Sioux City	+0.2	-0.5	-2.0	-3.0	-5.2	-1.7	-3.1
Kansas							
Concordia	-0.6	-1.8	-2.5	-5.0	-8.2	-3.2	-4.8
Dodge City	-1.6	-3.2	-4.9	-8.3	-11.5	-5.3	-4.8
Wichita	-0.6	-1.5	-2.0	-5.2	-9.0	-3.2	-5.4
Kentucky							
Louisville	+3.6	+1.7	-1.0	-2.9	-4.0	0	-2.7
Louisiana							
New Orleans	+4.0	+1.8	-0.3	-0.6	+1.0	+1.5	-3.5
Shreveport	+2.7	+0.3	-2.5	-6.1	-6.0	-1.7	-3.9
Maine							
Eastport	+3.5	+2.6	+1.8	+1.4	+1.0	+2.2	-1.6
Portland	+4.6	+3.1	+0.8	+0.1	-1.1	+1.9	-1.6
Maryland							
Baltimore	+2.7	+1.1	-0.8	-1.8	-1.9	+0.2	-3.7
Massachusetts							
Boston	+3.1	+2.1	-0.2	-0.8	-2.0	+0.8	-1.4
Michigan							
Alpena	+1.6	+1.7	+1.1	+0.7	-0.9	+1.0	-1.3
Detroit	+2.4	+1.6	+0.1	-0.9	-2.9	+0.4	-2.2
Grand Haven	+2.2	+2.0	+1.4	+0.5	-1.4	+1.1	-1.4
Lansing	+2.1	+1.8	+1.0	+0.4	-2.5	+0.8	-1.6
Minnesota							
Duluth	+1.0	+0.8	+0.4	+0.7	-0.9	+0.5	-1.7
Minneapolis	+0.7	+0.4	-0.4	-0.9	-3.5	-0.5	-2.4
Moorhead	+0.6	+0.5	-1.0	-1.9	-3.0	-0.7	-2.9
Mississippi							
Meridian	+5.4	+2.4	-1.1	-2.1	-0.8	+1.3	-2.3
Vicksburg	+4.8	+2.0	-1.8	-3.6	-2.6	+0.4	-2.6

TABLE NO. 1 (Continued)
PRECIPITATION MINUS EVAPORATION

Station	Dec Jan Feb	Mar Apr Nov	May Oct	Jun Sep	Jul Aug	Average Annual	Critical Year Corrections (1) (Deduct from Avg Annual)
	Rate of Gain or Loss in c.f.s. per 1000 Acres of Designated Period						
Missouri							
Columbia	+1.0	+0.6	-0.2	-0.1	-3.8	-0.3	-4.6
Kansas City	+0.5	+0.2	-0.9	-2.2	-5.7	-1.3	-4.8
St Joseph	+0.2	-0.4	-1.3	-1.6	-6.2	-1.6	-3.7
St Louis	+1.7	+1.4	-0.5	-2.3	-4.9	-0.5	-3.0
Springfield	+1.4	+0.9	0	-1.1	-3.2	-0.1	-3.9
Montana							
Havre	0	-1.5	-3.6	-5.4	-9.3	-3.4	-3.2
Helena	-0.3	-1.7	-2.6	-5.0	-9.6	-3.4	-1.4
Kalispell	+1.1	-0.8	-2.2	-4.2	-8.5	-2.4	-1.2
Miles City	+0.1	-1.5	-4.0	-7.0	-11.5	-4.1	-3.7
Nebraska							
Lincoln	-0.2	-1.3	-2.5	-3.9	-6.6	-2.5	-4.6
N. Platte	-0.8	-2.3	-3.8	-6.2	-8.2	-3.8	-3.7
Omaha	+0.1	-1.1	-2.4	-2.9	-5.8	-2.1	-4.3
Valentine	-0.4	-1.5	-3.1	-6.0	-8.0	-3.3	-3.2
Nevada							
Reno	-0.2	-3.4	-6.0	-10.1	-13.7	-6.9	-2.2
Winnemucca	-0.2	-2.9	-5.6	-9.8	-14.6	-6.8	-3.6
New Hampshire							
Concord	+2.9	+2.2	+0.8	+1.1	+0.2	+1.6	-1.6
New Jersey							
Trenton	+3.0	+1.7	+0.3	-0.2	+0.2	+1.2	-1.8
New Mexico							
Roswell	-3.0	-5.9	-7.9	-9.1	-9.6	-6.7	-2.8
Santa Fe	-1.2	-3.7	-5.9	-8.9	-8.0	-5.0	-2.2
New York							
Albany	+2.2	+1.6	-0.2	-0.7	-2.2	+0.4	-1.6
Binghamton	+2.2	+1.6	+0.7	+0.2	-0.9	+0.9	-1.8
Buffalo	+3.3	+2.0	+0.8	-0.5	-1.6	+1.1	-1.3
Canton	+2.9	+2.5	+1.4	+0.8	-1.0	+1.5	-1.6
New York	+3.0	+1.6	+0.2	-0.5	-0.8	+1.0	-1.3
Oswego	+3.0	+2.1	+1.2	+0.1	-1.4	+1.3	-1.5
Syracuse	+2.7	+1.8	+0.1	-0.5	-1.9	+0.7	-1.4
North Carolina							
Asheville	+2.2	+0.8	0	+0.1	+0.4	+0.8	-2.5
Charlotte	+2.9	+0.1	-1.7	-2.4	-1.0	-0.1	-2.4
Raleigh	+2.9	+0.4	-1.2	-0.7	+0.1	+0.5	-2.8
Wilmington	+2.1	-0.1	-0.8	+0.8	+2.7	+1.0	-2.1
North Dakota							
Bismarck	0	-1.0	-2.8	-4.5	-7.6	-2.7	-3.1
Devils Lake	+0.4	-0.3	-1.9	-2.6	-4.6	-1.5	-1.9
Williston	+0.2	-1.0	-3.0	-4.5	-7.6	-2.7	-2.3

TABLE NO. 1 (Continued)
PRECIPITATION MINUS EVAPORATION

Station	Rate of Gain or Loss in c.f.s. per 1000 Acres of Designated Period						Critical Year Corrections (1) (Deduct from Avg Annual)
	Dec Jan Feb	Mar Apr Nov	May Oct	Jun Sep	Jul Aug	Average Annual	
Ohio							
Cincinnati	+2.7	+1.4	-0.8	-2.4	-3.3	-0.1	-2.9
Cleveland	+2.2	+1.5	-0.3	-1.1	-2.1	+0.3	-1.8
Columbus	+2.6	+1.6	-0.7	-2.0	-2.7	+0.2	-2.9
Dayton	+2.8	+1.8	-0.1	-1.5	-3.0	+0.4	-3.4
Toledo	+2.2	+1.5	-0.5	-1.4	-3.4	0	-2.2
Oklahoma							
Oklahoma City	-0.8	-1.7	-2.7	-6.8	-10.7	-4.0	-4.7
Oregon							
Baker	+0.6	-1.1	-3.0	-5.3	-9.0	-3.0	-1.7
Portland	+7.0	+3.5	-0.2	-2.9	-6.0	+1.1	-2.1
Roseburg	+5.5	+2.4	-0.5	-3.8	-7.4	0	-2.4
Pennsylvania							
Erie	+2.1	+2.0	+0.7	-0.6	-1.9	+0.7	-1.6
Harrisburg	+2.3	+0.8	-0.6	-1.1	-2.4	+0.1	-2.4
Philadelphia	+2.8	+1.4	-0.8	-1.5	-1.7	+0.4	-1.6
Pittsburg	+2.4	+1.2	-0.8	-1.9	-2.6	0	-2.0
Reading	+2.7	+1.4	-0.4	-1.1	-1.3	+0.6	-2.3
Scranton	+2.6	+1.8	+0.3	+0.1	-0.7	+1.1	-2.0
Rhode Island							
Providence	+3.5	+2.2	+0.2	-0.1	-1.3	+1.2	-1.7
South Carolina							
Charleston	+1.0	-1.0	-1.7	-1.4	+0.5	-0.4	-1.0
Columbia	+1.9	-1.1	-2.8	-2.9	-1.5	-1.0	-2.3
South Dakota							
Huron	+0.1	-0.7	-2.5	-4.6	-6.6	-2.4	-3.1
Pierre	-0.3	-1.5	-3.6	-7.0	-9.3	-3.8	-2.8
Rapid City	-0.9	-1.6	-2.7	-6.5	-9.6	-3.8	-3.1
Yankton	+0.3	-0.9	-1.6	-2.7	-4.8	-1.7	-2.3
Tennessee							
Chattanooga	+4.9	+2.6	-0.7	-2.7	-1.8	+1.0	-2.9
Knoxville	+4.5	+2.0	-1.0	-1.7	-1.6	+0.9	-2.5
Memphis	+4.4	+2.1	-1.4	-3.9	-4.2	0	-2.6
Nashville	+4.1	+2.2	-0.8	-2.2	-2.7	+0.6	-1.8
Texas							
Abilene	-2.4	-5.4	-5.4	-10.9	-14.6	-7.1	-4.4
Amarillo	-2.9	-5.3	-6.6	-10.3	-12.0	-6.9	-5.3
Brownsville	-0.7	-3.0	-3.1	-2.7	-6.7	-3.0	-1.0
Corpus Christi	-0.2	-1.6	-2.5	-3.1	-6.5	-2.5	-2.1
Dallas	+0.2	-1.7	-3.2	-7.4	-10.6	-3.9	-2.8
Del Rio	-3.2	-6.2	-7.1	-10.7	-14.0	-7.7	-3.7
El Paso	-4.3	-8.4	-11.0	-13.2	-11.9	-9.2	-2.4
Houston	+2.4	-0.2	-1.2	-3.1	-3.2	-0.7	-4.0
Palestine	+1.9	0	-2.2	-5.3	-7.6	-2.0	-3.5
San Antonio	-1.4	-3.2	-4.4	-7.4	-10.0	-4.8	-3.3

TABLE NO. 1 (Continued)
PRECIPITATION MINUS EVAPORATION

Station	Dec Jan Feb	Mar Apr Nov	May Oct	Jun Sep	Jul Aug	Average Annual	Critical Year Corrections (1) (Deduct from Avg Annual)
	Rate of Gain or Loss in c.f.s. per 1000 Acres of Designated Period						
Utah							
Modena	-0.3	-3.2	-6.3	-12.0	-13.2	-6.1	-2.9
Salt Lake City	+0.6	-1.1	-3.8	-9.5	-13.2	-4.5	-3.0
Virginia							
Norfolk	+2.1	+0.3	-1.1	-1.3	-0.3	+0.2	-2.0
Richmond	+2.5	+0.5	-0.8	-1.2	-0.7	+0.3	-2.5
Wytheville	+2.4	+0.8	+1.2	+0.9	+0.7	+1.3	-2.8
Vermont							
Burlington	+1.8	+1.7	+0.9	+0.5	-1.2	+0.9	-1.7
Northfield	+2.5	+2.2	+1.2	+1.3	+0.8	+1.7	-1.0
Washington							
North Head	+9.2	+5.5	+2.6	+1.0	-0.8	+4.1	-2.2
Seattle	+5.2	+2.3	+0.1	-2.1	-4.1	+0.9	-1.8
Spokane	+1.7	-0.7	-2.9	-6.1	-9.9	-2.9	-1.7
Walla Walla	+1.2	-0.8	-3.4	-7.2	-12.6	-3.8	-2.5
West Virginia							
Elkins	+3.9	+2.6	+1.7	+2.1	+2.0	+2.6	-2.9
Parkersburg	+2.8	+1.3	-0.4	-1.3	-1.6	+0.5	-3.9
Wisconsin							
Green Bay	+1.1	+1.2	+0.2	-0.9	-2.7	0	-1.7
LaCrosse	+0.9	+0.6	+0.1	-0.1	-1.9	+0.1	-2.3
Madison	+1.2	+1.2	+0.3	-0.5	-2.9	+0.1	-1.6
Milwaukee	+1.4	+1.4	+0.4	-0.9	-3.5	0	-1.8
Wyoming							
Cheyenne	-1.7	-1.8	-2.9	-6.2	-7.9	-3.7	-2.5
Lander	-0.1	-1.1	-2.4	-7.3	-10.6	-3.7	-1.5
Sheridan	-0.1	-1.4	-2.9	-5.9	-10.3	-3.6	-2.6
Yellowstone	+0.6	-0.4	-1.8	-4.4	-7.2	2.2	+1.6

(1) The following values are amounts to be deducted from the average annual gain or loss to obtain the critical year values. To obtain the gain or loss for a shorter critical period, multiply the critical year correction by the constant for the period, as given below, and add this correction algebraically to the value for the period.

Period	Constant	Period	Constant
Dec Jan Feb	0.25	Jun Sep	1.15
Mar Apr Nov	0.45	Jul Aug	2.35
May Oct	1.35		

TABLE NO. 2
RATES FOR GAIN OR LOSS - REGION OF WASHINGTON, D.C.
VALUES IN C.F.S. PER 1000 ACRES OF RESERVOIR AREA

Month	Average Year	Critical Year Correction	Critical Year
January	+2.8	-1.0	+1.8
February	+2.8	-1.0	+1.8
March	-0.7	-1.8	-1.1
April	+0.7	-1.8	-1.1
May	-0.2	-5.4	-5.6
June	-0.7	-4.6	-5.3
July	-1.0	-9.4	-10.4
August	-1.0	-9.4	-10.4
September	-0.7	-4.6	-5.3
October	-0.2	-5.4	-5.6
November	+0.7	-1.8	-1.1
December	+2.8	-1.0	+1.8
Annual	+0.6	-4.0	-3.4

Table No. 2 gives as an example the monthly values of gain or loss for an average annual and a critical year for Washington, D.C. Average annual values are taken direct from Table No. 1 and the monthly values for the critical year are computed by using the critical year correction value and the constants indicated in footnote 1 in Table No. 1.

EXHIBIT II

COMPUTER PROGRAM ABSTRACTS

The Hydrologic Engineering Center

U.S. Army Corps of Engineers

609 Second Street

Davis, California 95616

(916) 440-2105

FTS 448-2105

THE HYDROLOGIC ENGINEERING CENTER LIST OF COMPUTER PROGRAMS

	NUMBER
Flood Flow Frequency Analysis	723-X6-L7550
Regional Frequency Computation	723-X6-L7350
HEC-4, Monthly Streamflow Simulation	723-X6-L2340
HEC-1, Flood Hydrograph Package	723-X6-L2010
HEC-2, Water Surface Profiles	723-X6-L202A
Gradually Varied Unsteady Flow Profiles	723-G2-L7450
Spillway Rating and Flood Routing	723-G1-L7100
Spillway Rating - Partial Tainter Gage Openings	723-G1-L2120
Spillway Gate Regulation Curve	723-G1-L2360
Reservoir Area-Capacity Tables by Conic Method	723-G1-L233A
Reservoir Yield	723-G2-L2400
HEC-3, Reservoir System Analysis for Conservation	723-X6-L2030
HEC-5C, Simulation of Flood Control and Conservation	723-X6-L2500

COMPUTER PROGRAM ABSTRACTS

Flood Flow Frequency Analysis (723-X6-L7550)

The purpose of this program is to perform frequency computations of annual maximum flood peaks according to the Water Resource Council Guidelines for Determining Flood Flow Frequencies, Bulletin 17, March 1976. The program computes a long-Pearson Type III frequency curve. The mean, standard deviation and computed skew coefficient are computed by the method of moments. The adopted skew is based upon a weighting of the computed skew and a generalized skew, provided as input. The program develops preliminary information based on systematic record, then if required, automatically adjusts for zero flood years, incomplete records, or low outliers. Incorporation of high outliers and historical information is then made. Final frequency curve is automatically calculated with the expected probability. The program is dimensioned for 50 historic events. The sum of the historic events and the systematic events must not exceed 130. Any number of stations may be sequentially analyzed. The skew coefficient cannot be greater than 2.0 nor less than -2.0.

Regional Frequency Computation (723-X6-L7350)

The purpose of this program is to perform frequency computations of annual maximum hydrologic events necessary to a regional frequency study. Frequency statistics are computed for recorded events at each station and for each duration. Missing events are computed so that complete sets of events are obtained for all years at all stations while preserving all intercorrelations. The mean, standard deviation, and skew coefficients of the logarithms are computed for each station and each duration. An approximate Pearson Type III distribution is assumed. Missing events are computed by a regression equation which includes a random component whose influence is proportional to the unexplained error. The flows are then arranged in order of magnitude and tabulated with median plotting positions. Statistics are then adjusted, standard deviation may be smoothed and regional skew may be specified, and frequency curves computed. The program is dimensioned for a maximum of 10 stations, but more stations can be interrelated by saving key stations from previous runs. The number of durations times the number of years cannot exceed 400, but the number of durations cannot exceed 8.

HEC-4 Monthly Streamflow Simulation (723-X6-L2340)

This program will analyze monthly streamflows at a number of interrelated stations to determine their statistical characteristics and will generate a sequence of hypothetical streamflows of any desired length having those characteristics. It will reconstitute missing streamflows on the basis of concurrent flows observed at other locations. It will also use the generalized simulation model for generating monthly streamflows at ungaged locations based on regional studies. The mean, standard deviation, and skew coefficients of the logarithms are computed for each station and each month. Each flow is converted to a normalized standard deviate using an approximation of the Pearson Type III distribution. Missing and generated values are computed by a multiple regression equation which includes a random component whose influence is proportional to the unexplained error. The previous month is one of the independent variables so as to preserve the serial correlation. The program is dimensioned for a maximum of 10 stations, but more stations can be intercorrelated by multipass operations. Input is limited to 100 years of monthly flows. Station numbers should be 3 digits or less (can be 4 digits by changing input format) and generated values cannot exceed 999,999 units.

HEC-1 Flood Hydrograph Package (723-X6-L2010)

All ordinary flood hydrograph computations associated with a single recorded or hypothetical storm can be accomplished with this package. Routines include rainfall-snowfall-snowpack-snowmelt determinations, computations of basin precipitation, unit hydrographs, and of hydrographs, routing by reservoir, storage-lag, multiple-storage, straddle stagger, Tatum and Muskingum methods, and complete stream system hydrograph combining and routing. Best-fit unit hydrograph, loss-rate, snowmelt, base freezing temperatures and routing coefficients can be derived automatically. Automatic plot routines are also provided. Unit hydrograph derivation is done by the instantaneous unit hydrograph method and Snyder coefficients are obtained. Snowmelt determinations are made by either the degree-day method or the energy budget method. Loss rates are computed using either an initial and uniform loss rate or by a variable loss rate function. Derivation of unit hydrograph and loss rate coefficients or routing coefficients is accomplished by means of an optimization subroutine utilizing the Univariate Method. The program is dimensioned for a variable number of locations, depending upon the number of alternative development plans or stream system computations and the maximum number of hydrographs retained in memory at any one time. Maximum number of flow values is 150 and maximum number of hydrographs is 270.

Water Surface Profiles (HEC-2) (723-X6-L202A)

The program computes water surface profiles for steady, gradually varied flow in rivers of any cross section. Flow may be subcritical or supercritical. Various routines are available for modifying input cross section data, for example, for locating encroachments or inserting a trapezoidal excavation on cross sections. The water surface profile through structures such as bridges, culverts and weirs can be modeled. Variable channel roughness and variable reach length between adjacent cross sections can be accommodated. Printer plots can be made to the river cross sections and computed profiles. Input may be in either English or Metric units. The method used is the step method which is generally like method 1, U.S. Army Corps of Engineers, Engineering Manual EM 1110-2-1409, 7 December 1959 - Backwater Curves in River Channels. Friction losses can be calculated from a choice of four different equations. Bridge losses are based on energy and momentum principles, and weir and orifice formulas. Critical depth is based on minimum energy.

Gradually Varied Unsteady Flow Profiles (723-G2-L7450)

This program simulates one-dimensional, unsteady, free surface flows. It calculates water surface elevations, discharges, velocities, and flow direction as functions of time at each cross section. Discharge hydrographs, stage hydrographs, or rating curves may be used for boundary conditions. Local (tributary) inflow can be accommodated. Solution of the one-dimensional equations of continuity and momentum (the St. Venant equations) is accomplished by numerical integration using an explicit, centered difference computation scheme.

Spillway Rating and Flood Routing (723-G1-L7100)

The main purpose of this program is to compute a spillway rating curve for a concrete ogee spillway with vertical walls for an assumed design head, then make a flood routing of the spillway design flood to determine the maximum water surface. The rating can also be for a broad-crested weir and can also include the discharge from a conduit or sluice. The routing can be for a gated or uncontrolled spillway. Rating curves for spillway based on WES Hydraulic Design Criteria. Rating curves for conduits based on $Q = CA\ 2gH/K$. Reservoir routing for uncontrolled spillway by modified puls. Reservoir routing for gated spillway by emergency release diagram discussed in EM 1110-2-3600. The program uses FORTRAN II.

Spillway Rating-Partial Tainter Gate Openings (723-G1-L2120)

This program was developed to compute the discharge for ogee-type weirs with partial tainter gate openings. Precise ratings can be obtained in a convenient table form for use in reservoir regulation sections or a limited volume of output can be obtained that is useful during the planning and design stages of a project. Partial gate opening ratings can be determined for any planned or existing ogee-spillway having radial-type gates. In general, the computational procedure shown on WES Hydraulic Design Charts 311-1 to 311-5 is followed with the primary difference being in the determination of G_0 (effective gate opening).

Spillway Gate Regulation Curve (723-G1-L2360)

This program will compute the gate regulation schedule curves for a reservoir utilizing the area capacity curves, the induced surcharge envelope curve, and a constant T_s which represents the slope of the recession leg of an inflow hydrograph. These curves are used to operate a gated spillway while the reservoir pool is rising under emergency conditions when communications have failed and in determining dam height for design purposes. The method of computation is based on EM 1110-2-3600, "Reservoir Regulation". FORTRAN II.

Reservoir Area-Capacity Tables by Conic Method (723-G1-L233A)

This program will compute reservoir area-capacity tables for an elevation increment of 1.0, .1 or .01 foot. The conic procedure employed is considered more suitable than the frequently used "average end area method" for determining reservoir capacities. Written in FORTRAN II.

Reservoir Yield (723-G2-L2400)

This program will perform a simulated operation study for a single reservoir with controls at the reservoir and one downstream control point. Operation is for water supply, power, water quality and water rights, taking account of flood control and other storage restrictions at the reservoir, quantity and quality of inflow to the reservoir, evaporation, quantity and quality of local inflows downstream and channel and outlet capacities, as well as project requirements. Operation interval can vary, but usually a monthly interval would be used. Translatory and channel storage effects are ignored. Water quality routing assumes thorough mixing of the inflow and reservoir quantities and pure-water evaporation before releases are made. Power is computed as a function of average head, efficiency, outflow and hydraulic losses. Written in FORTRAN II.

HEC-3 Reservoir System Analysis (723-X6-L2030)

Program will perform a multipurpose, multireservoir routing of a reservoir system. All requirements are supplied from reservoirs so as to maintain a specified balance of storage in all reservoirs, insofar as possible. Power is computed as a function of average head for each period, efficiency, outflow and hydraulic losses. In the case of conservation functions, a monthly computation interval is usually used, and economic benefits are computed based on a fixed relationship between the hydrologic quantity for a specified calendar month and location, and associated economic benefit for that month. In the case of flood control studies, the computation interval can be any length, time translations are accomplished by translating all input flows by the time required to travel to a common location, and damages are computed as a function of peak flow only. Program will accept system power demands that override individual power plant requirements, but does not provide for channel routings or percolation losses. It can assign economic values to all outputs and summarize and allocate these in various ways.

Simulation of Flood Control and Conservation Systems (HEC-5C) (723-X6-L2500)

The program is designed to simulate the sequential operation of a reservoir-channel system of any configuration. The program can be used to evaluate existing and proposed systems using defined flow sequences. Operating time intervals can be varied throughout a discharge sequence to best define the essence of the modeled operation. Expected annual damages, system costs, and net benefits for flood damage reduction can also be determined. This program represents a major expansion of the capabilities of the HEC-5 program for flood control operation. The input for HEC-5 is generally compatible with the requirements for this program. Discharge hydrographs are provided to the modeled locations in the system. Hydrographs are routed through channels by any of five hydrologic routing techniques. Reservoir releases are made to evacuate flood control storage as rapidly as possible without causing flooding, to provide for two levels of minimum flow requirements, and to provide defined hydropower requirements. Diverions can also be simulated within the system. The program is dimensioned for 15 control points, 10 reservoirs, 9 power plants, 11 diverions and 50 time periods. (Any number of time periods can be used with the program.) The dimensions can be varied for larger or smaller computer systems.

VOLUME IV: EXISTING FACILITY INTEGRITY

SECTION 1: INTRODUCTION AND OVERVIEW

Scope

This volume presents guidelines for investigating and evaluating, at the feasibility level, existing dams and appurtenant structures to determine their structural suitability for the addition of small hydroelectric facilities. These guidelines are applicable for dams of heights up to about 100 feet. The more common modern and older types of dams are covered herein, since they are the most likely to be considered for the addition of small hydroelectric facilities. For dam types not covered herein, the principles discussed should provide sufficient guidance for their investigation.

For the purposes of this report, dams are classified into the following four basic categories:

1. Earth and rockfill
2. Concrete
3. Masonry
4. Miscellaneous (includes stonewall-earth and rockfill timber crib).

Appurtenant works included in this document are those that exist prior to the addition of new hydroelectric facilities; they consist of spillways and outlet works for the most part, though occasionally they may include existing penstocks, powerhouses, and flumes.

Section 2 presents a classification and description of the types of dams and appurtenant works that are expected to be encountered most often during investigations for the addition of small hydroelectric facilities. Also included in Section 2 are detailed discussions of reservoir conditions from an integrity viewpoint, common deficiencies and failure modes of dams and appurtenant works, and the adverse effects that power additions may have.

Section 3 presents recommended methodology, guidelines, and personnel requirements for performing the investigations.

Section 4 discusses methods of rehabilitating the various components of existing facilities if found to be deficient.

Section 5 presents guidelines for estimating the costs for performing the various stages of the integrity investigations and for performing rehabilitation works. It must be pointed out that dams are unique structures, each fitting specific site and material conditions; that costs will therefore vary widely depending on site-specific conditions; and that considerable judgment is required for estimating realistic costs for both engineering and rehabilitation work.

Objectives of Integrity Investigations

The installation of hydroelectric facilities at existing dams will require significant capital investments, and there are liabilities to consider should the dam fail and cause loss of life and destruction of property downstream. It is therefore prudent to determine if the existing facilities are structurally sound and will perform adequately if modified and operated as a hydroelectric facility prior to committing the necessary investment for modifications. Studies of failure probability at any time during the life of a dam indicate that, where failures occur, the incidences of failure during initial years of operation are relatively high; thereafter the probability of failure at any time decreases for a number of years, and then increases rather dramatically as the dam deteriorates after many years of existence. Thus, the fact that a dam has performed adequately for a number of years is not proof that it is structurally sound. In fact, the opposite may be true: it may have deteriorated to the point that it is about to experience difficulties and may even fail.

The integrity studies discussed herein serve a different purpose than do surveillance and inspections of dams conducted by local, state, and federal regulatory agencies, which are concerned with the safety and protection of downstream inhabitants and property. The prior determination by an agency that a dam has been safe under existing operating conditions, while indicating that it is more likely to be sound than a dam not so studied, does not assure that the facility will continue to be sound under hydroelectric operating conditions, or that excessive maintenance will not be required for the life of the project. Several types of dams, such as some in the New England area, that are likely to be considered for installation of hydroelectric facilities do not meet the legal definition of a dam accepted by dam safety regulatory agencies, and they have not been evaluated for safety by appropriate governmental agencies.

The primary objectives of the type of integrity investigation discussed herein are to determine the structural integrity of existing facilities for hydroelectric power operational conditions, to evaluate the cost of remedial work, if required, and to assess maintenance requirements and the expected longevity of the existing facilities.

Major Technical Components of a Feasibility Level Investigation

From technical, cost effective, and administrative viewpoints, the integrity investigations of existing facilities to determine the feasibility of adding hydroelectric power units may be best accomplished by a program that has potentially one to three stages:

Stage 1 - Evaluation of existing data and site inspection

Stage 2 - Developing new data and performing additional evaluations

Stage 3 - Developing designs for rehabilitation and construction cost estimates.

The Stage 1 investigation consists of collecting, reviewing, and evaluating available data and information pertaining to the facilities; making a detailed site inspection; evaluating the existing facilities for the intended use; and making recommendations for additional investigations, if appropriate. If the dam and facilities are relatively new and if sufficient data are available, it may be possible to determine that no additional investigations are required and that the dam and facilities are adequate for the addition of hydroelectric facilities. If the dam is old, and/or adequate data are not available, or the site inspection reveals the possibility of potential problems, a Stage 2 investigation will be required. The final step of the Stage 1 investigation is preparation of a report which presents the evaluation of the existing facilities and recommendations as to any problems related to adding power facilities; the report would also include a program and cost estimate for a second-stage investigation, if required.

Stage 2 consists of implementing the program of additional investigations developed during Stage 1 if the dam is not deemed suitable for direct hydro addition. This consists of developing additional required data by drilling, sampling, and testing; evaluating the new data; analyzing portions of the facilities that are questionable; and preparing a report that describes the work that was performed and presents the data developed, results of analyses, evaluation of the facilities, recommendations, and a Stage 3 program, if recommended. The Stage 2 program must be kept flexible and the investigation modified to accommodate or evaluate unanticipated conditions that are revealed during the investigation, so that the required data are obtained and the work is cost effective. If the Stage 2 investigation indicates that the dam and facilities are suitable for the addition of hydroelectric facilities, Stage 3 would not be required. If rehabilitation of the dam and/or facilities is required, a program for Stage 3 would be developed.

Stage 3 consists of developing methods of rehabilitating the dam and/ or facilities and estimating the cost of construction to make the dam and facilities suitable for adding hydroelectric facilities.

If, at any time during the various stages of investigation, it becomes obvious that the existing facilities are technically unsuitable or that rehabilitation costs will be excessive, the investigations should be terminated. In addition, if significant integrity problems exist at a dam, the dam owner or operator should be notified so that the deficiencies can be examined further and corrected so as to minimize hazard to life and property.

Intended Use of These Guidelines

This volume, as well as the other volumes of the manual, is intended for the use of owners of potential power sites, governmental agencies, private consultants, and research and educational institutions. The use of this volume by non-technical individuals (i.e., persons other than engineers and geologists) should be limited to selecting people competent to perform the work, planning for investigation costs, and administering theinvestigation program. The investigation of existing structures to determine the feasibility of utilizing them for the addition of hydroelectric facilities must be performed by engineers and engineering geologists with experience related to dams and other hydraulic structures.

The addition of hydroelectric facilities represents a significant financial investment and, as attested by past dam failures, human lives and property downstream of a dam can be in jeopardy. Since the dam is the focus of and essential to the project, it is imperative that the suitability of the existing facilities be firmly established before capital is committed to the addition of hydroelectric facilities and the liability for the consequences of a dam failure is assumed. Dams and appurtenant structures are complex, and knowledge covering a wide range of geology, hydrology and various applicable engineering fields is required to adequately evaluate existing dams. It is beyond the scope of this volume to provide basic engineering and geologic information to technically educate all readers. It is therefore presumed that persons utilizing this volume for evaluating the technical suitability of existing dams and appurtenant structures have a basic understanding of geology, soils engineering, structural engineering, hydrology, and hydraulics, with sufficient experience to apply engineering judgment during the feasibility evaluations.

This volume can be best utilized to obtain an awareness of the more common problems that have been historically associated with dams and that the investigator must be alert for; and beyond that, the volume provides a means of organizing and performing the investigation, and offers guides to methods of rehabilitation and cost estimating. This volume cannot be used as a substitute for proper educational background, experience and engineering judgment, which are essential to the proper conduct of this type of investigation. A recognized expert in any areas of concern should be consulted before a project is adopted as feasible.

SECTION 2
CLASSIFICATION OF DAMS
AND PRINCIPAL AREAS OF CONCERN

Classification and Description of Principal Dam Types

Dams are generally classified on the basis of materials used for their construction. The basic dam classifications are (1) concrete, (2) masonry, (3) earth and rockfill, with the remainder grouped as (4) miscellaneous. These basic categories can be further subdivided based on geometric configuration or internal zoning. Composite dams consist of a combination of two or more different types of structures.

Concrete Dams. Concrete dams include gravity, arch and buttress types. Examples are shown in Figures 2-1 through 2-4. Concrete gravity dams depend on their mass for stability, and may be either straight or (sometimes) curved in plan. The curved plan takes advantage of the arch action for added strength. Gravity dams generally require sound rock foundations but may be founded on alluvial foundations. Concrete gravity sections are often used as overflow sections for composite dams. Concrete gravity dams founded on pervious foundations require special design considerations to control seepage, prevent excessive uplift pressures, and maintain the integrity of the foundation.

Figure 2-2. Variable radius arch dam with radial spillway gates, control for low level river outlet, and intake structure for tunnel outlet

Concrete arch dams utilize arch action to transmit most of the water load from the reservoir to the dam foundation and/or abutments. In order to transmit the arch action to the abutment walls, the arch dam must act as a monolithic structure and thus be free of open cracks and other structural discontinuities. Arch dams are much thinner in section than gravity dams. The U.S. Bureau of Reclamation classifies arch dams with a b/h ratio (ratio of the base thickness of the crown cantilever to the structural height of the dam) of 0.2 and less as thin, between 0.2 and 0.3 as medium thick, and 0.3 and greater as thick. Arch dams may be either single or double curvature and the radius of curvature may be constant or variable.

Figure 2-1. Concrete gravity dam.

Since the majority of the reservoir force is transmitted to the abutments of an arch dam, the abutments must be capable of withstanding the arch thrust. Narrow canyons with steep, strong walls are generally best suited for arch dams.

Buttress dams utilize a sloping membrane, generally of concrete, to transmit hydrostatic forces to a series of structural buttresses placed at right angles to the dam axis. There are several types of buttress dams, including flat-slab or Ambersen, multiple arch, multiple dome, roundhead, diamondhead, and cantilever buttresses. The most common and important buttress dams are the flat-slab and multiple arch. A few timber and steel deck buttress dams have been constructed. Buttress dams generally require considerably less concrete than gravity dams of the same size, but require more formwork. Reinforcement is required in buttress dams with thin slabs or arches.

Buttress dams are best suited to wide valleys with gradually sloping abutments; they can be founded on rock or sound alluvium. Depending on the foundation, the buttress may be cast directly into excavation into rock or supported by spread footings, or a continuous slab foundation may be utilized for alluvial foundations. Flexible joints are normally provided between adjacent slabs and buttresses, allowing each section to act independently. Thus, minor settlements and deformations are not critical.

Figure 2-3. Multiple arch dam.

Figure 2-4. Multiple arch dam (buttresses have been strengthened and cross channel stability increased by added diaphragm)

Masonry Dams. Masonry dams consist of rubble or stone laid in mortar (Figure 2-5). Both gravity and arch masonry dams have been constructed. These are similar in section and foundation requirements to concrete dams. Masonry dams were a principal type of dam constructed before rising labor costs and better utilization of mass concrete made them uneconomical. Most masonry dams were constructed before the twentieth century and there have been few, if any, significant masonry dams constructed in the United States for several decades.

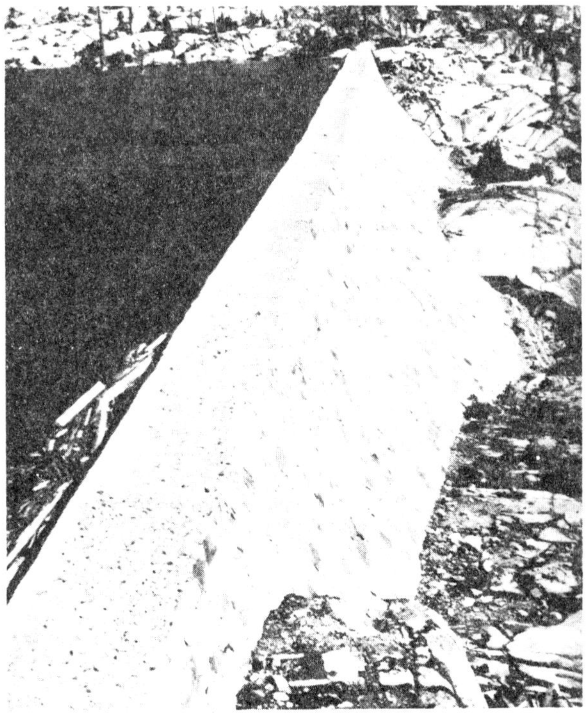

Figure 2-5. Gunite-faced masonry dam

Earth and Rockfill Dams. Earth and rockfill dams utilize natural materials for construction. The development of large, rapid performing, earth-moving equipment in recent years has made these types of dams extremely cost competitive. In addition, earth dams can be constructed on foundations that are unsuitable for other types of dams. Examples of earth dams are shown on Figures 2-6 through 2-8.

In the past earth dams were constructed by loose fill and hydraulic or semi-hydraulic methods. The rolled fill (or compacted fill) type of construction is now used almost exclusively. There are several types of earth dams, including homogeneous, zoned, and diaphragm types.

Homogeneous dams are constructed of essentially one type of material. In order to control the level of saturation and water pressure within embankments, drainage zones are often used. Homogeneous dams modified in this way can have either rockfill toe drains, blanket drains, chimney drains or combinations of these drains. These drain zones are constructed of essentially freedraining sand or gravel, protected with suitable filter zones to prevent migration of fine-grained soils into the coarser drain zones.

Zoned earth dams consist of an impervious zone surrounded by more pervious shells. The shells generally consist of sand, gravel, or cobbles; they support the dam core and increase dam stability by controlling the phreatic surface within the embankment. Piping of the fine-grained core material into the coarser shells is prevented with selected graded filter zones. The dam core can be either central or sloping upstream.

Figure 2-6. Zoned earth dam under construction

Diaphragm type earth dams consist of pervious material (sand or gravel) with a thin impervious membrane or diaphragm which acts as an impermeable barrier. The diaphragm may be centrally located or may form the upstream face of the dam. Earth, cement concrete, and asphaltic concrete, asphaltic membranes, and plastic and rubber sheets have been used as diaphragm material.

Rockfill dams are constructed of rock with an impervious membrane. The rockfill provides stability while the impervious membrane serves to retain the water. The impervious membrane is protected by filter or bedding zones and transition zones. The membrane may be an upstream facing or a thin interior core. The upstream

Figure 2-7. Constructing crest of earth dam (spillway control structure in foreground)

facings of rockfill dams have been constructed of impervious soil, concrete slabs, asphaltic concrete and steel plates. Interior cores are generally of earth. The rockfill sections have been constructed with dumped rock (sluiced or dry) and by placing and compacting the rock in horizontal lifts. Rockfill dams usually require either a rock or a competent sand or gravel foundation.

Figure 2-8. Earth dam (line of stakes marks failure scarp)

Miscellaneous Dams. The more important dams in the miscellaneous category include timber dams (Figure 2-9) and stonewall-earth dams (Figure 2-10). A number of small timber dams have been constructed in the west and northwest. Most timber dams are less than 20 feet high, but a few have been constructed that are over 60 feet in height. These dams have a relatively short life due to rotting of the timbers. However, with maintenance, these dams have performed successfully for a number of years. There are several types of timber dams. These include rockfilled crib, frame and deck, crib and deck, and beaver type. Other than the rockfilled crib dams, most of these dam types are rarely over 10 feet in height.

Figure 2-9. Timber crib dam

Figure 2-10. Stonewall-earth dam

There are at least several hundred stonewall-earth dams located in the northeastern U.S. These dams generally consist of an upstream earth section with a downstream vertical to near-vertical wall of stacked stone. Mortar has generally not been used in the stonewall. These dams are generally less than 30 feet high with most of the dams under 10 feet high. Most are 100 to 200 years old.

Beyond these discernible categories, there are structures which combine various design principles and construction materials. One example is shown in Figure 2-11.

Figure 2-11. Unusual structure which combines concrete gravity abutment segments with central timber slab and buttress segment

The Reservoir

There are primarily two types of reservoirs where hydroelectric facilities may be installed at existing impoundments. One is a run-of-the-river type installation where the heads are low and reservoir capacity small, and where there would not be much fluctuation in the reservoir level. The other type of facility consists of a storage reservoir where the reservoir level will fluctuate significantly.

For the run-of-the-river type installation, reservoir conditions would not vary significantly from existing conditions and therefore the addition of hydroelectric facilities would not have a significant effect on the reservoir. Past performances of these types of reservoir would offer an excellent guide to the reservoir conditions that would be expected after the hydroelectric facilities were added.

At the storage type of reservoir, the installation of hydroelectric facilities may not change the reservoir operating conditions significantly, or it could result in major changes in the operation of the reservoir. If hydroelectric power is generated from normal releases of water (such as for irrigation purposes), then reservoir operating conditions would not change significantly and past performance would fairly indicate the reservoir conditions that would be expected after the hydroelectric facilities were added. If the reservoir had been performing well for a number of years, the investigation would be directed primarily towards signs of progressive deterioration.

If the reservoir had been operating at a relatively constant level (such as essentially full, as for a recreation facility), and if power releases resulted in a significant reduction in the reservoir level, some of the previous benefits could be impaired and the change in operating conditions could significantly affect the performance of the reservoir. Drawdown conditions have long been recognized by earth dam designers as one of the most critical conditions affecting stability of the upstream slope of a dam. Reservoir drawdown conditions could also create adverse slope stability conditions in the reservoir just as in an earth dam. Therefore, if a reservoir has not been subjected to drawdown, past performance is not an adequate guide for this operating condition. In this case, the potential for dam slope and reservoir side slope slides would have to be investigated to evaluate the possibility of impairing the functioning of the dam, appurtenant structures, and reservoir; endangering facilities above the reservoir level; or, if no facilities are involved, of producing scars along the reservoir rim that are not acceptable.

If storage in the reservoir is increased beyond previous operating levels (as for a flood storage facility), the benefits could also be impaired and the performance of the reservoir could be significantly different. Reservoir slope stability could be reduced because of additional saturation and drawdown; and reservoir seepage could be a problem because the upper level of the reservoir might not be as impervious as the lower levels.

Appurtenant Works

The appurtenant works are the structural facilities associated with a dam by means of which the purposes of the dam and reservoir and the use of the water are achieved, usually by controlling the flows of water which enter and leave the reservoir. The most common and critical appurtenances are spillways and outlet works. If power plant components are closely integrated with a dam, hydraulically or structurally, the plant can be considered an appurtenance. Navigational locks, fish ladders, and log sluices are other kinds of appurtenances. Inlets also are associated with dams which store water from a remote source.

Spillways. Spillways are required at storage dams for releasing incoming flood waters that are in excess of available reservoir capacity. They are also required at diversion dams to bypass stream flows in excess of the diversion capacity.

Separately identifiable spillway components are the approach or entrance channel, the control structure, the discharge channel, the terminal structure, and the return or outlet channel. Some examples are shown on Figures 2-12 and 2-13. The topography, geology, dam type and spillway type determine which components are needed.

The entrance channel conveys water from the reservoir to the control structure and is usually required except for concrete dam overflow spillways. The channel profile and cross sections are sized and configured to minimize channel head losses, to provide uniform head, and to optimize the discharge coefficient for the control structure crest.

Figure 2-12. Spillway control structure

The control structure governs the reservoir outflow. Control structures may be an overpour crest in the form of a shaped weir or a sill, orifice-like openings, or conduit entrances. They may be either unregulated or regulated by gates, flashboards, and valves. Control structures are configured, positioned, and shaped in many different ways. Siphon control crests may sometimes be in use at older dams.

The discharge channel conveys and returns the water to the stream beyond the dam or into other topographic depressions beyond the reservoir basin. The channel may be on the face of a concrete dam; an open channel, lined or unlined, in natural formations; a conduit through or beneath the dam; or a tunnel through an abutment. Free falling outflows from overpouring crests require no discharge channel. Profiles, cross sections, alignments, and lengths are dimensioned and positioned in a variety of forms.

The terminal structure prevents undue erosion of the stream channel or damage to adjacent structures and the dam from the high-energy-laden spillway discharges. Stilling basins, roller buckets, baffled impact-type basins, and dentated aprons are commonly used for dissipating the energy as the flow returns to the stream. For efficient performance, their position in elevation with respect to tailwater elevation is critical. Where erosion resistant bedrock is present, releases may sometimes be made directly back to the stream at a distance from vulnerable structures. If the jet impingement can be predicted and controlled, the terminal structure can be a cantilevered or flip bucket extension of the discharge channel, provided that the impingement region and associated plunge pool will not endanger nearby structures or the bucket substructure.

Figure 2-13. Spillway discharge channel, terminal structure, and return channel

Figure 2-14. Radial gates at spillway control structure

The return channel conveys the flow back to the stream. The profiles and cross sections are dimensioned to avoid creating a hydraulic control that would adversely affect the energy-dissipating characteristics of the terminal structure and to provide velocities that minimize scour.

The spillway type and location are dependent upon the type of dam, site topography, geologic and foundation characteristics, and magnitude of expected floods. For example, spillways can be safely located on concrete dams, but it is not considered good practice to place them on major embankment type dams. A topographic saddle distant from the dam may provide a favorable location.

Spillway types are categorized by some distinctive characteristics of their components. The more common types are free-drop overflow, ogee overflow, open channel chute, side channel, conduit, tunnel, drop inlet, shaft, and siphon. Common classifications by use or operation are service, auxilliary, emergency, fuse-plug, controlled, and uncontrolled.

Controlled spillways often afford economic and operational advantages but they can also be hazardous at embankment dams if they fail to operate as planned. Redundant operational and alternative overflow relief features and seasonal variation of the level of the reservoir or the position of the control devices can sometimes reduce the probability of overtopping the dam should the system fail to function as planned.

Common control devices are radial gates, drum gates, Bascule-type gates, slide and wheeled gates, flashboards, stoplogs, soldier beams, and bulkheads. Two common types of gates are shown in Figures 2-14 and 2-15. The electrical, mechanical, and operational control systems for operating, installing, and removing headwater control devices range from simple to complex, and include local or remote, automatic or manual. Power sources may be hydraulic, commercial electrical, locally generated electrical, internal combustion engines, or a combination of these.

Outlet Works. Outlet works regulate the release of water from a reservoir and are sized and designed to meet the water demands and other purposes of the project. Releases of water are required for irrigation, municipal, industrial, and power generation use; for flood control regulation; for stream flow maintenance; and to satisfy prior or other downstream water rights.

Outlet works are usually classified according to (1) purpose such as canal outlets and pressure pipe outlets which divert water into canals and pipelines; river outlets which release water directly into the stream channel; flood control outlets which release water beyond the dam; and power outlets which admit water into tunnels and penstocks serving detached and integral power plants; (2) structural configuration - such as open channels or closed conduits; and (3) hydraulic operation - pressure or free-flow.

Outlet works are also used to lower the reservoir stage or empty the reservoir for inspection, maintenance, and precautionary reasons. Controlled, conduit type outlet works should not be considered as a spillway or part of the spilling capability for passing the spillway inflow design flood because of the uncertainties of availability and of absolute operational reliability during extreme floods.

Figure 2-15. Roller gates for intake control structure

Figure 2-16. Intake structure of low-level outlet works

The type and location of the outlet works are dependent upon the purpose of the outlet, the type of dam, topographic, geologic and foundation characteristics, and the point of downstream release. Outlet works components which can be separately identified include the entrance channel; the intake structure; the waterway; the control structure; the terminal structure; and access shafts, bridges, and tunnels to operation and maintenance stations. Figures 2-16 and 2-17 show typical intake and terminal structures. The required components and their features are determined by the type, purpose, and location of the outlet works. A dam may have several outlet works for different purposes and they may be at different elevations.

The entrance channel conveys water to the intake structure of the outlet works. The intake structure establishes the ultimate drawdown level, guards against entry of trash, and may incorporate water control devices for flow regulation or closure devices for unwatering the outlet works during inspection and maintenance. Intake structures may be vertical or inclined towers; drop inlets; or submerged, box-shaped structures. Intake elevations are determined by the head needed for discharge capacity, storage reservation for siltation, the required amount and rate of withdrawal, and the desired extreme drawdown level.

The waterway conveys the released water from the intake structure to the point of downstream release. Waterways may be open channels, steellined sluiceways or ports through concrete dams, lined or unlined tunnels in abutments or from the reservoir basin elsewhere, or closed cut-and-cover conduits beneath the dam. Closed waterways may be designed for pressure and non-pressure flow. Pressure pipelines and penstocks may be extended through non-pressure conduits and tunnels, affording access and pressure relief.

The control structure regulates the flow of water through the outlet works and may be located at the upstream or downstream limits of the waterway, at intermediate positions, or at several positions. They house and support control devices which proportion or shut off outflow. Types of valves and gates used for control devices include slide gates; commercial gate valves; butterfly valves; ring follower, fixed-wheel, and roller train leaf gates; needle tube, jet-flow, hollow-jet, and Howell-Bunger valves; and bottom-seal and top-seal radial gates. For satisfactory performance, the type of valve or gate must be matched to service conditions such as maximum head, flow velocity, in line or free discharge, fully open or closed or partially open, and unbalanced or balanced head operation. The operational control systems are similar in principle to those discussed above for spillways.

The terminal structure delivers the flows to the point of downstream release. Any need for and the type of terminal structure is determined by the purpose of the outlet works. For river and flood control outlets the terminal structures can be similar in principle to those for spillways or the outlet releases may be conveyed through the spillway terminal structure. Normally, terminal structures are unnecessary for pipe, canal, and power outlets.

Figure 2-17. Terminal structure of outlet works

The access component may be a bridge from the dam crest to a tower intake structure, a vertical cast or drilled shaft from the ground surface to a valve chamber in a conduit or tunnel, an inclined tunnel, roadways and bridges to control structures and houses at the downstream end of the waterway, or galleries and ladders at concrete dams.

Power Plant. A power plant may be hydraulically and/or structurally coupled to a dam. The intake and penstock water passages for the plant may be formed within the body of a concrete dam, or the plant penstock may be appended to the face of the dam. The plant enclosures may be formed by the buttresses and face elements of a buttress-type dam, or the plant substructure and enclosure may create a water barrier auxiliary to or in conjunction with the actual dam. An underground plant may be located within the dam abutment mass. Economic advantages from reduced head losses and hydraulic transient control associated with plant operation are possible with these close coupled arrangements.

It is unlikely that these coupled arrangements will be encountered at existing impoundments under study for small hydropower. If they are, the existing plant elements must be evaluated for their impoundment integrity. If they are not, it is unlikely that any proposed power plant facilities would replace portions of the existing dam as a primary water barrier. Instead, the dam modifications would likely be those to accommodate power outlets for structurally detached power plants.

Navigational Locks. A navigational lock may form a portion of a dam. It is subject to the same gravity, seismic, seepage and hydraulic forces as a dam. The lock is usually joined to adjacent earth, rockfill, or gravity sections of the actual dam.

The facilities controlling admission and discharge of water for the lock chamber and the lock gates are hydraulically and structurally similar in many ways to the control devices and gates for outlets and spillways. The same engineering design and performance principles can be applied in their integrity investigation.

Fish Ladders and Log Sluices. Facilities for fish and log passage through or over dams are also similar to outlets and spillways in their hydraulic performance and manner of control. Their integrity for safe impoundment is structurally investigated employing the same techniques used for outlets and spillways. Examples are shown in Figures 2-18 and 2-19.

References. Several of many excellent references in the literature of dam engineering amplify, discuss in detail, and present examples of the principles and descriptions discussed in this portion (Golze, 1977; Justin et al., 1945; U.S. Department of Interior (USBR), 1974/1; *et passim*).

Figure 2-18. Fish ladder at flashboard and buttress dam

Common Deficiencies or Failure Modes Associated with Small Existing Dams

Many existing dams may have significant deficiencies or may be subject to failure, even though they have performed satisfactorily for decades. The state of the art in dam design and construction has advanced considerably from the past and many older dams do not meet present standards. Furthermore, dams, appurtenant works, and their foundations are subject to aging and deterioration; or potential problems may develop when the operating conditions of the dam change. Deterioration of the dam and related structures may be readily apparent, or the deterioration may be very subtle and not manifest itself until substantial damage has occurred and failure becomes imminent. Several typical dam deficiencies or failure modes are listed below.

1. Dam overtopping
2. Piping
3. Uncontrolled and excessive seepage
4. Foundation instability
5. Embankment slope instability
6. Deterioration of slope protection on embankment dams
7. Deterioration of concrete
8. Excessive hydraulic uplift pressures
9. Spillway and outlet failure or inadequacies
10. Erosion.

One of the most common modes of small dam failure is recognized to be due to overtopping. This is due to the inadequacy of, or lack of, spillways in many old, existing structures. The effects of overtopping on different types of dams can vary considerably. Overtopping can cause serious distress and even the total failure of earthfill structures. In the case of concrete dams, uncontrolled overtopping can damage or destroy the abutments and/or structures or appurtenant works immediately downstream of the dam and can cause erosion and scour downstream of the structure. Many old, stonewall earth dams have been subjected to reported minor overtopping without significant damage to the structure.

Reservoir slides can cause large waves that can overtop the dam. Reservoir slide debris can also block outlets and spillways, leading to overtopping.

Spillway and outlet failures and malfunctions can lead to dam overtopping. Spillway and outlet facility criteria are discussed in Section 3.

A form of erosion known as piping is caused by the movement of soil particles to unprotected exits due to uncontrolled seepage. Piping failures are recognized to be a very common mode of failure of dams. Piping can occur through embankments or through a dam's foundation or abutments. Areas adjacent to conduits are particularly susceptible to piping because of the difficulty in properly compacting backfill around these conduits. Piping problems have developed at dams with many years of satisfactory performance due to solution of soluble materials (such as gypsum) within the dam abutments and foundations. Similar problems have resulted from animal burrows and rotted tree roots. Differential settlement cracks can also provide paths for uncontrolled seepage with attendant erosion.

Figure 2-19. Log chute (debris skimmer)

Such uncontrolled foundation seepage can lead to high uplift pressures or uplift pressure distributions not anticipated during the design of the dam. Such excessive pore water pressures can lead to the formation of boils and springs and, by reducing the shearing resistance, to the failure of abutments and slopes.

Problems that are associated with adverse foundation conditions include slides, differential settlements, and excessive seepage. Foundation slides have occurred where foundation materials have low shear strength or where seams of weak material exist in an otherwise competent foundation. Differential settlements of compressible foundations can lead to excess cracking in the dam. Pervious seams and adverse bedding planes can provide paths for uncontrolled seepage. Foundation problems can also result from deteriorating grout cutoff curtains and plugged relief wells or drains.

Embankment slope instability can lead to catastrophic failure of a dam. The most common cause of slope instability is the development of excess pore water pressures due to unfavorable seepage conditions. Dams with many years of satisfactory performance can develop slope stability problems when operating conditions change, such as drawdown of the reservoir, or when embankment material properties change due to aging.

Embankment dam slopes are subject to erosion from wave action on the upstream slope and from surface runoff on the downstream slope. Riprap slope protection can suffer degradation from wave action, slaking, and decomposition. When riprap is placed directly on embankment surfaces without suitably graded bedding or filters, the underlying embankment materials can be washed out, causing sinkholes and riprap sloughing.

Numerous concrete and masonry structures have exhibited substantial deterioration of structural concrete and grout. Deterioration of concrete from alkali-aggregate reaction, sulfate attack, freeze-thaw deterioration and leaching of soluble substances from the cement are typical problems that may develop in concrete dams. This deterioration (or, sometimes, poor construction practices) can lead to vuggy (cavitated) concrete, "popcorn" concrete, areas with mortar and no aggregate, and areas of aggregate and no mortar.

Alkali-aggregate reaction can lead to concrete deterioration well within the interior of structures, greatly reducing the concrete strength. Freeze-thaw deterioration is generally concentrated on concrete surfaces.

Over-stressing and differential displacements of concrete structures create areas of distress and cracking where freeze-thaw action or leaching of the concrete can lead to further deterioration of the structure.

Concrete deterioration and cracking can lead to exposure of steel reinforcement. Corrosion of steel reinforcement with subsequent loss of strength, along with loss of concrete strength, has been a problem with some buttress dams.

Dams located in seismically active regions may be subjected to severe shaking or foundation displacements due to fault movement. Seiches (seismically induced water waves) may cause overtopping. Seismic shaking causes an increase in pore pressures in impervious or semi-pervious materials and a resulting decrease in shear strength. The decrease in shear strength, coupled with the seismically induced shear stresses, can lead to the failure of the dam. Other earthquake effects may be cracking of embankments or excessive settlements, the former providing direct paths for water flows with resultant erosion and possible breaching, and the latter leading to loss of freeboard and possible overtopping.

The more unusual types of dams can be subject to unique problems. Timber dams are subject to deterioration of the wood timbers. The rate of deterioration depends on the type of wood used and on dam operation. Redwood timbers generally have a longer life than cedar. Timbers that are repeatedly wetted and dried deteriorate at much greater rates than timbers kept continuously wet. Stonewall earth dams have been known to fail due to frost heave.

Adverse Effects of Power Additions

Any modification of existing dams, appurtenant structures, reservoir conditions, or the area near these facilities will modify stresses within the components. Some modifications resulting from the addition of hydroelectric facilities would have an adverse effect on the integrity of the existing facilities. Changes in reservoir operating conditions and the possible effects on the reservoir area are discussed above, and the effects of adding hydroelectric facilities to the existing facilities are discussed briefly below. Methods of investigating and rehabilitating existing facilities are discussed in Sections 3 and 4 of this volume.

The most common problems that have been encountered in adding hydroelectric facilities to existing structures have been associated with utilizing existing outlet conduits or installing new water passages and making excavations for the power facilities downstream of the dam.

The existing outlet facilities form the most obvious waterway from the impoundment to the powerhouse, and also the least expensive to construct. However, there are several pitfalls in using these facilities as a penstock. Conduits with controls at the upstream end or at intermediate points were probably not designed for and are not capable of withstanding the full hydrostatic head created by the reservoir. If the controls are moved to the downstream end (as is normally done when power facilities are added), full reservoir pressure will exist in the conduit when power is not being generated, and when the powerhouse is in operation the internal pressures will be higher than normal for a dam conduit. At several facilities, it has also been found that the outlet conduits can be inadequate for other reasons: the small diameter of the conduit may cause excessive energy loss, or cavitation may occur because of abrupt changes in alignment.

If the existing outlet facilities are inadequate for power generation, a new water passage must be constructed or the inadequate portion replaced or modified. Small diameter conduits buried under earth and rockfill dams are not amenable to modification, and generally the only practical means of adding a new water passage to an earth or rockfill dam is by tunneling through an abutment. On the other hand, it is practical to construct a water passage through a concrete dam. However, a structural analysis of the changed condition should be performed, since the introduction of an opening in the concrete may result in overstressing portions of the structure. If blasting is required to drive a tunnel through or remove concrete from an existing structure, the charges must be controlled so that the structure and foundation are not damaged.

Each existing facility is unique and good engineering judgment must be used to evaluate and solve problems in adapting water passages for use in hydroelectric generation. For example, at one site, a diversion tunnel which had been constructed through one abutment during initial construction was successfully adapted for hydropower use. The tunnel had been plugged with concrete near the axis of the dam, with a small diameter steel conduit extending through the plug to a point downstream of the dam. It was concluded that excessive energy losses would occur if the samll diameter conduit was used for power generation. The solution was to blast an opening in the plug, reseal the plug around a larger diameter steel pipe, and extend the pipe through the tunnel to the powerhouse. Construction specifications limited the energy release during blasting and instruments were used to record accelerations when blasting was performed.

As well as structural and hydraulic problems, environmental problems may occur when existing outlet facilities are utilized. At one such facility, where only a low level intake existed, it was determined that, at certain times of the year, water releases from that level were deficient in dissolved oxygen and aquatic life for some distance downstream of the powerhouse would be destroyed. Either a multiple level intake structure needed to be added or the water aerated before release from the powerhouse area.

The addition of power facilities downstream of the dam will require regrading of the area. Building up the area at the toe of the dam will generally not reduce the stability of the dam if drains or other seepage outlet paths are not blocked off. Excavations within the influence area of the dam will result in weakening the existing facilities and increasing the potential for sliding, foundation failures, and piping problems. As a general rule, for planning and feasibility purposes extensive excavations which are over 5 feet deep should not be performed close to the downstream toe of the dam. A distance of at least one half the height of the dam should be maintained between the downstream toe and the upstream edge of an excavation in solid rock. This distance should be increased to at least the full height of the dam when the excavation is in soft rock or soil. Where this is impractical, rock stabilzation techniques may have to be used. Where foundation conditions are poor or questionable, a subsurface exploration program should be conducted and an evaluation made of the effects of excavation in the vicinity of the dam. Prior to final design and construction, a subsurface exploration program must be conducted and an evaluation made of the effects of excavation in the vicinity of the dam regardless of the type of foundation material.

The addition of power facilities at the toe of the dam may require relocation of the spillway so that it does not discharge in the powerhouse area. If spillway discharges can be directed to an adjoining valley, this may be the best solution to the problem. Otherwise flow from the spillway must be carried past the powerhouse and the power facilities will have to be protected from backwater during flood discharges. The area downstream of the dam and power facilities must be protected from erosion by discharges from the spillway and tailrace.

SECTION 3
INTEGRITY INVESTIGATION

General

As introduced in Section 1, the objectives of a feasibility level integrity investigation of an existing impoundment for the addition of hydroelectric facilities are (1) to determine the structural conditions and hydraulic performance characteristics of the dam, reservoir, and appurtenant works; (2) to assess their capability of being utilized safely for small hydroelectric power generation; (3) to determine the nature and to estimate the cost of any remedial measures necessary for such safe utilization; and (4) to estimate their longevity and future maintenance needs while serving that purpose.

These objectives are most readily achieved by conducting the investigation in from one to three stages. In Stage 1 the dam, reservoir, and appurtenant works and all existing records pertaining to them are examined, reviewed, and evaluated. In Stage 2 supplemental data and analyses are acquired and evaluated and conclusions are made concerning the integrity of the impoundment and any need for remedial repairs or alterations. In Stage 3 the alteration and repair schemes for any necessary rehabilitation are conceived and their costs estimated. The remaining useful life of the facilities and the associated annual maintenance costs are also determined. The need and scope of Stages 2 and 3 are determined by the Stage 1 findings. Should addition of power facilities prove feasible, additional detailed investigations and analyses are carried out at the design level, but discussion of these is not within the scope of this volume.

Stage 1 - Review of Existing Data and Site Reconnaissance

General. The purpose of Stage 1 is to make an initial evaluation of the integrity of the existing facilities by maximum utilization of all available records and by detailed on-site examinations. One of the objectives of Stage 1 is to determine whether or not Stage 2 is needed for a final evaluation and to establish the scope of that stage. Rarely, it may be possible to proceed directly to Stage 3. Occasionally the Stage 1 findings may dictate that the investigation be terminated.

Review Existing Data. The investigation logically and purposely commences with assembling, organizing and reviewing all information that is already available concerning the facilities. Among the important questions which should be addressed are the following:

- How were the facilities designed?
- What were the loading assumptions?
- What engineering properties were assigned to the construction materials and the foundation?
- Were they based on laboratory and field tests?
- What were the test procedures?
- Were they reliable and representative of actual service conditions?
- What criteria were imposed for stress and stability analyses and what were the actual results?
- How were the flood-producing characteristics of the drainage basin evaluated?
- What runoff records were then available?
- How was the inflow design flood developed?
- How do the peak flow and volume of the hydrograph compare with the envelope values for other hydrologically similar basins in the region?
- What have been the record flows at the facility since completion?
- What kinds of construction procedures and methods were used?
- What were the corresponding technical provisions of the contract specifications?
- What were the specified construction materials properties and characteristics?
- How was quality control maintained and measured?
- What engineering inspections were made during construction?
- What were the actual conditions encountered in exposing the foundations?
- What design changes were made to conform to those conditions?
- What has been the performance record to date, as revealed by instrumental observations and reports of past inspections?
- Have any repairs or alterations been necessary? Why?
- How were they made?
- Has the dam ever been raised? How?

Answers to the foregoing questions and others can be obtained in varying degrees from records, if they were made and can be found. Depending upon their quality and completeness, they can be of great value in initially evaluating the structural and hydraulic suitability of the facilities. In any event, advance study of whatever records are available (such as previous inspection reports) will provide selective guidance to the inspecting personnel during their on-site examinations. Those records may also provide basic data such as material test results and foundation exploration information for use in the engineering analyses to be made in Stages 2 and 3. The need for and nature of additional basic data for Stage 3 will also be determined by the kind and quality of the data found in the records.

The records on existing dams vary considerably in completeness, quality, and usefulness. Their existence and character will vary with the age of the facilities, the type of ownership, and the project engineer, if there was one. In many cases, records (especially of design and construction) may be totally nonexistent, fragmentary, or inaccurate. It is important, however, that a diligent search be made for all records, because the information therein may be vital and unavailable from any other source, e.g., treatment of unusual or difficult foundations.

The search for records should include the files of the owner, of his engineer (in-house or retained), and of supporting specialists such as geotechnical engineering firms and consultants. Rarely records may be available from construction contractors. State agencies administering effective dam safety programs will have accumulated past records and they maintain current records. Their files may ease the investigator's search for records and be highly informative. Useful information may be reported in volumes of periodicals such as *Engineering News Record,* especially for older dams.

Answers to questions like those mentioned previously and other disclosures essential to the investigation will be found in engineering design and construction records often descriptively and conventionally titled with regard to their original purpose and use. Of course, the quality and accuracy of the engineering reported by the record must be examined and used by the investigator with discrimination and not uniformly accepted at face value. (For example, the drawings may not show actual, as-built conditions.) A reasonably comprehensive list of records and reports categorically grouped is presented herein. Such complete records will be rare for the dams being investigated.

---

1. Design records -

Contract plans and specifications

Geologic report

Site and materials exploration report

Design report or design bases (methods of analyses, analyses assumptions, assigned materials and foundation properties, stress and stability summaries, spillway design flood, flood routing summary, etc.)

Materials testing and appraisal report

Site seismicity report

Designers' operating criteria

Stress model reports

Hydraulic model reports

Technical record of design

2. Construction records -

Photographs - especially of foundation surfaces and preparation

Daily inspector's reports and construction progress reports - especially for descriptions of foundation and construction materials quality, unusual treatment and preparation, contractor's compliance with technical provisions of the specifications, etc.

Record of foundation drilling and grouting, and contraction joint grouting

Quantified materials quality control record of embankment and concrete engineering properties

Weekly, monthly or other periodical or special interim reports

Final construction report

Final geology report

Final grouting report

Instrumentation installation report and record of measurements during construction to establish baseline data

3. Reservoir operation records -

Chronological reservoir stages - especially for unusual stages, noteworthy spillway and outlet discharges, taxed spillway capacity, etc.

Standard operating procedures - especially for unusual, difficult, or uncertain functioning of gates, valves, controls, etc.

4. Performance record -

Hydraulic performance records of the separate spillway and outlet components at different stages and discharges.

Instrumentation design, layout and records, observation program, schedule, chronological plots, etc.

5. Maintenance record -

Reports of previous inspections, including photos of both normal and unusual conditions.

Recent evaluation reports of structural and hydraulic conditions and recommendations for remedial work or operational requirements and restrictions.

6. Records of significant past repairs, raises or alterations -

Correspondence files over the life of the facilities commencing with the design period may contain clues concerning the integrity of those facilities.

Basic Data Studies. Besides studying records relating to the dam in question, available data relating to the area and site (which may or may not have been available or used in the original or subsequent work on the dam) should also be reviewed. It must be determined how (if at all) this current knowledge modifies the conditions that must be considered in the dam's operation as a hydroelectric facility. This study of available hydrologic, meteorologic, geologic and seismic data should be performed prior to the dam inspection to form a frame of reference for the inspection.

Conduct Site Inspections. Inspections of existing impoundments are most intelligently made when the inspector is armed with the knowledge obtained from the record; guided by an understanding and familiarity with the way structures behave under various loads and water flows; informed on the way materials and natural formations react to their environment; and acquainted with actual modes of accidents and failures and their underlying causes.

Experience has revealed general classes of concerns meriting integrity investigation, especially of older impoundments in the size ranges appropriate for the addition of small hydropower. A whole host of specific conditions creating the concerns have been identified within these classes. For a comprehensive investigation, the principles expressed by these general classes must not become obscured while concentrating on specific details. The following general classes of concern prevail at all types and sizes of dams:

1. Ability to handle expected inflow floods
2. Stability of the dam and other water barrier structures under all anticipated forces and modes of operation.
3. Stress ranges in the dam and other structures critical to impoundment and operation of the reservoir.
4. Hydraulic capability of the outlet works
5. Load supporting capability of foundations
6. Control of seepage, leakage, and erosion in dam, foundation, and the confining boundaries of the reservoir
7. Deterioration of materials and foundation
8. Reliable service and operation of spillway and outlet control devices.

Failure modes and causes have been reported and discussed extensively in engineering literature (ICOLD, 1973; ASCE, 1975; Biswas, 1971). They are discussed in a general manner in Section 2.

The inspecting party should be comprised of a group of qualified, professional personnel, educated and experienced in dam design, construction and inspection. The number and discipline of the members are determined by the type and complexity of the structure and the reservoir environs. A civil engineer and engineering geologist would be a minimum-sized party. A mechanical engineer would be included dependent upon the type and complexity of the installed mechanical equipment. The individual responsible for making the final integrity evaluation should be a member of the party whenever possible. In many instances he would be the civil engineer member. A civil engineering specialist (a soils engineer, a concrete specialist, a structural engineer, a hydraulic design engineer or others) may be needed. The owner's project operation and maintenance personnel familiar with the facilities should be present to assist the party and supply information from their experience and knowledge. If the owner has an engineering staff, an individual from the staff should be present.

Checklists are often helpful to the inspection party for guidance and as "memory joggers". In principle, a checklist tabulates separately identifiable components of the dam, appurtenant works and other project features that merit observation for structural and hydraulic behavior, durability of materials, stress, strain, stability, seepage, leakage, drainage, erosion, operational capability and reliability, cavitation, temperature response, performance, instrumentation capability and serviceability, and maintenance. Checklists also are reminders for obtaining general information associated with the inspection itself, such as participants, project access, and communications. A checklist can be prepared in advance and tailored to each specific impoundment, using the information obtained from the review of the record, while keeping in mind the general classes of concern discussed earlier in this section. Reference to a universal checklist (Exhibit I of this volume) will help make the specific checklists complete.

While a checklist may be a useful tool in the hands of a knowledgeable person, it may mislead, confuse, or inhibit unqualified or inexperienced personnel by limiting the scope and detail of the inspection. Checklists are of little value unless the party members know what to look for, how to interpret what is visible, and how to make an evaluation based on indirect as well as direct evidence. Interpretation and evaluation of the observations are done by the application of engineering principles and judgment. Completion of a checklist should not be regarded as a selfsufficient measure of evaluation.

Evaluation of Data and Formulation of Conclusions.

General. The preliminary evaluation of the integrity of the impoundment is made by collectively considering all pertinent information revealed by the record, all conditions observed at the site, and the results of those engineering analyses that can be made by the investigator with the existing record data and by his checks of any recorded analyses. Engineering judgment by individuals experienced in dam design and construction is essential in the process.

If the Stage 1 preliminary evaluation is favorable in all respects, the feasibility study described in Volume I may proceed without Stages 2 and 3 of the integrity investigation. If the preliminary evaluation is favorable and can identify positively the specific rehabilitation needs, Stage 3 may follow directly. If the preliminary evaluation is uncertain but promising, Stage 2 should follow. If Stage 1 or Stage 2 evaluations are clearly unfavorable, the investigation should be terminated at the completion of those stages after consultation with the responsible project managers.

The engineering analyses portions of the evaluation process are discussed in this section on Stage 1, even though all the needed analyses may not be possible during Stage 1. Presentation here, then, not only includes analyses to be made during both Stage 1 and Stage 2 but also serves to identify additional data required for Stage 2 (See discussion below on Stage 2).

Standards and Engineering Criteria. In order to decide whether a dam and its appurtenant works can, in fact, safely store and control flows of water, an investigator must apply some measure of adequacy. Because there are many associated considerations, both direct and indirect, these decisions can seldom be made based solely on the application of rigid "standards" and engineering criteria. A "standard" as used here is considered to be a definite rule established by authority (usually governmental regulation), while a criterion is considered a test of quality by the application of engineering principles. The statutes, codes, and regulations of governmental agencies having various kinds of jurisdiction over such matters as water rights, public safety, environmental protection, or occupational health and safety may be controlling or contradictory. How, then, do they apply in the case of this type of study? What is the pertinence of the state of the art as practiced when the dam was designed and constructed compared to that existing today, especially if the facility has ably performed over the years? What are the relative hazards (source of potential danger created by the existence of the dam and reservoir) when compared to the degree of risk (the probability of failure and the chance of loss of life and property) that exists at the time of the study? What is the influence of public opinion and the public's demonstrated unwillingness to accept involuntary risks (Starr, 1969) and how are they relevant to the study's conclusions? How does the owner knowingly view his liability? What is the liability of the evaluator? What are today's commonly accepted practices for designing, constructing, operating, and maintaining dams, and how are they influenced by conflicting schools of thought among different groups of engineers? Yesterday's standards may prevail when evaluating liability for a failure of an old structure but new standards will prevail for an altered structure. All these questions are legitimate and must be considered in evaluations such as are covered by this volume. The blind application of standards or criteria is not adequate.

The use of standards and criteria as measures of adequacy can be dangerous, biased, or restrictive when numbers and specific values are generated by an engineering analysis and then compared as a pass or fail test. Of far greater importance than the numbers that are generated by the analysis is the evaluator's understanding of the degree of accuracy of the values and the assumptions going into the analysis, the limitations of the analysis, and the true representation of actual conditions. The interpretation and application of the numbers from the analysis must be tempered with common sense, understanding, experience, and judgment.

Instead of basing his evaluation on just barely meeting some imposed minimum standard, the investigator should make his evaluation based on demonstrated sound engineering practices generally endorsed by the collective dam engineering profession, coupled with his own experience and convictions. Competent, conscientious investigators will usually match or exceed the so-called standards without being unduly conservative.

Methods of Analyses. Many analytical techniques - mathematical, graphical, and physical (models) - have been developed for investigating and predicting the behavior and response of dams, other hydraulic structures, and their foundations in different physical environments and service conditions under all kinds of loading. These techniques are used to help find dependable answers to the general classes of concern introduced above.

These techniques are available in prolific detail with examples from many sources - university textbooks for fundamentals; professional engineering society publications such as United States Committee on Large Dams (USCOLD), International Congress on Large Dams (ICOLD), and American Society of Civil Engineers (ASCE) for practical specific applications; design manuals, monographs, handbooks, and design standards of federal and state agencies engaged in water resource development for methodical, production-basis use — for example, publications of the U.S. Army Corps of Engineers (COE), especially the Hydrologic Engineering Center; technical publications of product manufacturers and construction materials associations such as the Portland Cement Association (PCA), the American Concrete Institute (ACI), the American Institute of Steel Construction (AISC), and the American Concrete Pipe Association (ACPA), the Stress Steel Corporation, ARMCO Drainage and Metal Products, the American Asphalt Institute, etc. for detailed analytical methods of hydraulic structure components where their products are used (ACPA, 1957; ACPA, 1959; ARMCO, 1955). Several publications (Golze, 1977; Justin, 1945) are outstanding. Some (USBR, 1974/2; NRC, 1939) are also especially suited as well to the size class of dams having potential for small hydropower. Private engineering firms specializing in hydraulic project planning and design have developed manual-like compilations for their in-house use.

Analyses most frequently and conventionally made for reservoirs, dams, and appurtenant structures in size ranges which may be candidates for small hydro investigations are:

1. Inflow design flood hydrograph (COE, v.3 April 1975; and v.5 March 1975).
2. Reservoir flood routing (COE, v.4 October 1973, and v.7 February 1976).
3. Spillway discharge rating curve (USBR, 1974/2, Sections 195-200, 211-214.)
4. Open channel water surface profile (USBR, 1974/2, Sections 203-204).
5. Tailwater elevation-discharge curve (USBR, 1974/2, Section B-8, B-9).

6. Outlet discharge rating curve (USBR, 1974/2, Sections 222, 232-236, B-3).

7. Hydraulic jump characteristics of stilling basin (USBR, 1974/2, Sections 205-210).

8. Water surface profile in the trough of side-channel spillways (USBR, 1974/2, Section 202).

9. Trajectory of overflowing nappe (USBR, 1974/2, Section 211) or free falling jet.

10. Plunge pool scour depth (USBR, 1974/2, Section 210).

11. Conduit (penstock) pressure surge.

12. Buoyancy resistance, stability, stresses for free-standing dry-type intake tower.

13. Active, passive, at rest earth pressure.

14. Retaining wall, spillway gate pier, spillway control structure stability, stresses, deflections.

15. Stresses, deflections, reactions in spillway and outlet control devices (radial gates, flashboards, etc.) and in anchorages.

16. Stresses in conduits (Beggs, 1968; USBR, 1965).

17. Stability of embankment and foundations (ASCE, 1969; Janbu in Hirschfeld, 1973, pp. 47-86).

18. Stability of natural formation confining the reservoir.

19. Stability of hillside adjacent to abutments.

20. Seepage flow nets for pore pressures, hydraulic gradients, and escape gradients in embankments and foundations (Cedergren in Hirschfeld, 1973, pp. 21-45).

21. Consolidation, subsidence, compression, and expansion of foundations.

22. Stability and stresses in concrete gravity sections (dams, locks) (Golze, 1977, pp. 385-393, 437-445, 583-587; Justin, 1945, pp. 247-423; USBR, 1976, Chapters 2 and 3).

23. Stresses in arch dams (Golze, 1977, pp. 385-437; Justin, 1945, pp. 425-553; USBR, 1977, Chapters 3 and 4), arch barrels (Justin, 1950, pp. 584-587), facing slabs (Justin, 1945, pp. 558-599).

The determination of the particular analyses that must be made will of course depend upon the type of dam, its age, observable conditions, performance history, watershed, stream, and reservoir characteristics, geologic setting, etc. For example, reservoir flood routings serve no useful purpose if the reservoir capacity is small and the drainage basin is large. Dynamic stability analyses are unnecessary in regions of low seismicity. The cost and time required will be reduced if simpler methods of analyses can be used. Refined procedures and precise results are not always needed in order to make a decision. The more experienced the analyzer, the fewer the analyses that may be needed. Analyzers who are generally knowledgeable but inexperienced need more data and studies to make evaluations. Evaluations by unknowledgeable persons will not only be inaccurate but can lead to false conclusions as well as dangerous expectations.

The detailed developments, explanations, instructions, applications, and examples of these analyses, some of which can be made by several different accepted methods, will be found in the selected references.

Of these many analyses the ones usually considered most critical for integrity investigations are: (1) those concerning adequate spillway capacity or, more generally stated, the ability to safely handle expected inflow floods; and (2) those concerning the stability of the dam and foundation for safely impounding the water in the reservoir. Because of their importance, these topics are discussed in more detail below.

Ability to Safely Handle Expected Inflow Floods. The ability of the impoundment to safely handle expected inflow floods first requires preparation of an inflow flood hydrograph or peak inflow value on some acceptable frequency or probability-of-occurrence basis. If detention storage capacity is operationally reserved for that purpose, the inflow flood is routed to determine the residual freeboard protecting non-overpour structures. The hydrologic techniques for flood estimating and routing are discussed in Volume III. Criteria for the flood magnitude and residual freeboard are discussed in a later segment of this section, "Suggested Engineering Criteria."

Certain investigations of spillway capability can be made by analytical methods. The spillway rating curve is calculated for use in the flood routing study. Usually the capacity will be established by the control structure but any other components that might become capacity-controlling, usually at higher discharges, must not be overlooked. For example, at a double side-channel spillway, the hydraulic control may shift to locations in the side-channel trough or to the juncture of the trough and the discharge channel. The control may shift from free-surface flow to orifice flow to pressure flow at shaft and drop-inlet spillways. The water surface profile in an open channel can be calculated to investigate wall overtopping. Cross-channel wave patterns created by channel convergence or curvature can be determined (at least qualitatively) for the same reason. The hydraulic jump characteristics or nappe and jet trajectories at the terminal structure can be calculated to investigate energy dissipating capability. The tailwater rating curve can be calculated from a known downstream hydraulic control to investigate the effect of the tailwater elevation on flows and on the terminal structure.

An impoundment may not have a spillway and it must then be investigated for ability to temporarily store the inflow volume and dependably draw off that volume through available release facilities before succeeding floods occur. In such cases the investigation of the capacity, structural integrity, and operational reliability of all components of the release facilities used for that purpose becomes of great importance.

Stability of a Dam and Foundation. Certain investigations for the stability of a dam and its foundation can be made by analytical methods, dependent upon the dam type.

Embankment-Type Dams. Various methods of slope and foundation stability analyses are available. The more common ones are two-dimensional and are based on limiting equilibrium. These analyses are known by a variety of titles, including slip circle, Swedish circle, Fellenius method, method of slices, sliding block, etc. There are differences in assumptions and force resolutions in the different methods. When forces representing earthquake effects are included, the analysis is often termed pseudostatic. The analysis is made by assuming some form and location of failure surface such as a circular arc, compound curved surface, or a series of connected plane surfaces. The configuration and positioning of the surface depend upon the kind of embankment dam, the internal zoning, and the foundation geologic structure. For example, connected plane surfaces are often used for an inclined or sloping core rockfill dam. The trial failure surfaces are positioned judgmentally to pass through weaker or more highly stressed regions. For example, a plane surface may be positioned in shallow weak clay or in shale layers in the foundation; or a circular surface may be positioned in a confined fluvial foundation susceptible to high pore pressure. The most critical surface is defined as the one having the least computed factor of safety which is considered to be the ratio of forces or moments resisting the movement of the mass above the surface being considered to the forces or moments tending to cause movement. Both embankment slopes are analyzed for the specific service conditions expected. The most critical case for the downstream slope is usually full reservoir with steady seepage; for the upstream slope it is usually either rapid drawdown or reservoir partially full with seismic loading. Seismic cases for these methods of analyses assume horizontal loads determined from constant horizontal seismic coefficients whose values are arbitrarily selected on the basis of ground motions anticipated at the site. The engineering properties and strength values used in these analyses must be selected to duplicate as closely as possible the actual field conditions expected. For example, if drainage during the application of forces is not possible, shear strengths should be based on quick or consolidatedundrained laboratory tests. These analytical methods can also be used to examine reservoir and abutment hillside slide potentials. More realistic but extensive and costly dynamic analysis methods are available for investigating the effects of earthquakes on stability. These methods are based on limiting strains and permanent displacements rather than factors of safety. Only in very special situations would such analyses be employed in small hydro investigations. Instead, simplified procedures (Makdisi, 1978) for estimating the earthquake-induced deformations are available if needed.

Allied analyses are used during stability studies to determine seepage patterns and amounts, pore pressures, uplift forces, hydraulic gradients, and escape gradients in the embankment zones and the foundation by the application of the principles of flow through porous media and the graphical or mathematical modeling of flow nets (Cedergren, 1977).

Concrete and Masonry Dams. The stability of gravity dams and the buttresses of buttress-type dams can be numerically evaluated for resistance to sliding and overstressing from water, weight, uplift, earth and silt, temperature, seismic, and ice loads. The resistance values are calculated on critical surfaces in the dam, on the foundation, and below the foundation level. The resistance to overturning can also be calculated, but any indicated instability will most likely be manifested by local crushing of the concrete or the foundation due to overstressing, rather than a physical toppling of the intact mass. The principles and procedures of these analytical methods are also applicable to lock walls, spillway control structures, and retaining walls.

The stresses in the arches and slabs of buttress-type dams and arch dams can be numerically evaluated for the same kinds of loads as for gravity dams.

Single-arch dams may be further characterized as being constant radius, constant angle, variable radius, or double curvature. The arch rings may be cylindrical and of uniform thickness or of irregular form and of variable thickness.

Depending upon the height, geometry, complexity, and importance of an arch dam, the stresses can be approximately determined by the cylinder theory (NRC, 1939; Justin, 1945, pp. 425-553) or by the application of the theory of elasticity using graphical and mathematical summation methods (Justin, 1945, pp. 425-553). Various assumptions and considerations can be included or omitted that will affect the complexity of the calculations and the relative validity of the resulting stresses. Two examples are deformations due to shear and the effect of Poisson's ratio. The arch rings may be considered fixed or hinged at the abutments. The abutments may be considered rigid or elastic. Contraction joints may be considered grouted or ungrouted.

The more realistic and exacting methods of trial-load analysis and two- and three-dimensional finite element analyses are available. Stress patterns in the abutment and foundation mass of concrete dams can be determined by the finite element methods. Such analyses will not usually be necessary in small hydro investigations; however, where special and critical situations exist, these types of analysis may be justified.

Suggested Engineering Criteria. "There was unanimous agreement that it would be unwise to publish recommended design criteria as standards to be adopted and used universally... Consequently, it could be extremely dangerous to publish design criteria and thereby imply that by following these criteria an engineering organization could assure that a safe structure will result" (ASCE, 1967). Although these words were written about the design of "large" dams they are equally applicable to the investigation of the integrity of smaller dams. This referenced joint ASCE-USCOLD committee report summarizes the practices for dam design and construction of major engineering organizations in the United States and provides excellent criteria statements for use here.

Criteria are sometimes stated on the basis of dam size and the related hazards and risks. There is no universally accepted definition of a "large" dam. Hazard is a function of dam size and physical condition. Risk is a function of potential project damage, monetary loss, and of population location and density. Definitions that have been suggested indicate that dams appropriate for small hydro are of small and intermediate size.

The U.S. Army Corps of Engineers has recommended (COE, 1977) expected inflow flood magnitudes for use in the National Program of Inspection of Dams. Those recommendations which are appropriate for guidance here are excerpted from that reference and presented as Tables 3-1, 3-2, and 3-3. There are some differences in terminology for floods, hazards, and risks but the interpretations are obvious.

TABLE 3-1
RECOMMENDED SPILLWAY DESIGN FLOODS[a]

Hazard[b]	Size[c]	Spillway Design Flood (SDF)[d]
Low	Small	50 to 100-yr freq
	Intermediate	100-yr to 1/2 PMF
Significant	Small	100-yr to 1/2 PMF
	Intermediate	1/2 PMF to PMF
High	Small	1/2 PMF to PMF
	Intermediate	PMF

[a] Source: COE, 1977.

[b] See Table 3-2 for definitions

[c] See Table 3-3 for definitions

[d] The recommended design floods in this column represent the magnitude of the spillway design flood (SDF), which is intended to represent the largest flood that need be considered in the evaluation of a given project, regardless of whether a spillway is provided; i.e., a given project should be capable of safely passing or storing the appropriate SDF. Where a range of SDF is indicated, the magnitude that most closely relates to the involved risk should be selected.

100-yr = 100-Year Exceedence Interval. The flood magnitude expected to be exceeded on the average of once in 100 years. It may also be expressed as an exceedence frequency with a one-percent chance of being exceeded in any given year.

PMF = Probable Maximum Flood. The flood that may be expected from the most severe combination of critical meteorologic and hydrologic conditions that are reasonably possible in the region. The PMF is derived from probable maximum precipitation (PMP), which information is generally available from the National Weather Service, NOAA. Most Federal agencies apply reduction factors to the PMP when appropriate. Reductions may be applied because rainfall isohyetals are unlikely to conform to the exact shape of the drainage basin and/or the storm is not likely to center exactly over the drainage basin. In some cases local topography will cause changes from the generalized PMP values, therefore, it may be advisable to contact Federal construction agencies to obtain the prevailing practice in specific areas.

TABLE 3-2
HAZARD POTENTIAL CLASSIFICATION[a]

Hazard Category	Loss of Life (Extent of Development)	Economic Loss (Extent of Development)
Low	None expected (No permanent structures for human habitation)	Minimal (Undeveloped to occasional structures or agriculture)
Significant	Few (No urban developments and no more than a small number of inhabitable structures)	Appreciable (Notable agriculture, industry or structures)
High	More than few	Excessive (Extensive community, industry or agriculture)

[a] Source: COE, 1977.

TABLE 3-3
SIZE CLASSIFICATION[a]

Category	Storage (Ac-Ft)	Impoundment Height (Ft)
Small	50 to 1,000	to 40
Intermediate	1,000 to 50,000	40 to 100

[a] Source: COE, 1977.

The U.S. Bureau of Reclamation has published (USBR, 1974/1) design criteria for concrete arch and gravity dams. Those criteria are appropriate for use here. The subject matter is organized and presented in brief, systematic fashion by first discussing each basic consideration and then making the criterion statement. Loads and load combinations, safety factors and their application limitations, assumptions and uncertainties of analyses and materials properties, limiting stresses, and minimum stability factors are all presented.

Two excellent publications (John Lowe III, in ASCE, 1969, pp. 1-35; Nilwer Janbu, in Hirschfeld, 1973, pp. 47-86) comprehensively discuss the state of the art and the mechanical principles for embankment-type dam stability analyses by limiting equilibrium methods. Although minimum factors of safety criteria are not presented, an appreciation and understanding of the advantages and limitations of the methods of analyses can be obtained from which the investigator for small hydro can better understand why it is no simple matter to declare universally applicable minimum factors of safety.

Calculated minimum safety factors used by many engineers and organizations are listed in Table 3-4. These are presented herein for guidance only. The reviewer must establish minimum requirements based on site-specific conditions and his best judgment.

Users of this volume of the small hydropower guide manual should obtain copies of the referenced literature and technical reports for advice on the analytical procedures and criteria contained therein while investigating the integrity of an impoundment. Many other excellent and widely recognized publications, organizational or otherwise, are equally suited for those purposes.

TABLE 3-4
CALCULATED MINIMUM SAFETY FACTORS[a]

Case	Loading	Slope	Minimum Factor of Safety
I	Steady seepage, reservoir at normal pool	Downstream	1.5
II	Drawdown from normal to minimum pool elevations	Upstream	1.2
III	Earthquake		
	a. Case I with seismic loading	Downstream	1.0
	b. Reservoir at intermediate pool with seismic loading	Upstream	1.0

[a] Source: COE, 1977.

Stage 2 - Development and Evaluation of Data

General. As discussed previously, all available records, visual site examinations, and numerical engineering analyses that can be made with the available data are fully exploited in an effort to reach a dependable evaluation of impoundment integrity. Often additional data will be needed to augment the Stage 1 investigation before a dependable evaluation is possible, especially if the impoundment promises to be favorable for small hydro use.

The type of information and numerical data needed will concern structural, geological, and performance features unobtainable by direct visual examination. Some kind of exploration will be required for sample extraction; for providing access for direct observation; and for instrumental measurements of forces, stresses, deformation, seepage, etc. Data may also be obtained by non-destructive testing. Laboratory tests will be required to determine engineering properties of the materials of the dam and appurtenances and of the foundation for use in analyses and to assess their state of preservation.

The kinds of data, the techniques for acquiring them, and their applications in the integrity evaluation are discussed in this section.

Subsurface Exploration. The integrity of facilities may be questioned if foundation or embankment conditions are unclear, or if saturation levels and seepage levels are of concern. In such cases, subsurface exploration will be required to develop additional data and to provide samples for laboratory testing to determine engineering properties. There are many exploration tools and techniques available to obtain and develop data in the evaluation of existing impoundment structures. Some of the commonly used exploration tools are described below.

Geologic Mapping. The geologic map and geologic cross-sections are essential tools for planning a subsurface exploration program, specifically in evaluating foundation conditions of the impoundment under investigation. If geologic maps are available, they should be updated to show existing features such as slope instability, groundwater seeps, etc., adjoining the impoundment and appurtenant works, as well as within the reservoir area.

Drilling. Information which can be obtained from drilling will be required for earth and rock dams if the original site conditions, design criteria and analyses, and construction records are unavailable or if visual inspection or performance records indicate that the facilities may not be performing adequately. The purpose of drilling is to obtain subsurface information which is used to construct a three-dimensional picture. Samples at depth can be secured, down-hole testing can be performed, water levels can be determined, and instrumentation such as piezometers and slope indicator casings can be installed. A large number of drilling and sampling systems are available to achieve the above purposes. Factors affecting the type of drilling and sampling used include type of impoundment structure, the materials constituting the embankment, abutments and foundation, and accessibility.

Core drilling with diamond drill equipment is the exploration method used most commonly for concrete or masonry structures and for relatively hard bedrock foundations. The core drilling program provides means of investigating and evaluating the structure and its foundation, construction joints, and cracking (if any) in the concrete or masonry. It also provides core samples for laboratory testing. The current practice of core drilling uses the rotary method almost exclusively because of the higher quality of samples obtained. The core barrel and diamond drill bit constitute the sampling device in which the cylinder or core of the sampled material is retained. Core barrels are available in a variety of sizes that produce cores with nominal diameters ranging from a fraction of an inch up to 48-inch or larger. However, NX-size cores (nominal diameter 2-1/8-inch) appear to be the minimum size core that would be meaningful for strength testing and visual inspection. In coring, the primary objective is maximum percent core recovery, so that the maximum amount of subsurface information is obtained. Unlike mining exploration, where the objective is almost solely maximum core recovery, drilling exploration for engineering evaluations requires that all available data during the drilling operations be collected and recorded. These data are of two types: permanent and fugitive. Permanent data are the cores obtained. Fugitive data are those which, if not observed and recorded in the hole log by the field geologist during the drilling operation, are lost forever. They include the time necessary to cut the core, the actions of the drill with the depths at which they occur, the driller's opinions, changes in the operation of the drill made by the driller with the reasons therefor, the color of the drilling water return, drill fluid "take" by the hole, and any other data of similar nature that may be requested by the project geologist. These data (which are not always obtainable from the drill core) are used to evaluate cracking, jointing, and other aspects of the material being sampled. It is the "non-core" information that may be most critical. Thus, it is most important that the geologist be at the drill and be carefully observing the details of the drilling operation at all times when the drill crew has the core barrel in the hole.

Core drilling is basically a sampling procedure for hard materials. In earth embankments or soil foundations the sampling procedure is entirely different, although the method of advancing the drill hole may be the same. The most commonly used method of drilling exploration in earth embankments is the straight rotary drilling method. Other drilling systems (such as augering) may be applicable, but they have certain limitations such as shallow depths, inability to utilize larger sampling devices (in the case of the hollow-stem auger), and difficulty in sampling below the water table. Percussion drilling is commonly used in alluvium containing cobbles and boulders. For many applications rotary drill rigs have several advantages. They can drill to greater depths than can be reached by other methods; they are extremely versatile; and they can accommodate different types of soil and rock samplers. Rotary drills conventionally use a circulating fluid (air, water or bentonite slurry) which is used to cool the cutting bit and remove cuttings by carrying them upwards to the surface.

Sampling, logging, and groundwater observation are the prime objectives of nearly all exploratory drilling. Sampling is essential for detailed examination and laboratory testing. Every precaution should be taken to guarantee that representative and uncontaminated samples are recovered. Basically, there are two types of soil samples: disturbed and undisturbed. Both types can be obtained using a variety of mechanical sampling devices. Sample tubes or barrels may be advanced into the soil by three basic methods: pushing, driving, or drilling. Pushing is usually preferred; however, in firmer material it often becomes necessary to drive or drill the sampler into the ground.

The most common type of sampler for obtaining disturbed samples is the split spoon. The split spoon is available in various sizes; however, the 1-1/2-inch-diameter sampler is popular because of correlations that have been developed between the number of blows required to drive the sampler into the soil strata and the relative density of cohesionless soils or the shear strength of cohesive soils. The sample obtained can be used for identification tests such as visual classification, water content, grain size analysis, Atterberg limit tests, etc.

Undisturbed samples preserve as closely as possible the natural structure and density of the in-situ material and are therefore suitable for strength tests as well as the identification tests that can be performed on disturbed samples. The open, thin-wall (Shelby) tube sampler is the most commonly used undisturbed sampler. The thin-wall tubes are pushed by the hydraulic or screw-fed system of the drill rig and are primarily used for sampling soft to stiff cohesive soils. The Shelby tubes are available in various diameters and lengths, but the most commonly used are 2 and 3 inches in diameter and 24 and 36 inches long. In embankment materials with gravel components, larger diameter (up to 6-inch) tubes should be used. A modification of the open thin-wall tube sampler is the closed-tube sampler in which a piston located at the lower end of the thin-wall tube is either released or withdrawn when the drive is started. The thin-wall stationary piston sampler and the Osterberg Piston sampler are examples of this type of sampler, which is used for sampling soft to stiff cohesive soils. The piston prevents shavings and "cave" material from entering the sampling tube and creates a partial vacuum between the piston and the sample which helps collect and retain the sample. When drilling with water or drilling below the groundwater table, the piston sampler offers the potential of more representative moisture contents and less contamination of the sample, because the sample tube is relatively dry when it reaches the bottom of the hole. When the material to be sampled is too soft to be cored or too hard to be sampled by pushing a thin-wall tube, modified push drill samplers are used. The Denison and Pitcher samplers are examples of this type. This type of sampler is primarily used for stiff to hard cohesive soils and dense cohesionless soils, and alternating hard and soft layers. They differ from the double-tube core barrel used for rock coring in that the stationary inner barrel (a thin-wall tube) extends ahead of the bit, thus preventing washing of weak materials. In the case of the Pitcher sampler, the inner tube is spring mounted; thus, the lead distance ahead of the bit depends upon the firmness of the material being sampled.

The U.S. Bureau of Reclamation's *Earth Manual* (USBR, 1974/3) contains a more detailed and comprehensive discussion of drilling and sampling equipment and should be used as a basic reference. Another good reference is *Basic Procedures for Soil Sampling and Core Drilling* (Acker, 1974).

Trenching or Test Pitting. These methods of exploration open a wider area of shallow subsurface materials to detailed examination than does drilling. The excavation can be done by backhoe or bulldozer or by hand. In-place field density tests and disturbed or undisturbed samples can be obtained from the exploratory trenches or test pits. Undisturbed samples could be handcarved either in blocks, in random shapes, or into sample tubes. Trenches which are deep enough for a person to be buried if a wall were to cave should not be entered unless the sides are determined to be stable naturally or there is adequate wall support provided. All normal safety precautions and regulations should be observed.

Other Surveys. Ambient vibration surveys on concrete or masonry structures measure the natural mode of vibration, and their shapes and periods of vibration.

Underwater surveys can be conducted to visually evaluate physical conditions of upstream earth embankment slopes or concrete dams under water. For example, concrete or masonry dams can be inspected for deterioration or cracking and earth embankment slopes for slides or deformation of the slope.

Laboratory Testing. Laboratory test results are performed to obtain data which is used for both rational evaluation of conditions and to obtain numerical data for use in engineering analyses. For convenience, laboratory tests are divided here into two categories (1) soils and (2) rock and concrete. All testing should be performed at established laboratories by experienced personnel. Therefore, only the types of test, their purposes, and the use of test results are discussed herein; test procedures are beyond the scope of this work. Test descriptions and procedures are available in several sources, with the *Annual Book of ASTM Standards* (ASTM, Annual) as the best of these. Parts 10 and 11 of the ASTM Standards cover concrete and soils respectively.

Soils. Laboratory testing of soils and soft rock consists of two types of test - (1) classification and physical properties testing and (2) engineering properties testing.

The Unified Soil Classification System is the system most commonly used for classifying soils. This system is based on a recognition of the various types and significant distribution of soil constituents, considering gradation characteristics, and plasticity of materials. Grain size distribution data and the results of Atterberg limits tests provide the information, except for the determination of organic content, to properly classify the material. The USBR's *Design of Small Dams*, (USBR, 1974/2), describes the Unified Soil Classification System and presents general properties of materials for each of the soil classification groups.

Other tests commonly performed which are not truly classification tests or engineering properties tests include natural water content, dry density, and specific gravity.

Engineering properties can be roughly estimated by experienced soil engineers if the soil classification is known. The estimated engineering properties may be adequate for preliminary evaluations or if the structure is obviously adequate. However, if detailed engineering analyses are to be performed, the engineering properties must be determined by laboratory tests. Tests that are commonly performed to determine engineering properties of soils include compaction tests to determine the moisture-density relationships of materials containing a significant percentage of fines; relative density tests to determine the maximum and minimum densities for relatively clean sands and gravels; consolidation tests; permeability tests; and shear strength tests. The engineering properties tests listed above are relatively straight forward except for the shear strength tests. Shear tests can be performed in direct shear apparatus or in triaxial shear apparatus. Normally all shear tests will be performed on saturated samples. In the direct shear apparatus, reliable pore pressure measurements cannot be made; the only pore pressure control available is to run the test slowly enough for pore pressures to dissipate, or rapidly enough so that the pore pressures build up in the sample to simulate field conditions where high pore pressures are expected to exist. Triaxial tests afford the opportunity to make good pore pressure measurements and the results from one triaxial test series can be utilized to determine both effective and total stress strengths or, in the terminology of the U.S. Army Corps of Engineers, "S" and "R" strengths respectively (S standing for slow rate of failure in the direct shear apparatus where pore pressures are allowed to dissipate, and R standing for rapid failure where pore pressures are not allowed to dissipate).

All engineering properties tests on materials from existing embankment and foundation materials that are to be left in-place should, if possible, be performed on undisturbed samples. The sample size should be large enough to permit testing of representative samples without having individual particles control the test results.

Concrete and Rock. Tests on concrete and hard rock samples are normally limited to determining the unconfined compressive strength. A standard method of determining the unconfined compressive strength of rock is contained in the *Annual Book of ASTM Standards,* Part 11. The method presented for unconfined compressive strength of rock core specimens is also applicable for concrete core specimens.

Analyses and Interpretations of Results.

The final numerical analyses and the methods employed in Stage 2 are the same as those described and enumerated by reference above in the discussion of Stage 1 - Methods of Analyses. However, they are more extensive, definitive, and refined. They are specifically tailored to represent the actual physical conditions disclosed by the investigations. Particular care should be taken to study suspicious or uncertain appearing features and conditions. The engineering data and information to be used in the analyses are those specifically obtained for that purpose during Stage 2. For example, if the spillway capacity appears inadequate for any reason, such as experienced near capacity discharges or high regional flood comparisons, a new flood estimate should be made and the existing spillway and impoundment components should be analytically tested for their ability to safely handle the updated flood. Or, for example, if the stability of an embankment-type dam appears marginal for any reason (such as apparently over-steep slopes, unusual saturation patterns, low strength soils, or indications of high foundation pore pressures) a stability analysis and companion seepage analyses should be made using soil strengths and permeability rates obtained by sampling and testing for use in those specific analyses.

In many cases, the final analyses will be the only analyses, rather than extensions and refinements of Stage 1 analyses.

As valuable as they are, numerical analyses cannot provide total and absolute answers upon which to base the final evaluation. Many physical conditions and reactive mechanisms cannot be mathematically analyzed, even qualitatively.

When all the objective factors that may influence the evaluation have been gathered, interpreted, analyzed, and discussed, the investigator must decide if the impoundment can be safely used to serve a small hydropower installation in its present condition, that it cannot, or that it is engineeringly feasible to rehabilitate it so that it can.

There are no clear-cut rules by which these decisions can be made. Instead the decisions are made by a value judgement process employing empirical reasoning and objective assessments by trained engineers. Comparisons with successfully performing similar impoundments are made. Criteria generally accepted and proclaimed by reputable practitioners and by professional engineering societies are applied as general tests in measuring adequacy. Throughout the decisions process, the general classes of concern enumerated for site inspections must dominate the mind of the evaluator.

The type, size, complexity and regional setting of impoundments are highly variable. For that reason, Exhibit II, "Considerations and Procedures for Impoundment Integrity Evaluations", is included at the end of this volume to provide a comprehensive list of actions, studies, and reviews that constitute the evaluation process. Obviously, all items are not applicable to all impoundments, nor are all the items of equal importance.

Stage 3 - Rehabilitation Methods and Cost Estimate

When an evaluation decision finds that an existing impoundment is suitable for a small hydropower installation, it will be possible to proceed directly with the feasibility study described in Volume I, provided no deficiencies were disclosed by the integrity investigation. However, it can be expected in some cases that the investigation will identify structural or hydraulic weaknesses in the dam or appurtenances, or even the reservoir confines, which would require remedial treatment before the impoundment could be safely used for a hydropower installation, whether or not such installation is close-coupled to the impoundment. In such cases, it would be necessary to formulate repair or alteration schemes for rehabilitating the particular component of the impoundment and to estimate the associated construction costs.

The required repairs or remedial measures may be simple or extensive and their costs will vary accordingly. Alternative designs and construction procedures are often feasible and their physical and cost advantages should not be overlooked. It may be possible to combine or coordinate the rehabilitation repairs or remedial measures with any alterations that might be needed to accommodate the hydropower installation. The need to maintain stream flows or continue operation for existing project purposes during repair must be considered and in some cases may control or influence the design and the construction schedule for the repairs.

Some of the deficiencies most likely to be encountered and examples of corrective repairs and reconstruction are presented and discussed in Section 4. The associated cost estimating procedures and the use of the cost estimates in making decisions regarding rehabilitation of the impoundment are presented and discussed in Section 5.

Program Administration and Personnel

General. As discussed in Section 1, it is imperative that the integrity of the impoundment be positively established because of the potential for capital investment loss and public liability should the dam fail. The feasibility of the power project depends on a sound dam or one that can economically be made sound. The integrity investigation must be conducted in a comprehensive, orderly manner by an individual or team educated and experienced in several technical and scientific disciplines essential to dam engineering, construction, and operation. The team will function most effectively if it is properly structured and managed to accomplish specific objectives on an established schedule. The size of the team and the different disciplines required will vary with the type and complexity of the impoundment and the breadth of each individual's expertise. The teamcomposition may also vary somewhat with the particular stage of the investigation.

Establishing and Administrating an Investigative Program. The organization and direction of the integrity investigation program should be assigned to an engineering program manager who has broad and extensive experience in design and construction. The program manager should be an engineer, usually in the field of general civil engineering. Non-technical personnel should not be assigned as program manager.

An initial schedule should be established for the overall program with each of the three stages separately identified. Target dates for the fundamental decisions of each stage should be established, while recognizing that the investigation may be terminated at the end of any stage or that Stage 2 or even Stage 3 might not be required, as previously discussed. The schedule should recognize and provide for sequential or simultaneous conduct of activity. For example, the site inspection should not precede the acquiring of existing data because familiarization with that data will provide special guidance for the inspection. Where data are to be acquired in Stage 2, other Stage 2 analyses independent of that data can proceed simultaneously.

The schedule must provide for flexibility so that as the objective of each stage nears achievement and the initial and final integrity decisions are made, the next appropriate activity can proceed without delay. A sample schedule for a relatively simple investigation is presented below. Additional inspections for specific or more detailed observations will be advisable as the investigation proceeds and schedule allowances should be made for that purpose.

Sample Schedule

Stage 1: June 1 - August 5	
Collect and evaluate available data	1 - 2 weeks
Site inspection	1 week
Evaluate integrity of existing facilities, develop Stage 2 program (if required), and prepare report	1 - 2 weeks
Stage 2: August 10 - October 31	
Administration and coordination	1 week
Subsurface exploration	1 - 2 weeks
Laboratory testing	1 - 2 weeks
Evaluate exploration and laboratory data, perform engineering analyses and evaluations	2 - 3 weeks
Stage 3: October 15 - November 30	
Prepare rehabilitation design	1 - 2 weeks
Compute construction quantities	1 week
Prepare construction cost estimates	1 week

If, during Stage 1, it becomes apparent that Stage 2 will be necessary, the specific drilling, sampling, and testing objectives and procedures must be planned in detail and their manner of accomplishment decided. Time requirements, costs, scheduling, and instructive procedures must be considered and established. If the services are to be provided by others, service agreements or contracts must be arranged. Rights of entry may be necessary. The dam owner's permission and liability clearances must be negotiated and obtained. The exploration must be coordinated with existing project operation schedules or requirements. Jurisdictional authorities may have legal controls that must be satisfied.

If more than one impoundment is under investigation by the group, management may have to establish priorities. Management should also recognize any advantages in staff utilization by coordination of activities for a multi-project program.

Should a difficult, complex, or unusual engineering problem arise, it may be advisable for the investigating group to retain a consultant or individual expert for advice, and management must arrange for those services. Such need might occur, for example, while deciding upon the manner of exploration or test of a suspected unsafe foundation condition during Stage 2. Or advice might be needed on the magnitude and severity of expected earthquake ground motions and foundation displacements at the site for use in studying the dynamic stability of the dam during Stage 2.

As Stage 2 nears completion and it is decided that Stage 3 is in order, management must schedule the study of rehabilitation methods and preparation of cost estimates. The study should provide for alternative plans to determine the possibilities of cost advantages.

Personnel Qualifications and Composition of Investigative Team. The minimum integrity investigation will include records review, site examination, and judgemental evaluations. Since evaluations at this stage are largely judgmental, it is important that an individual or group experienced in all phases of dam engineering perform the investigation. As a minimum, the individual or group must have scientific knowledge and experience in the fields of geotechnical engineering, structural engineering, and hydrology and hydraulics as related to water retention structures.

When Stage 2 and Stage 3 are to be performed and the investigations are relatively straight forward, the individual or group that performed the Stage 1 investigation can perform the additional work with support from lower level staff. The amount of work shown in the sample schedule represents a relatively simple Stage 2 and Stage 3 investigation that could be performed by an individual or small group, except for the drilling and laboratory work, which requires special equipment. This type of program would cost in the order of $15,000 at 1978 prices. If the Stage 1 investigation reveals questionable integrity such as the need to perform a complete seismic analysis of an earth dam, additional expertise and substantial costs (in the order of $100,000 at 1978 prices) will be required. This type of investigation would economically be practical if the anticipated revenue is high for a small hydroelectric project but would not be justified if the project was considered to be economically marginal.

Peer Review. Evaluation decisions seldom can be based solely on the results of mathematical analyses or simply on the external appearances seen at the time of the site examination. Decisions are made mainly by empirical analyses and judgmental evaluation, supplemented by the mathematical analyses and site examination.

The report of the Los Angeles County Coroner's Jury after the failure of St. Francis Dam in 1928 notes: "...public safety... demands that the construction and operation of a ...dam should never be left to the sole judgment of one man, no matter how eminent, without check by independent expert authority, for no one is free from error, and checking by independent experts will eliminate the effect of human error and ensure safety."

The statements in the two preceding paragraphs emphasize the reasons for and the purposes of peer review, especially when evaluations are being made in difficult or unusual circumstances. The wise investigator will recognize when and why he should seek peer review. Peer review is available from individual consultants and from other engineering firms engaged in dam design.

SECTION 4
REHABILITATION METHODS

General

The integrity evaluation will find that an impoundment is (1) safe in its present state; (2) unsafe and obviously cannot be rehabilitated economically for hydropower use (in which event removal of the dam by the owner would seem to be in order under some dam safety regulatory process); or (3) defective in some manner, but may be restorable economically for hydropower use.

This section discusses various ways in which defective dams have been successfully restored and used for the safe storage and control of water. Engineering feasibility is emphasized. The possible alternative solutions, considered with their costs, can then be incorporated into the feasibility study.

Most defects in an existing dam and in the appurtenances are usually associated in some way with one or more of the following physical circumstances:

1. Reaction of the foundation formation and construction materials to their environment.
2. Resistance to forces and loads
3. Control of seepage
4. Hydraulic capacity and flow performance characteristics
5. Serviceability of mechanical/electrical components and systems.

Many general examples will immediately come to mind, e.g., alkali-aggregate reaction in concrete structures in the case of (1); slides in earth embankments for (2); emerging seepage under pressure from drains in concrete dams or along the toe of an earth dam for (3); stream bed erosion and undercutting of a spillway terminal structure for (4) combined with (2); seizing of an outlet slide gate or a neglectful dismantling of a spillway radial gate hoisting system for (5).

The defects arise either because of original poor design, shoddy construction, lack of maintenance, changed operational demands, or from the application to the original design of more dependable, present-day analytical methods and accumulated hydrologic and seismic records.

The defect may be extensive and seriously threaten the structural integrity of the dam unless promptly counteracted by extensive repair or even replacement. The defect may be in an early stage of development, and if so can be successfully arrested by intensified maintenance. The true nature of a suspected defect may not be immediately determinable and a period of operational monitoring instrumentally or visually may be needed for diagnosis. It may be possible to eliminate or mitigate the defect by reducing the storage level permanently or by operating the reservoir in a different manner.

Decisions on alternatives are thus influenced not only by differing physical designs and methods but also by differing funding arrangements. An extensive replacement such as a new, relocated spillway to replace one historically threatened by obstruction from slides, ice formation, or drift accumulation requires a capital investment. Alternatively, an improved, more attentive maintenance program for continual patrolling and removal of obstructions would require increased annual maintenance funding. Or the useful remaining life and cost of a repair such as patching rotted portions of a timber facing of a rockfill dam might be compared with the life and cost of total removal and replacement with a reinforced gunite facing. The reduced benefits resulting from operating the reservoir at a lower stage for increased flood detention capacity might be compared with the capital cost of enlarging the spillway discharge capability.

Rehabilitation of Dams

In this section and the one that follows the rehabilitation of dams and their foundations are discussed separately. However, it cannot be emphasized too strongly that a dam and its foundation must perform together as an integral unit. This is especially significant along the immediate interface. Many defects simultaneously implicate the dam and the foundation, especially in the case of embankment dams.

Earth and Rockfill, Stonewall-Earth, and Rockfilled Timber Crib Dams. The more common defects encountered are:

1. Insufficient control of seepage and of the accompanying pore pressures and escape gradients.
2. Overly steep slopes of marginal stability, incipient slides, loss of freeboard from crest settlement.
3. Severe erosion and benching of the upstream slope, deep gullying of the downstream face and groins—all tending to reduce the embankment cross section at the most critical elevations.
4. Transverse cracking of the embankment from differential settlement of the fill and consolidation upon saturation of the foundation.
5. Crushing, cracking, parting of waterstops in concrete face slabs of rockfill dams from settlement and deformations of the fill.
6. Excessive large tree growth with large root systems near or on the dam crest creating a breaching potential from uprooting during high winds or root deterioration after the tree dies. Rodent holes can cause similar problems.

7. Utility pressure conduits penetrating or traversing the dam.

Examples of successful remedial measures for these defects are described in the same order. The reader must recognize that in every case the specific details will be different and that the construction methods must be adapted to the actual conditions.

1. Seepage through so-called homogeneous earth dams, where permeability is relatively high or where leakage may concentrate through anomalous regions or transverse cracks, can be controlled by placing a compacted, more impervious zone on the stripped face of the existing dam. The reservoir must be emptied. If the normal drawdown operation of the reservoir cannot be limited or if the new slope cannot be made sufficiently flat, a pervious zone surmounting the added impervious zone may be necessary. If the defect includes excessive seepage through the foundation or along the interface with the dam (often the result of inadequate foundation preparation originally) the new impervious zone can be extended into a cutoff trench excavated into the bedrock formation across the valley section and into the abutments along the upstream toe of the dam or upstream as a blanket. Time must be allowed for accumulated silt deposits to dry; or excavating by dragline may be possible.

If seepage emerges uncontrolled along the toe or over the lower portion of the downstream face, a berm or mildly sloping zone of sand and gravel or cobbles and rock fragments may be added to that face. The grading of the materials positioned immediately against the dam and abutment hillsides must be much more pervious than the material upstream and also prevent movement (piping) of fines from the dam or foundation. If pervious material of the requisite grading is scarce or costly, the main body of the added mass can be comprised of other types of materials, if they are enveloped by pervious materials at all interfaces. With variations, this treatment also improves downstream slope stability.

If the seepage is largely concentrated along the toe or groins, a drain pipe of clay tile, sewer tile, or asbestos-bonded CMP, successively enveloped by gravel and by sand, can be installed in a trench excavated into the foundation along the toe of the dam. If the drain can be safely installed on an alignment upstream of the toe, it will be more effective, especially for slope stability.

2. Actual slope failures can be repaired by first removing all or critical portions of the disturbed mass. If the strength of the materials within the mass has been permanently reduced or if the internal deformations adversely affect the function of a particular zone, then reconstructing to new configurations and zoning suited to the engineering properties of the construction materials is called for. The materials used in reconstruction may be either derived from new sources or reused from the slide volume.

Upstream slope failures are most likely to result from drawdown. Reconstruction requires lowering or even emptying the reservoir. The configuration of the slope to be reconstructed is established by analysis using the engineering properties of the available materials and applying the proposed reservoir operation plan. Construction of a drawdown zone of free-draining rock or cobbles should be considered.

If slide movement has not actually occurred but is considered possible, the slopes can be strengthened by various combinations of seepage control for reduction of destabilizing pore pressures and by adjustments of the exterior slopes. Slopes may be flattened or bermed in lower elevations or unweighted in upper elevations. Free-draining buttress or reverse filter blankets can be added over the ground beyond the toes to counteract instability from high pore pressures in confined, buried aquifers.

The design elevation of the dam crest can be restored by simply stripping the surface and placing and compacting more soil on it. If the crest is narrow, local steepening of both slopes may be acceptable for accommodating the restored elevation, or even for an increased elevation when greater freeboard is needed. Reinforced concrete parapet walls can also be used for either restoring or increasing freeboard.

The near-vertical downstream face of a stonewall-earth dam can be strengthened by adding a downstream zone of compacted, free-draining rock on a slope somewhat flatter than the natural angle of repose of the added rock. The filtering capability at the original interface between the upstream earth zone and the rock wall must be carefully investigated. If piping has occurred, or is likely to occur, a properly graded transition zone should be placed between the existing rock wall and the added rock. The transition zone must be terminated in a non-pipable formation across the channel section and up the abutments, so that all seepage is forced to pass through the filter. Sink holes in the earth zone can sometimes be excavated, shaped, and backfilled with filter materials and compacted earth, and a new compacted earth zone placed on the existing upstream slope to improve the long-term suitability of the impervious zone. If indications of piping, sink holes, and slope disruptions are extreme, rehabilitation by these methods may be inadequate.

Extensive restoration of decayed timber elements of a rock-filled timber crib dam is generally not feasible. Depending upon the degree of disruption and the quality of the rock originally retained by the crib, it may be possible to rehabilitate the dam by adding transition and filter zones and an impervious earth zone upstream and utilizing the old dam as a downstream shell element.

3. Upstream slopes severely benched by erosion can be restored by surface stripping and replacement with

compacted fill. A cushion or transition bedding of correctly graded sand and gravel or small rock is placed on the restored slope beneath the riprap stone. This bedding is essential for adequate performance of the slope protection. Soil-cement properly proportioned, placed, and compacted has been used to restore a slope and to protect it against wave action at the same time.

Gullying of the downstream face and groins can be mended and recurrence prevented by excavating to provide working room, refilling the eroded and excavated areas, then placing a protective course of crushed or angular rock. A system of concrete surface drains, cast or preformed, installed on narrow berms and coupled with a nurtured cover of local grasses has also been successful.

4. Transverse cracking can be repaired if the causative forces have stabilized or have attenuated with time. One method has been discussed in (1) above. When the cracks are limited to the higher elevations in the dam, as they usually are, a narrow trench can be excavated from the dam crest and backfilled with impervious plastic soils. The reservoir may have to be drawn down or even emptied during repair. The strength of the backfill materials must be adequate, otherwise a critical failure plane may be induced by the backfilled trench. Reinforced plastic fabrics, anchored or planted along their perimeters, placed on a smooth prepared surface on the upstream slope and covered by a protective element, can be considered.

5. Excessive leakage caused by disruption of the concrete face elements of a rockfill dam can be reduced or eliminated by selective removal and replacement of damaged panels, if the waterstops from adjacent panels are serviceable. If the embankment is still settling at a significant rate, the repair process will have to be repeated several times. The damaged panels can be covered with courses of redwood tongue and groove planking for increased flexibility during the active settlement period. Anchored butyl rubber sheets have been successfully used on the surface of the panels to waterstop the panel joints.

A rockfill dam can be modified to include an inclined earth core by using the existing dam for the downstream shell and constructing transition zones, filter zones, impervious zones, and shell elements upstream. The opportunity for improved control of foundation seepage, if necessary, is available in such an alteration.

6. The upper crest sections of embankments that are riddled with tree roots or rodent holes can be restored by complete removal of the infested portions and by replacement with compacted fill securely bonded to the unaffected portions.

7. A utility pressure conduit located longitudinally on or near the dam crest can be totally relocated, or it can be rerouted at normal pool level on the upstream face if the reservoir is usually operated full. A longitudinal or transverse conduit can be totally encircled by a larger diameter pipe, or partially encircled underneath by a semi-circular pipe segment of sufficient capacity to safely transport water from a ruptured conduit away from the dam. Automatic shutoff valves controlled by pressure sensing devices can be installed in the conduit beyond both ends of the dam. Transverse conduits can be either relocated away from the dam or replaced using the proven design principles and upstream gating arrangements that are employed for safe outlets.

Concrete and Masonry Dams. The more common defects encountered are:

1. Concrete deterioration from alkali-aggregate reaction, frost action, and poor concrete and construction methods originally.
2. Excessive uplift on the base, on foundation planes at depth, and on horizontal construction joints.
3. Marginal stability for reasons other than excessive uplift.
4. Overstressing, especially in buttress type dams.

Successful remedies and repairs are discussed in the same order.

1. Concrete deterioration appears to be the most prevalent concrete dam defect. The great advances in cement and concrete technology and manufacture and in concrete placement methods are most likely responsible for the improved resistance to deterioration now being observed in newer dams, and it would be expected that this would be confirmed by future performance as the dams become older.

Alkali-aggregate reaction once started cannot be totally stopped by any means now known. If deterioration is advanced, the defective concrete can be removed. For example, in an arch dam if the concrete is less severely affected at lower elevations, its useful and safe service life can be extended at a reduced storage capacity by removing the upper portions and converting the lowered crest to an overpour spillway. The defective concrete can also be replaced. If the entire dam is badly deteriorated but the reservoir basin, detached appurtenances, available yield, and power head provide sufficient benefits, a new dam can be constructed in close proximity to the existing dam or even on the same site by removing the old dam. If the site topography is suitable, the old dam can even be incorporated into a new embankment-type dam.

Alkali-aggregate reaction can be slowed and the useful life of the dam extended by the application of protective upstream coatings and by densifying the concrete itself by grouting, all in order to reduce the severity of the wet environment which helps promote the reaction.

Deterioration of an upstream dam face from alternating freezing and thawing action can be repaired by scaling and chipping the surface to fresh concrete. Steel forms or precast concrete panels can be positioned to the restored face configuration and the intervening space filled by the preplaced aggregate concrete process. Once the panels or forms are installed, the repair can be completed with water in the reservoir. This method restores the full dam cross section. Gunite or shotcrete directly applied with the reservoir empty, of course, can be used if the dam cross section has not been diminished significantly. Seal coats of materials such as neoprene rubber compounds and asbestile can be applied to the prepared concrete surface.

2. Excessive uplift results from inadequate control of seepage. If there are foundation drains and formed drains in the dam which have become plugged with chemical deposits, they can sometimes be reamed and their effectiveness restored if they are accessible from drain galleries or from the dam crest. New foundation drains can be drilled. If water losses are excessive, the foundation can be regrouted from galleries, if they exist but the more effective way to reduce uplift is the addition of drainage.

3. Marginal or inadequate stability in a concrete gravity dam can be counteracted by installing post-tensioned stress tendons through the concrete section and into the foundation. The resisting capabilities of gravity thrust blocks for an arch dam can be increased by the addition of concrete or by post tensioning into the foundation. Post tensioning of a gravity section is especially suited where the horizontal lift surfaces cannot transmit shear because they were not cleaned of laitance during construction.

The stability of a gravity dam can be increased by building concrete buttresses against and bonded to the downstream face. Reservoir water load during construction, temperature control of the new concrete, preparation of the old weathered concrete surfaces, and details of the joint between the two require special design considerations and construction sequences for proper transmittal of shearing stresses and achievement of load sharing. Stress magnitude and distribution, as well as stability, can be improved in both gravity and single arch dams by increasing the cross section with added mass concrete downstream. Slots are left between the new and old concrete for later filling when the new concrete temperatures have equalized.

Stability of buttress-type dams is discussed under (4) below.

4. Buttress-type dams most likely to be encountered during small hydropower feasibility studies are concrete multiple arch dams and concrete or timber slab and buttress dams. Dams having concrete buttresses and removable timber flashboards may also be found.

Sliding stability will seldom be a problem if the angle between the upstream face and a vertical plane is substantial. Because of historically changing construction costs, most of the buttress dams of concrete will be quite old, 50 years or more. Consequently, defects will not only be associated with the inherent low quality and deterioration of vintage concrete but also with stresses in the members comprising the dam. Characteristically, very little reinforcing steel was used in these older dams.

High tensile stresses in the upstream regions of the buttresses of a dam can be reduced by installing tendons or high strength steel rods along the groin at the face of each buttress between the arch barrels, anchoring them into the foundation and then stressing them a predetermined amount while the reservoir is at a low stage. The tendons are covered with protective concrete. The same technique can be used along the intrados of the arch barrels on both sides of the buttress. Lateral rigidity of an individual unreinforced buttress can be increased by attaching reinforced bond beams on both sides of the buttress or by attaching vertical pilasters.

Indicated high stresses in arch barrels attributable to loss of effective thickness from concrete deterioration can be remedied by scaling and chipping the extradosal surface and then restoring, or even increasing, the thickness with gunite or shotcrete reinforced with steel bars or mesh.

Cross channel stability during earthquakes may be low or lacking. The arch barrels and architectural struts between buttresses supply very little resistance. This defect can be overcome by converting alternate pairs of buttresses into single, tower-like supports. This can be done by adding a series of steel or reinforced concrete truss members or vertical concrete diaphragms between the two buttresses. The joint details are extremely important for safe load transfer, especially if the existing buttresses are only nominally reinforced. The buttress can be stiffened by bond beams or pilasters.

Defects in timber buttresses and decks are mainly associated with rotting, corrosion, or other deterioration of the materials forming the members and joint fasteners. Reconstruction with new materials must be undertaken.

Rehabilitation of Dam Foundations

The importance and consequences of foundation defects will vary with the type of dam and the degree and methods of rehabilitation must be planned accordingly. For example, a geological defect such as an open joint at the surface of a rock foundation beneath an embankment dam is of much greater concern than it is beneath a concrete gravity dam. The physical features of a foundation defect usually are not directly observable, because they are hidden by the dam. The presence of characteristics of the defect must be deduced from indirect as well as direct evidence obtained instrumentally or from extracted cores and from study of visual manifestations, such as dissolved solids in seepage water, or movements and strains in the dam itself.

Foundation rehabilitation is often difficult and in some cases may not be possible.

Some of the more common defects encountered are:
1. Insufficient control of seepage and consequent piping, dissolution, or softening of the foundation materials; and displacement of rock masses.
2. Insufficient supporting strength.
3. Inelastic deformations.
4. Loss of local supporting capability from undercutting due to rock plucking.
5. Presence of faults.
6. Excessive or differential consolidation and subsidence.

Some remedial measures that have been used for these defects are described in the same order. Because the foundation must support the dam without excessive deformations or displacements in either the dam or foundation and must control seepage as well, it is obvious that the foundation defects which influence stress and stability in the dam and their rehabilitation cannot be considered independently of the dam.

1. Seepage through foundations can be controlled by grouting, blanketing, new cutoffs, drainage, and pressure relief wells.

A grout curtain can be installed beneath the impervious zone of an embankment dam by drilling through the dam. Care must be used to avoid hydraulic fracturing of susceptible fills with the drilling fluid. Injection of grout between the foundation surface and the base of the embankment should be done carefully. Different techniques are available. A new grout curtain can be installed or an existing curtain supplemented beneath an arch or gravity dam from existing foundation galleries, along the upstream toe, with the reservoir emptied, or even by drilling from the dam crest. A grout curtain can be installed beneath a thin arch dam by slant drilling from the downstream face.

An impervious blanket of compacted earth or a commercially available liner can be placed on the floor of the reservoir. The blanket must be joined to the impervious element of the dam and to the abutments, and must terminate in a satisfactory manner.

The construction of a new cutoff and an impervious facing is described under item (1) of the subsection "Earth and Rockfill, Stonewall-Earth, and Rockfilled Timber Crib Dams." A new cutoff can also be formed in alluvial deposits with a slurry wall. The wall must be joined to the impervious element of the dam. A horizontal impervious zone (blanket) can sometimes be used.

Embankment toe drains and drain blankets are described in the same subsection referenced above. The toe drain or part of the blanket drain can also be installed at depth in the foundation for dual service.

Pressure relief wells or trenches backfilled with drain rock and filter material can be drilled or excavated along or beyond the toe of an embankment dam to control the escape gradients of seepage flowing through the foundation.

Drain holes beneath gravity dams are described under item (2) in the subsection "Concrete and Masonry Dams."

Drain holes can be drilled along the downstream toe of an arch dam, greatly reducing the possiblity of high uplift pressures in the rock structure which tend to displace a foundation rock mass at the abutments.

A drain tunnel can be driven into the foundation from an abutment hillside at an embankment dam, or even from an existing gallery in a concrete dam.

2. The strength of a foundation beneath an existing dam is difficult to increase directly. Tensioned rock bolts or steel tendons may increase the strength of rock foundations, and consolidation grouting may increase the strength of sand and gravel foundations. However, the forces that must be resisted can be changed, or additional resistance can be provided. For example, the imposed shearing stresses on a weak clay seam or bed in a horizontally stratified sedimentary formation can be reduced by flattening the slopes of an embankment dam or by adding buttressing fills if the weak bed outcrops on an abutment hillside. Beneath a concrete dam, the resistance to sliding can be increased by casting concrete shear keys across the bed from trenches or drifts; but it is a difficult and expensive process.

Loose to medium-dense sandy alluvial foundations lose strength during prolonged ground motions from earthquakes. Increased drainage, consolidation of loose materials, and increased confining pressure would all improve the strength of the materials during earthquakes. However, drainage and consolidation may be difficult to achieve and the increase of confining pressure may result in additional dynamic stresses and may actually decrease the stability. The imposed shearing stresses are also difficult to reduce by exterior adjustments of the dam configuration. The most positive way of increasing the stability is to remove the susceptible soils in preparing the foundation beneath flattened slopes or buttressing zones.

3. Irrecoverable deformations in hard rock foundations, which are of concern primarily for concrete dams, occur on first loading, when the mass modulus of elasticity is lower than for subsequent loadings. For existing dams of the moderate sizes under consideration here, it may not be practical or even necessary to attempt treatment of the foundation if it can be demonstrated that the dam is not presently overstressed and that irrecoverable deformations are not continuing.

The deformation characteristics of limited masses of rock defined by geological structural features can be altered by a combination of consolidation grouting and tensioned rock bolts or steel tendons.

4. Local losses of hard-rock foundation may be caused by overpour along the downstream toe of gravity sections, along buttresses, and along the contact between the abutment and extrados of arches. The resulting cavity can be filled with concrete and the resulting interfaces between the rock foundation and the dam concrete then grouted after the concrete mass has cooled to ambient temperature. A plunge basin deeply eroded and retrogressing into the adjacent foundation of an arch dam can be unwatered, cleaned out, and covered with mass concrete anchored to the rock, coupled with treatment of the cavity beneath the dam as just described.

5. Treatment methods for inactive faults or large shear zones beneath existing dams are limited because they are not directly accessible. If a transverse-trending fault is transmitting seepage, it can be locally mined out to practical depths near the toes of the dam and plugged with concrete upstream and filled with filtered, free-draining materials downstream.

Active faults cannot be treated. Instead, the ability of the dam to accommodate fault displacements without disastrous release of water must be evaluated, and if necessary the dam must be modified to accommodate expected movements without failure.

6. Excessive or differential consolidation and subsidence cannot be effectively arrested or controlled by any direct treatment of the foundation at depth. Instead, any continuing foundation movements and their effect on the dam are continuously monitored. The dam can then be repaired or modified accordingly. In some cases, the cause of the subsidence may be detected and corrected, especially if the subsidence is related to old mining works or fluid withdrawal from the substratum beneath the dam.

Rehabilitation of Appurtenant Works

Spillways. The more common defects encountered are:

1. Inadequate capacity to safely pass floods without overtopping the dam.
2. Unpredictable capacity.
3. Damaging hydraulic performance characteristics caused by extreme channel convergence or curvature, lack or mislocation of energy-dissipating terminal structure, excessive velocities, shifting hydraulic control sections, etc.
4. Obstructions to flow.
5. Controlled spillways without redundant features for embankment dams.
6. Spillways founded on fill materials or located over embankment dams.
7. Structural weaknesses in channel walls and inverts, gate piers and anchorages, retaining walls, conduits, etc.
8. Poorly maintained or inoperative mechanical/electrical components.
9. Concrete deterioration.

There are many types and configurations of spillways. Locations vary extensively and are influenced by many factors related to the site and to the type of dam. Consequently, only a few examples of remedial measures can be included here.

1. It may be physically impractical to increase the capacity of a spillway—one with a tunnel discharge carrier, for example; but its capacity can be usually supplemented by a second spillway separately located.

An open channel spillway capacity can be increased by raising the dam crest in different ways, including a parapet wall, even on an embankment dam. Approach channel and discharge channel freeboard must be investigated. If necessary, they can be increased by extending the walls or linings in various manners. A weir type control structure can be lengthened if a new transition to the discharge channel can be fitted in structurally and hydraulically. Usually a capacity increase can be made more efficiently by increasing the head rather than the length, because the capacity varies with the three halves power of the head.

The ability of an impoundment to safely pass floods newly estimated at greater magnitudes can be achieved without enlarging the spillway, if increased flood detention storage capacity can be economically dedicated and the project scrupulously operated accordingly.

Indicated overtopping by the new flood for infrequent limited durations may be acceptable at a concrete dam on an erosion-resistant foundation.

The existing spillway can be considered a service spillway and a new so-called emergency spillway constructed at a higher elevation designed to operate only during a very infrequent flood of the largest magnitude. Project damage, especially to the emergency spillway, can be economically accepted.

Fuse plug control devices in spillways are unpredictable and can create peak flows greater than those of the natural flood. They may also fail to work and thus not provide the intended protection from the inflow flood.

2. The capacity of a siphon spillway may not be reliably predictable. It is also vulnerable to obstruction by trash and ice. It discharges sudden flows at high rates. A battery of siphons can be converted to an open free-discharge crest by removal of the siphon hoods and reshaping of the crests. If additional freeboard is needed with the modified crest, it can be provided as discussed in (1).

3. Freeboard can be increased for an open discharge channel by raising vertical sidewalls or extending a sloping lining to contain overtopping waves or rideups created by excessive channel convergence or alignment curvature. A sloping lining can be extended with a vertical wall. A curved channel can be compartmented by several vertical training walls which will decrease the

rise of the water surface on the outside concave wall in proportion to the number of compartments.

Ill performance of a stilling basin set too shallow can be improved by imposing sufficient tailwater with an end sill or downstream weir.

A foreshortened stilling basin can be extended to compensate for jump sweep-out.

Existing retrogressive channel erosion can be arrested by adding a bucketed terminal structure positioned well above tailwater and supported on deep-seated, cast-in-place piling in drilled holes.

4. An incipient slide or overly steep slope endangering a spillway approach or discharge channel can be stabilized by methods similar to those discussed in the subsection "Earth and Rockfill, Stonewall-Earth, and Rock- filled Timber Crib Dams," item (2).

Persistent drift and trash can be held at bay and contained for periodic removal by installing a securely anchored trash boom fabricated from lengths of timber or other suitable floats such as styrofoam-filled, thin-walled steel pipe linked with chains.

5. Spillway control devices such as gates and flashboards that are ill-suited, poorly designed, or uncertain of operation are really nothing more than spillway obstructions. They pose a hazard, especially to dams that cannot withstand overtopping flows.

Where floods are seasonally predictable, the control devices can be kept clear of the waterway during the flood season.

The control devices can be eliminated, and the desired storage level established by raising the control section with a wall or sill and the required spilling capacity supplemented by methods described in (1).

Redundant spilling capacity over inoperative closed devices may already exist or can be provided.

Redundant operating systems can be installed that will be activated should the primary system fail or when operating personnel cannot or do not arrive at the control station. Radial gates can be modified by counterweighting and adding automatic operating control systems actuated by the rising reservoir stage that will open the gate at a compound rate sufficient to pass the estimated maximum flood. That system can be further backed up by installing buoyancy chambers on the face of the gate designed to force the gate open by water pressure alone in direct ratio to the rise in reservoir stage. The outflow capacity will be less for the backup system; but, if it is designed to pass the largest flood of a long period of record, the probability of an inoperable gate during the more critical large, routine floods will be greatly diminished without seriously affecting the capacity for unprecedented infrequent occurrences.

6. A spillway located on fine-grained fill materials without carefully designed and constructed invert cutoffs, water stops, and a filtered drain system can be withdrawn from service by closure with an earth embankment extended and bonded to an impervious foundation. A new spillway can be constructed at another, more secure location. Addition of the necessary seepage and piping control features at the existing spillway can also be considered, but the practicality and security of doing so may be quite uncertain. A spillway located over an embankment dam will settle, particularly during an earthquake. If the spillway components cannot conform to the settlement without significant structural damage or impairment as a watertight channel, it can be similarly decommissioned and replaced.

7. The stability of spillway control sections, gate piers, large retaining walls and channel walls can be increased by methods similar to those discussed under the subsection "Concrete and Masonry Dams," item (3).

A distressed reinforced concrete conduit discharge carrier can be strengthened with internal steel sets and a concentric concrete lining if the reduced discharge capacity is acceptable.

Damaged or overstressed radial gate anchorages can be replaced with new post-tensioned trunnion block systems.

8. Gates, valves, hoists, bulkheads, stoplogs, etc., can be removed and disassembled and then refurbished by sand blasting, welding, machining, and otherwise repairing each item. Replacement parts are available or can be custom manufactured. Gate seats and seals can be replaced. New improved gate lifts, hoists, engines, motors, etc., can be obtained. Standby emergency generators can be installed to back up the supply of commercial energy.

9. Concrete deterioration and remedial methods for spillway components are similar to those discussed in the subsection "Concrete and Masonry Dams," item (1).

Outlet Works. As with spillways, there are many types and locations of impoundment outlets. There are tunnels or conduits. There are openings and ports through concrete dams. Outlets disrupt the continuity of the dam or of the foundation. They are internally subjected to reservoir water pressure and can transmit that pressure to the dam or foundation anywhere along their alignment. They are a major source of potential weakness in the dam or foundation, especially in the case of an embankment dam. Some of the more common defects encountered are:

1. Inadequate capacity to lower or control the reservoir stage.
2. Unsafe location of control structure; dangerous or restrictive gating facilities.

3. Unsafe location of outlet conduit.
4. Inadequate control of peripheral seepage.
5. Structural weaknesses.
6. Damaging hydraulic performance characteristics, cavitation, lack of energy-dissipating terminal structures, or unsafe release points.
7. Obstructions to flow.
8. Poorly maintained or inoperative mechanical/ electrical features.
9. Deterioration of concrete and metal.

Because of location and surrounding physical constraints, it may be impossible to rehabilitate or modify an outlet. In such a situation, the only practical solution is to construct another one. The existing outlet can be safely removed from service in several ways, depending on the nature and endangerment of the defect and its relationship to the adjacent dam or foundation.

1. An outlet of inadequate capacity can be supplemented with a new one. A new outlet can be constructed on the foundation of an embankment dam by breaching the dam, installing or casting the conduit in place, and replacing the embankment. Proven design and contruction features similar to those for a new, modern project are employed. A tunnel outlet can be driven through an abutment. An opening can be broken through a concrete dam by drilling and pneumatic jacking; a steel conduit or liner installed; the annular space filled with concrete, mortar, or grout; and control facilities installed. The altered stress pattern about larger openings is investigated and reinforcing members added when needed.

If the only defect is inadequate capacity, the old outlet can remain in service. If the outlet is structurally defective, it can be reinforced and kept in service, or it can be plugged with concrete or mortar and grouted to remove it from service. The entire conduit can be filled or the plug can be of limited length and the conduit filled with drain material downstream. If the conduit is removed from service, it may or may not require replacement depending on the need for water service.

2. Outlets beneath embankment dams that are gated only at the downstream end are particularly hazardous because the surrounding embankment and foundation are subjected to full reservoir pressure when the gate is closed. Any leakage from the conduit can result in piping.

Upstream and downstream bifurcations and associated gates and valves can be added to an outlet conduit for safer, more dependable, more flexible control of outflow, and to facilitate otherwise neglected maintenance and repair. Guard gates can be added in line ahead of service or regulating gates.

3. An outlet conduit positioned in the fill of an embankment dam or on a yielding foundation is potentially unsafe, unless it is securely designed for flexibility, axial stretching, and watertightness, and unless the materials will not deteriorate. This potential is especially great for a conduit that crosses a deep embankment foundation cutoff. These outlets can be replaced and safely deactivated as described in (1).

4. Seepage appearing around the exterior of an outlet conduit must be intensively investigated for its source and travel path in order to determine the correct remedial measures.

The conduit can be exposed over a portion of its length near the downstream end and enveloped with drain and filter zones.

The interface and surrounding backfill can be chemically grouted through the walls of larger conduits.

A shaft can be sunk from the surface above the conduit alignment and cutoffs placed about the conduit exterior.

5. A distressed reinforced concrete conduit of larger size can be strengthened as discussed under "Spillways," item (7).

A bare steel conduit of doubtful strength or which may be badly corroded can be strengthened and rehabilitated by centering a smaller pipe or liner inside the conduit and pressure-filling the annular space with mortar. The alignment and grade of the conduit must be reasonably straight and the reduced discharge capacity must be acceptable. Construction details for proven techniques are available.

A dry-type intake tower of doubtful stability or of resistance to flotation can be converted to a more stable wet-type tower by modifying the piping and gating system.

Structural defects in other external outlet works components, such as open channels, intake structures, walls, and energy dissipators can be rehabilitated as described in the subsection "Concrete and Masonry Dams," item (3).

6. Cavitation of conduit surfaces in high velocity outlet works at flow-disrupting locations and at gates and valves can be repaired with resistant materials such as stainless steel liners or epoxy concrete. The fluidway boundary surfaces can be straightened and irregularities removed or smoothed. Air can be introduced where sub-atmospheric pressures are created in the water, especially at gates and valves. Spring points can be formed in the conduit walls for flow separation.

An energy-dissipating terminal structure can be added to control erosion at the outlet release point.

A conduit can be extended to a point of safe release.

A defectively designed or constructed stilling basin can be modified as discussed under "Spillways," item (3).

7. A silted intake structure can be vertically extended by constructing a riser on top of the existing intake.

An actual or incipient slide imperiling an entrance or return channel can be removed or stabilized.

8/9. Defects and rehabilitation measures for mechanical/electrical features and materials deterioration are similar in principle to those discussed under "Spillways," item (8).

Other Considerations

Defects that may be associated with the reservoir basin are:

1. Thin, weak, natural topographic and geologic barriers impounding the reservoir.
2. Large-volume incipient or potential slide masses that can move suddenly at high velocities into the reservoir pool and create water surges that overtop the dam.
3. Economic loss of stored water through pervious geologic structure.

A weak natural barrier of limited topographic expression and extent can be strengthened by seepage control, drainage, and stabilizing measures similar to those employed for embankment dams.

A reservoir that leaks over a large area probably cannot be sealed economically. If the leaking areas are of limited extent and can be selectively identified, it may be possible to reduce the water losses from the reservoir by blanketing those areas with compacted impervious soils, covered by a protective blanket of sand and gravel or fine rock. Reservoir leakage of this nature would not be expected to cause any loss of basin integrity or catastrophic release of storage, except where it might occur in the immediate proximity of the dam or thin natural barriers.

The stability of reservoir slides can be improved by unloading the upper portion of the slide, buttressing the base, drainage, and chemical treatment. The potential for such an event should be examined during the integrity investigation. The freeboard on the dam can be increased some judgmental amount to provide for slide volume and wave generation. The unusual topographic, geologic, and ground water conditions contributing to those very few cases where devastating slides have actually occurred would appear to be extremely rare.

Defects in the following project features, although not directly affecting impoundment integrity, can impede project operation and maintenance, especially during an emergency situation:

1. Impassable or inadequate access roads and bridges.
2. Lack of communication facilities.
3. Lack of emergency lighting at critical locations along spillways, outlets, and the dam crest.

Appropriate rehabilitation methods are obvious.

SECTION 5
COST ESTIMATING GUIDELINES

General

Cost estimating is a specialized field and can best be performed by persons who routinely make cost estimates for the service required or the type of construction that is to be performed. However, in order to initiate a small hydroelectric project, administration must have a viable means of acquiring reasonable cost data for the integrity investigation of the existing facilities; and if the facilities are amenable to the addition of small hydroelectric plants, the cost of remedial work and maintenance must be determined as input for the economic feasiblity determination.

This section presents guides for methods of estimating costs, sources of cost estimating information, and some ranges of cost at 1978 prices for some of the major common items. Because of unique site conditions, climatic conditions, location, quantities, and other factors, unit costs may vary widely (sometimes by several hundred percent) from site to site.

Costs associated with the integrity investigations and rehabilitation of existing facilities consist of engineering costs, construction costs, and administration costs. These functional costs are composed of labor, material, and equipment with all of their associated variation. Costs can readily be determined if the quantities and unit prices of all of the cost factors are known for the time that the work is to be performed. However, determining the quantities and unit prices for all the items involved with a sufficient degree of accuracy for the intended purpose can be a major challenge. There are at least two different types of cost estimates. They are approximate estimates and detailed estimates. For a feasibility investigation, an approximate estimate is normally adequate. A detailed estimate would not normally be required and is not feasible until the design plans and specifications have been prepared.

The unit costs associated with rehabilitating an existing structure will vary more widely depending on quantities and location and will generally be more expensive than costs for similar items for new structures. Basic reasons for these widely varying costs are that working room and access are limited, demolition or preparation of the portion of the structure to be rehabilitated will be required, mating new equipment to old equipment is difficult or parts may not be available as a shelf item, quantities are normally small resulting in high mobilization and unit costs, and the work is normally labor intensive. As an example, concrete in place could cost less than $50 per cubic yard in a massive structure where a plant and raw materials are readily available, whereas the repair of a structure requiring a few cubic yards of concrete may cost several hundred dollars per cubic yard if it is located in a remote area where access is poor and if several man-days of labor are required to chip out old concrete, construct forms, mix and place the concrete, strip the forms, and cure the concrete.

Estimating Integrity Investigation Costs

Stage 1. The work associated with the Stage 1 integrity investigation of existing facilities is similar to work performed under the Corps of Engineers Phase 1 dam safety inspections program. The Phase 1 inspections are generally being performed by private consulting engineering firms and are reported to cost generally in the range of $7,000 to $9,000 per dam inspected during 1978. The $7,000 to $9,000 1978 costs properly escalated would give a reasonable cost estimate for Stage 1 investigation by consulting engineers for "average" facilities. Unusual or complex facilities could cost considerably more. Of course the most reliable method of acquiring a cost estimate would be to get a quotation from an engineering firm that is qualified to perform the work, or to estimate the time and materials costs if the investigation is to be performed by in-house staff.

Stage 2. The costs for Stage 2 investigations are highly variable and dependent on the extent of the investigations, laboratory testing, and analyses and evaluations that are required. The cost for this stage of the investigations should be estimated as part of the Stage 1 investigation work, or an estimate of the costs could be obtained from the engineer that would be performing the Stage 2 work. The following 1978 unit costs are pre-

---

Description	Cost
Engineering	$25 - $75/hour
Drilling, Soil	$ 7 - $12/L.F.
Rock	$20 - $40/L.F.
Classification Testing of Soils	
Atterberg Limits	$48 ea.
Specific Gravity	$33 ea.
Sieve Analysis	$40 ea.
Hydrometer	$38 ea.
Unconfined Compressive Strength	$22 ea.
Compaction Properties	$95 ea.
Direct Shear	$55/point
Triaxial Shear (with pore pressure measurement)	$90/point

sented as a guide for estimating major Stage 2 cost items after an estimate of time and quantities have been made. Approximately 10 to 20 percent should be added to the labor cost for miscellaneous items such as printing, telephone, transportation, etc.

Stage 3. The engineering costs for this stage will be highly variable, depending on the extent of the work to be performed as defined in Stages 1 and 2. The costs for performing the Stage 3 engineering work can best be estimated by persons that performed the Stage 1 and Stage 2 work or the persons that are to perform the Stage 3 work. Advanced estimates of Stage 3 costs can be made only if the man-hours required for engineering and supporting help can be reasonably estimated. The cost of this type of engineering is much higher than for new works for reasons similar to those that make rehabilitation construction cost more than new construction.

Construction Rehabilitation Costs

Information Required for a Feasibility Cost Estimate. The Stage 3 investigation must be completed before a reasonable feasibility cost estimate can be prepared. As part of Stage 3, the rehabilitation work that is to be performed must be well defined as to scope and extent. Drawings to scale showing the dimensions of materials that are to be removed, replaced, or added are necessary to determine the volumes of all significant materials and number of major items.

The major cost items associated with the rehabilitation of existing structures will generally fall within the classifications of earthwork, concrete, structural steel, timber, and electrical and mechanical items.

Unit prices for earthwork within a job can vary widely, depending on the type of earthwork involved which will affect the amount of labor, equipment, and materials and supplies required for a unit volume of earthwork. Earthwork is generally broken down into excavation and fill.

Excavation is normally broken down further according to material type, e.g., soil or rock (soil being defined as a fine-grained material which can be readily excavated with scrapers, and rock being defined as a material which requires heavy ripping or blasting prior to excavation). In addition to the two types of materials discussed above, there are coarse-grained materials (sands and gravels) which would cost about the same as soil per unit volume of excavation, fine-grained and coarse-grained materials containing cobbles and boulders which increase the cost of excavation, and soft rock which is easily rippable and has an intermediate unit cost of excavation. Excavation quantities must be computed for each material type which would have a different unit cost. In addition to material type, unit costs for excavation will increase greatly if the material is to be excavated from below the water table; quantities to be excavated below and above the water table should be separated. Of course, access to the excavation area and working room can have a significant effect on unit cost and must be considered when separating quantities to be used in preparing a cost estimate.

Unit costs for fill materials are primarily dependent on material type, availability, in-situ conditions, haul distance, access, working room, and placement and compaction requirements. Where a variety of material types are to be placed for remedial work, the above factors must be considered and separate volume computations made for material types that may have significantly different unit costs.

Slopes of excavations and fills for rehabilitation of existing dams are typically irregular and the volume cannot usually be computed by volume formulas for standard shapes. The common method of determining excavation and fill quantities is by the average end area method by the formula $V = \ell/2\ (A_1 + A_2)$ where V is the volume (cubic feet) of the prismoid of length ℓ (feet) between cross-sections having areas (square feet) A_1 and A_2. End areas can be determined by drawing vertical cross-sections to scale and planimetering the areas; or quite often it is quicker to planimeter the areas of horizontal planes from the plan view on a contour map. A great deal of ingenuity is required to obtain some quantities quickly and accurately. As an example, thin layers of slope protection can be computed by planimetering the plan area and converting the planimetered area to the true surface area by multiplying the area times the square root of the sum of the squares of the horizontal and vertical distances along the slope divided by the horizontal distance, and then multiplying this area times the thickness to develop its volume.

Primary work items for concrete work are preparation of the area where concrete is to be placed, forming, placing reinforcing steel, placing and finishing the concrete, curing, and removal of forms and clean-up. To make an accurate estimate of concrete cost, the quantities for labor, materials and equipment must be determined. Sometimes, however, if the volume of concrete is determined, a unit price can be assigned by using unit prices determined from a previous job where similar concrete work was performed.

Structural steel is normally priced on the basis of weight. The American Institute of Steel Construction *Steel Construction Manual* (AISC, 1973) provides weights per linear foot for all standard shapes and sizes of structural steel members.

The cost of timber items can best be determined by computing the boardfeet of the various types of timber members required to replace rotted or damaged material or for a required addition.

Each major mechanical/electrical item, along with the necessary controls and leads to be replaced or added, must be identified and sufficiently specified so that proper replacements can be secured. The manufacturer's name and model or identification numbers are most helpful for replacement items, while specifications for new items such as valves, the type, size, head, type of controls, etc., must be identified.

Sources of Cost Information. After the volumes of materials or number of items have been identified for rehabilitation, unit or item costs must be applied. There are many sources of information for costs, several of which are discussed below.

The most reliable method of obtaining a good cost estimate is to have a professional estimator or local contractor that regularly performs the type of work being considered prepare the estimate. When small and difficult jobs are bid competitively by contractors, it is not unusual for the high bidder to be double the total price of the low bidder, with wide variations from the engineer's estimate. Therefore, even having a professional estimator or contractor estimate the cost does not assure that his estimate is what the cost will be if and when the project goes to construction. Reserves and contingencies must be used to protect the project in the event that the cost estimates prove to be inadequate due to circumstances beyond the control of the estimator.

Another method of obtaining reasonably reliable costs is to utilize adjusted unit costs from a similar project. Costs should be adjusted for inflation, difference in locale, site conditions, quantities, etc. Considerable judgment is required to determine if the work is similar and what adjustments should be made in unit prices for any differences.

Engineering News Record, published weekly by McGraw-Hill, gives quarterly statements of construction trends, cost indexes for common items for a number of years, equipment rental rates, and material rates. Periodically *Engineering News Record* prints unit prices bid for government projects. These prices can be indicative of costs. However, the projects are generally large and conditions at the sites are not defined; thus the unit prices are of questionable value for application to rehabilitation of existing small dams.

Unit costs for many construction items, equipment rental costs, equipment production rates, and labor rates are available in some annual publications. Two of these, *Dodge Guide to Public Works and Heavy Construction Costs,* (McGraw-Hill, annual) and Heavy Construction cost File (Engelsman, 1977), provide good unit cost data for use in estimating the civil works costs. The total unit cost, as well as the labor, material, and equipment unit costs, is presented. *Estimating Construction Costs* (Peurifoy, 1975) is an excellent general reference for methods of preparing detailed cost estimates.

Equipment rental firms will supply costs as well as information on equipment specifications and production. Local material suppliers will readily furnish costs for items which they have for sale, and costs of materials for concrete, steel, timber and such items are readily available. The cost for mechnical/electrical items such as gates, valves, hoists, etc. can best be determined by the supplier of the specific items. The local office of the U.S. Department of Labor will supply labor rates.

Cost Summary

The cost estimates for investigating and rehabilitating existing facilities for the addition of small hydroelectric facilities are intended for use in planning and in economic and financial feasibility analyses. (Volume II of this manual, *Economic and Financial Analysis,* discusses in detail the use of the cost data developed.) The costs developed, as discussed above in this section, must be summarized and documented in a form that is usable by the economic and financial evaluators. (See Figure 5-1 under "Examples," below).

The project and type of cost estimate should be identified in the title. The major work items should be identified by number, described briefly, and the units of measurement, quantity, unit prices and the total amount of the cost for each major work item should be given in tabular form. The cost of major work items should be totaled and an appropriate contingency factor applied to account for minor items not included in the cost estimate and for additional work which may be required by conditions revealed during final design investigations and analyses or during construction. The contingency factor that should be applied depends on the level of the study at the time the cost was prepared (i.e., conceptual, feasibility, final design, construction), whether site conditions are well defined or not, the extent to which minor items are included in the estimate, and the reliability of quantities and unit prices. The contingency factor should never be less than 10 to 15 percent for this type of work and could be as high as 30 to 40 percent or more if site conditions are not well defined and the work is in a preliminary stage.

The cost should be based on the prevailing costs at the time that the estimate is made and the date of the estimate should be identifed. Volume II discusses in detail methods of applying escalation factors. It should be left to the people performing the economic and financial feasibility analyses to escalate costs for all phases of the project.

Cost Examples

Unit Costs for construction Items. Typical 1978 unit costs for the more common rehabilitation construction items are presented in Table 5-1. These unit costs must be adjusted for escalation and specific site conditions. Manufactured items are not included in Table 1 because they can be readily checked by a telephone call to a supplier.

TABLE 5-1
TYPICAL UNIT COSTS

Item	Units	Unit Cost
Common excavation (dry)	C.Y.	$ 1.50
Common excavation (wet)	C.Y.	3.00
Rock excavation	C.Y.	3.50
Earth fill	C.Y.	2.00[a]
Rock fill	C.Y.	4.50[a]
Filter-drain material	C.Y.	15.00
Concrete (reinforced)	C.Y.	200.00

[a] Includes excavation, haul, placement, and compaction.

BLUE RIVER HYDROELECTRIC PROJECT
FEASIBILITY COST ESTIMATE
FOR
INVESTIGATION AND REHABILITATION OF
EXISTING FACILITIES

Item No.	Description	Units	Quantity	Unit Price	Amount
1	Foundation Excavation, Soil	C.Y.	5,000	$ 2.50	$ 12,500
2	Foundation Excavation, Rock	C.Y.	2,000	10.00	20,000
3	Zone 1 Fill	C.Y.	4,000	3.50	14,000
4	Zone 2 Fill	C.Y.	50,000	2.75	137,500
5	Slope Protection	Sq.Ft.	5,000	8.00	40,000
6	Remove and Replace Concrete Spillway Walls	Sq.Ft.	500	250.00	125,000
7	Furnish and Install 18" Butterfly Valve	Ea.	1	3,000.00	3,000
			Subtotal		$352,000
	Contingency fl 25%				88,000
			Subtotal		$440,000
	Investigations				30,000
	Engineering				50,000
	Administration				10,000
			TOTAL		$530,000

Note: This estimate is based on current (1978) unit prices. Unit prices must be escalated for work performed after 1978.

Figure 5-1. Sample feasibility cost estimate summary

Feasibility Cost Estimate. A sample cost estimate summary is presented in Figure 5-1 as a guide for summarizing and documenting the cost estimate data.

Actual Costs for Repair of a Non-standard Dam. An example of a timber crib dam was shown in Figure 2-9. Figure 5-2 is a close-up of a segment of the same dam, showing downstream wood planking in disrepair. The dam is approximately 400 feet long and 24 feet high. Interior timber cribs were rehabilitated and refilled with rock where necessary, and the downstream face was replaced with gunite over wire mesh. The cost of the repair work in 1978 was approximately $300,000, or about $15 per square foot of facing. This example demonstrates the high cost for types of repair work that contractors are not used to performing. Figure 5-3 shows a segment of the dam after rehabilitation.

Utilization of Cost Information in Decision-Making Processes

As discussed above, the cost information will be utilized to evaluate the economic justification and financial feasibility of the project. In addition to the cost of investigating and rehabilitating the existing facilities, many other factors such as cost of installing hydroelectric equipment, power production capacity, marketing, financing, etc. have major effects on the feasibility of adding hydroelectric facilities to existing structures. However, the existing facilities are different from the other aspects to be evaluated in that they are there and something must be done with them.

If they are suitable for their existing use and are not a financial liability to the owner, leaving them in their current state would not adversely affect the owner. However, if the investigations should reveal that the existing facilities are unsafe under existing operating conditions, the facilities would have to be rehabilitated and operated in a different manner which would be safe, or breached and abandoned. Any of the above courses of action would have a financial impact on the owner, and this should be considered in the decision making process.

Figure 5-2. Close-up of broken, rotting timbers and rock washed out at toe of dam

Figure 5-3. One segment of dam after repairs to timbering, replacement of rock fill, and placement of gunite facing

REFERENCES

Acker, W. L., III, 1974, *Basic Procedures for Soil Sampling and Core Drilling:* Acker Drill Co., Inc., P. O. Box 830, Scranton, PA, 18501, 246 pp.

American Concrete Pipe Association (ACPA), 1959, *Concrete Pipe Handbook.*

_____, 1957, Hydraulics of Culverts.

American Institute of Steel Construction (AISC), 1973, *Manual of Steel Construction,* 7th ed.

American Society for Testing and Materials (ASTM), *Annual Book of ASTM Standards.*

ARMCO Drainage and Metal Products, Inc., 1955, *Handbook of Drainage and Construction Products.*

American Society of Civil Engineers (ASCE), 1967, *Current United States Practice in the Design and Construction of Arch Dams, Embankment Dams, Concrete Gravity Dams,* Joint publication of ASCE and International Congress on Large Dams (ICOLD).

_____, 1969, *Stability and Performance of Slopes and Embankments,* Soil Mechanics and Foundations Division Conference, Berkeley, August 22-26.

_____, 1975, *Lessons from Dam Incidents - USA,* ASCE-ICOLD joint publication.

Biswas, A. K., and Samar Chaterjee, 1971, Dam Disasters - An Assessment, *Engineering Journal* (Canada), v. 54 no. 3, pp. 3-8, March.

Cedergren, Harry R., 1977, *Seepage, Drainage and Flow Nets,* 2nd ed., Wiley, pp. 148-280.

Dodge Guide for Estimating Public Works Construction Costs, Annual, McGraw-Hill.

Engelsman, Coert, 1977, *Heavy Construction Cost File,* Van Nostrand Reinhold.

Engineering News Record, Weekly, McGraw-Hill.

Golze, Alfred R. (ed.), 1977, *Handbook of Dam Engineering,* Van Nostrand Reinhold.

Hirschfeld, Ronald C., and Steve J. Poulos (eds.), 1973, *Embankment Dam Engineering,* Wiley.

International Congress on Large Dams (ICOLD), 1973, *Lessons from Dam Incidents,* Committee on Failures and Accidents to Large Dams, Paris, France.

Justin, Joel D., William P. Creager and Julian Hinds, 1945, *Engineering for Dams,* 3 volumes, Wiley.

Longyear Company, 1972, *Soil Sampling Methods and Equipment* (commercial catalog with good descriptions of equipment and applications).

Makdisi, Faiz I., and H. Bolton Seed, 1978, *Simplified Procedure for Estimating Earthquake Induced Deformations,* ASCE Journal of the Geotechnical Engineering Division, v. 104 no. GT7, July, pp. 849-867.

National Resources Committee (NRC), 1939, *Low Dams,* Subcommittee on Small Water Storage Projects, Water Resources.

Peurifoy, R. L., 1975, *Estimating Construction Costs,* 3rd ed, McGraw-Hill.

Starr, Chauncey, 1969, *Social Benefit versus Technical Risk,* Science, v. 165, September 19, pp. 1232-38.

U.S. Department of Interior, Bureau of Reclamation (USBR), 1965, *Stress Analysis of Concrete Pipe.*

_____, 1968, *Beggs Deformeter Stress Analysis of Single Barrel Conduits,* 2nd ed., rev.

_____, 1974/1, *Design Criteria for Concrete Arch and Gravity Dams.*

_____, 1974/2, *Design of Small Dams,* 2nd ed., rev.

_____, 1974/3, *Earth Manual.*

_____, 1976, *Design of Gravity Dams.*

_____, 1977, *Design of Arch Dams.*

_____, 1978, *Manual for Safety Evaluation of Existing Dams* (Preliminary): Engineering and Research Center.

U.S. Department of the Army, Corps of Engineers (COE), 1977, *National Program of Inspection of Dams Report,* v. 1, Appdx. D - Recommended Guidelines for Safety Inspection of Dams, Office of the Chief of Engineers.

_____, *Hydrologic Engineering Methods for Water Resources Development,* Hydrologic Engineering Center. Volume 3 Hydrologic Frequency Analysis, April 1975; v. 4 Hydrograph Analysis, October 1973; v. 5 Hypothetical Floods, March 1975; v. 7 Flood Control by Reservoirs, February 1976.

EXHIBIT I
UNIVERSAL CHECKLIST FOR INSPECTIONS
(Adapted from USBR, 1978, Appendix D)

1. GENERAL

The integrity of dams and appurtenant works is controlled by (1) their designs, (2) the characteristics of their constituent materials, (3) the nature of their foundations, and (4) their regional settings.

The objective of the inspection is to visually examine the structural conditions and hydraulic performance characteristics in relationship to these "performance controllers."

2. CHANGES IN THE CHARACTERISTICS OF MATERIALS

2.1 *General* - Observe for defective, inferior, unsuited, or deteriorated materials. A variety of different materials makes up the different types of dams and appurtenances. The quality and durability of these materials must be determined for each specific structure.

2.2 *Concrete* - (1) alkali-aggregate reaction, pattern crazing and cracking, (2) leaching, (3) frost action, (4) abrasion, (5) spalling, (6) general deterioration, (7) strength loss.

2.3 *Rock* - (1) disintegration, (2) softening, (3) dissolution.

2.4 *Soils*- (1) degradation, (2) dissolution, (3) loss of plasticity, (4) strength loss, (5) mineralogical change.

2.5 *Soil-cement* - (1) loss of cementation, (2) crumbling.

2.6 *Metals*- (1) electrolysis, (2) corrosion, (3) stress-corrosion, (4) fatigue, (5) tearing and rupture, (6) galling.

2.7 *Timber*- (1) rotting, (2) shrinkage, (3), combustion, (4) attack by organisms.

2.8. *Lining fabrics*- (1) punctures, (2) seam partings, (3) light deterioration, (4) disintegration of boundary seals, (5) loss of plasticity and flexibility.

2.9 *Rubber*- (1) hardening, (2) loss of elasticity, (3) light deterioration, (4) chemical degradation.

2.10 *Joint sealers* - (1) loss of plasticity, (2) shrinkage, (3) melting.

3. GENERIC OCCURRENCES

3.1 *General* - Observe generic occurrences for their characteristics, locations, and recency. These occurrences are of a universal nature, regardless of structure type or foundation class. The details of what to look for in observing these generic occurrences, actual or evidential, must be observed at all structures and locations.

3.2 *Seepage and leakage* - (1) Discharge-stage relationship, (2) increasing or decreasing, (3) turbidity and piping, (4) color, (5) dissolved solids (6) location and pattern, (7) temperature, (8) taste, (9) evidence of pressure, (10) boils, (11) recency and duration.

3.3 *Drainage*- (1) obstructions, (2) chemical precipitates and deposits, (3) unimpeded outfall, (4) sump pump facilities, (5) bacterial growth.

3.4 *Cavitation*- (1) suface pitting, (2) sonic evidence, (3) implosions, (4) vapor pockets.

3.5 *Ice action* - (1) evidence of ice forces decreasing stability of structures, lifting gate hoists, obstructing gate leaves and operational and mechanical installations.

3.6 *Stress and strain* - evidence and clues - (1) in concrete cracks, crushing, displacements, offsets, shears, creep; (2) in steel - cracks, extensions, contractions, bending, buckling; (3) in timber - compression, buckling, bending, shears, extensions, compressions; (4) in rock and soils - cracks, displacements, settlement, consolidation, subsidence, compression, zones of extension and compression.

3.7 *Stability*- evidence and clues - (1) in concrete and steel structures - tilting, tipping, sliding, overturning; (2) in embankment structures, cutslopes, natural slopes - bulging, sloughing, slumping, sliding, cracks, escarpments; (3) in rock cutslopes, foundation, and unlined tunnels - slumps, slides, rockfalls, bulges, cracks.

4. OPERATION AND MAINTENANCE

4.1 *Service reliability of outlet, spillway, sump pump mechanical/ electrical features* - (1) broken or disconnected lift chains and cables, (2) test operation including auxiliary power sources, (3) reliability and service connections of primary power sources, (4) ease and assurance of access to control stations, (5) functioning of lubrication system.

4.2 *Gate chambers, galleries, tunnels, and conduits*- (1) ventilation and heat control of damp, corrosive environment of mechanical/electrical equipment.

4.3 *Accessibility and visibility* - (1) obscuring vegetal overgrowth; (2) galleries, access ladders, lighting; (3) access roads and bridges; (4) communication and remote control lines, cables, and telemetering systems.

4.4 *Control of vegetation and burrowing animals* - (1) harmful vegetation on embankments - oversize, dead root channels; (2) harmful vegetation in structural concrete joints; (3) obstructing vegetal growth in hydraulic flow channels; (4) ground squirrels, muskrats, and beavers.

5. BEHAVIOR

5.1 *General* - Resident operational personnel can often supply valuable information and may have been the only observers (during earthquakes, for example).

5.2 *Warning, safety, and performance instrumentation* - (1) piezometers, flow recorders, accelerometers, seismoscopes, joint meters and gage points, strain meters, stress meters, inclinometers, direct and inverted plumblines, surface reference monuments, stage recorders, extensiometers; (2) serviceability; (3) access to readout stations; (4) type and location suitable for condition being observed; (5) need for recalibration; (6) faulty readings, sources, and reasons; (7) alarm systems operable and at appropriate set points; (8) random check readings during inspections.

5.3 *During and after large floods* - (1) driftmarked high waterlines; (2) evidence of taxed spillway capacity; (3) undesirable or dangerous spillway flow patterns directly observed or deduced from flow stains, erosion trails, swept vegetation, deposition of solids.

5.4 *During and after large outlet releases* - (1) undesirable or dangerous discharge flow patterns, dynamic pressures, vibrations, cavitation sonics.

5.5 *After earthquakes* - (1) cracks, displacements, offsets in structural features; (2) cracks, slumps, slides, displacements, escarpments, settlements in embankments, cutslopes, and fill slopes; (3) broken stalactites in galleries, tunnels, chambers; (4) toppled mechanical equipment; (5) sand boils, (6) changes in seepage patterns and rates.

6. CONCRETE AND MASONRY DAMS

(Any of these observations are applicable also to reservoir-impounding power intake structures, spillway control structures, lock walls.)

6.1 *Stress and strain* - evidence and clues - (1) cracks, crushing, displacements, offsets in concrete monoliths, buttresses, face slabs, arch barrels visible on exterior surfaces and in galleries, valve and operating chambers, and conduit interior surfaces; (2) typical stress and temperature crack patterns in buttresses, pilasters, diaphragms, and arch barrels; (3) retention of design forces in post tensioned anchorages and tendons.

6.2 *Stability* - evidence and clues - (1) excessive or maldistributed uplift pressures revealed by piezometers, pressure spurts from foundation drain holes, construction joints, and cracks; (2) differential displacements of adjacent monoliths, buttresses, and supported arch barrels or face slabs; (3) disparities in regions near the interface between arches and thrust blocks; (4) movement along construction joints; (5) uplift on horizontal surfaces revealed by seepage on downstream face and in galleries at construction lift elevations.

6.3 *Hillsides and river channels adjacent to the abutments and river section foundation along the downstream toe of the dam* - (1) leakage, (2) seepage, (3) stability, (4) boils.

6.4 *Special attention to stability and seepage control at discontinuities and junctures* - (1) embankment wraparound sections, (2) waterstops in monoliths and face slabs, (3) reservoir impounding backfill at spillway control sections and retaining walls.

6.5 *Foundation* - (1) piping of weathering products from old solution channels and rock joint structure; (2) efficiency of foundation seepage control systems - drains, drainage holes, grout curtains, cutoffs, drainage tunnels; (3) history of shear zones, faults, cavernous openings; (4) zones of varying permeability; (5) orientation of stratification and bedding planes effect on permeability, uplift, foundation stability; (6) subsurface erosion and piping; (7) thin weaker interbeds - effect on stability.

7. EARTH AND ROCKFILL, STONEWALL-EARTH, AND ROCKFILLED TIMER CRIB DAMS

7.1 *Stress and strain - evidence and clues* - (1) settlement; (2) consolidation; (3) subsidence; (4) compressibility; (5)cracks, displacements, offsets, joint opening changes in concrete facings on rockfills; (6) loss of freeboard from settlement; (7) zones of extension and compression visible along dam crest or elsewhere; (8) crushing of rock points of contact; (9) differential settlement of embankment cross sectional zones visible along dam crest, indicating stress transfer along region of zone interface (increases possibility of hydraulic fracturing).

7.2 *Stability* - evidence and clues - (1) cracks, displacements, openings, offsets, sloughs, slides, bulges, escarpments on embankment crest and slopes and on hillsides adjacent to abutments; (2) sags and misalignments in parapet walls, guardrails, longitudinal conduits or other lineaments parallel to embankment axis; (3) irregularities in alignment and variances from smooth, uniform face planes; (4) bulges in ground surfaces beyond toes of slopes.

7.3 *Inadequate seepage control* - evidence and clues - (1) wet spots; (2) new vegetal growth; (3) seepage and leakage; (4) boils; (5) saturation patterns on slopes, hillsides, and in streambed; (6) depressions and sinkholes; (7) evidence of high escape gradients.

7.4 *Erosion control* - (1) loss, displacement, and deterioration of upstream face riprap, underlayment, and downstream face slope protection; (2) beaching.

7.5 *Foundation* - (1) see 6.5 also, (2) consolidation, (3) liquefaction potential.

7.6 *Other endangerments* - (1) utility pressure conduits on, over, or through embankments; (2) diversion ditches along abutment hillsides.

8. SPILLWAYS

8.1 *Approach channel* - (1) obstructions; (2) slides, slumps, and cracks in cutslopes.

8.2 *Log booms* - (1) submergence, (2) uncleared accumulated drift, (3) parting, (4) loss of anchorage, (5) inadequate slack for low reservoir stages.

8.3 *Hydraulic control structure* - (1) stability, (2) retention of capacity rating, (3) erosion at toe, (4) installations on crest, raising storage level and decreasing spilling capacity, (5) gate piers, (6) trash control systems, (7) nappe and crotch aeration, (8) siphon prime settings.

8.4 *Headwater control (gates, flashboards, fuse plugs,*

fabric dams) - (1) position, (2) wedging, (3) gate trunnion displacements, (4) loss of gate anchorage post tensioning, (5) undesirable eccentric loads from variable positions of adjacent gates, (6) gate-seal binding, (7) erosive seal leakage, (8) failure of lubrication system, (9) availability of bulkhead facilities for unwatering, and of cranes and lifting beams.

8.5 *Operating deck and hoists* - (1) broken or disconnected lift chains and cables; (2) unprotected exposure of electrical/mechanical equipment to weather, sabotage, vandalism; (3) structural members and connections.

8.6 *Shafts, conduits, and tunnels* - (1) vulnerability to obstruction; (2) evidence of excessive external overloading - pressure jets, contorted cross sections, cracks, displacements, circumferential joints; (3) serviceability of linings (concrete and steel), materials deterioration, cavitation, erosion; (4) rockfalls; (5) severe leakage about tunnel plugs; (6) support system for pressure conduits in walk-in tunnels.

8.7 *Bridges* - (1) possibility of collapse with consequent flow obstruction, (2) serviceability for operational and emergency equipment transport.

8.8 *Discharge carrier (open channel or conduit)* - (1) vulnerability to obstruction; (2) evidence of excessive external sidewall loading - large wall deflections, cracks, differential deflections at vertical joints; (3) invert anchorage and foundation support - drummy soundings, buckled lining, excessive uplift; (4) observation or evidence of dangerous hydraulic flow patterns - cross waves, inadequate freeboard, wall climb, unwetted surfaces, uneven distribution, ride-up on horizontal curves, negative pressures at vertical curves, pressure flow, deposition; (5) drain system serviceable; (6) air ingestion and expulsion; (7) tendency for jump formation in conduits; (8) buckling, slipping of slope lining; (9) erosion of unlined channels.

8.9 *Terminal structures* - (1) inadequate dissipation of energy, (2) jump sweep out, (3) undercutting, (4) retrogressive erosion, (5) loss of foundation support for flip bucket substructure, (6) unsafe jet trajectory and impingement, (7) erosive endangerment of adjacent dam or other critical structures.

8.10 *Return channels* - (1) impaired outfall; (2) obstructions; (3) slides, slumps, cracks in cutslopes; (4) erosion of deposition creating dangerous tailwater elevations or velocities; (5) evidence of destructive eddy currents.

9. OUTLETS

9.1 *General* - Many of the observations made of outlet components are similar in nature and purpose to those made for spillway components, stilling basins for example.

9.2 *Approach channels (may seldom be directly visible and may require underwater inspection)* - (1) siltation, (2) underwater slides and slumps.

9.3 *Intake structures (including appended, inclined, and freestanding towers, both wet and dry* - (1) lack of dead storage; (2) siltation; (3) potential for burial by slides and slumps; (4) damage or destruction of emergency and service bulkhead installation facilities; (5) availability of bulkhead, cranes, lifting beams; (6) serviceability of access bridges.

9.4 *Trashracks and raking equipment* - (1) clogging of bar spacing, (2) lodged debris on horizontal surfaces, (3) collapse.

9.5 *Gate chambers, gates, valves, hoists, controls, electrical equipment, air demand ducts* - (1) accessibility to control station under all conditions; (2) ventilation; (3) gate or valve positions; (4) binding of gate seals; (5) seizing; (6) erosive seal leakage; (7) failure of lubrication system; (8) drainage and sump pump serviceability; (9) vulnerability to flooding under reservoir pressure through conduits, bypasses, and gate bonnets surfacing in chamber.

9.6 *Conduits and tunnels* - (1) see 8.6 also, (2) seepage or leakage along external periphery of conduit, (3) extension strains in conduits extending through embankments, (4) capacity and serviceability of air relief and vacuum valves on conduits.

9.7 *Terminal structures* - See 8.9.

9.8 *Return channels* - See 8.10.

10. ENVIRONS

10.1 *Reservoir* - (1) stage at time of inspection; (2) indications of recent noteworthy stages; (3) depressions, sinkholes in exposed reservoir basin surfaces; (4) massive water-displacing slide potentials - leaning trees, escarpments, hillside distortions; (5) flood pool encroachments; (6) siltation adversely affecting loading on dam, and forming approach channel and waterway obstructions.

10.2 *Reservoir linings - compacted, PCC (Portland Cement Concrete) and AC (Asphaltic Concrete), fabric* - (1) depressions, sinkholes; (2) erosion; (3) animal disruption.

10.3 *Downstream proximity* - (1) tailwater stage at time of inspection, (2) reservoir-connected springs; (3) endangering seepage or leakage regardless of source, (4) river obstructions creating unanticipated tail- water elevations or interference with outfall channel capacities of the spillway and outlets.

10.4 *Watershed* - (1) surface changes that might materially affect runoff characteristics.

10.5 *Regional vicinity* - (1) subsidence indications - sinkholes, trenches, subsidence surveys, settlements of buildings, highways, other structures in the region; (2) assessment of land forms and regional geologic structure; (3) records of mineral, hydrocarbon, and groundwater extractions, locations, producing horizons, accumulated production, and current rate of production.

10.6 *Downstream flood plain* - (1) limits of natural, improved, or leveed channel; (2) areas of potential inundation - for spillway design flood, for hypothetical failure; (3) proximity of developed areas, (4) habitation, population, communication and transportation corridors.

11. EXAMPLE - SPECIFIC CHECKLIST

An example of a specific checklist for a zoned earth dam follows. Lists for other types of dams, for reservoirs, and for appurtenant works can be similarly prepared with the aid of the universal list.

Check List For Inspection of Zoned Earth Dam

Upstream face - (1)slides; (2) settlement, cracks, and displacements; (3) vegetative growth; (4) slope protection for erosion, beaching, grading, durability, loss of bedding.

Downstream face - (1) slides; (2) settlement, cracks, and displacements; (3) seepage, saturation, wetness; (4) vegetation; (5) slope protection for furrowing, durability; (6) rodents.

Regions adjacent to abutments and foundations - (1) seepage; (2) cracks, slides; (3) vegetation; (4) groins for erosion; (5) formation joints, fractures, bedding planes; (6) boils; (7) depressions; (8) sinkholes, (9) rodents.

Crest - (1) cracks; (2) settlement; (3) lateral movements; (4) camber; (5) parapet walls for sags and misalignment.

Performance Instrumentation - piezometer gauge house and equipment; (2) surface positions of observation wells, piezometers, deflectometers, cross-arm settlement devices; (3) surface settlement and deflection monuments; (4) reference monuments.

Adjacent endangerments - (1) utility pressure conduits; (2) diversion ditches along abutment hillsides.

EXHIBIT II

CONSIDERATIONS AND PROCEDURES FOR IMPOUNDMENT INTEGRITY EVALUATIONS

(Adapted from USBR, 1978, Appendix C)

Note: The term "review" as used in this Exhibit means a study of project records or project-related publications; or an appraisal or analysis of a condition, apparent or suspected, based on available information or supplemental data acquired during Stage 2.

1. GEOLOGY

1.1 Review geologic mapping, plans, and cross sections showing exploration features and summarizing drill logs and geologic interpretations for the dam, appurtenant structures, materials sources, and the reservoir geology. Particular attention should be paid to geologic features such as: shear zones; faults; open fractures; seams, joints, fissures, or caverns; landslides; variability of formations; compressible or liquefiable materials; weak bedding planes, etc.

1.2 Review exploration logs for lithologic and physical conditions, water test data, standard penetration or other resistance testing results.

1.3 Review geophysical data.

1.4 Review groundwater level records in the vicinity of the reservoir.

1.5 Review petrographic or chemical studies of foundation materials and natural construction materials.

1.6 Review geologic portions of all reports relevant to the site.

1.7 Review aerial photographs of site and reservoir.

1.8 Review published or unpublished regional geologic studies that are relevant to the dam and reservoir setting.

1.9 Inspect the pertinent features of the areal geology at the dam and appurtenant sites, borrow and quarry sites, and, to the extent practicable, in the reservoir basin. Inspect representative core recovered from exploration, particularly from zones indicated on the logs as being badly broken, weathered, or highly pervious.

1.10 On the basis of general geologic setting, is this an acceptable site for the type of dam? Are attitudes of bedding and joints particularly favorable or unfavorable to seepage, slope stability, foundation stability, acceptance of dam and reservoir loads and pressures, and sliding?

1.11 Review any effect of raised groundwater levels on the stability of abutment and reservoir slopes.

1.12 Review potential chemical activity - reactivity of aggregate, quality of surface and groundwater, type of cement.

1.13 Was foundation improved by treatments such as pressure grouting slurrying grouting, blanket grouting, drainage, dental concrete, and deeper or more extensive excavation?

1.14 Was the actual treatment of the geologic conditions adequate?

2. SEISMICITY

2.1 Review seismic and tectonic history of region.

2.2 Review seismic history of site.

2.3 Determine location and relative influence of active and potentially active faults which could affect the project site.

2.4 Consider all potential earthquake effects which could influence the project site such as:

- Surface rupture
- Ground tilting
- Elevation changes
- Shaking
- Landsliding
- Slumping
- Liquefaction
- Settlement
- Seiches

2.5 Review design earthquake - location, magnitude, and recurrence interval.

2.6 Were expected baserock motions for design earthquake developed? What are they and how were they developed? Are design accelerograms available?

2.7 Were pseudostatic "g" factor(s) recommended for design? How were they determined?

2.8 Review aerial photographs and space imagery of site and region.

3. HYDROLOGY AND SPILLWAY DESIGN FLOODS

3.1 Review summary hydrologic data contained in project reports.

3.2 Review design reports, operations and maintenance manuals, and contract plans and specifications regarding spillway design and operation.

3.3 Review design flood criteria:

- Hazard potential of impoundment.
- Downstream risk evaluation.
- Appropriate flood magnitude.

3.4 Review design storm precipitation, duration, and runoff values:

- Storm distribution with time.
- Assumed snowpack conditions.
- Watershed characteristics - antecedent moisture, vegetation type, topography, land use, etc.

3.5 Review flood routing studies:

- Reservoir area - capacity curve.
- Spillway rating curve (gated or uncontrolled).
- Flood routing analysis.
- Assumed reservoir water surface prior to design flood inflow.
- Maximum flood surcharge level.
- Residual freeboard between crest of dam and maximum flood surcharge level.

3.6 Review flood control and other storage operation plans.

- Seasonal storage requirements.
- Seasonal flood potentials.
- Potential operational conflicts.
- Normal outlet releases.

3.7 If spillway is gated:

- Review seasonal gate operation procedures and schedules.
- Do the flood routing studies consider gate malfunctions and any redundant provisions for passing floods?

3.8 Review downstream flood plain conditions:

- Limits of improved channel and/or flood levees.
- Areas of potential inundation for spillway design flood discharges.
- Proximity of developed areas.

3.9 Is a spillway capacity reevalutation needed in light of the present state-of-the-art and post-construction hydrological records?

4. GENERAL CONSIDERATIONS FOR DAMS OF ALL TYPES AND APPURTENANT STRUCTURES

4.1 Review contract plans and specifications and design reports.

4.2 Review basic design including dam layout, cross-sections and zoning, specified foundation treatment, and grouting. Note any unusual aspects or omissions.

4.3 Review exploration, geology, and seismicity data for dam and reservoir, and evaluate. Note potential adverse effects of known geologic features.

4.4 Review laboratory test procedures and results.

4.5 Assess unforeseen conditions and their treatment for relationship to safety and performance of dam and appurtenances.

4.6 Review construction photographs.

4.7 Review construction control test results. Compare these with the design-phase exploration and test results and with the design assumptions.

4.8 Compare materials and foundation properties determined during construction with general criteria used for design. Assess adequacy of criteria and specifications provisions from safety standpoint with regard to specific items such as seepage control, capacity, and clogging potential of foundation and interior drains, piping potential, etc.

4.9 Evaluate design criteria and methods of analyses and their relationships to present state-of-the-art.

4.10 Are there any activities in the region such as mining or oil or water extraction which could adversely affect the dam or appurtenance?

4.11 Evaluate whether construction specifications, procedures, and materials were compatible with general design assumptions and known site conditions.

4.12 Review instrumentation installations and assess adequacy of instrumentation for monitoring probable operational performance in general or for specifically identified behavioral patterns.

4.13 Review instrumentation records and evaluate significance of results.

4.14 Conduct detailed inspection of site and environs. Note any unusual or suspect conditions. Observe selected drill cores, if available.

4.15 Was design and construction in accord with the state-of-the-art at the time?

4.16 How would design and construction compare with present state-of-the-art?

5. EARTH AND ROCKFILL, STONEWALL-EARTH, AND ROCKFILLED TIMBER CRIB DAMS

5.1 *General*

5.1.1 See Section 4 of this exhibit.

5.1.2 Review adopted foundation and embankment materials design properties and compare with exploration and field and laboratory test results for appropriateness. Evaluate compatibility of the dam and foundation.

5.1.3 Review stability analyses, including the loading and operational conditions analyzed. Note any apparent deficiencies and/or unusual appearing results. Were currently acceptable methods of analyses employed?

5.1.4 Review as-built drawings and data including foundation configuration, grouting summaries, drainage provisions, construction changes, type and depth of cutoff, foundation discontinuities, special foundation treament, etc., and assess their potential effects on performance.

5.2 *Materials Properties - Placement, Testing, and Control*

5.2.1 Classification, gradation, Atterberg limits.

5.2.2 Laboratory maximum densities for fine-grained materials, relative density for coarse-grained materials. Optimum moisture.

5.2.3 Freeze-thaw (riprap durability).

5.2.4 Consolidation and settlement.

5.2.5 Dispersive clay tests, solubility tests.

5.2.6 Filter and drain materials, gradation, permeability, etc.

5.2.7 Petrographic and minerological descriptions.

5.2.8 Lift thickness, compactive effort, method of compaction.

5.2.9 Number and distribution of control tests. Variation of density and moisture.

5.2.10 Select material and placement methods at abutments and around structures.

5.2.11 Variability of material in borrow areas.

5.2.12 Relative settlement of adjacent zones.

5.2.13 Dynamic and static strength properties (friction angle and cohesion).

5.3 *Foundation*

5.3.1 Methods used in determining the strength and behavioral characteristics of the foundation mass.

5.3.2 Extent of foundation investigation - area covered - number and type of exploratory holes.

5.3.3 Summary of grouting - depth, take, pressures, additives, and mixes.

5.3.4 Drain holes, seepage, and uplift control systems.

5.3.5 Strike and dip of joint system.

5.3.6 Specified foundation treatment.

5.3.7 Size and location of seams and shears.

5.3.8 Characteristics of any joint fillings.

5.4 *Analytical Data*

5.4.1 Method of analysis - finite element, slip circle, wedge, etc. What materials, engineering properties (strength, etc.) were used? Were they valid? What were assumptions for foundation strengths and interaction with the dam?

5.4.2 What loading conditions were adopted?

5.4.3 Results of analysis - stresses, strain, displacements, stability factors, foundation pressures.

5.4.4 Was any analysis made of pore pressure distribution within the dam and foundation?

5.4.5 Was analysis made of seepage distribution within the dam and foundation?

5.4.6 Were the abutments analyzed?

5.4.7 Compare computed and measured deformations in dam and foundation.

5.4.8 Was uplift and fracturing caused by grouting considered and monitored?

6. CONCRETE AND MASONRY DAMS

6.1 *General*

6.1.1 See Section 4 of this exhibit.

6.1.2 Review adopted foundation and concrete materials design properties and compare with exploration and field and laboratory test results for appropriateness. Evaluate compatibility of the dam and foundation.

6.1.3 Review results of stress analyses or stability analyses, including loading and operational conditions analyzed especially for any apparent deficiencies and/or unusual appearing results. Were currently accepted methods of analyses used?

6.1.4 Evaluate possible effects of freezing and thawing on structural response and operational performance of the impoundment.

6.2 *Material Properties - Placement, Testing, and Control*

6.2.1 Strength and durability of concrete employed - 90-day strength, etc.; size of cylinders (design vs. construction values), coefficient of variation - high and low values - number of cylinders.

6.2.2 Modulus of rupture and elasticity of concrete.

6.2.3 Have any cores been taken from dam and tested? How do the results compare with design criteria?

6.2.4 Type of cement, cement factor, admixtures, and water-cement ratio. What tests were conducted on the cement used? Proportions of concrete mix? Was the creep property of concrete determined?

6.2.5 Lift height and method of placement.

6.2.6 Treatment of vertical or contraction joints and lift surfaces.

6.2.7 Concrete placement and joint grouting schedule - as performed.

6.2.8 Heat generation characteristics of the concrete mixes.

6.2.9 Physical, chemical, and mineralogical characteristics and sources of aggregates used.

6.3 *Foundation*

6.3.1 Methods used in determining the strength and behavioral characteristics of the rock mass.

6.3.2 Extent of foundation investigation - area covered - number and type of exploratory holes.

6.3.3 Summary of grouting - depth, take, pressures, additives, and mixes.

6.3.4 Drain holes, seepage, and uplift control systems.

6.3.5 Strike and dip of joint system.

6.3.6 Specified foundation treatment.

6.3.7 Size and location of seams and shears.

6.3.8 Characteristics of any joint fillings.

6.4 *Analytical Data*

6.4.1 Method of analysis - trial load - finite element - number of cantilevers - arches, etc.

6.4.2 How was the foundation deformation considered?

6.4.3 What loading conditions were adopted?

6.4.4 What temperature variation was assumed?

6.4.5 When were construction joints grouted relative to construction sequence?

6.4.6 How much cooling occurred prior to grouting?

6.4.7 Results of analysis - stresses, thrust, movements, stability factors, shear-friction safety factors, foundation pressures.

6.4.8 Was any analysis made of pressure distribution within the foundation?

6.4.9 Abutments radial or nonradial?

6.4.10 Shear keys in vertical or contraction joints?

6.4.11 Was the effect of cracked sections included?

6.4.12 Were the abutments analyzed?

6.4.13 Impact forces of water in plunge pool (arch dams only)

6.4.14 Compare computed and measured stresses and deformations in dam and foundation.

7. APPURTENANT STRUCTURES

7.1 *General*

7.1.1 See Section 4 of this exhibit.

7.1.2 Review basic design, including plans, section, details, assumptions, and criteria. Note any unusual aspects and omissions.

7.1.3 Review laboratory and hydraulic model test procedures and results.

7.1.4 Review adopted foundation, concrete and steel reinforcement design properties, and compare with exploration, field and laboratory test results, and generally accepted practice, for appropriateness. Evaluate compatibility of the structure with its foundation and environment.

7.1.5 Review results of stress and stability analysis, including loading and operational conditions analyzed. Note any apparent deficiencies and/or unusual appearing results.

7.1.6 Evaluate possible effects of freezing and thawing on structural and operational service of structures.

7.2 Spillway

7.2.1 Hydraulic evaluations - Evaluate spillway capability to pass all design floods without endangering the dam. If the spillway has control gates, evaluate redundant provisions for safely passing floods should the gates fail to fully operate for any reason. Review provisions (log booms, etc.) for keeping spillway entrance free of obstructions.

7.2.2 Structural evaluations - Review and evaluate the following relevant to the security of the dam:

- Geologic data regarding the spillway foundation and compatibility with structural design.
- Design criteria in comparison with generally accepted standards. The evaluation would include review of the various combinations of loading for which components of the spillway facility might be subjected, such as:

  Earth loads
  Hydrostatic loads
  Uplift forces
  Dynamic water forces
  Earthquake forces

- Design of seepage cutoffs and drainage provisions behind spillway walls and beneath floor slabs.
- Energy dissipation features.

7.3 Outlet Works Structures and Controls

Review and evaluate the following items relevant to the security of the dam:

7.3.1 Design criteria with regard to hydraulic and structural requirements.

7.3.2 Operational criteria including capability of outlets to reduce or completely withdraw reservoir storage in event of emergency.

7.3.3 Geologic conditions and any potentially adverse effects on structural or operational requirements.

7.3.4 Backup systems available in event of operation malfunctions.

7.3.5 Energy dissipation features.

7.4 Materials Properties for Spillways and Outlets - Placement, Testing, and Control

7.4.1 Strength and durability of concrete employed - 90-day strength etc., size of cylinders (design vs. construction values), coefficient of variation - high and low values - number of cylinders.

7.4.2 Modulus of rupture and elasticity of concrete.

7.4.3 Type of cement, cement factor, admixtures, and water-cement ratio. What tests were conducted on cement? Proportions of the concrete mix?

7.4.4 Methods of concrete placement.

7.4.5 Treatment of construction and contraction joints.

7.4.6 Physical, chemical, and mineralogical characteristics and sources of aggregates.

7.4.7 Properties of steel reinforcement.

7.4.8 Do the properties of the materials actually used conform with design assumptions?

7.5 Foundations of Spillways and Outlets

7.5.1 Methods used in determining the strength and behavior characteristics of the supporting rock.

7.5.2 Extent of foundation investigation - area covered - number and type of exploratory holes.

7.5.3 Summary of grouting - depth, take, pressures, mixes, additives.

7.5.4 Drain holes, seepage, and uplift control systems.

7.5.5 Strike and dip of joint systems.

7.5.6 Specified foundation treatment.

7.5.7 Size and location of seams and shears.

7.5.8 Characteristics of any joint fillings.

7.6 Analytical Data for Spillways and Outlets

7.6.1 Were methods of analysis adequate and appropriate?

7.6.2 How were foundation characteristics handled?

7.6.3 Were adopted loading conditions adequate and appropriate?

7.6.4 Results of analyses - stresses, stability factors.

7.6.5 Evaluate anticipated hydraulic performance of energy dissipation features, channel or conduit flow patterns, and scour resistance.

8. OPERATION AND MAINTENANCE

8.1 Review reservoir topography and geology and assess reservoir landslide potential.

8.2 Review any established designers' operating criteria and standard operating procedures or similar documents, for the project. Note particularly the operational capability of outlets to reduce reservoir storage in an emergency, the redundant systems available to operate gated spillways and outlets works during power and operational malfunctions. Identify project operation and maintenance factors relating to the safety of the dam.

8.3 How often are operators required at the dam?

8.4 Identify any adverse or difficult operational aspects related to dam impoundment integrity.

VOLUME V: ELECTROMECHANICAL FEATURES

SECTION 1: INTRODUCTION AND OVERVIEW

Scope of Electromechanical Report

This volume covers the selection, spatial requirements, cost, and representative manufacturers of the major equipment items and systems which comprise the electromechanical functions of a small hydroelectric power plant. The historical record of cost increases in equipment is presented for escalation of the cost component of the foregoing items to a time period beyond the July 1978 base used in this volume. The upper limits of small hydroelectric projects, as defined herein, are operating heads up to 100 feet and flow rates producing an output of up to 15,000 kW of power. The lower limits are a function of available equipment and the economics of developing power at the site.

Definition of Electromechanical Equipment

Electromechanical equipment is considered to be the equipment and systems required to develop the energy, either potential or kinetic, available in impounded or flowing water, to convert it to electric energy, to control it, and to transmit it to a regional power grid. The major equipment items are the hydraulic turbine, the electric generator, and a switchyard consisting of a transformer, circuit breaker and switchgear. Included are supporting systems which control and protect these major equipment items. Maintenance facilities such as a crane for lifting, which may be required, are also considered in a broad definition of electromechanical equipment.

Dependent upon the type and capacity of the hydraulic turbine and electric generator, the cost of the electromechanical equipment can vary from one quarter to one half of the total small hydroelectric power additions cost. When the cost of transmission lines and rehabilitation of the impoundment structure is included to the total project cost, the ratio decreases.

The selection of some equipment, primarily the generator, is dependent on the type of hydraulic turbine selected. Other equipment, such as transformer, switchgear and electrical protection systems are examples of equipment not dependent on the type of hydraulic turbine used.

Limitation of Data

The data provided herein regarding cost and dimensions were obtained from manufacturers, federal agencies, engineering consultants and contractors. The data was analyzed and factored to represent reasonable costs to be used for the intended purpose of this volume, which is feasibility level cost estimates. Generally, the cost data presented should be considered the mid-point in a band of costs varying as much as plus or minus 10 percent.

There are significant factors which can cause the costs of equipment to vary that are not controllable. It is not unusual to have competitive bids for turbines and generators vary by 25 percent. These variations can occur for several reasons. Whenever standardized units are proposed by a manufacturer, their cost should be less then a custom-made unit, because a custom made unit usually includes some development engineering costs. The exchange rate of the dollar is directly related to the cost of foreign-made equipment. The final selection of the type of turbine/generator and other mechanical equipment should be made by totalling firm bid prices from manufacturers and the estimated powerhouse civil/structural cost with due consideration to guaranteed hydraulic efficiences and anticipated life.

Power Equation

Power can be developed from water whenever there is available flow which may be utilized through a fall in water level. The potential power of the water in terms of flow and head can be calculated with the following equations:

$$hp = (Q \times H)/\ 8.815$$

where: hp is theoretical horsepower available
Q is quantity of water flowing through the hydraulic turbine in cubic feet per second
H is available head in feet

In terms of electrical output the above equation becomes:

$$kW = (Q \times H \times E)/11.81$$

where: E is the overall efficiency of the hydropower plant. For general estimating purposes, E is normally taken to be 0.85.

Functional Differences Between Large and Small Hydroelectric Power Plants

It is the general practice in the design of modern hydropower plants to include adequate controls in the generation equipment to enable the unit to maintain the system frequency in the event the power plant and its local distribution system become electrically separated from the regional power grid. For small hydroelectric plants which are typically very small in comparison to the generation capability of the regional grid, there are cost savings in the governoring system if the turbine-generator speed control system does not include frequency regulation of the electric system. Under these latter conditions an electric separation from the regional grid would shut down the small hydroelectric power plant.

Furthermore, for smaller hydropower plants, the duplication of transmission lines leading from the power plant to the connection with the grid is not necessary. The loss of one small power plant would not normally cause any impact on the system but the savings realized by not duplicating the transmission line could significantly effect the economic feasibility.

Smaller hydroelectric power plants can also be designed with less flow control than larger plants. The flow of water to most turbines is controlled by a set of gates called wicket gates. The wicket gates, which are controlled by signals from the governor, regulate the flow of water into the turbine and control the amount of power produced. Where it is not important to control the amount of power produced or regulate the flow for hydrological reasons, wicket gates can be eliminated, reducing the first cost of the turbine by about 10 percent, and also providing a reduction in maintenance costs.

Figures 1-1 and 1-2 show a comparison of a new and an old small hydroelectric power plant project.

Figure 1-1. McSwain Power Plants located on the Merced River, California with a capacity of 10,000 KW. Constructed in 1969 (Courtesy of Merced Irrigation District).

Figure 1-2. South Power Plant located on Bottle Creek, California with a capacity of 4,000 kW. Constructed in 1910. (Courtesy of Pacific Gas and Electric Company).

SECTION 2
ELECTROMECHANICAL FEATURES

General

The major electromechanical components of a power plant are the inlet valve, turbine, draft tube, draft tube gates, generator, control and protection equipment, and substation for transformation of the power to the transmission line voltage. In terms of spatial requirements and costs, the major items are the turbine and generator. Other miscellaneous plant equipment include the crane, station light and power systems, fire protection systems, heating and ventilating equipment, potable water system, and sanitary facilities. Most of the traditional miscellaneous equipment for larger hydroelectric projects can be either eliminated or reduced in scale for smaller, unattended hydropower projects.

General Considerations for Selection

Economic vs. Actual Life. In the selection of electromechanical equipment, differences between economic and actual life are important in the determination of the project feasibility. The economic life is the period of time to retire the bonds or to retrieve the capital required for construction. Bonds with a forty year retirement period would indicate an economic plant life of forty years. However, this does not properly credit the actual life of a hydro facility. A turbine/generator unit, properly maintained, may last over 75 years and consideration of a capital investment credit at the end of economic life should therefore be given.

Technical Considerations. Small hydro plants have generated the interest of a number of turbine and generator manufacturers in both the foreign and domestic market. The advantage of having foreign suppliers is the competitive aspect they introduce to the domestic marketplace as well as innovative technology. The disadvantage is the communication gap that may exist when spare parts are needed or technical problems in operation must be resolved. Although the economics of foreign equipment may prove attractive, consideration for future problems must be evaluated in the selection of equipment.

Another technical consideration is the speed of rotation of the turbine and generator. Under low head conditions, the turbine speed is generally below 450 r/min, which is considered a low speed. A low speed generator is larger in diameter than a high speed machine of the same capacity. By selecting a gear driven speed increaser, it is possible to couple a low speed turbine to a generator operating at several times the speed of the turbine. The higher speed generator would cost less to manufacture, weigh less, reduce the structural requirements and decrease the building size. Although this reduces the cost of procurement and construction, the speed increaser represents a loss in efficiency of one to two percent and a potential increase in maintenance cost.

Reaction turbine runners are subject to pitting caused by cavitation. For further discussions on cavitation refer to USBR Engineering Monograph No. 20-1976. Damage to the runner from cavitation can be avoided by proper selection of the speed of the turbine and the distance the runner is set above or below the tailwater surface. Selection of these parameters primarily have an effect on the cost of the turbine/generator and powerhouse excavation. This variation in cost for a typical installation is within the accuracy of the estimated costs presented in this volume. The dimension and cost data presented in this volume is based upon the centerline of the runner being set at approximately the minimum tailwater elevation. If site excavation costs are unusual it is suggested that turbine manufacturers be contacted for recommendations of speed and setting elevation to avoid cavitation.

Design Trends for Small Hydropower Installations. Industry is responding to the needs of the small hydropower plant market. Equipment manufacturers are standardizing the sizes and capacities of small units. Such applications are particularly applicable for the bulb-type and the tube-type predesigned turbine and generator units.

For a manufacturer, standardization is the preparation of functional control diagrams and physical layouts which may be readily adapted to a range of job site conditions. The purpose is to reduce engineering costs and establish the criteria applicable to a specific range of power plant sizes. Another application is the establishment of predetermined modes for unit start-up and the reduction of requirements for control equipment such as governors and synchronizing equipment. Standardization represents cost savings and an increase of project feasibility.

Methodology for Selection of Unit and Determination of Cost

The methodology by which the turbine/generator and accessory electromechanical equipment for a small hydropower plant is selected and its cost determined is described in the following paragraphs. The steps are shown graphically on Figure 2-1. The selection process is a trial-and-error process and two or more selections may be carried through this procedure at the same time. The final selection is made by combining all costs, including civil features of the powerplant and improvement of the impoundment, then comparing the annual cost to the amount of energy generated.

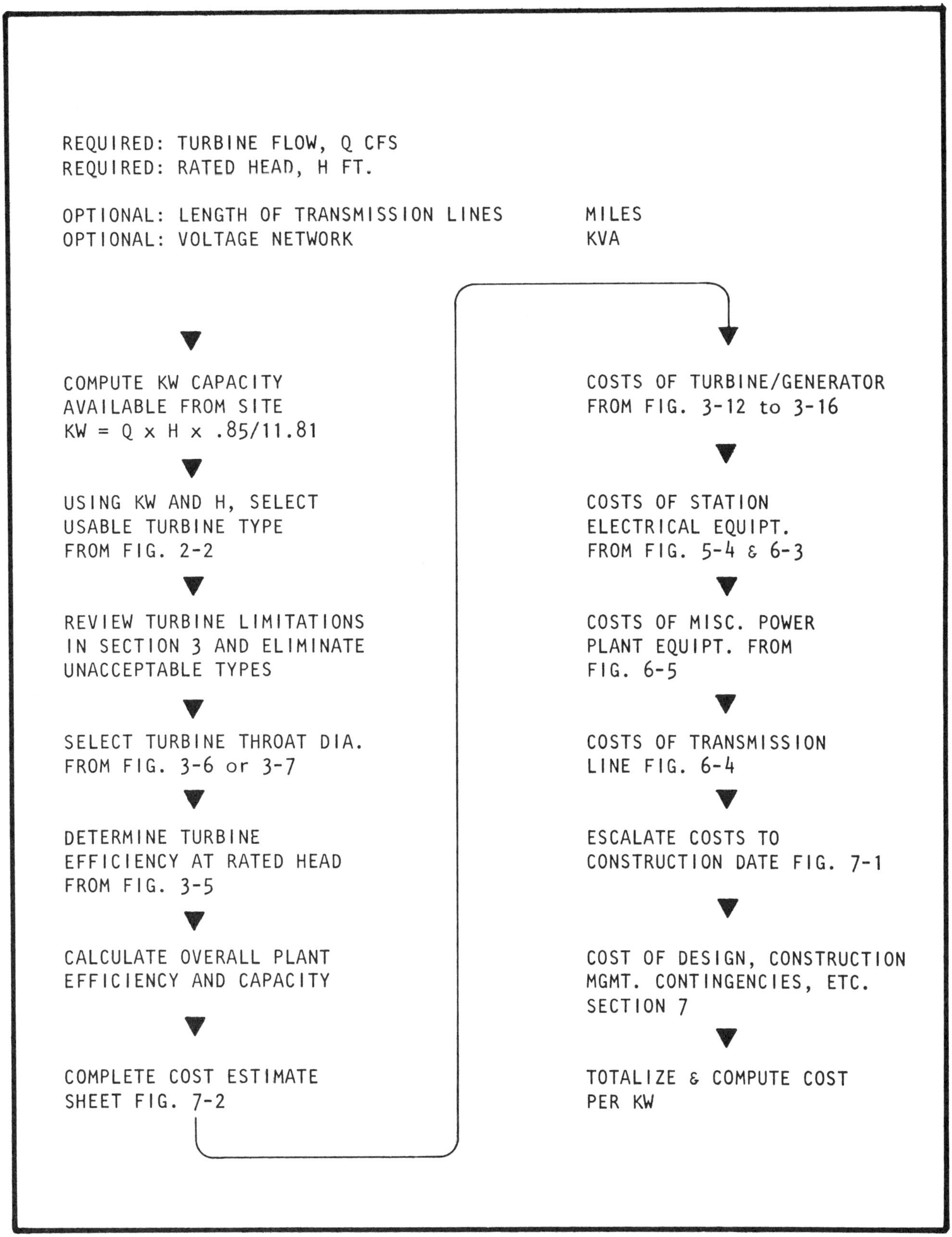

Figure 2-1. Turbine selection methodology

Collect Site Data. The basic data which must be considered in the selection of a hydraulic turbine are the design flow of water and the net head on the turbine. If the results of the study are to include transmission line cost, the other data needed to complete the electromechanical cost estimate include location and voltage of the nearest transmission line with the available capacity to handle the power from the project. Additional data which would be desirable include condition of the water, variation of water level in the impoundment, variation of tailwater level with flow and climatic conditions for the site. The collection of this data is further described in Volume III, Hydrologic Studies.

Obtain Effective Head. The effective head is the static head, the difference between the level of water in the impoundment and the tailwater level at the outlet, less the hydraulic losses of the water passage. The effective head must be used for all power calculations. The hydraulic losses can vary from essentially zero from flume-type turbine installations to amounts so significant for undersized outlet conduits that the energy potential of the site is seriously restricted. The hydraulic losses in closed conduits can be calculated using the principles set out in general hydraulic text books. In addition to conduit losses, an allowance for a loss through the intake structure should also be included. In general a hydraulic loss of one velocity head (velocity squared divided by 64.4) or greater would not be uncommon. The hydraulic losses through the turbine and draft tube are accounted for in the turbine efficiency curves.

Select Turbine/Generator. From the turbine design flow and maximum effective head, the kilowatt capacity of the unit can be computed by the power equations. Note that with the installation of multiple turbines, the turbine design flow should be divided by the number of units to give the flow per unit. The installation of multiple turbines should be considered in order to obtain higher efficiency over a wide range of flows. If multiple units are selected, all of the units should be equivalent, same capacity and same manufacturer, in order to reduce the required spare parts inventory.

The efficiency to be used in the preliminary sizing should be 85 percent. Based upon the kW capacity of the unit and turbine net head, the type of turbine (or turbines) can be determined from Figure 2-2. This graph was developed from data available from turbine manufacturer and information contained in public and private utilities publications. Section 3 contains pertinent information relative to the type of turbines shown on the figure, including general information and the limitations of operation of turbines relative to various flows and head.

Select Turbine Throat Diameter and Other Dimensions. Modern reaction turbine design has evolved through the trials of various dimensions and shapes in models which are tested in hydraulic laboratories. The critical dimension which dictates the amount of discharge capable from a turbine is the throat diameter, i.e. the diameter immediately below the runner of a Francis turbine or the tip diameter of a Propeller turbine. The throat diameter is a function of the type of turbine, effective head and capacity and may be estimated by use of the appropriate chart in Section 3. All other dimensions of the turbine may be estimated by use of the figures given in Volume VI - Civil Features. These dimensions will vary with different manufacturers' designs; however the data given are suitable for preliminary sizing.

Determine Cost of Turbine/Generator. Given the net head and kW capacity, the cost of the turbine/generator, including transportation and installation, can be determined by using the procedures described in Section 3. In employing cost data, reference is made to Section 1 of this volume relative to the limitations of data. The data may be further modified to reflect the user's experience with respect to items that may be special to the area, such as unusual labor rates or special transportation charges.

Select Performance Curves for the Overall Plant Efficiency. The overall plant efficiency is the turbine efficiency times the generator efficiency times transformer efficiency times an efficiency factor to account for station use and average station deterioration. The turbine efficiency is explained in Section 3. The generator and transformer efficiency is described in Sections 4 and 5. The efficiency often selected for average station use and deterioration is 98 percent. For a small hydroelectric project, where the flow and head remain relatively constant, and overall plant efficiency of 85 percent is a reasonable value.

For plants where the flow and head vary over a wide range, monthly, weekly, or possible daily operation schedules with use of the performance curves of the selected units may be required to obtain a reasonable estimate of the annual power production. If the flow varies over a wide range, two or more units are often more cost effective because of the improved efficiency characteristics of a multi-unit installation.

Select Control and Protection Equipment and Determine Cost. There are options for the selection of control and protection equipment, including type of governor and degree of operational control for the unit, and amount and degree of protection for the equipment. These options and the cost differences are discussed in Section 5.

Select Miscellaneous Power Plant Equipment and Determine Cost. The selection of miscellaneous power plant equipment is discussed in Section 6. The options include a crane, heating and/or ventilation equipment, sanitary facilities and a potable water supply. The latter two items depend upon whether the station is attended or unattended. The selection criteria may include the remoteness of the site and the owners preference for providing such facilities. Consideration may be given to a separate building to house these facilities not required for actual plant operation.

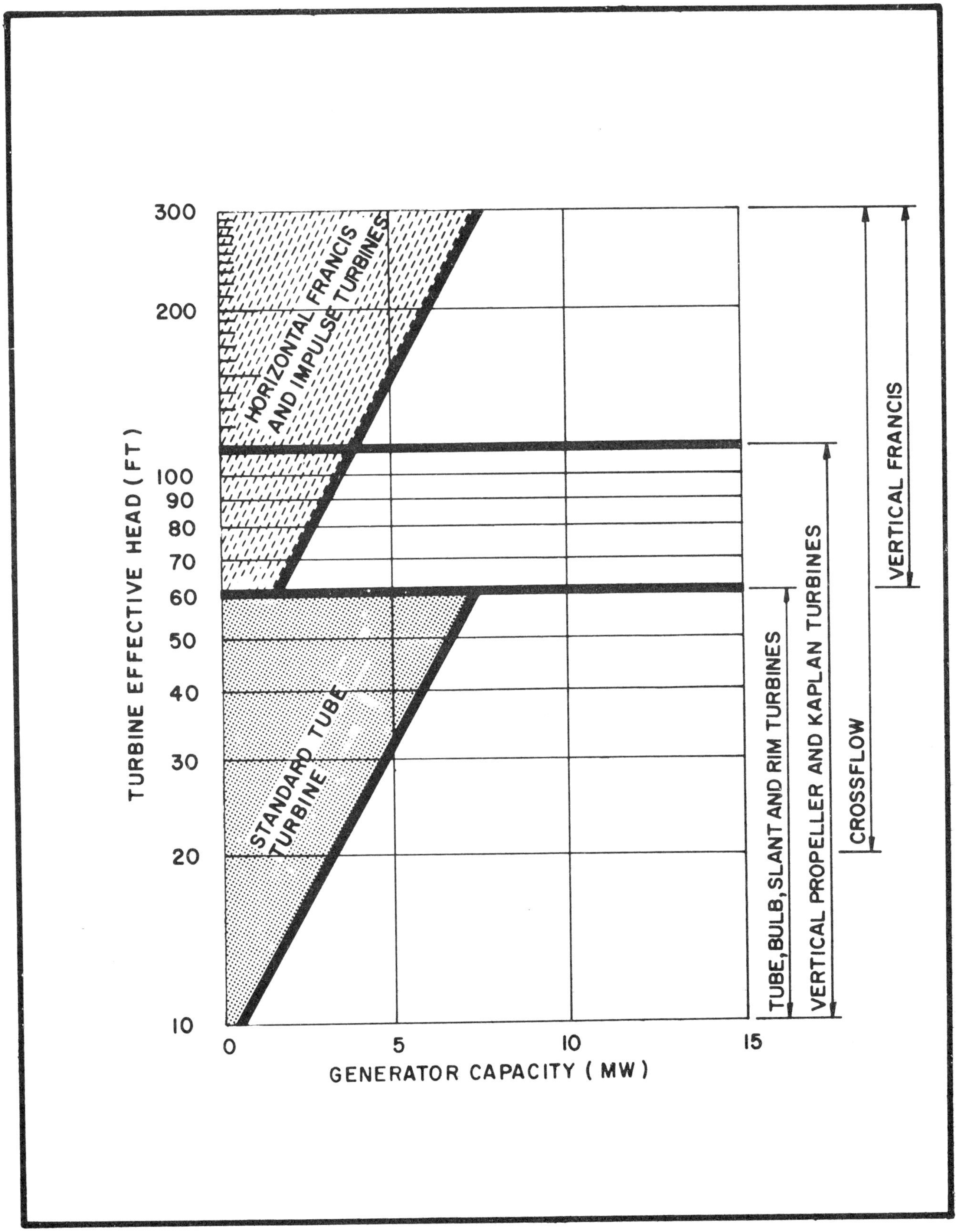

Figure 2-2. Turbine operating range

Select Transformer, Switchyard Equipment and Transmission Line Sizes and Determine Costs. Section 6 includes parameters for the selection and cost determination of the station transformer, switchyard equipment and transmission line to the connecting grid. The generator voltages vary from 480 volts from small units to 13,800 volts for larger units. Where the transmission voltage is uncertain, it is suggested that a voltage step-up to 34,500 volts be assumed. Consideration should be given to the type of terrain travelled as noted in the cost data.

Addition of Indirect Costs. In the preparation of an estimate for the electromechanical equipment, additional indirect costs need to be considered as outlined in Section 7. These indirect costs are escalation of construction and equipment prices and development costs. The costs of construction and equipment provided herein are at the July 1978 cost level and should be escalated to date required for either the reconnaissance or feasibility study. The development costs include expenditures for license and permit applications, preliminary and final design, construction management and administration. The development costs are provided as a proportion of the direct electromechanical equipment cost. Section 7 also provides a proportion of the estimated cost as related to the electromechanical equipment for annual operation and maintenance of the hydropower plant.

SECTION 3
HYDRAULIC TURBINE SELECTION AND COST GUIDELINES

Classification of Turbines

General. The net head available to the turbine dictates the selection of type of turbine suitable for use at a particular site. The rate of flow determines the capacity of the turbine.

Hydraulic turbines have two general classifications, impulse and reaction. Reaction turbines are by far the most widely used within the head range addressed by this volume.

Reaction turbines are classified as Francis (mixed flow) or Propeller (axial flow). Propeller turbines are available with both fixed blades and variable pitch blades (Kaplan). Both Propeller and Francis turbines may be mounted either horizontally or vertically. Additionally, Propeller turbines may be slant mounted. Trade names have been applied to certain Propeller turbine designs such as Tube, Bulb and Straflo. The runner design principals, however, are the same.

Impulse turbines may have some application for small hydropower installations. However, there are very few manufacturers interested in developing a standardized product line. In general the cost to manufacture a reaction turbine of comparable head and capacity is less.

Proprietary turbines (i.e., rim and crossflow) are available and discussed further in this section. These turbines have unique characteristics which may be beneficial for some projects.

Cross-sections of the various types of turbines commercially available are shown in Figure 3-1.

Francis Turbines. A Francis turbine is one having a runner with fixed buckets (vanes), usually nine or more, to which the water enters the turbine in a radial direction, with respect to the shaft, and is discharged in an axial direction. Principal components consist of the runner, a water supply case to convey the water to the runner, wicket gates to control the quantity of water and distribute it equally to the runner and a draft tube to convey the water away from the turbine.

A Francis turbine may be operated over a range of flows from approximately 40 to 105 percent of rated discharge. Below 40 percent rated discharge, there can be an area of operation where vibration and/or power surges occur. The upper limit generally corresponds to the generator rating. The approximate head range for operation is from 60 to 125 percent of design head. In general, peak efficiencies of Francis turbines, within the capacity range of 15 MW, will be approximately 88 to 90 percent. The peak efficiency point of a Francis turbine is established at 90 percent of the rated capacity of the turbine. In turn, the efficiency at the rated capacity is approximately 2 percent below peak efficiency. The peak efficiency at 60 percent of rated head will drop to near 75 percent.

The conventional Francis turbine is provided with a wicket gate assembly to permit placing the unit on line at synchronous speed, to regulate load and speed, and to shutdown the unit. The mechanism of large units are actuated by hydraulic servomotors. Small units may be actuated by electric motor gate operations. It permits operation of the turbine over the full range of flows. In special cases, where the flow rate is constant, Francis turbines without wicket gate mechanisms may be used. These units will operate at a fixed load dependent upon the net head. Start up and shut down of turbines without wicket gates is normally accomplished using the shut off value at the turbine inlet.

Francis turbines may be mounted with vertical or horizontal shafts. Vertical mounting allows a smaller plan area and permits a deeper setting of the turbine with respect to tailwater elevation without locating the generator below tailwater. Generator costs for vertical units are higher than for horizontal units because of the need for a larger thrust bearing. However, the savings on construction costs for medium and large units generally offset this equipment cost increase. Horizontal units are often more economical for small higher speed applications where standard horizontal generators are available.

The water supply case is generally fabricated from steel plate. However open flume and concrete cases are often used for heads below 50 feet. Concrete and open flume cases are discussed in a subsequent section. Closed concrete and steel cases are also known as spiral cases.

Francis turbines are generally provided with a 90 degree elbow draft tube which has a venturi design to minimize head loss. Conical draft tubes are also available, however the head loss will be higher and excavation may be more costly.

Propeller Turbines. A propeller turbine is one having a runner with four, five or six blades in which the water passes through the runner in an axial direction with respect to the shaft. The pitch of the blades may be fixed or movable. Principal components consist of a water supply case, wicket gates, a runner and a draft tube. Figure 3-2 illustrates a stay ring and wicket gate assembly and Figure 3-3 illustrates a fixed blade propeller runner.

The efficiency curve of a typical fixed blade Propeller turbine forms a sharp peak, more abrupt than a Francis turbine curve. For variable pitch blade units the peak efficiency occurs at different outputs depending on the blade setting. An envelope of the efficiency curves over

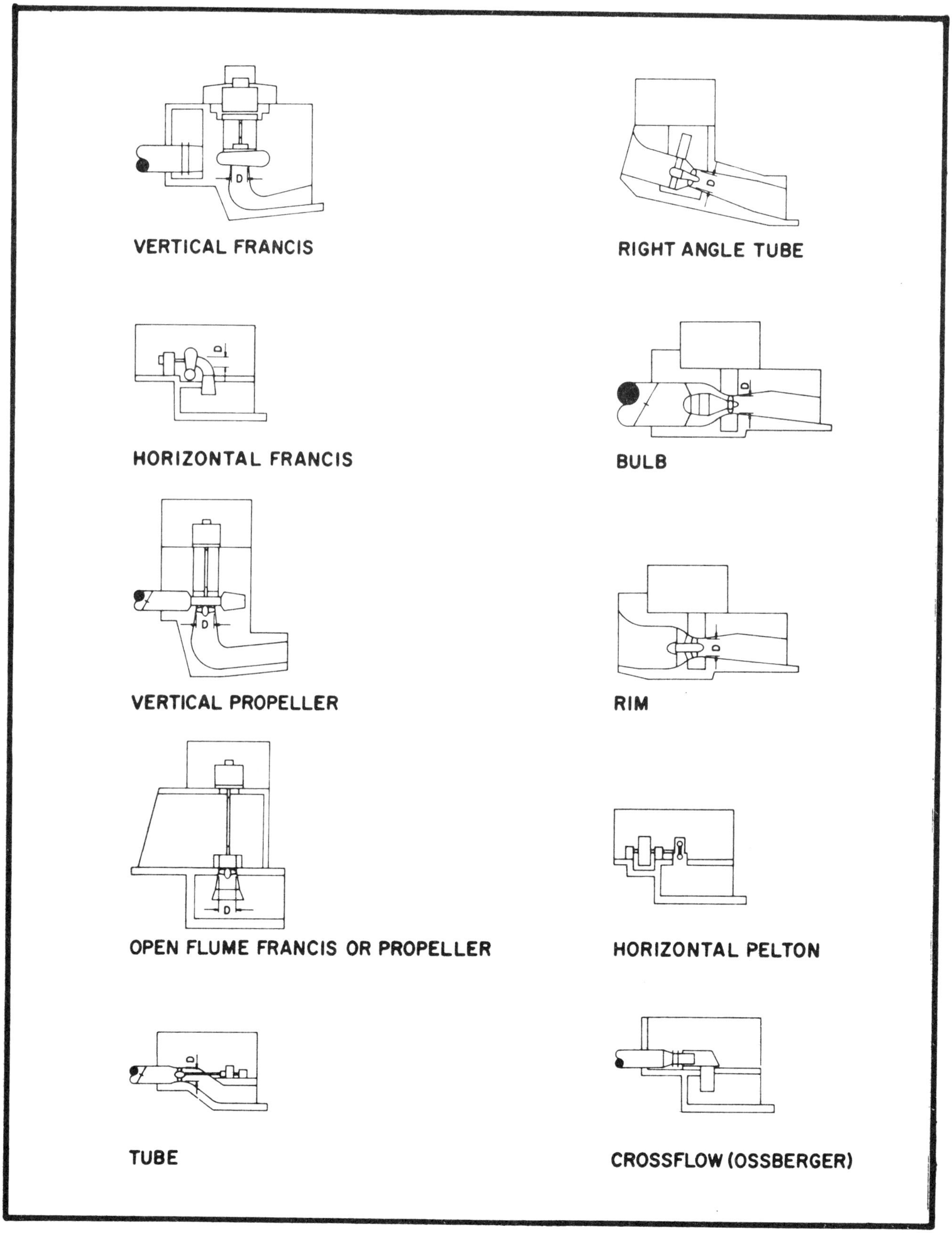

Figure 3-1. Turbine cross sections

the range of blade pitch settings forms the variable pitch efficiency curve. This efficiency curve is broad and flat. Fixed blade units are less costly than variable pitch blade turbines; however, the power operating ranges are more limited.

Turbine manufacturers have developed runner designs for a head range of 15 to 110 feet. Four blade designs may be used up to 35 feet of head, five blade designs to 65 feet and six blade designs to 110 feet. In general, peak efficiencies are approximately the same as for Francis turbines.

Propeller turbines may be operated at power outputs with flows from 40 to 105 percent of the rated flow. Discharge rates above 105 percent may be obtained; however, the higher rates are generally above the turbine and generator manufacturers' guarantees. Many units are satisfactorily operated beyond these limits; however, for purposes of feasibility studies, it is suggested that these limits be maintained. Head range for satisfactory operation is from 60 to 140 percent of design head. Efficiency loss at higher heads drops 2 to 5 percentage points below peak efficiency at the design head and as much as 15 percentage points at lower heads.

The conventional propeller or Kaplan (variable pitch blade) turbines are mounted with a vertical shaft. Horizontal and slant settings will be discussed separately. The vertical units are equipped with a wicket gate assembly to permit placing the unit on line at synchronous speed, to regulate speed and load, and to shutdown the unit. The wicket gate mechanism units are actuated by hydraulic servomotors. Small units may be actuated by electric motor gate operators. Variable pitch units are equipped with a cam mechanism to coordinate the pitch of the blade with gate position and head. The special condition of constant flow, as previously discussed for Francis turbines, can be applied to propeller turbines. For this case, elimination of the wicket gate assembly may be acceptable. Variable pitch propeller turbines without wicket gates are discussed in a subsequent section.

The advantages and disadvantages discussed above with regard to vertical versus horizontal settings for Francis turbines apply also to propeller turbines.

The water supply case is generally concrete. Either an open flume or a closed conduit type of construction may be used. Open flume construction may be economical when heads are below 35 feet. At higher heads the turbine shaft length becomes excessive. Also open flume construction is disadvantageous with regard to maintenance costs. The wicket gate assembly and guide bearing are water lubricated causing additional maintenance particularly when silt or debris is in the water. At

Figure 3-2. Wicket gate and stay ring assembly for an open flume turbine. (Courtesy of James Leffel and Company)

capacities above 1500 kW, wicket gate and guide bearing loading are such that an open flume may not be a satisfactory choice. For closed conduits, spiral cases of steel or concrete may be used. The concrete case is generally less costly. The cross-section of a concrete case, taken in a direction radial to the shaft is usually rectangular.

The draft tube designs discussed for Francis turbines apply also to propeller turbines.

Tubular Turbines. Tubular or tube turbines are horizontal or slant mounted units with propeller runners. The generators are located outside of the water passageway. Tube turbines are available equipped with fixed or variable pitch runners and with or without wicket gate assemblies.

Performance characteristics of a tube turbine are similar to the performance characteristics discussed for propeller turbines. The efficiency of a tube turbine will be one to two percent higher than for a vertical propeller turbine of the same size since the water passageway has less change in direction.

The performance range of the tube turbine with variable pitch blades and without wicket gates is greater than for a fixed blade propeller turbine but less than for a Kaplan turbine. The water flow through the turbine is controlled by changing the pitch of the runner blades.

When it is not required to regulate turbine discharge and power output, a fixed blade runner may be used. This results in a lower cost of both the turbine and governor system. To estimate the performance of the fixed blade runner, use the maximum rated power and discharge for the appropriate net head on the variable pitch blade performance curves.

Several items of auxiliary equipment are often necessary for the operation of tube turbines. All tube turbines without wicket gates should be equipped with a shut off valve automatically operated to provide shut- off and start-up functions. Tube turbines may also be equipped with an air clutch between the turbine and generator when the generator is not designed for turbine runaway speed. The clutch is normally set to disengage at 125 percent of design speed and is used to prevent damage to the equipment if a runaway condition occurs. This aspect is further discussed in Section 5.

Tube turbines can be connected either to the generator or to a speed increaser. The speed increaser would allow the use of a higher speed generator, typically 900 or 1200 r/min, instead of a generator operating at turbine speed. The choice to utilize a speed increaser is an economic decision. Speed increasers lower the overall plant efficiency by about one percent for a single gear increaser and about two percent for double gear increaser. (The manufacturer can supply exact data regarding the efficiency of speed increasers.) This loss of efficiency and the cost of the speed increaser must be compared to the reduction in cost for the smaller generator.

The required civil features are different for horizontal units than for vertical units. Horizontally mounted tube turbines require more floor area than vertically mounted units. The area required may be lessened by slant mounting, however, additional turbine costs are

Figure 3-3. Fixed blade propeller runner. (Courtesy of James Leffel and Company)

incurred as a larger axial thrust bearing is required. Excavation and powerhouse height for a horizontal unit is less than that required for a vertical unit.

Standardized tube turbines are available from a domestic turbine manufacturer. Ten sizes are currently available with up to 7000 kW of capacity and for heads up to 60 feet. Standardization should provide lower costs and shorter delivery periods. Figure 3-4 shows the shop assembly of a standardized tube turbine.

Bulb Turbines. Bulb Turbines are horizontal units which have propeller runners directly connected to the generator. The generator is enclosed in a water-tight enclosure (bulb) located in the turbine water passageway. The bulb turbine is available with fixed or variable pitch blades and with or without a wicket gate mechanism. Performance characteristic are similar to the vertical and tube type turbines previously discussed. The bulb turbine will have an improved efficiency of approximately two percent over a vertical unit and one percent over a tube unit because of the straight water passageway.

Due to the compact design, powerhouse floor space and height for Bulb turbine installations are minimized. Maintenance time due to accessability, however, may be greater than for either the vertical or the tube type turbines.

Standardized bulb turbines are offered by some foreign manufacturers.

Rim Type Turbines. A rim type turbine is one in which the generator rotor is mounted on the periphery of the turbine runner blades. This turbine has been developed by Escher Wyss Ltd. of Zurich, Switzerland and given the name "Straflo". The concept was developed 40 years ago and approximately 75 units are now in service. Capacities range from 1000 to 1900 kW at heads of 26 to 30 feet. All units built to date have fixed blade propeller runners. The existing seal design, to prevent leakage of water into the generator annulus, is a rubber "lip" seal type. This design is not satisfactory for variable pitch runner nor for capacities over 2000 kW. A new seal design has been developed which will permit Escher Wyss to offer units with runner diameters up to

Figure 3-4. Shop assembly of a standardize tube turbine. (Courtesy of Allis Chalmers Corporation).

32 feet and head up to 130 feet. The old lip seal design will be used on units with runner diameters of 11.5 feet or less at heads of less than 50 feet.

Performance characteristics of the Straflo turbine are similar to those of the bulb unit. Rim turbines are offered with or without wicket gates, and are also available with partial closure wicket gates, which require shut-off valves as discussed previously. The compact design of the Straflo turbine provides the smallest power house dimensions of all the turbine types considered in this volume. The "Straflo" design is attractive because of simplicity and compactness, however, the design for large units has limited application experience.

Crossflow Turbines. A crossflow turbine may best be described as an impulse type turbine with partial air admission. This type of turbine is offered by Ossberger Turbine Fabrik Co. of Weissenburg, Germany and has the name "Ossberger Turbine."

Performance characteristics of this turbine are similar to an impulse turbine, and consist of a flat efficiency curve over a wide range of flow and head conditions. The wide range is accomplished by use of a guide vane at the entrance which directs the flow to a limited portion of the runner depending on the flow. This operation is similar to operation of multi-jet impulse turbine.

Peak efficiency of the Crossflow turbine is less than that of other turbine types previously discussed. Guaranteed maximum efficiency is 83 percent and expected peak efficiency is 85 percent.

At the present time, the largest size runner produced by Crossflow is 4 feet in diameter. This limits the unit capacity but multi-unit installations are often used. Allowable heads range from 20 to 600 feet.

Crossflow turbines are equipped with a conical draft tube creating a pressure below atmosphere in the turbine chamber. Therefore the difference between the turbine centerline elevation and the tailwater is not lost to an Crossflow turbine as is the case for an impulse turbine. Air is admitted into the chamber through an adjustable air inlet value used to control the pressure.

Crossflow turbines are free from cavitation, but are susceptible to wear when excessive silt or sand particles are in the water. Runners are self-cleaning and, in general, maintenance is less complex than for the other types of turbines discussed in this volume.

Floor space requirements are more than for the other turbine types, but a less complex structure is required and a savings in cost might be realized.

Impulse Turbines. An impulse turbine is one having one or more free jets discharging into an aerated space and impinging on the buckets of a runner. Efficiencies are often 90 percent and above. Application of the impulse turbine within the capacity and head range of this volume is limited. In general, an impulse turbine will not be competitive in cost with a reaction turbine below 1000 feet of head. However, certain hydraulic conditions or surge protection requirements may warrant investigation into the suitability of an impulse turbine in the 100 foot range.

Single nozzle impulse turbine have a very flat efficiency curve and may be operated down to loads of 20 percent of rated capacity with good efficiency. For multi-nozzle units, the range is even broader because the number of operating jets can be varied.

Control of the turbine is maintained by hydraulically operated needle nozzles in each jet. In addition, a jet deflector is provided for emergency shut down. The deflector diverts the water jet from the buckets to the wall of the pit liner. This feature provides surge protection for the penstock without the need for a pressure release valve because load can be rapidly removed from the generator without changing the flow rate.

Control of the turbine may also be accomplished by the deflector alone. On these units the needle nozzle is manually operated and the deflector diverts a portion of the jet for lower loads. This method is less efficient and normally used for speed regulation of the turbine under constant load.

Runners on the modern impulse turbine are a one-piece casting. Runners with individually attached buckets have proved to be less dependable and, on occasion, have broken away from the wheel causing severe damage to powerhouse. Integral cast runners are difficult to cast, costly and require long delivery times. However, maintenance costs for an impulse turbine are less than for a reaction turbine as they are free of cavitation problems. Excessive silt or sand in the water however, will cause more wear on the runner of an impulse turbine than on the runner of most reaction turbines.

Draft tubes are not required for impulse turbines. The runner must be located above maximum tailwater to permit operation at atmospheric pressure. This requirement exacts an additional head loss for an impulse turbine not required by a reaction turbine.

Impulse turbines may be mounted horizontally or vertically. The additional floor space required for the horizontal setting can be compensated for by lower generator costs on single nozzle units in the lower capacity sizes. Vertical units require less floor space and are often used for large capacity multi-nozzle units.

Selection of Turbine Efficiency Curves

General. A calculation of the annual energy must be made in order to determine the feasibility of a hydroelectric power installation. The calculation consists of the product of multiplying flow, head and efficiency over a specific period of time. There are several methods for estimating the energy, including flow duration curves and systematic routing studies that simulate the operation of the plant. The simulated operation studies can be performed either by hand calculation or more commonly by a digital computer. There are many computer programs available to simulate the operation of a hydroelectric power plant. (Reservoir System Analysis for Conservation - HEC Program)

If the head and discharge rate are relatively constant,

an overall turbine, generator, station use and deterioration, and transformer efficiency of 85 percent can be used for estimating the energy from a flow duration curve. Whenever the flow rate and/or the head varies, a more precise analysis of the efficiency of the hydraulic turbine is required. A value of 95 percent may be used for all other losses including generator, station use and deterioration, and transformer. Therefore, the total efficiency to be used in the power operation studies is the product of the turbine efficiency and all other losses (0.95). If the turbine and generator are coupled together with a speed increaser, the losses, other than the turbine, may be estimated as 93 percent.

Turbine Efficiency Curves. Typical efficiency curves of the various types of turbines are shown for comparison in Figure 3-5. These curves are shown to illustrate the variation in efficiency of the turbine through the load range at the design head. Performance of the various types of turbines when operated at heads above and below design head are discussed below. Approximate efficiencies at rated capacity for the reaction turbines are shown for a turbine with a throat diameter of one foot. Rated efficiency will increase as the size of the turbine increases. The bottom curve shows the relationship of efficiency to throat diameter. The rated efficiency for turbines with throat diameters larger than one foot may be calculated in accordance with this curve. This curve was developed from model test comparisons with field-tested prototype units. It is common practice to apply the step-up value to all efficiency values throughout the operating range.

The efficiency curves shown are typical expected efficiencies. Actual efficiencies vary with manufacturer and design.

To find the approximate efficiency for a reaction turbine, determine the approximate throat diameter from Figure 3-6 or 3-7, and find the size step up factor in the bottom curve. Add this value to the rated efficiency values given for the appropriate turbine type. Size step up efficiency factors do not apply to impulse or cross flow type turbines. The values asshown may be used. Note, that these curves can only be used when the head on the turbine does not vary and less precise results are warranted. For more precise results, see the following section on turbine performance curves, USBR Engineering Monograph No. 20 or consult with turbine manufacturers.

Turbine Performance Curves. Figures 3-8 and 3-9 show performance characteristics for Francis, Kaplan (variable pitch blade Propeller with wicket gates) Propeller (fixed blade with wicket gates) and Tube (variable pitch blades without wicket gates) type turbines. These curves were developed from typical performance curves of the turbines of a specific speed that was average for the head range considered in this volume. The data used was obtained from turbine manufacturers' data and USBR Engineering Monogram No. 20. Comparison of performance curves of various specific speed runners were made and the average performance values were used. The maximum error occurs at the lowest kW_R and was approximately three percent. These curves may be used to determine the power output of the turbine and generator when the flow rates and heads are known. These curves are proposed herein because they are easily adapted for use in a digital computer instead of the more conventional "contour-type" performance curve often found in literature on turbines. The curves show percent turbine discharge, $\%Q_R$, versus percent generator rating, $\%kW_R$, throughout the range of operating heads for the turbine.

Following determination of the selected turbine capacity as discussed in Section 2, the power output at heads and flows above and below rated head (H_R) and flow (Q_R) may be determined from the curves as follows:

Calculate the rated discharge Q_R using the efficiency values discussed previously.

$$Q_R = (11.81 \times kW_R) / (H_R \times E_R \times E_g), \text{ (cfs)}$$

Compute the percent discharge, $\%Q_R$ and percent head, $\%H_R$, for the various flow and head requirements of the site.

$$\%Q_R = (Q/Q_R) \times 100, (\%)$$
$$\%H_R = (H/H_R) \times 100, (\%)$$

Enter the curves with the $\%Q_R$ and find the $\%kW$ on the appropriate H_R line. Calculate the power output.

$$P = (\%kW_R) \times (kW_R), \text{ (kW)}$$

The heavy lines at the border of the curves represent limits of satisfactory operation within normal industry guarantee standards. The top boundary line represents maximum recommended capacity at rated capacity. The turbine can be operated beyond these gate openings, however, cavitation guarantees generally do not apply beyond these points. The bottom boundary line represents the limit of stable operation. The bottom limits vary with manufacturer. Reaction turbines experience a rough operation somewhere between 20 to 40 percent of rated discharge with vibration and/or power surges. It is difficult to predict the magnitude and range of the rough operation as the water passageway configuration of the powerhouse affects this condition. Where operation is required at lower output, straightening vanes can be placed in the draft tube below the discharge of the runner to minimize the magnitude of the disturbance. These modifications reduce the efficiency at higher loads. The right-hand boundary is established from standard generator guarantees of 115 percent of rated capacity. The head operation boundaries are typical, however, they do vary with manufacturer. It is deemed that these typical performance curves are satisfactory for preliminary feasibility assessments.

When the $\%Q_R$ for a particular selection is beyond the curve boundaries, generation is limited to the maximum $\%kW_R$ for the $\%H_R$. The excess water must be bypassed. When the $\%Q_R$ is below the boundaries, no

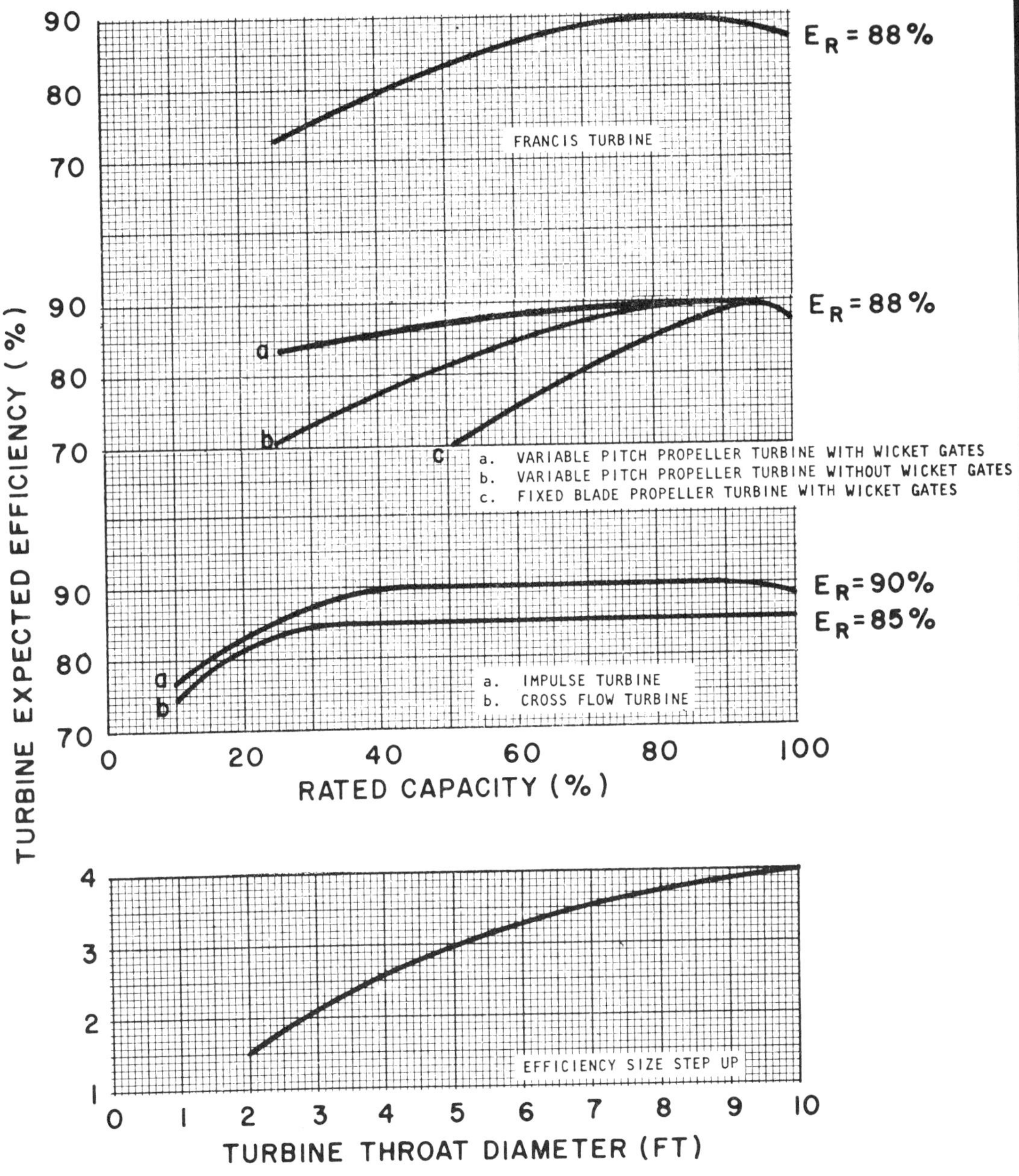

NOTES:

1. E_R = Turbine Efficiency at rated output, kW_R and head, H_R
2. The values shown are typical for a turbine with a 1 foot diameter runner. The values shown in the size step up curve may be added to the E_R values for larger units. Values apply for Francis, fixed and variable pitch propeller, tube, slant, bulb and rim turbines. Do not apply step up on impulse or cross flow turbines.

Figure 3-5. Turbine efficiency curves

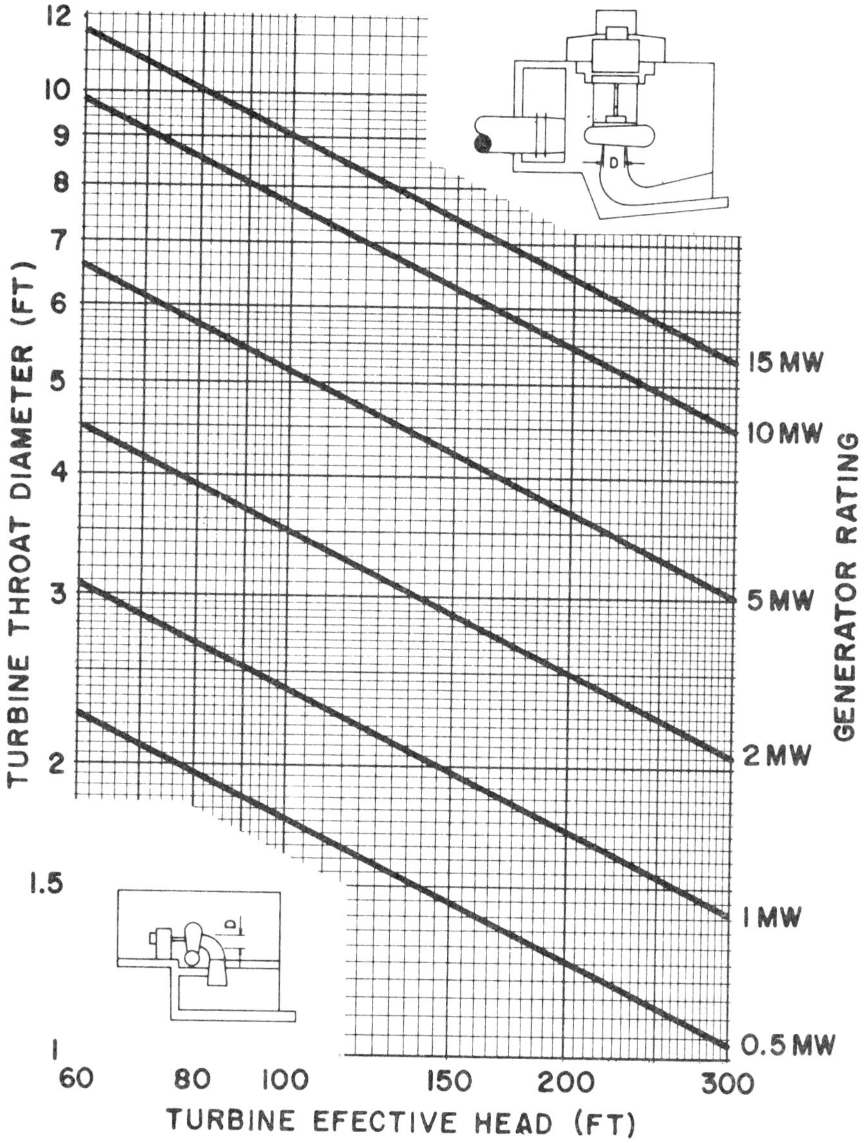

NOTES:

1. The approximate throat diameters are based upon typical values for the turbine set with the centerline of the runner at minimum tail water elevation. Actual diameters vary with turbine manufactuers.
2. The estimated diameters may be used for vertical or horizontal Francis turbines.

Figure 3-6. Francis turbine throat diameters

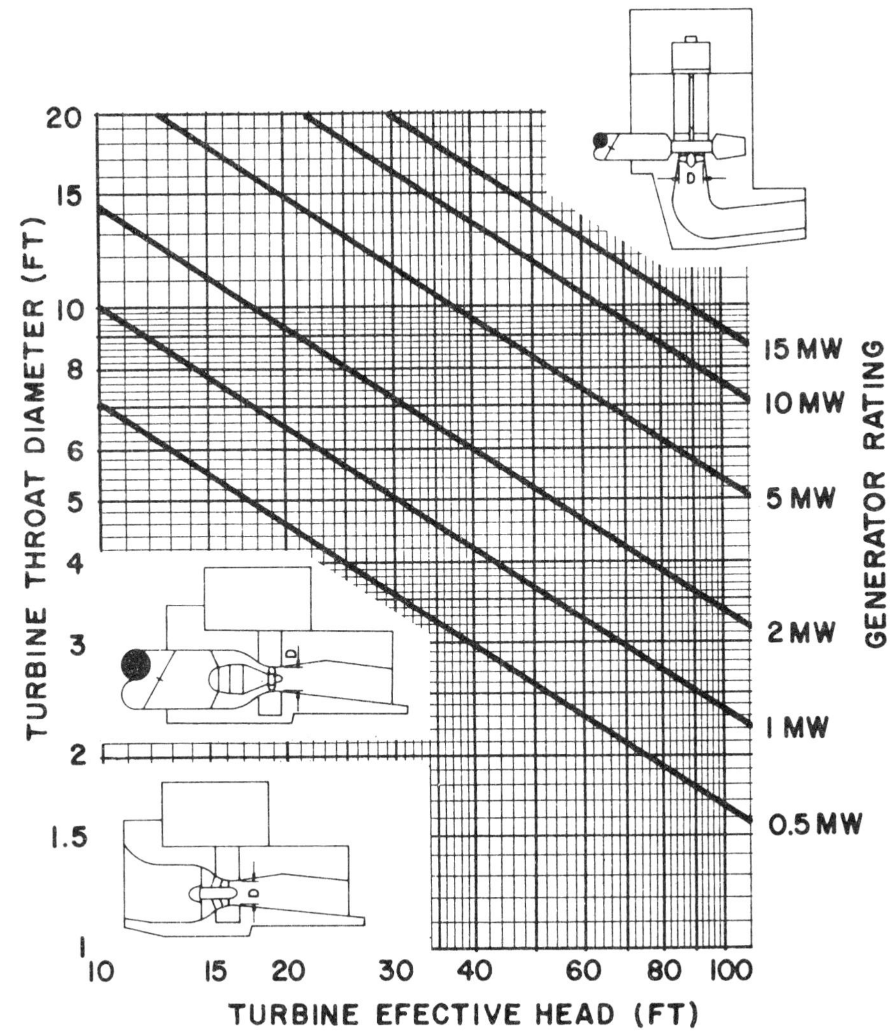

NOTES:

1. The approximate throat diameters are based upon typical values for the turbine set with the centerline of the runner at minimum tail water elevation. Actual diameters vary with manufacturers.
2. The estimated diameters may be used for both fixed and variable pitch propeller turbines, vertical, tube, slant, bulb and rim types. Dimensions for standardized tube turbines are shown on a separate sheet.

Figure 3-7. Propeller turbine throat diameters

VERTICAL OR HORZONTAL FRANCIS TURBINE

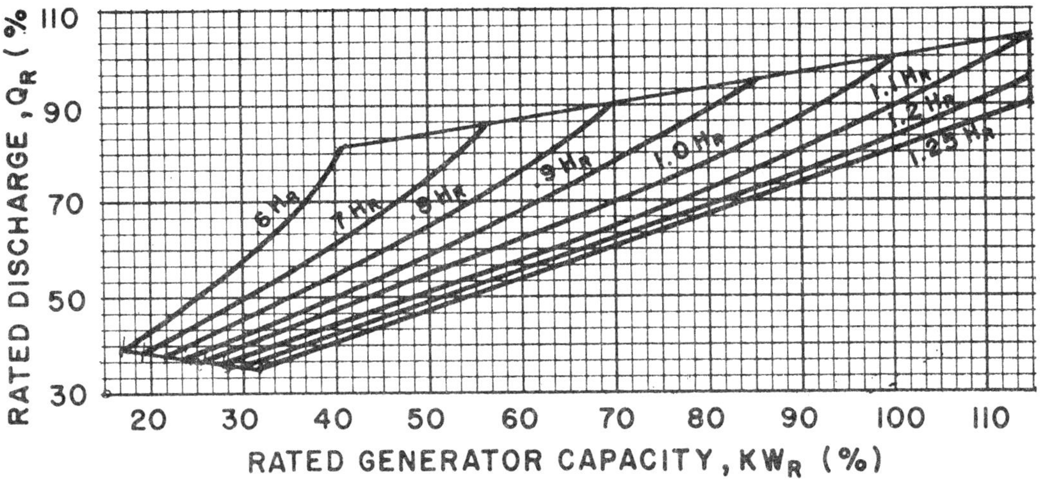

VARIABLE PITCH PROPELLER WITH WICKET GATES
VERTICAL, TUBE, BULB OR RIM

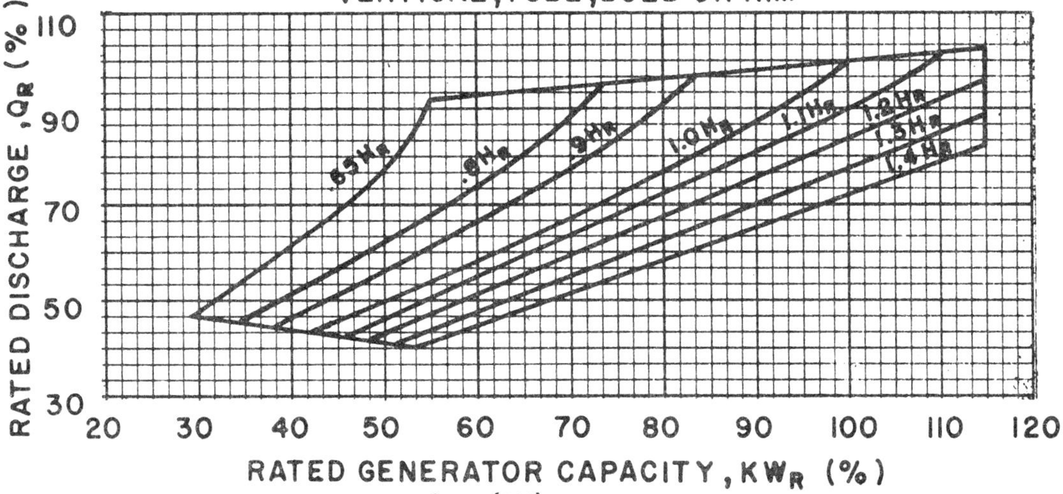

$kW_R = H_R\ Q_R\ E_R\ E_G/11.81$, (kW)

where:

kW_R = Rated capacity at H_R

H_R = Selected Design Head, (ft.)

Q_R = Turbine Discharge at H_R & kW_R, (cfs)

E_R = Turbine efficiency at H_R & kW_R, (%)

E_G = Generator efficiency, (%)

Figure 3-8. Francis and Kaplan performance curves

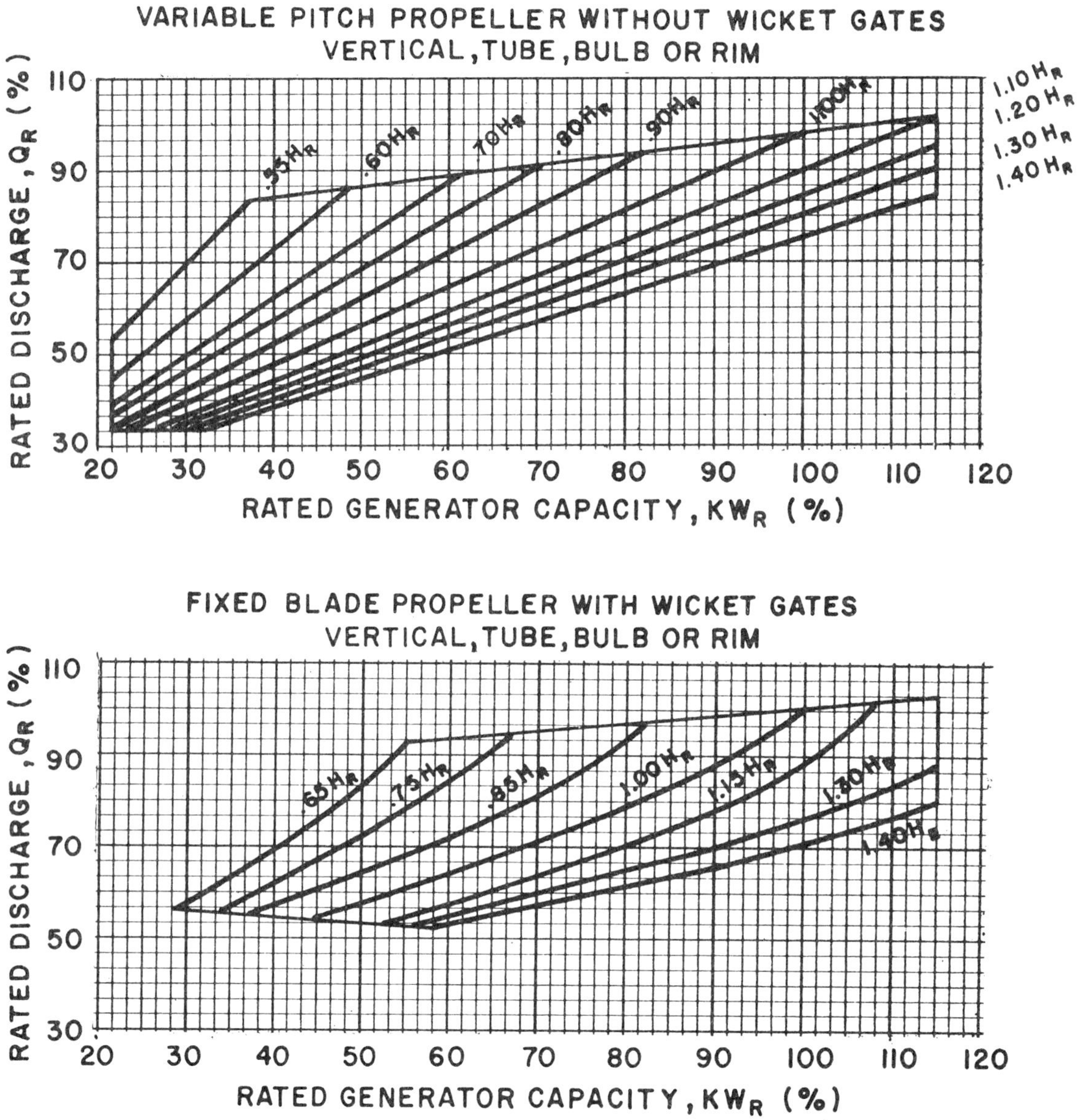

$kW_R = H_R\ Q_R\ E_R\ E_G/11.81$, (kW)

where:

kW_R = Rated capacity at H_R

H_R = Selected Design Head, (ft.)

Q_R = Turbine Discharge at H_R & kW_R, (cfs)

E_R = Turbine efficiency at H_R & kW_R, (%)

E_G = Generator efficiency, (%)

Figure 3-9. Propeller turbine performance curves

power can be generated. When the $\%H_R$ is above or below the boundaries, no power can be generated.

The optimum number of turbines may be determined by use of these curves for an annual power computation. If power is being lost because the $\%Q_R$ is consistently below the lower boundaries, the annual power produced by lowering the kW rating of each unit and adding a unit should be computed. If the power increase is substantial, an approximate turbine cost of the alternatives may be approximated from the turbine costs curves and the incremental increase in cost per kWh compared. If the total construction cost of the powerhouse is assumed to roughly equal the cost of the turbine and generator, the cost per kWh derived above can be doubled and compared with the financial value of the energy. If the selection of more turbines seems favorable from this calculation, it should be pursued in further detail with more accurate studies. Conversely, the first selection of the number of turbines may be compared with a lesser number of units and compared on a cost per kWh basis as described above.

Following the establishment of the number of units, the rating point of the turbines can be optimized. This generally is done after an estimate of the total project cost have been made. Annual power production of turbines having a higher rating and a lower rating should be calculated and compared to the annual power production of the turbine selected. With the annual cost estimate, a cost per kWh may be calculated for the selected turbine. Total project costs for the lower and higher capacity ratings may be estimated by correcting the turbine/generator costs from the cost charts and correcting the remaining costs on a basis of a constant cost per kW capacity. Rates of incremental cost divided by incremental energy generation indicate economic feasibility. As an example, if a contemplated capacity increase would produce 1,000,000 kWh per year and would cost $20,000 per year in debt service and operation and maintenance, the incremental cost of energy is 20 mills/kWh. If energy is worth more than this, the capacity increase is justified.

The rated head of the turbine can be further refined by optimization in a similar manner. The annual power production is computed for higher and lower heads with the same capacity rating. The rated head yielding the highest annual output should be used.

The boundaries established on these curves are typical. Should energy output of a particular site be curtailed, it is suggested that turbine manufacturers be consulted as these boundaries can be expanded under certain conditions.

Standardized Tube Turbine. Performance curves for the Allis-Charmers units are shown on Figure 3-10. The same procedure for selection of turbines previously described is applicable for tube turbines. Following selection of the size, Figure 3-11 may be used for estimating power over the range of flow and head.

Dimensions of Turbines

General. The size of reaction turbines may be estimated after the capacity, kW_R, and effective head have been established. Figure 3-6 shows the approximate throat diameter of Francis turbines, both vertical and horizontal. Figure 3-7 shows the approximate throat diameter for Propeller turbines and may be used for both fixed and variable pitch blade units, vertical, tubular, bulb or rim types. Other dimensions of the turbine may be found in Volume VI, Civil Features. These dimensions are suitable for feasibility assessments. Actual dimensions vary with manufacturer and should be obtained from the manufacturers for final sizing studies.

Dimensions of Standardized Tube Turbine. Dimensions of the Allis-Chalmers units are shown in Figure 3-11.

Dimensions of Impulse and Crossflow Turbines. These turbine dimensions are not provided. Manufacturers should be contacted and dimensions requested when the application is suitable for these types of turbines.

Turbine/Generator Costs

General. Charts have been prepared for the various types of turbines and generators considered in this volume. The data used to prepare these charts were obtained from turbine, generator and governor manufacturers over the past five years and escalated to a July 1978 price level. Price lists are not available on turbines as most turbines are custom design. In general, turbine and generator costs per installed kW decrease as the capacity of the unit increases. However, the effective head available to the turbine has the greatest influence on the cost. The lower the head, the higher is the cost per installed kW. This increase is due to the larger size and lower synchronous speed turbine required for the low head application. The cost curves are suitable to indicate the feasibility of a project. However, it is recommended that prices be requested from manufacturers when final feasibility studies are made. It is also suggested that firm bids be received on the turbines and generators prior to the final design of the powerhouse. The bids would permit competition between the various types available and an evaluation of the overall generation cost be made including civil costs and average annual energy generation.

Turbine/Generator Cost Curves. The cost curves included in this volume are as follows:

Turbine Type	Figure No.
Vertical & Horizontal Francis	3-12
Vertical Kaplan (Variable Pitch Propeller)	3-13
Standardized Tube	3-14
Bulb and Rim	3-15
Cross Flow	3-16

Turbine Manufacturers. Exhibit I is a list of manufacturers which design small hydroelectric turbines within the capacity range of this volume.

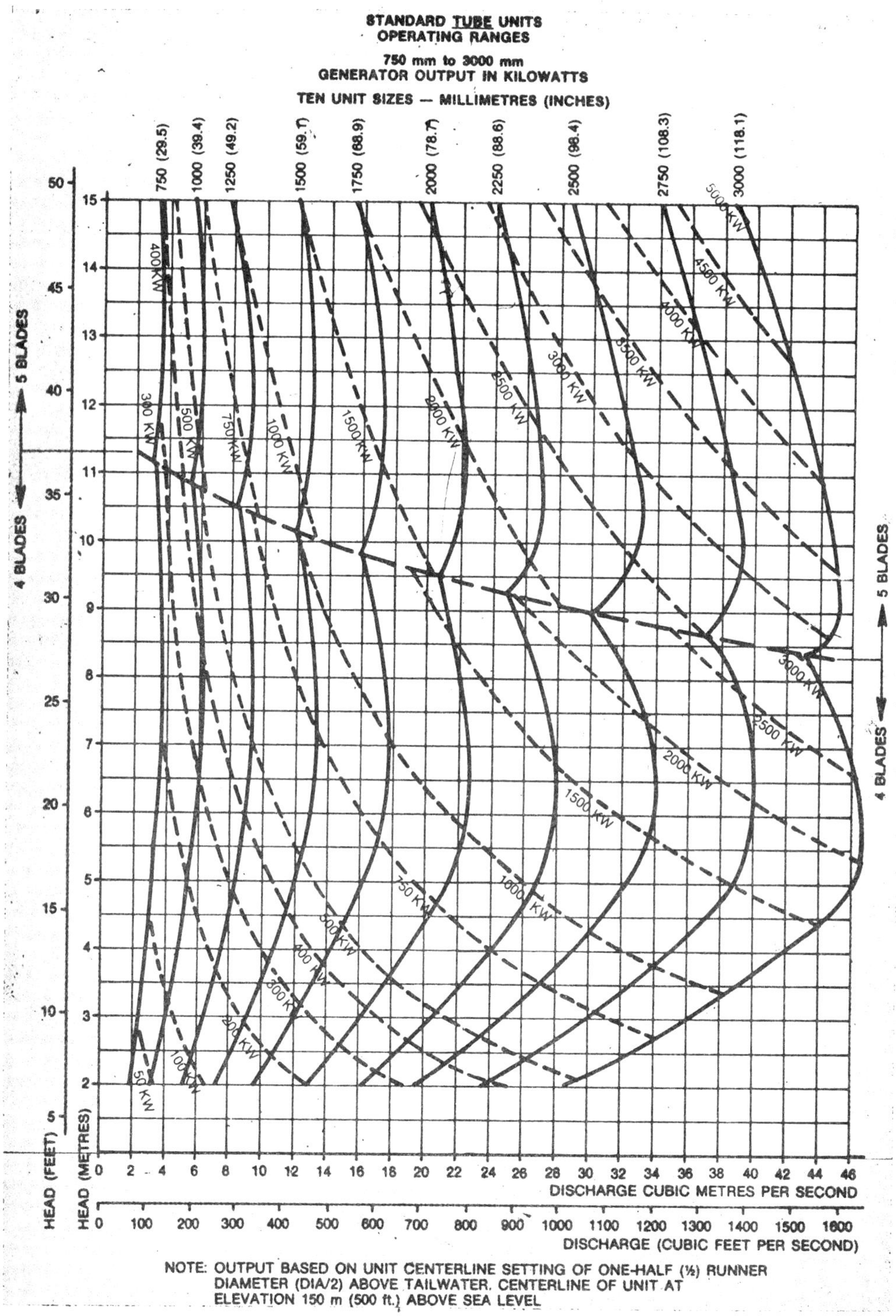

Figure 3-10. Allis Chalmers standardized Tube turbine performance curves

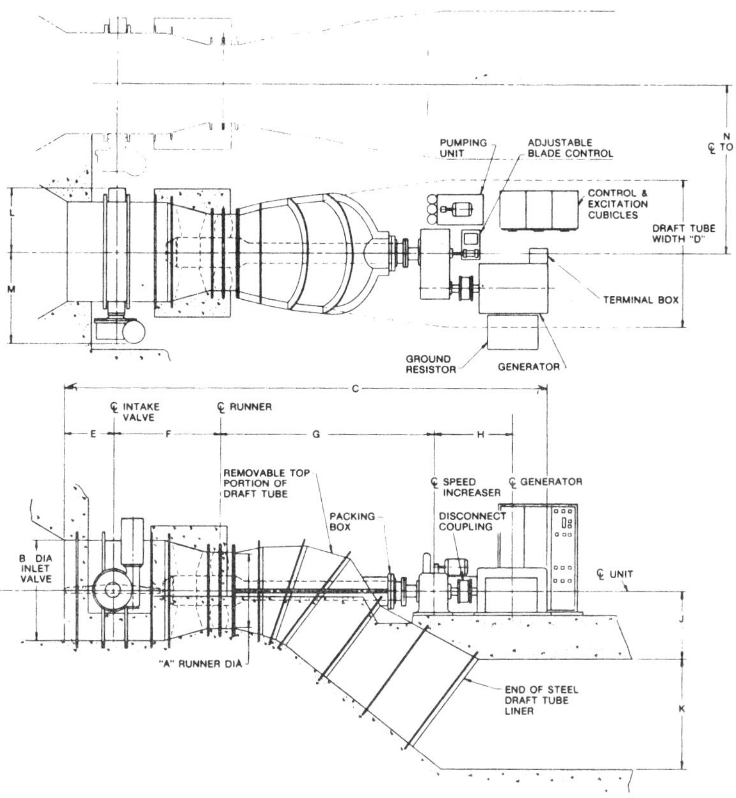

Figure 3

BASIC DIMENSIONS

A = Runner Diameter in millimeters (inches) = 1.00

All Other Dimensions Are In Proportion From Runner Diameter

A	750 (29.5)	1000 (39.4)	1250 (49.2)	1500 (59.6)	1750 (68.9)	2000 (78.7)	2250 (88.6)	2500 (98.4)	2750 (108.3)	3000 (118.1)
B	1.43	1.37	1.34	1.32	1.31	1.30	1.29	1.28	1.27	1.22
C	9.28	8.66	8.28	7.80	7.50	7.34	7.19	7.08	7.03	6.97
D	2.00	2.00	2.00	2.00	2.00	2.00	2.00	2.00	2.00	2.00
E	0.73	0.70	0.70	0.70	0.66	0.65	0.64	0.64	0.64	0.63
F	1.75	1.60	1.52	1.50	1.46	1.45	1.42	1.40	1.38	1.47
G	3.07	3.04	3.02	2.93	2.88	2.87	2.86	2.85	2.92	2.87
H	1.33	1.32	1.32	1.13	1.13	1.12	1.22	1.13	1.10	1.07
J	0.93	0.93	0.93	0.93	0.93	0.93	0.93	0.93	0.93	0.93
K	1.49	1.49	1.49	1.49	1.49	1.49	1.49	1.49	1.49	1.49
L	0.93	0.90	0.88	0.87	0.80	0.78	0.77	0.77	0.76	0.73
M	1.33	1.30	1.28	1.25	1.26	1.22	1.22	1.22	1.22	1.17
N	3.00	2.75	2.60	2.50	2.43	2.38	2.33	2.30	2.27	2.25

Figure 3-11. Allis Chalmers standard tube turbine dimensions

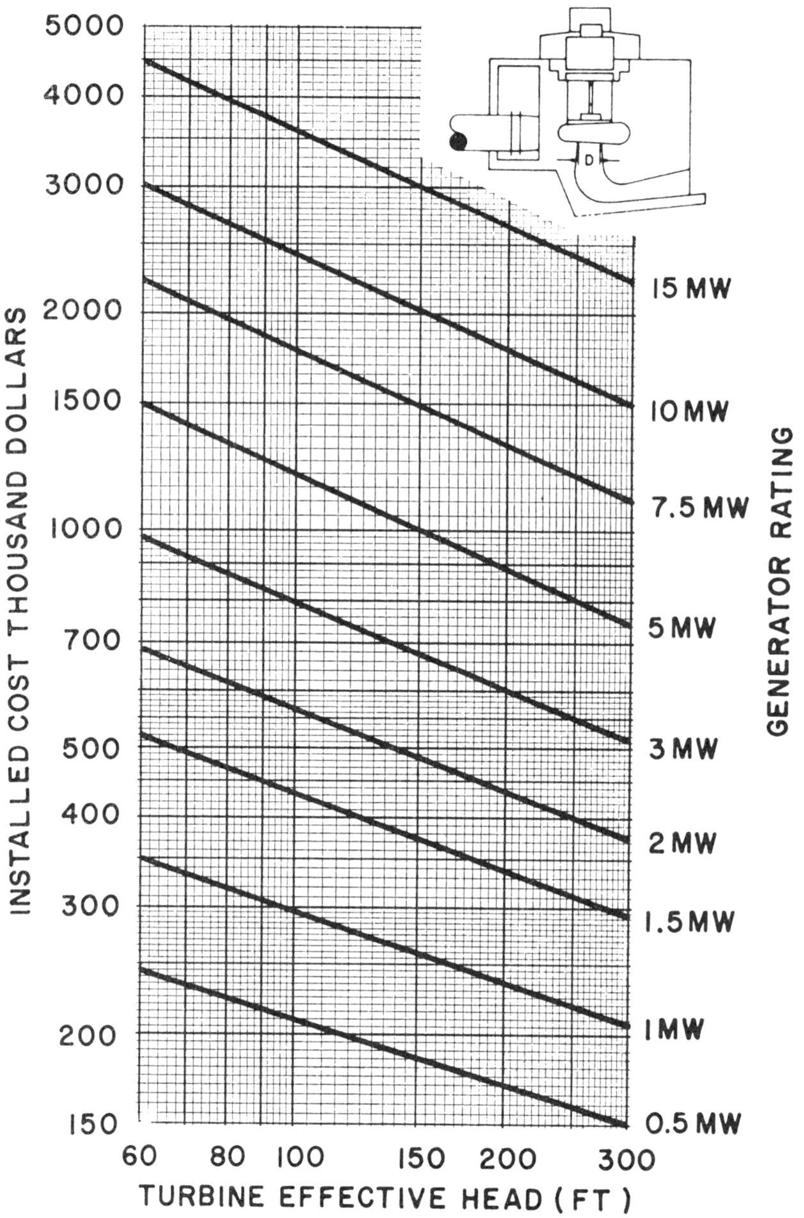

NOTES:

1. Estimated costs are based upon a typical vertical turbine direct coupled to the generator.
2. Cost includes turbine, generator, exciter, inlet valve non speed regulating governor and installation.
3. Installation costs estimated at 15% of total equipment cost.
4. Add $60,000 for speed regulating governor.
5. Horizontal Mounting Deduct 7%.
6. Cost Index is July 1978.

Figure 3-12. Francis turbine costs

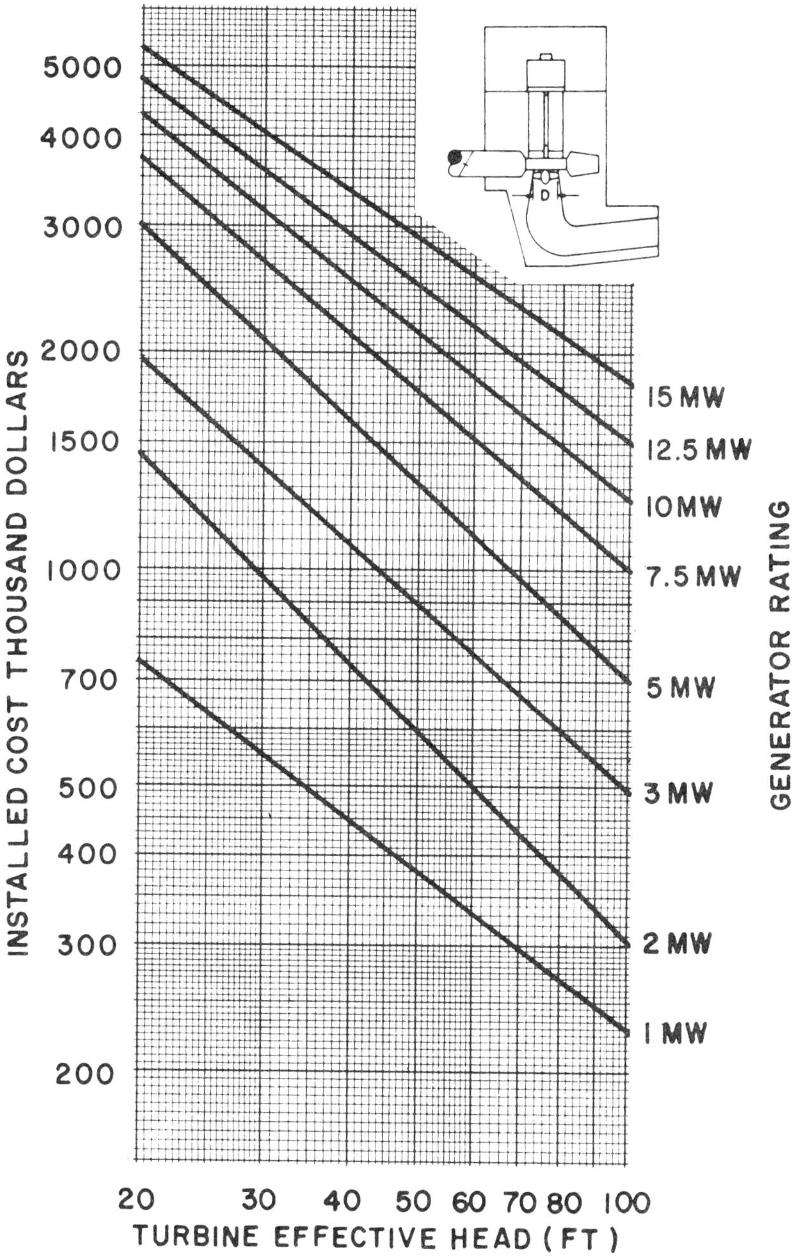

NOTES:

1. Estimated costs are based upon a typical vertical Kaplan turbine with concrete spiral case direct coupled to the generator.
2. Cost includes turbine with adjustable blades and wicket gates, generator exciter, speed regulating governor and installation.
3. Installation costs estimated at 15% of the total equipment cost.
4. For fixed blade Propeller turbine deduct 10%.
5. Cost Index is July 1978

Figure 3-13. Vertical Kaplan and Propeller costs

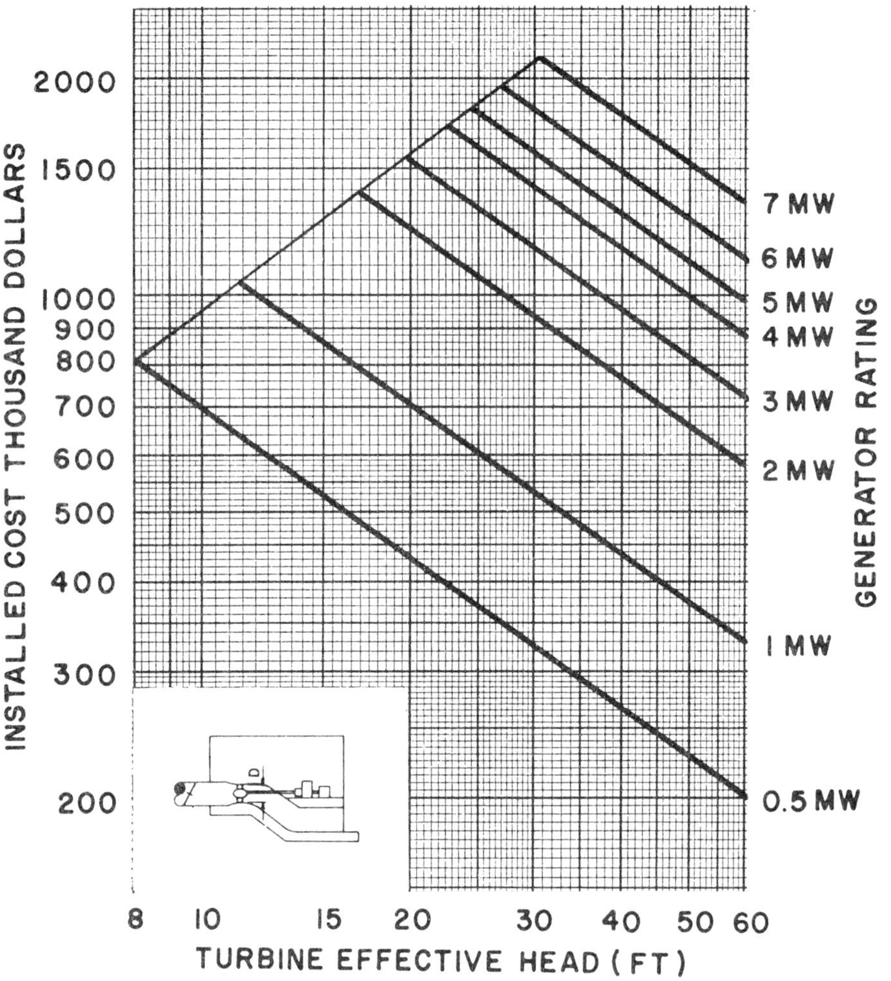

NOTES:

1. Estimated costs are based upon standardized tube turbines coupled to the generator through a speed increaser.
2. Costs include turbine with variable pitch blades and fixed guide vanes, inlet Butterfly Valve, air clutch, speed increaser, generator with exciter, speed regulating governor, controls and installation.
3. Installation costs were estimated at 15% of total equipment costs.
4. Deduct $27,000 for fixed blade type.
5. Cost Index is July 1978

Figure 3-14. Standard Tube turbine costs

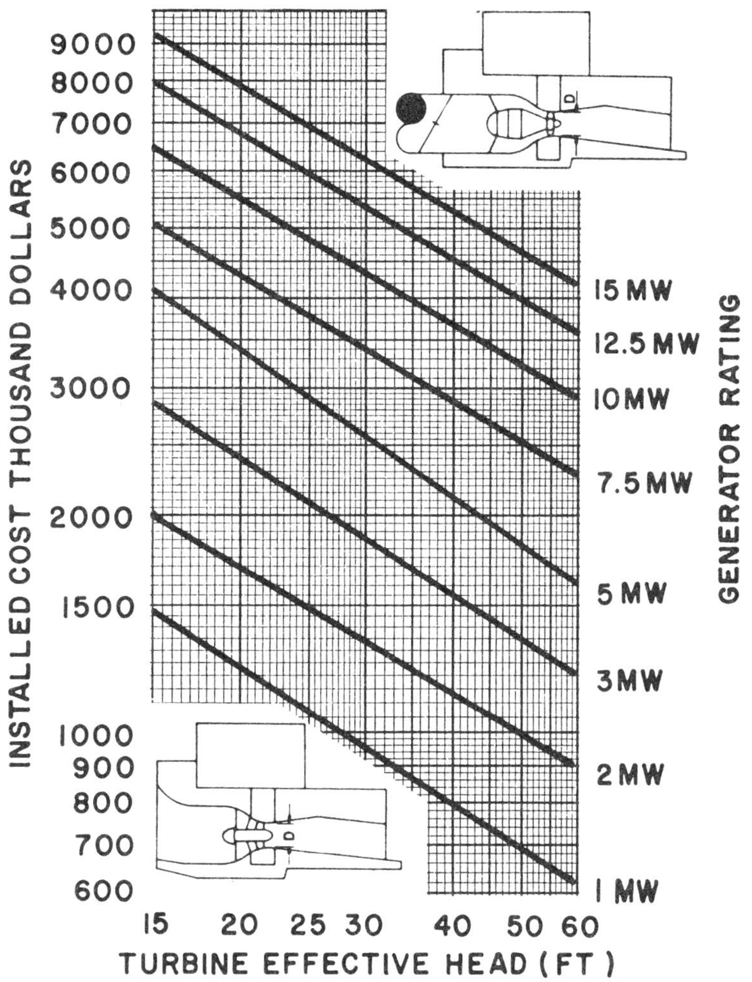

NOTES:

1. Estimated costs are based upon typical-horizontal bulb turbines with variable pitch blades and wicket gates direct coupled to the generator
2. Cost includes turbine, generator, exciter, speed regulating governor and installation.
3. Installation costs estimated at \$250,000 for the large units, to \$75,000 for the small units.
4. For fixed blade units, deduct 10%.
5. Cost of rim turbines are approximately the same as bulb turbines and the above chart may be used for preliminary costs of same.
6. Cost Index is July 1978

Figure 3-15. Bulbs and Rim turbine costs

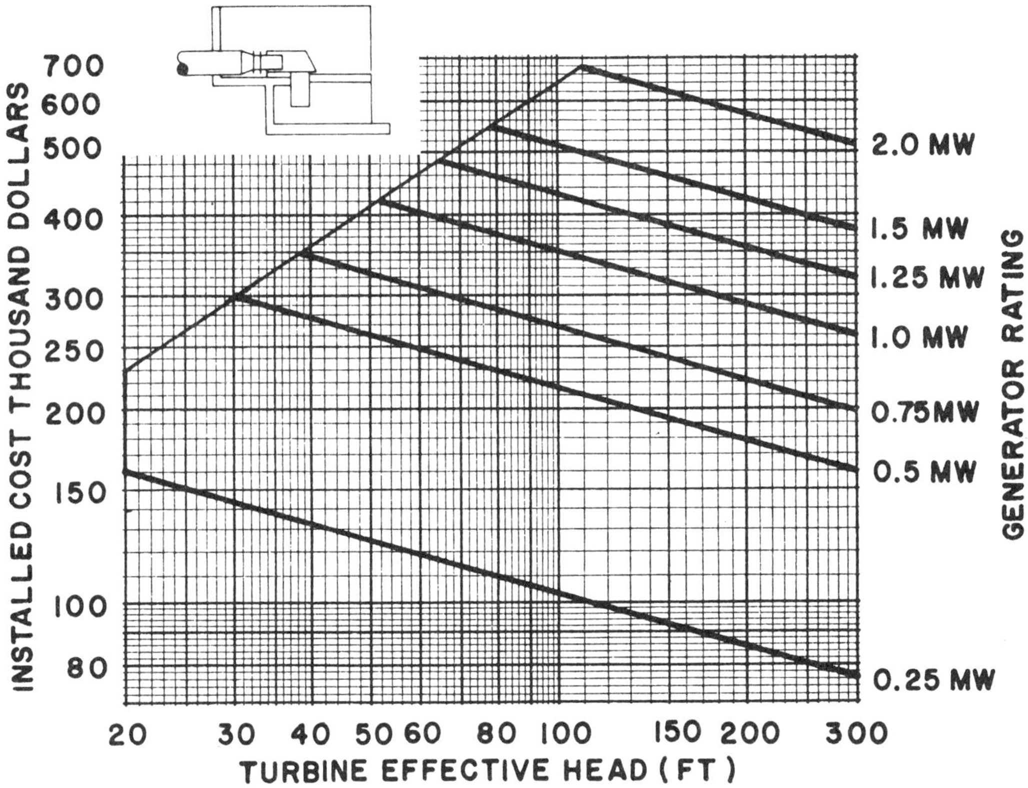

NOTES:

1. Estimated costs are based upon a typical single turbine direct coupled to the generator at high heads and coupled through a speed increaser at low heads.
2. Costs include turbine, generator, exciter, inlet valve, non speed regulating governor and installation.
3. Add $60,000 for speed regulating governor.
4. Cost Index is July 1978

Figure 3-16. Crossflow turbine costs

SECTION 4
ELECTRIC GENERATORS SELECTION AND COST GUIDELINES

Classification of Generators

General. The electric generator converts the mechanical energy of the turbine into electrical energy. The two major components of the generator are the rotor and the stator. The rotor is the rotating assembly to which the mechanical torque of the turbine shaft is applied. By magnetizing or "exciting" the rotor, a voltage is induced in the stationary component, the stator. The principal control mechanism of the generator is the excitor-regulator which sets and stabilizes the output voltage. The speed of the generator is determined by the turbine selection, except when geared with a speed increaser. In general, for a fixed value of power, a decrease in speed will increase the physical size of the generator.

The location of the generator is influenced by factors such as turbine type and turbine orientation. For example, the generator for a bulb type turbine is located within the bulb itself. A horizontal generator is usually required for a tube turbine and a vertical shaft generator with a thrust bearing is appropriate for most Francis turbine installations.

Conventional cooling on a generator is accomplished by passing air through the stator and rotor coils. Fan blades on the rotating rotor assist in the air flow. Depending on the temperature rise limitations of the winding insulation of the machine, the cooling may be assisted by passing air through surface air coolers which have circulated cold water as the cooling medium. On both indoor and outdoor installations the generator and associated cooling equipment and piping are enclosed in a housing, usually fabricated of steel, with entrance hatches and with a top hatch for an emergency exit (outdoor only). Indoor installations provide additional plant security but add an additional cost to the structure. Outdoor housings are generally accommodated with crane rails on the generator deck to provide for removal of the machine during maintenance. A photograph of a typical vertical, hydroelectric generator is shown in Figure 4-1.

Synchronous. A synchronous generator is so named because it is synchronized to the system voltage and frequency before the breaker device which connects the generator to the system is closed and, when connected, continues to operate at synchronous speed.

The excitation of the generator is achieved by impressing a direct current (dc) source across the rotor field coils and creating a magnetic field within the stator which induces a voltage potential in the stator coils. Present day designs employ a static excitation device which converts an alternating current (ac) source to a dc source via solid state circuitry. The static system has replaced the shaft-driven dc excitation generator and comparatively costs less, has a quicker response time and accommodates discharge of the field energy without a field discharge resistor upon a sudden disconnect of the unit from the system. However, for generators of 5,000 kW or less, a brushless shaft driven exciter may still be used in lieu of a static excitation system. The brushless exciter is a rotating ac generator with rectifiers on the main shaft to produce dc current for the field.

The voltage regulator functions as an automatic control device. It senses machine voltage and compares it to a set point. As the generator load changes, the voltage regulator adjusts the machine excitation to hold the generator voltage constant.

The exciter-regulator generally consists of one modular unit. It primarily affects generator reactive power output, power factor and voltage levels. The equipment is used in conjunction with the synchronizing equipment in the starting sequences of placing the generator on-line. Once the exciter-regulator brings the machine voltage up to system voltage and the synchronizing equipment matches frequency and phase with the system, the generator may be connected to the power grid. Small machines are frequently started and brought up to rated speed without excitation, the breaker closed and excitation applied to pull the generator into step with the system. This procedure eliminates the cost of the synchronizing equipment.

Induction. The major difference between an induction and a synchronous generator is that the induction generator obtains its excitation from the power grid.

The general method of getting the plant on-line is to start the generator as a motor with the turbine runner spinning "dry" and then open the wicket gates of the turbine to load the unit. The generator then begins to operate as a generator.

Present-day costs for induction generators are somewhat less than for synchronous generators of the same rated output. However, commercially available induction generators are generally limited to capacities of less than 2,000 kW. For the purposes of the preliminary feasibility study, synchronous machines should be used.

The choice of generator, synchronous or induction, is a function of application. An induction generator has a fixed power factor which if operating into a small power system can be a disadvantage because other generators in the system will be required to provide the reactive component for the operation of the induction generator. Synchronous generators can vary the power factor and contribute reactive power into the system. The proper adjustment of reactive components of synchronous generators can be utilized to reduce losses in the system.

Selection should be based on a case by case analysis of the power grid into which the generator will contribute power.

Procedure for Selection of Generator

General. With a Francis turbine, a vertical or horizontal configuration is possible. The orientation becomes a function of the turbine selection and of the power plant structural and equipment costs for a specific layout. As an example, the Francis vertical unit will require a deeper excavation and higher power plant structure. A horizontal machine will increase the width of the power plant structure yet decrease the excavation and overall height of the unit. It becomes apparent that generator orientation and setting are governed by compatibility with turbine selection and an analysis of overall plant costs.

Dimensions. Three factors affect the size of generator. These are orientation, kVA requirements and speed. The turbine choice will dictate all three of these factors for the generator.

Figure 4-2 lists dimensional information on vertical generators rated at 4160 volts which is the rated voltage

Figure 4-1.
Vertical indoor generator, Kern Canyon Power Plant on the Kern River, California. Capacity 8,480 kW. (Courtesy of Pacific Gas and Electric Company).

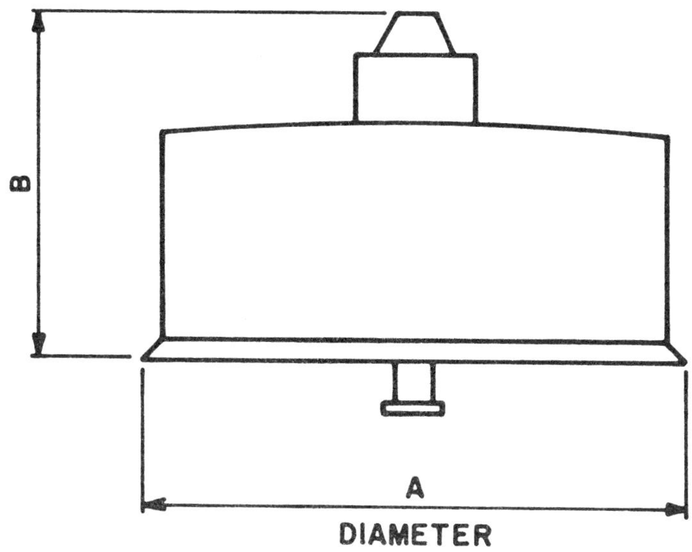

RATING (MW)	A (in)	B (in)
0.5	80	40
1	85	50
2	95	75
5	125	120
10	145	145
15	180	180

NOTES:

1. All values represent median sizes for varying head conditions (0-300 ft.).
2. Data based on 3-phase generators, 60 cycles, 0.9 power factor and 4160 volts.
3. For units above 5MW, several extra feet should be added to the diameter if surface air coolers are used.

Figure 4-2. Dimensions for generators, vertical configuration

commercially available. All dimensions shown represent synchronous generating equipment.

The size of the generator for a fixed kVA varies inversely with unit speed. This is due to the requirements for more rotor field poles to achieve synchronous speed at lower r/min.

Generator Efficiencies. The efficiency of an electrical generator is defined as the ratio of output power to input power. Efficiency values for commercially available generators are included in Section 3. There are five major losses associated with an electrical generator. Various test procedures are used to determine the magnitude of each loss. Two classes of losses are fixed and therefore independent of load. These losses are (1) windage and friction and (2) core loss. The variable losses are (3) field copper loss, (4) armature copper loss and (5) stray loss or load loss. (Fink and Carroll, 1968)

Windage and friction loss is affected by the size and shape of rotating parts, fan design, bearing design and the nature of the enclosure. Core loss is associated with power needed to magnetize the steel core parts of the rotor and stator. Field copper loss represents the power losses through the dc resistance of the field. Similarly, the armature copper loss is calculated from the dc resistance of the armature winding. Stray loss or load loss is related to armature current and its associated flux. Typical values for efficiency range from 96 to 98 percent. This efficiency value is representative throughout the whole loading range of a particular machine; i.e., the efficiency is approximately the same at 1/4 load or at 3/4 load.

Cost Data. Generator costs vary with kVA capacity, speed and configuration. Cost will increase with an increase of kVA or a decrease in speed. Vertical generators are more costly than horizontal generators due to the addition of a thrust bearing for vertical units.

A refinement in costs would compare indoor versus outdoor installations. The indoor generator has a lower generator housing cost; however, this reduction is outweighted by the additional building structure costs. This examination is considered in the Civil Features discussed in Volume VI.

Cost data for the generators is included with the turbine cost data in Section 3.

Generator Manufacturers. The source of generator data concerning dimensions, cost and operating characteristics is the generator manufacturers. The manufacturers are continually adapting designs to new market criteria. With the increased interest in this area, generator manufacturers have come forth with new generator designs to accommodate small hydro conditions. Activity on both foreign and domestic markets is presently very active.

A partial list of generator manufacturers for small hydroelectric generators is shown in Exhibit II at the end of this volume.

SECTION 5
GENERATION CONTROL, PROTECTION EQUIPMENT AND TRANSFORMERS

Hydroelectric Plant Control

General. The governor is the primary controller of a small hydroelectric plant. The governor may be actuated by manual operation, by float level control in the waterway or by the flow of water in a conduit. Each method provides control for starting and loading the unit.

The generator is controlled through the excitation and voltage regulation equipment. In coordination with the synchronizing equipment, these systems allow for unit start-up, and voltage and power generation control when the unit is on the line.

The central location for plant control is the main control board. From it emanates the complete operation and monitoring of all plant equipment. Together with the plant switchgear and storage energy system, normally batteries, the above are the predominant control systems for small hydroelectric plant operation.

There are several differences in the control systems required for large versus small hydroelectric installations. A comparison of the two sizes of installation indicates that the primary descriptions of the above control systems are representative for both classes. In large plants, the complexity of the regulation equipment (e.g., governor, synchronizing equipment, excitation gear) will be greater since slight increments in turbine gate position or generator field adjustment may result in large increments of power swing relative to the power grid. Small hydroelectric units do not create such an impact on the system and thereby require less costly and complex equipment. The other area where a difference occurs is in the horsepower of the auxiliary pumps, storage battery capacities and protective systems. Large hydroelectric plants employ larger sized control systems simply because the auxiliary systems are larger.

Main Control Board. The function of the main control board is to control and monitor all plant functions. In small hydroelectric plants, often unattended, a primary function of the control board is to give indication of plant function status after a remote alarm has occurred so that an operator may be dispatched to determine the nature of the alarm. From this display the operator can determine the nature of the malfunction, and can then follow established operating instructions for handling the plant malfunctions and often restart the unit.

The control board consists of indicating meters, control switches, lights, annunciators, mimic arrangements, interposing relays, protective relays and recording instruments. The indicating meters provide information on voltage levels, current levels, watts, vars, temperature and unit speed. Indicating lights show status, such as "pump on - pump off" or "valve open - valve closed". The annunciators display specific alarm or malfunction conditions throughout the plant. Generally the annunciator points are grouped by function. One layout often used has the annunciators partitioned into generator, turbine and transmission line functions. The annunciator may be accompanied by a local alarm and facilities for initiating remote alarm. Protective relays are mounted in a separate area of the control board and are visible from the front for inspection of relay targets. Interposing relays are often mounted behind the control board for ease of interwiring within the board. Recording wattmeters, varmeters and voltmeters in addition to flow and water level recording meters are mounted in the vicinity of the protective relays.

The configuration in design and layout is arbitrary and dependent on operator logic and system conformance. Note that the control switches are mounted with corresponding indicating lights, indicating meters and the annunciator sections. This arrangement is quite common. Relays and recording devices are grouped together. The typical control board layout includes a walk-in configuration. The whole panel layout may be set up for front view if this is deemed feasible. Figures 5-1 and 5-2 illustrate an old and new control board arrangement. Both boards are still in operation.

Generator Control

Synchronization and Voltage Regulation. The synchronizing equipment allows the generator breaker or line breaker to be closed when the generator voltage is in phase and frequency with the system voltage. This function may be performed manually with use of a synchroscope or automatically employing speed matching and automatic synchronizing relays. For small units automatic synchronizing equipment may be eliminated.

The voltage regulator works jointly with the static excitation equipment. After the field has been excited to achieve system voltage and the generator is synchronized to the system, the voltage regulator assures that the set point voltage is automatically maintained. A voltage adjustment device is provided to set the desired generator voltage.

Generator Breakers and Line Breakers. Generator breakers and line circuit breakers are the link that connects the generator to the power grid. The generator breaker closing occurs when the unit is in synchronism with the power grid. These breakers also act as an interrupting or tripping mechanism to disconnect the unit from the system when an abnormal condition or a normal shutdown takes place.

Breakers are classified by type, voltage class, continuous rated current and interrupting capacity. Types of breakers include magnetic, air blast, gas, oil and vacuum and are indicative of the medium in which the electrical arc is extinguished. The distinction between the generator breaker and the line circuit breaker is that for multiple unit arrangements with a single step-up transformer, a separate low voltage breaker is required for each generator. Some single unit plants eliminate the generator breaker and connect the plant to the system with the line circuit breaker. See Figure 5-3 for a typcial one-line diagram of a single unit plant.

Generator breakers for small hydroelectric installations are commonly air blast or vacuum type, metal-clad units rated at 4.160 kV. The interrupting capacity is dependent on fault calculations which determine system and generator contribution to a fault. Metal-clad units can be supplied with associated metering and instrument transformers.

The line circuit breaker is located on the high voltage side of the step up transformer in the switchyard. Vacuum and gas type are being installed more frequently due to decreased maintenance costs. However, many utilities still standardize designs for small installations around the oil type unit. Standard voltage levels are 15,500, 38,000, 48,300, 72,500 and 121,000 volts.

Figure 5-1. Control board for South Power Plant, installed in 1910 and still in operation. (Courtesy of Pacific Gas and Electric Company)

Figure 5-2. Control board for the New Exchequer Power Plant which was installed in 1969. (Courtesy of Merced Irrigation District)

For all aforementioned breakers, control cabinets and consoles are available for the circuitry required to close and trip the breaker. Bushings come with provisions for instrument transformers. Options include relaying equipment and key interlocks.

Unit Starting. The method for starting the unit is regulated by the governing system. The governor controls speed and loading of the turbine. After allowing an initial flow of water through the turbine to achieve breakaway of the turbine runner, the speed regulation and matching equipment begins a feedback network to adjust governor speed and to check system speed. Once the speed adjustment is achieved and the voltage is regulated, the unit is connected to the system. The function of the governor is then to control load on the unit by positioning the gates to regulate flow of water to the turbine runner.

Starting the Unit by Motoring. An alternative means of unit starting is to start the generator as a motor. Induction generators are started by this method. If a synchronous generator is to be started by this method it requires that a damper winding be designed into the machine to handle starting requirements. There are some additional requirements to this system which offset its advantages in economy and ease of operation.

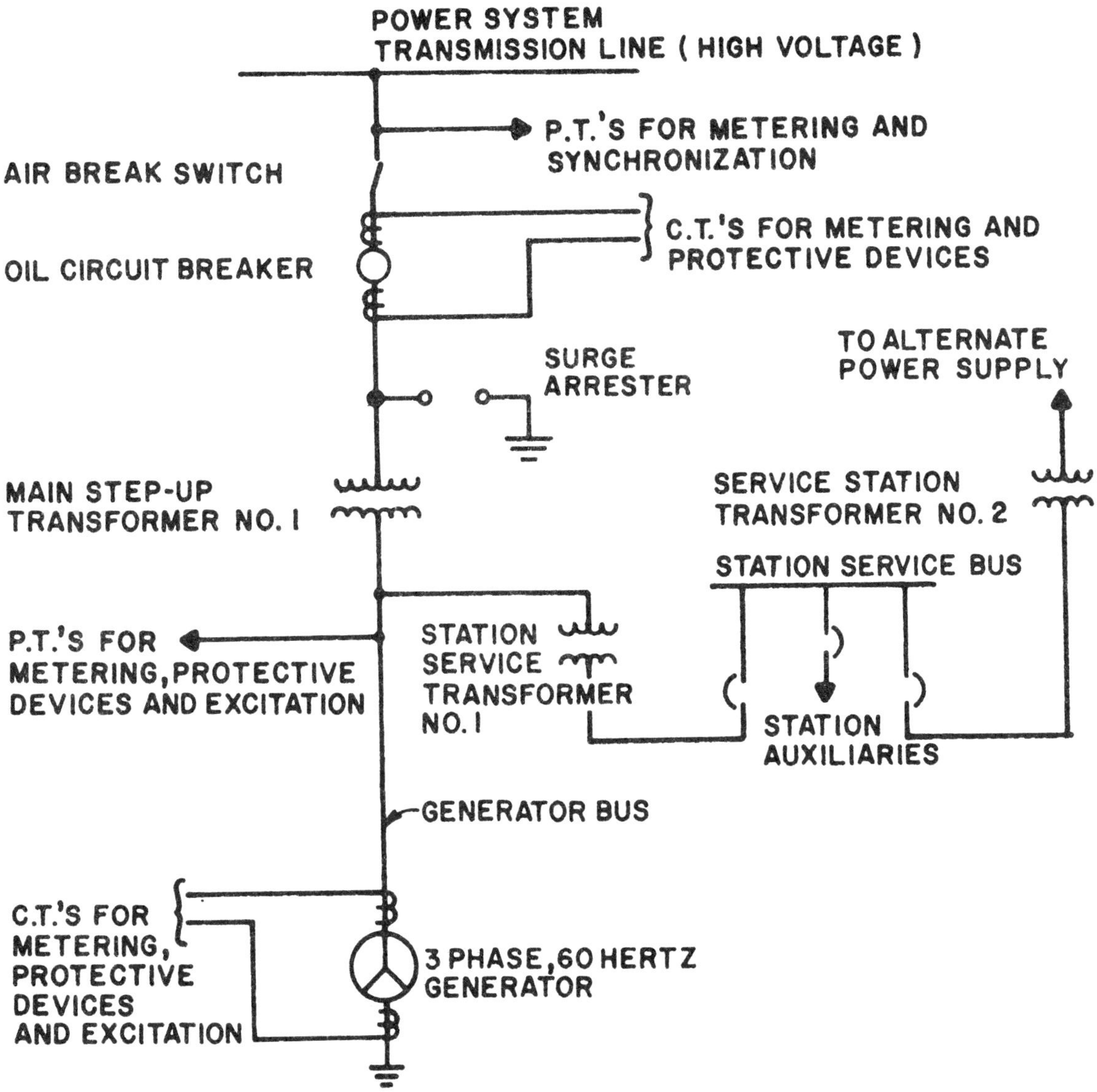

NOTES:

1. C.T. and P.T. refer to Current Transformer and Potential Transformer respectively.
2. For two or more units, a single main step up transformer may be employed; each unit will require a separate load side generator breaker (4.16kV).

Figure 5-3. Single unit one-line electrical diagram

These requirements include the need of a seal water system to cool the turbine seals during starting or motoring and additional metering to measure power flow into the generator. This measurement must be figured into the net power production of the plant.

Governor and Load Control Equipment. Large hydroelectric installations are equipped with hydraulic-mechanical or hydraulic-electric governors which regulate speed. These governors are capable of regulating the speed of the turbines with a gate control deadband of less than 0.02 percent. Small hydroelectric installations generally have little effect on the frequency of the power grid and may be installed without speed regulation governors which result in a cost savings. Gate control equipment is generally part of the equipment furnished by the turbine manufacturer and the estimated costs are included in the turbine generator cost curves.

For small hydroelectric installations, non-speed-regulating governors may be either hydraulic or electric-operated and their function is to bring the turbine to near synchronous speed for start-up, to regulate load after synchronous speed has been achieved and to shut down the unit during both normal and emergency conditions. The units must be equipped with mechanical speed switches and an independent energy source which will shut down the turbine in the event of load rejection or loss of station power. When hydraulic systems are used, an air-oil accumulator is used as an independent energy source. When electric operators are used, a dc battery system is used.

In cases where load regulation is not required, the turbine is equipped with an inlet valve which must be able to shut the unit down under emergency conditions. The power to close the valve can be provided by a hydraulic accumulator system, a battery system or a weight trip lever device.

Station Equipment and Protection Systems

Relaying Equipment and Surge Protection. An important part of hydroelectric plant operations deals with safety and protection. In particular, short circuits and ground faults within the plant must be monitored and corrective action must be initiated to prevent injury to personnel or damage to equipment.

Two types of protective devices are the protection relays and the surge protection arresters. The relays examine time-current relationships and operate when the voltage and current characteristics lie outside of the pre-calibrated settings. An example is the generator differential relay which senses a fault within the machine itself. Its operation initiates emergency shutdown of the unit. The purpose is to immediately arrest any further damage to the equipment and to alert the local operator or a remote control center to the condition. Surge protection is required to restrict any line surges from the system or any surges in voltage that have not been properly contained by the station lightning arresters. The surge arresters are physically located as close to the generator terminals as possible. The surge equipment prevents insulation damage and flashover on the generator windings.

Protection relays often cost in the range of several hundred dollars. They are flush mounted on the main control board to allow the display of relay targets. The targets indicate which phase of the three-phase system activated the relay. The settings of the protective relays are based on voltage-current information and time characteristics. Settings must be coordinated so that the closest relay to the fault activates first and the next one up the line represents the back-up. In most cases, this information is coordinated with the operator of the power grid.

Fire Protection. A $C0_2$ fire protection system is employed in the generator housing assembly and general plant area. The purpose of the generator $C0_2$ system is to extinguish fires that occur within the generator housing. A bank of cylinders for both initial and delayed discharge is actuated by $C0_2$ thermal switches. Portable extinguishers are positioned about the plant to contain local fire hazards. Steam or water may be used in place of $C0_2$, but both require that the generator be disconnected from the bus and the excitation system before the fire protection system is activated. A further advantage of $C0_2$ is the fact that it is harmless to the insulation. A common physical configuration is a bank of cylinders against a wall with a discharge header pipe to the generator housing.

Small hydroelectric installations may not warrant automatic fire systems. Local hand-operated $C0_2$ extinguishers may be suitable. The costs for fire protection is included as an item in Miscellaneous Power Plant Equipment, Section 6.

Cost Data. Station electric equipment includes station switchgear, battery system, station service transformer and equipment, lighting, protection system, control board, cable and conduit. These systems represent a fixed expenditure of plant cost regardless of turbine and generator selection. Figure 5-4 illustrates the dollars vs. MW for the range of small hydroelectric plants.

Transformers

General. The power transformer is a highly efficient device to step the voltage from generation level to transmission level. Efficiencies are generally in the range of 99 percent. For small hydroelectric installations, a single, two winding, oil-filled substation type transformer is required. The main tank is pressurized with nitrogen to monitor rupture of the vessel and with loss of pressure to cause an alarm to sound. The bus entry to and from the transformer is accomplished by porcelain bushings, which may be supplied with current and potential transformers for metering, relaying and instrumentation. A terminal cabinet is located on the side of the transformer. Its function is the termination of auxiliary devices such as sudden pressure relays, over-temperature devices, and cooling fans and pumps. The cooling system consists of fin-type radiators which

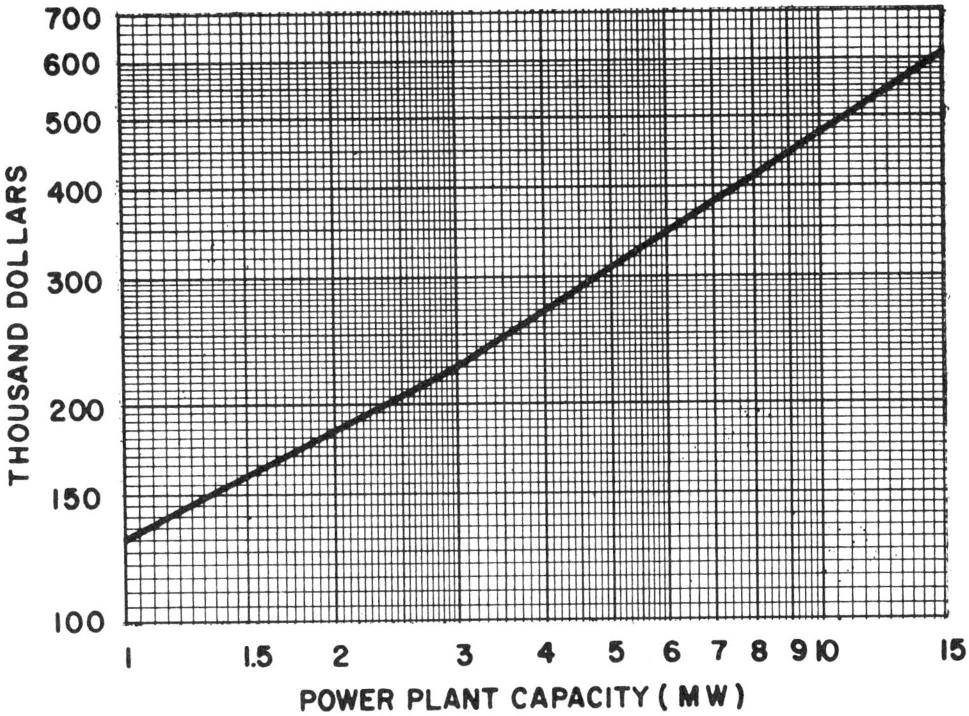

NOTES:

1. Major Equipment is listed below:
   a. Battery and Battery Charger
   b. Station Switchgear
   c. Station Service Transformers
   d. Bus, Cable, Conduit and Grounding
   e. Main Control Board
   f. Lighting System
2. All items applicable to FERC Account No. 334
3. Costs include freight and installation.
4. Costs shown are for a single unit plant. For multiple units a cost for generator breakers and additional controls must be added. Add $20,000 + $58,000 x (n-1) to the cost of a single unit plant of the same total kW capacity. (n = number of units).
5. Cost Index is July 1978.

Figure 5-4. Station electrical equipment costs

depend strictly on convection. To augment natural cooling, fans or fans combined with oil circulating pumps may be employed. A further refinement of cooling can be accomplished with oil-to-water heat exchangers. This method, however, requires that the coolers be operated at all times. Small hydroelectric installations are normally limited to open air transformers with forced air cooling only for extremely warm days or short term overload conditions.

Cost Data. The main variable for transformer cost is the capacity of the unit for power transfer (kVA). Voltage levels are the next variable in cost as higher voltage requires more insulation material. Each transformer is provided with a control cabinet and sudden pressure relay. It is assumed the low voltage of the hydroelectric generation will be 4160 volts. Several high side voltages are presented. Transformer costs are included in switchyard costs given in Section 6.

SECTION 6
SWITCHYARD, TRANSMISSION LINES, AND MISCELLANEOUS TOPICS

General

The switching and delivering of power to some distant point represents the final link to the power grid. Although sometimes disregarded in preliminary feasibility assessments, the length of transmission of the power may be an economic constraint that seriously affects project feasibility.

Switchyard

The switchyard is comprised of line circuit breakers, disconnect switches, transformers, structures, buswork and miscellaneous power plant equipment. The arrangement of this equipment should allow for the future movement of circuit breakers and other major equipment into position without de-energizing existing buses and equipment. For single unit small hydroelectric installations, the switchyard will consist of the generator bus, step-up transformer, a disconnect switch, a line circuit breaker and a take-off tower. Station transformers, excitation transformers, and surge and metering cubicles may also be included in the switchyard to decrease floor space requirements in the powerhouse structure. Another alternate arrangement would have the metal-clad (enclosed in cabinets) generator breakers located in the switchyard. A typical arrangement drawing for a single unit plant is shown in Figure 6-1. Multiple unit switchyards may be similarly arranged as long as electrical protection and a means for isola tion is maintained between individual generators by use of generator breakers.

The location of the switchyard with respect to the powerhouse is dependent on soil conditions, space requirements and topography. Where geographically feasible, the best location of the switchyard is close to the powerhouse structure. This eliminates costly extension of the generator bus and reduces power losses in the bus. A photograph of a typical switchyard for a hydroelectric power plant is shown in Figure 6-2. Note that the plant is shown under construction.

Cost Data. The costs presented are for single unit switchyards. Cost data reflects the installed cost for level of transmission up to a maximum of 115 kV. Cost data relative to site preparation and clearing, foundation, structures and fencing of the switchboard area is provided in Volume VI. See Figure 6-3 for cost data.

Transmission

Reduction of line losses is the key to optimum transfer of power. For the potential developer of a small hydroelectric plant, transmission facilities may be the responsibility of the purchaser of the power. Thus the developer will only need to coordinate the outgoing take-off structure with the positioning of the incoming transmission line.

However, some projects will require that the developer also be responsible for transmission to some point at which an intertie to the transmission grid can be made. This construction then must be included in a cost analysis to determine the economic viability of the project. Consideration of right-of-way for construction of the transmission line and siting of the line l also must be taken into account.

The physical equipment for a transmission line includes conductors, poles, supporting guys, insulators and connectors. Wood pole line construction is quite applicable for the range of transmission levels discussed herein. The nominal values of transmission voltage considered in this study are 13,800, 34,500, 69,000 and 115,000 volts. (Alcoa Aluminum Overhead Conductor Engineering Data, 1960).

Cost Data. Figure 6-4 illustrates cost data for transmission lines of up to 30 miles and at voltages typical for small hydroelectric plants. The cost data has been developed from data supplied by the U.S. Department of the Interior. These costs reflect open, level terrain and favorable foundation conditions. The figure also indicates a procedure for increasing the costs when more adverse conditions are present.

Miscellaneous Power Plant Equipment

Small hydroelectric installations are generally operated and monitored from a remote location and therefore designed to house only the generation equipment. Heating, ventilating and air conditioning and waste systems for human habitat are normally not required. During infrequent maintenance periods, bottled water and portable toilet facilities may be provided. The estimated costs for miscellaneous equipment contained in this volume reflect minimum equipment for average site conditions and consist of the following:

Ventilation. A central blower located in the roof or end walls with temperature control to actuate when ambient temperature rises above 74 degrees F is provided. Filtered air inlets near floor at generator level are also included.

Water System. Duplex pump system with strainers is provided for water-cooling requirements of the turbine and generator. The water is taken from the penstock or tailrace. The cooling water system should operate independently of the plant generating equipment. No water cooling system for generator ventilation is included in the costs data.

Crane. A permanent powerhouse crane is not recom-

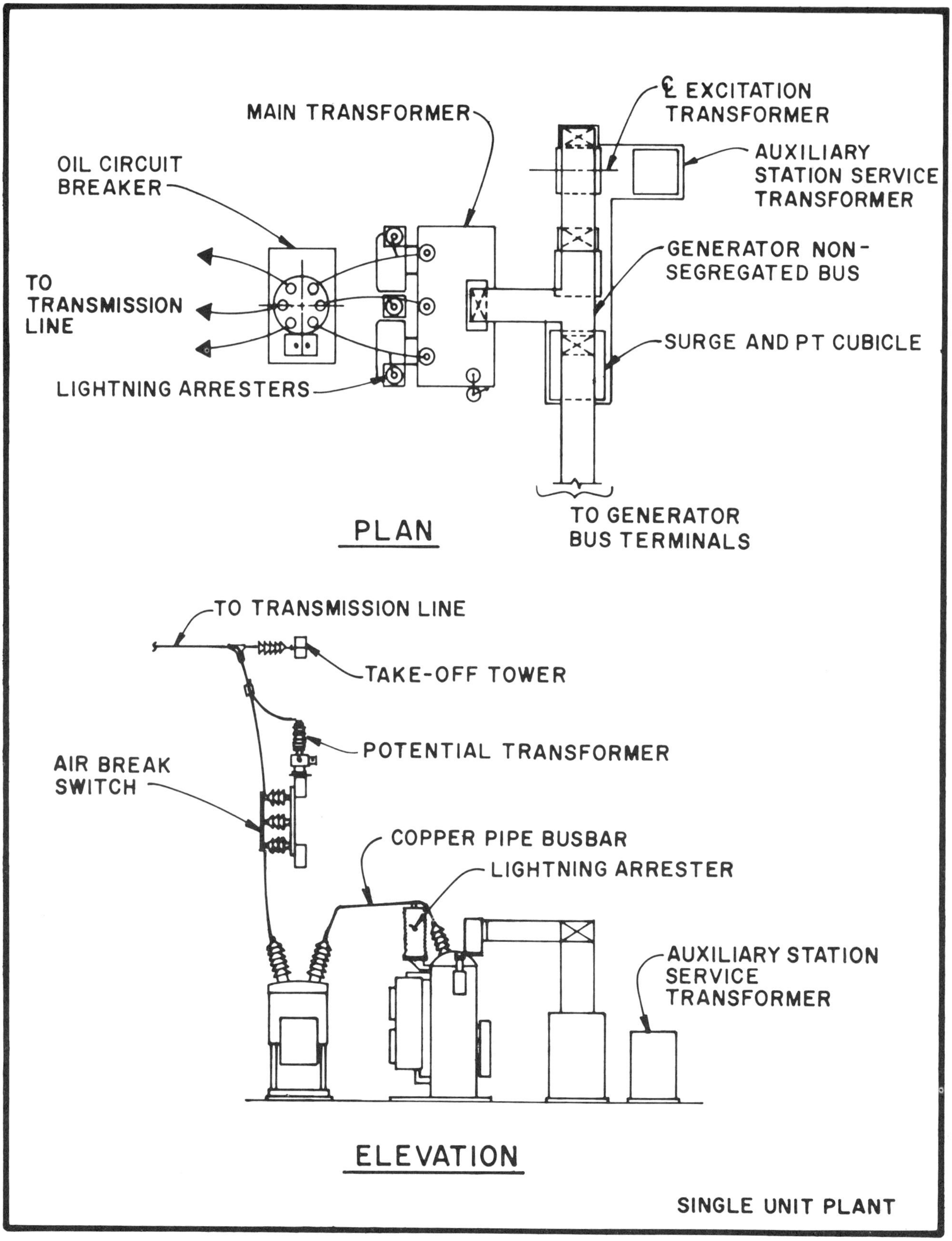

Figure 6-1. Typical arrangement of a switchyard

Figure 6-2. Switchyard for McSwain Power Plant. (Courtesy of Merced Irrigation District)

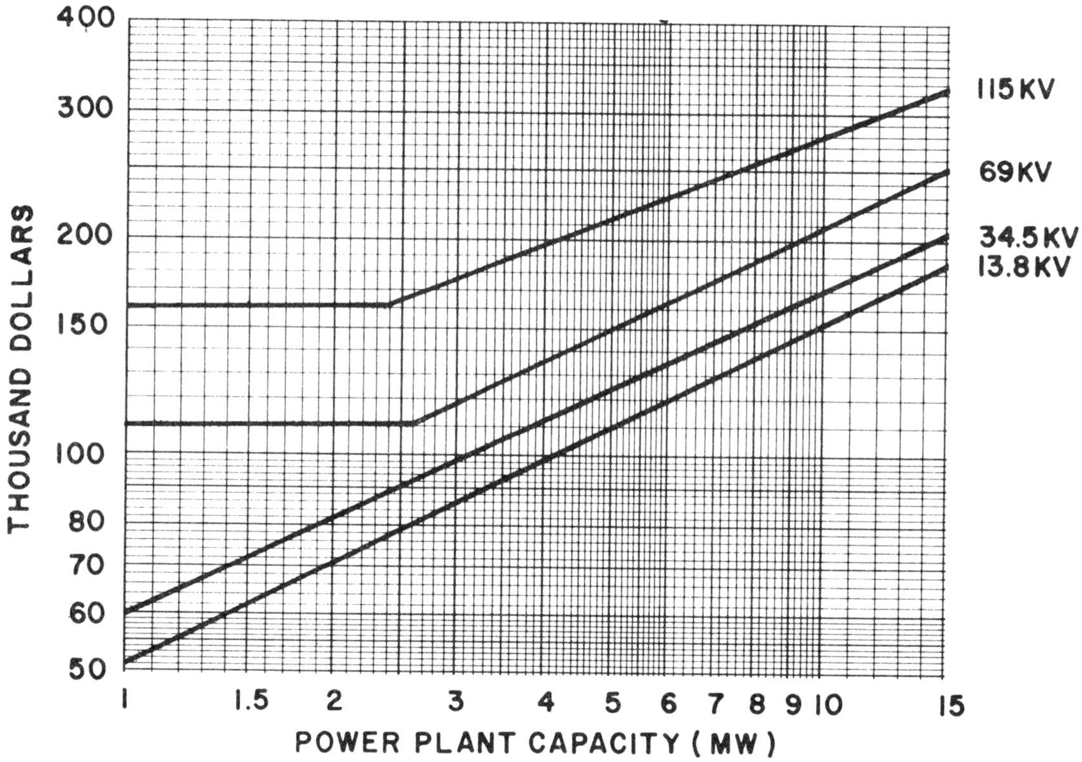

NOTES:

1. The major Equipment is listed below:
   a. Main Step-up Transformer
   b. Line Side Oil Circuit Breaker
   c. Lightning Arresters
   d. Air-Break Switches
   e. Bus work
2. Costs include 25 percent for freight and installation.
3. Foundations and Switchyard structures are covered in the Civil Features (Volume VI).
4. Above costs reflect a design of 45 feet of generator buswork. For extension beyond 45 feet, use a factor of $200 per foot for generator buswork.
5. Cost index is July, 1978.

Figure 6-3. Switchyard equipment costs

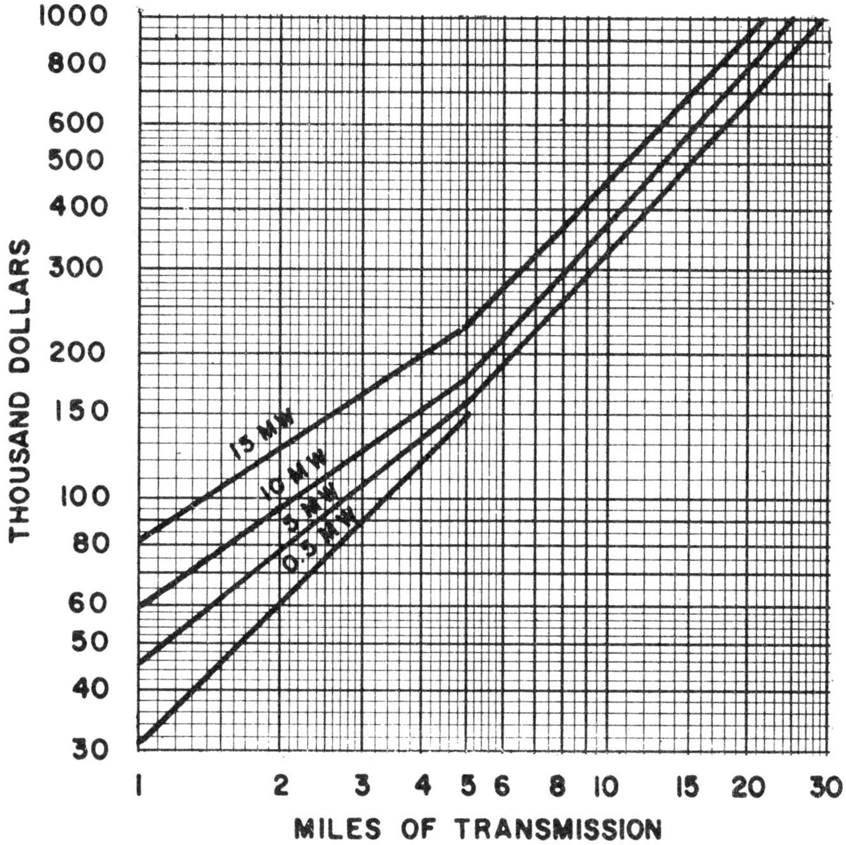

NOTES:

1. Direct Costs are for Prairie Construction and favorable foundation conditions.
2. Costs should be increased for areas with more difficult access (up to 50 percent for swampy or mountainous areas).
3. Costs should be increased for unfavorable foundation conditions (up to 50 percent for swampy or rocky areas).
4. Direct Costs include normal construction road costs but do not include any contingencies, land and rights, relocations or clearing and access roads.
5. It is not feasible to transfer 500 kW more than 5 miles because of line loss of power. A limiting distance of efficient transfer should be calculated for values up to 5000 kW to ensure that a design be feasible.
6. Cost index is July, 1978.

Figure 6-4. Transmission line costs

mended for small hydroelectric plants. Due to size and cost of equipment, it is considered more economical to bring in portable equipment for major plant overhauls. Provisions for a portable gantry crane for larger power plants should be provided. This would include crane rails embedded in the generator deck and a power connection. Appropriate hatches should be provided for access to all movable machinery.

Miscellaneous. An eye wash bath and a ventilating fan for battery area are required for safety of the workers.

Fire Protection System. The fire protection system included in the cost is for a detector operated CO_2 system for extinguishing fires in generator housings and for hand held portable extinguishers for other fire protection.

Cost Data. Figure 6-5 contains data for estimating the cost of miscellaneous power plant equipment. The costs contain only the minimal equipment as described in the previous paragraphs. For an attended station with facilities for operators and maintenance personnel, the costs should be increased by a factor of 2 to 3.

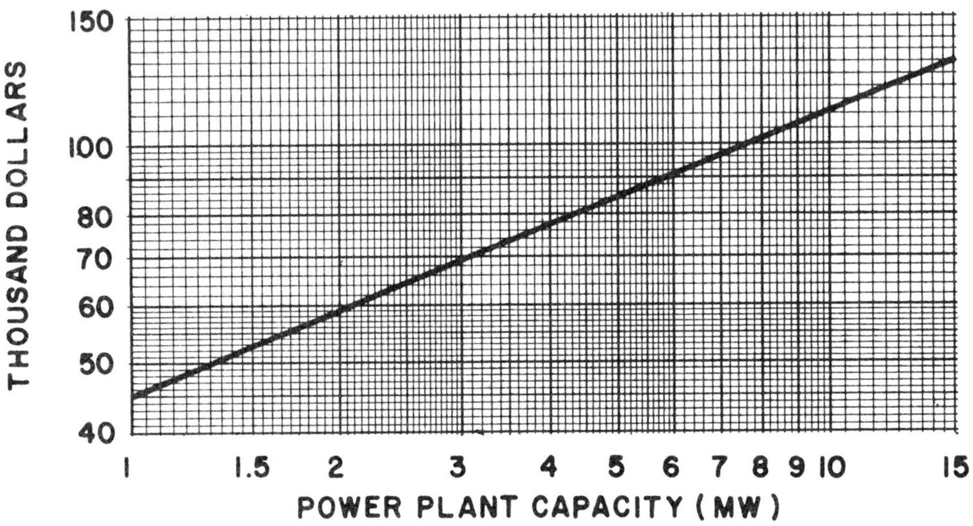

NOTES:

1. The Major Miscellaneous Power Plant Equipment is listed below:
   a. Ventilation Equipment
   b. Fire Protection Equipment
   c. Communication Equipment
   d. Generation Bearing Cooling Water Equipment
2. Communication equipment includes alarm and communication facilities for unattended operation of unit; further cost figures should be obtained for very remote locations or integration with complex communication networks.
3. All figures shown include a 15 percent factor for freight and installation.
4. Cost index is July, 1978.

Figure 6-5. Miscellaneous power plant equipment costs

SECTION 7
COST SUMMARY

Additional Cost Considerations

This section will provide a cost summary and describe additional factors which should be considered in the preparation of an estimate for the electromechanical equipment of a small hydropower plant. These factors include escalation, development costs, and annual operation and maintenance costs.

Escalation. The costs for the electromechanical features described in the previous sections were given as July 1978 bid prices. In order to determine the equipment costs after that date, the previously presented costs must be excalated to the desired future date. The bid price, escalated to the desired future date, may be determined by use of the U.S. Bureau of Reclamation indices. These indices are published quarterly and indicate constructuion cost trends. The quarterly publication is known as *Construction Cost Trends* and can be obtained from any USBR regional office. In the Sacramento region, the publication can be obtained by writing the Bureau at 2800 Cottage Way, Sacramento, CA 95825, Attn: M.P. 200. The indices are also published quarterly in *Engineering News Record.*

In order to determine escalated costs, it is first necessary to plot all up-to-date indices on the graphs provided in Figure 7-1. After plotting the indices, the resulting curve should be extrapolated to the desired future date. The extrapolation should be an extension of the latest trend in the indices. After extending the curve, the appropriate index should be picked off for the desired future date.

Once the index for the desired future date is known, it is then possible to estimate the equipment cost for that date. The equipment costs obtained from previous sections of the report should be multiplied by the index for the desired future date, then divided by the July 1978 index. The resulting figure is the facility bid price, escalated to the desired future date.

Contingency. A contingency allowance should be added to the escalated prices to cover the unknown items and items omitted which would normally be covered with a more detailed cost estimate. Contingency is also considered an allowance for possible cost increases do to unforeseen conditions. This allowance is normally taken to be 10 to 20 percent of the escalated prices. The percentage used should reflect upon the confidence level of the data used.

Engineering, Construction Management and Other Costs

Once the escalated construction cost has been determined, it is necessary to estimate the engineering, construction management and administration costs, sometimes referred to as development costs or indirect costs. These costs include expenditures for feasibility study, license and permit applications, preliminary and final design, construction management, and administration. A multiplier of 20 percent should be applied to the total escalated construction cost, including contingencies, to estimate these development costs.

For a more detailed breakdown of these development costs the following percentages, applied to escalated constructuion costs plus contingencies, may be used:

Feasibility Study	2%
License and Permit Applications	2%
Preliminary Design	3%
Final Design	6.5%
Construction Management	5.5%
Administration	1%

The above percentages are for electromechanical costs only, hence the multipliers should be applied only to electromechanical bid costs. Not included in the above development costs are interest during construction, legal fees and financing fees. These omitted costs will be covered in Volume II which describes economic and financial considerations.

Operation and Maintenance Costs

Operation and maintenance costs for small hydropower plants are difficult to forecast accurately. The costs are directly related to the site and owner's capability to perform the operation and maintenance function. The amounts which are suggested to be used in this report are based on those published by the U.S. Bureau of Reclamation and are updated to reflect recent experience.

Operation and maintenance costs as described herein, include the items listed below.

Insurance. The Government is basically a self-insurer, however, for a commercial installation, coverage is required for fire and storm damage, vandalism, property damage and public liability. Insurance can also be purchased for major mechanical or electrical damage. This latter insurance is not usually considered for small hydropower installations.

Routine Maintenance and Operation. An amount must be budgeted to cover the costs of manpower, wages, services, equipment and parts utilized in the normal operation and maintenance of the hydroelectric plant.

Interim Replacement. During the life of a hydroelectric project, miscellaneous equipment and facilities will wear out and require replacement. This replacement is in addition to those routine replacements

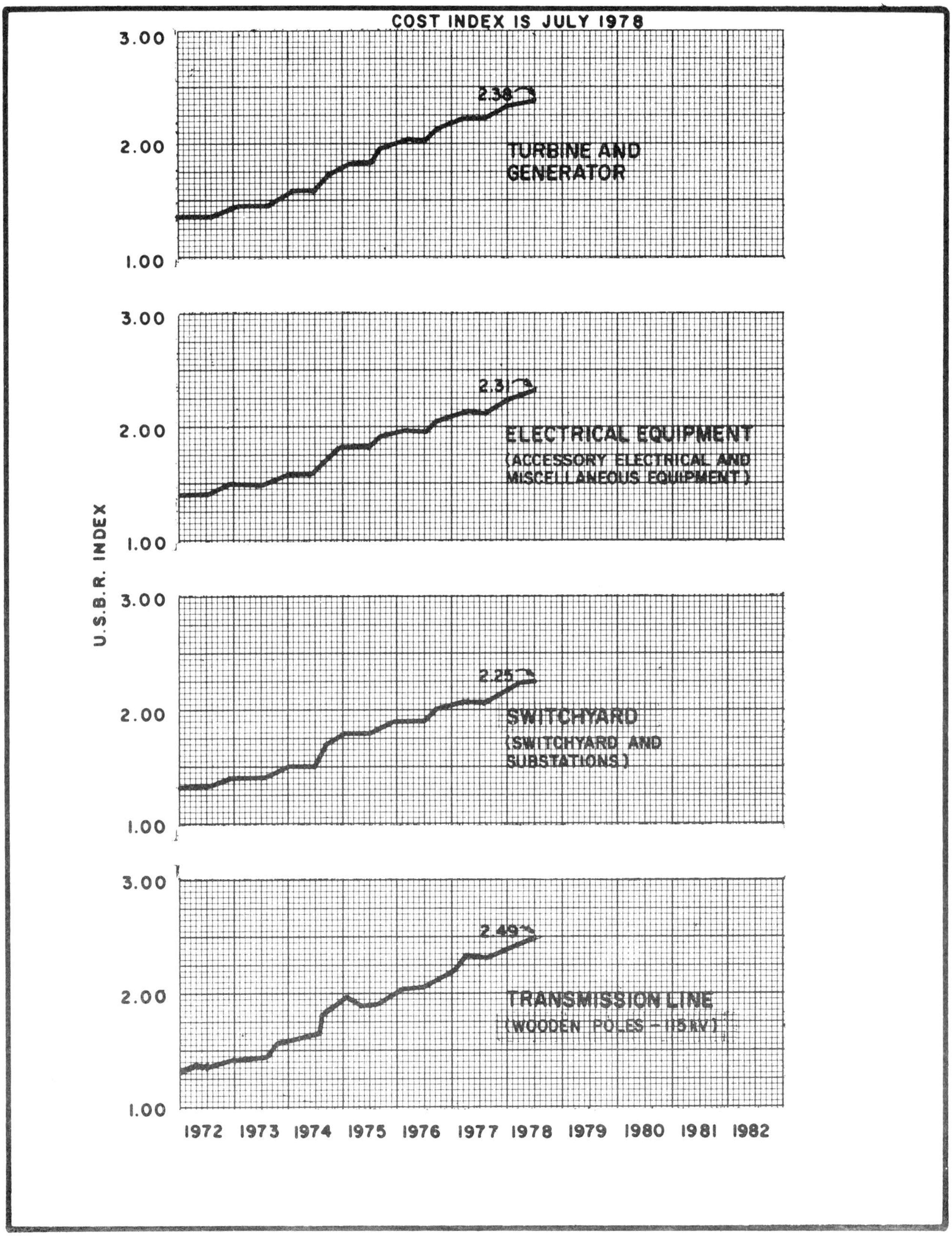

Figure 7-1. Escalation Costs

covered in normal plant maintenance. A sinking fund should be established for these interim replacements. For the most preliminary studies, the yearly deposit into this sinking fund should be taken as 0.1 percent of total construction cost plus contingencies.

General Expenses. The final portion of operation and maintenance costs are made up of those expenditures for administration fees and other miscellaneous costs required during project operation.

Total Annual Cost. The total annual cost of operation and maintenance expenses can be estimated by multiplying the investment cost, i.e., escalated bid price, contingencies and development costs, for plant facilities, by 1.2 percent. The resulting amount will be the estimated cost for operation and maintenance of the hydroelectric plant for the first year of operation. The annual operation and maintenance costs will increase with time, corresponding to inflationary trends. The current annual increase for operation and maintenance costs is taken to be 6-1/2 percent.

There are two final comments to be observed in determining the operation and maintenance costs of hydroelectric plant electromechanical facilities. First, the annual costs for operation and maintenance should never be estimated below a certain minimum amount, approximately $20,000 in 1978 dollars. Second, the multiplier given previously, 1.2 percent, should be used only if the owner can integrate the operation of the small hydropower facility with other related operations. If the operating entity will operate and maintain only the small hydropower facility under consideration, a multiplier of 2 to 4 percent should be used to determine annual O&M costs.

Manpower Allocation for Studies. The allotment of time for the preparation of feasibility studies for small hydroelectric power plants varies depending upon site conditions and degree of depth of the study. As a general rule of thumb, however, approximately ten man-days should be allocated for the electromechanical portion of the study. Of that total, approximately 10 percent would be spent by a lead Hydroelectric Engineer with experience in Hydropower or electromechanical projects. An experienced Mechanical Engineer would spend approximately 35 percent of the man-hour allocation, and an experienced Electrical Engineer would spend approximately 25 percent. The remainder of the man-hour allocations, approximately 30 percent, would be spent by a designer and draftperson, familiar with engineering calculations. For more complex sites, or if substantial text is required, the required manpower will increase. Guidelines have been established by various professional organizations which indicate that feasibility studies should cost between one and two percent of the total project costs. The larger percentage is applicable to more complex installations.

Cost Summary

Figure 7-2 is a cost estimate summary sheet for the electromechanical components of a small hydroelectric power plant. Account numbers have been assigned to correspond with the account number assigned by the Federal Energy Regulatory Commission. Each item is referenced to the chart as section of the report providing cost data. The cost data contained in this report is adequate for indication as to the feasibility of a potential hydroelectric power site. As indicated pre viously, actual prices on equipment should be obtained whenever possible for a final feasibility determination.

COST ESTIMATE SHEET

PROJECT ______________ PLANT CAP. ______________ MW DATE ______________

JOB NO. ______________ ANN ENERGY ______________ MKWH BY ______________

ACCOUNT NO.	DESCRIPTION	QUANTITY	UNIT COST	COSTS JULY 1978	Escalation Factor	COSTS DATE
333	WATER WHEELS, TURBINES AND GENERATORS TURB. TYPE GEN. TYPE RATING MW FT. D_2 FT. INSTALLAED COST (FIG 3-12 TO 3-16)					
334	STATION ELECTRICAL EQUIPMENT TRANSFORMER, LIGHTING ARRESTOR, AIR BREAKER SWT, GEN. BREAKER AND LINE OCB (FIG. 6-3)...................... BATTERY SYS., STA. SWT GEAR, STA. SER. TRANS.,BUS,CABLE CONDUIT, GRD., CONTROL BD., LIGHTING SYS., FREIGHT AND INSTALLATION (FIG. 5-4)............................					
335	MISC. POWER PLANT EQUIPMENT VENTILATION, FIRE PROTECTION, COOLING WATER, COMMUNICATION SYS., FREIGHT AND INSTALLATION (FIG. 6-5)...					
350	TRANSMISSION LINE LENGTH MI VOLT. KV (FIG. 6-4).............					
COST PER INSTALLED KW				SUB TOTAL		
	CONTINGENCY (%)					
				SUB TOTAL		
	ENGINEERING, CONSTR. MG. & OTHER COSTS (20%)					
	TOTAL INSTALLED COST					

Figure 7-2. Cost summary sheet

REFERENCES

Alcoa Aluminum Overhead Conductor Engineering Data, 1960, Section 5, p. 79-83.

Fink, D. G., and Carroll, J. M., 1968, *Standard Handbook for Electrical Engineers,* Section 6.

U. S. Army Corps of Engineers, *Reservoir System Analysis for Conservation HEC-5C,* Computer Program 723-X6-L2500.

United States Department of Interior, Bureau of Reclamation, *Selecting Hydraulic Reaction Turbines, Engineering Monograph No. 20,* 1976.

EXHIBIT I
TURBINE MANUFACTURERS

Name/Address/Offices	Type Manufactured
Allis Chalmers Corp. Hydro Turbine Div. P.O. Box 712 York, PA 17405 (U.S.A.)	Francis, Impulse, Kaplan, Propeller, Tube, Slant and Standardized Tube
Axel-Johnson Corp. 1 Market Plaza San Francisco, CA. 94105 (Sweden-U.S.A.)	Francis, Impulse, Kaplan, Propeller, Tube and Slant
Barber Hydraulic Turbines Ltd. 65 Queen St. West Toronto, Ontario, Canada 195H2195 (Canada)	Vertical, Horizontal and open flume Francis-Packaged Hydro Units
Bell Hydropower 3 Leather Stocking St. Cooperstown, N.Y. 13326 (U.S.A.)	Crossflow
Charmilles Krupp International Inc. 550 Manaronek Ave. Harrison, NY 10528 (Switzerland)	Francis, Impulse, Kaplan, Propeller, Tube, Slant
Dominion Engineering P.O. Box 220 Montreal, Canada (Canada)	Francis, Kaplan, Propeller, Bulb and Tube
Fuji Electric Co. Nissho-Iwai American Co. 700 So. Flower St. Los Angeles, CA 90017 (Japan)	Francis, Kaplan, Propeller and Standardized Bulb
General Electric Co. 55 Hawthorne St. San Francisco, CA 94105 (U.S.A.)	Francis, Kaplan and Propeller
Hitachi America Ltd. 100 California St. San Francisco, CA. 94111 (Japan)	Francis, Kaplan, Propeller and Bulb
Independent Power Developers Rte. 3 Box 285 Sandpoint, Idaho 83864 (U.S.A.)	Impulse and Propeller
Kvaerner-Moss Inc. 800 - 3rd Ave. New York, N.Y. 10022 (Norway)	Francis, Impulse, Kaplan, Tube and Slant-Standardized Units

James Leffel & Co. 426 East St. Springfield, Ohio 45501 (U.S.A.)	Francis, Impulse, Kaplan, Propeller and Open Flume.
Mitsubishi International Corp. 50 California St. San Francisco, CA 94111 (Japan)	Francis, Kaplan, Propeller, Bulb, Slant and Tube
Neyrpic Alsthon-Atlantic 50 Rockefeller Plaza New York, N.Y. 10022 (France)	Francis, Impulse, Kaplan, Propeller, Bulb, Slant and Tube
Noble Automated Systems Inc. 226 Phelan Ave. San Jose, CA. 95112 (U.S.A.)	Impulse
Ossberger Turbines F.W.E. Stapenhorst Inc. 285 Labrosse Ave. Pt. Claire, Quebec H9R1A3 (Canada)	Crossflow
Siemens-Allis Utility Sales Operation P.O. Box 89000 Atlanta, GA. 30338 (U.S.A.)	Bulb
Sulzer Bros. Inc 1255 Post St. San Francisco, CA. 94109 (Switzerland)	Francis, Impulse, Kaplan, Propeller, Bulb, Slant and Tube
Voest-Alpine 1923 Magellan Dr. Oakland, CA. 94611 (Austria)	Francis, Impulse, Kaplan, Propeller, Bulb, Slant and Tube
Westinghouse Electric One Maritime Plaza San Francisco, CA. 94111 (U.S.A.)	Bulb

EXHIBIT II
GENERATOR MANUFACTURERS

ASEA Inc.
4 New King Street
White Plains, New York 10604
(Sweden)

Beloit Power Systems
555 Lawton Avenue
Beloit, Wisconsin 53511
(U.S.A.)

Electric Machinery Manufacturing Company
800 Central Avenue
Minneapolis, Minnesota 55413
(U.S.A.)

General Electric Company
1 River Road
Large Motor and Generator Department
Schenectady, New York 12345
(U.S.A.)

Hitachi America, Ltd.
100 California Street
San Francisco, California 94111
(Japan)

Ideal Electric
330 East First Street
Mansfield, Ohio 44903
(U.S.A.)

Kato Engineering Company
1415 First Avenue
Mankato, Minnesota 56001
(U.S.A.)

Westinghouse Electric Corporation
Hydro Generator Department
700 Braddock Avenue
East Pittsburgh, Pennsylvania 15112
(U.S.A.)

Siemens - Allis
P.O. Box 2168
Milwaukee, Wisconsin 53201
(U.S.A.)

VOLUME VI: CIVIL FEATURES
SECTION 1: INTRODUCTION AND OVERVIEW

Scope of Civil Features Volume

The objective of the Civil Features volume is to provide guidance in the preliminary civil facilities layout and to establish cost guidelines for the civil work required to add small hydroelectric powerplants to existing impoundments. The basic civil work includes:

1. Site preparation
2. Hydraulic conveyance facilities
3. Powerhouse and appurtenant facilities

Site preparation includes grading, foundation excavation, drainage and erosion control, access roads and parking facilities, and construction noise abatement and dust control.

Hydraulic conveyance facilities include penstocks, tunnels, canals, valves and gates, outlet works, and tailraces.

Powerhouse and appurtenant facilities include all structures for the powerhouse and equipment handling facilities, foundations for both the powerhouse and switchyard, and fencing around the project area.

Construction costs in this volume are for July, 1978. Unless otherwise noted, the construction costs used are from the engineering files of Tudor Engineering Company. These costs were developed from numerous small hydroelectric project feasibility cost studies. The historical record of cost increases in construction is presented in Section 6 for use in updating the costs beyond July 1978.

Small hydroelectric power projects, as defined in this volume, have operating heads of 100 feet or less and plant capacities of 15,000 kW or less. The lower limits on hydroelectric development are a function of the available equipment and the economics of developing power at the site.

Civil Features; Proportion of Project Costs

To determine the typical range of civil costs to total project costs for small hydroelectric installations, three main cost categories were compared; civil, electrical/mechanical and development or indirect costs.

Only the costs associated with the on-site features were included in the comparison. Transmission line costs from the plant switchyard to the power grid were not included. Neither were any right-of-way costs included, because these costs are unique to every project and do not have a common or predictable value.

The civil costs are described in this volume. The electromechanical costs are presented in Volume V. These costs were further categorized as turbine/generator, accessory electric equipment and miscellaneous power plant equipment costs for the purpose of this comparison. The indirect costs include the costs associated with engineering, administration, construction management, and legal and financial consulting. Interest during construction is included in the indirect costs. A contingency percentage is not shown as it would be applied to all cost categories to cover any unexpected increases in costs.

Figure 1-1 illustrates the range of proportional costs possible for small hydroelectric development. The top graph shows the case where the civil costs are minimum. The bottom graph shows where the civil costs are maximum. The proportion of the indirect cost for both cases is the same, at approximately thirty percent. The upper graph of Figure 1-1 includes those projects which usually have an existing outlet work which permits the power house to be designed with minimal penstock and other water way passage costs. The lower graph of Figure 1-1 includes the projects that require long penstocks and either major alteration of the existing outlet work or bypassing it by constructing new civil facilities.

Types of Sites Suitable For Hydroelectric Development

Hydroelectric power may be developed at any site where there is a flow of water between two bodies of water at different elevations. Besides sites at impoundment facilities, there may be possible hydroelectric sites at analagous sites such as a drop structure between two reaches of a canal. (Figure 1-2 shows a vertical drop of an irrigation canal.) Another example is a water aqueduct facility where an outlet, either to a lower pressure feeder main or to a storage facility, discharges water under pressure.

Basic Civil Feature Differences Between Small and Major Hydroelectric Installations

Differences between small and major hydroelectric installations are primarily related to the plant size and the importance of the installation to the power grid. The small hydroelectric plant usually has major equipment of a physical size that can be readily handled with portable lifting devices. Large, permanently installed cranes are required only for major installations. Similarly, small hydroelectric installations, not being critical to the power grid requirements, can have longer overhaul periods than would be appropriate for larger installations. This minimizes the area and facilities that are required for maintenance operations. Often, centralized shop facilities can be used, as the equipment items may be of a size that are easily transportable.

Typically, small hydroelectric installations are unattended and remotely operated. This reduces the require-

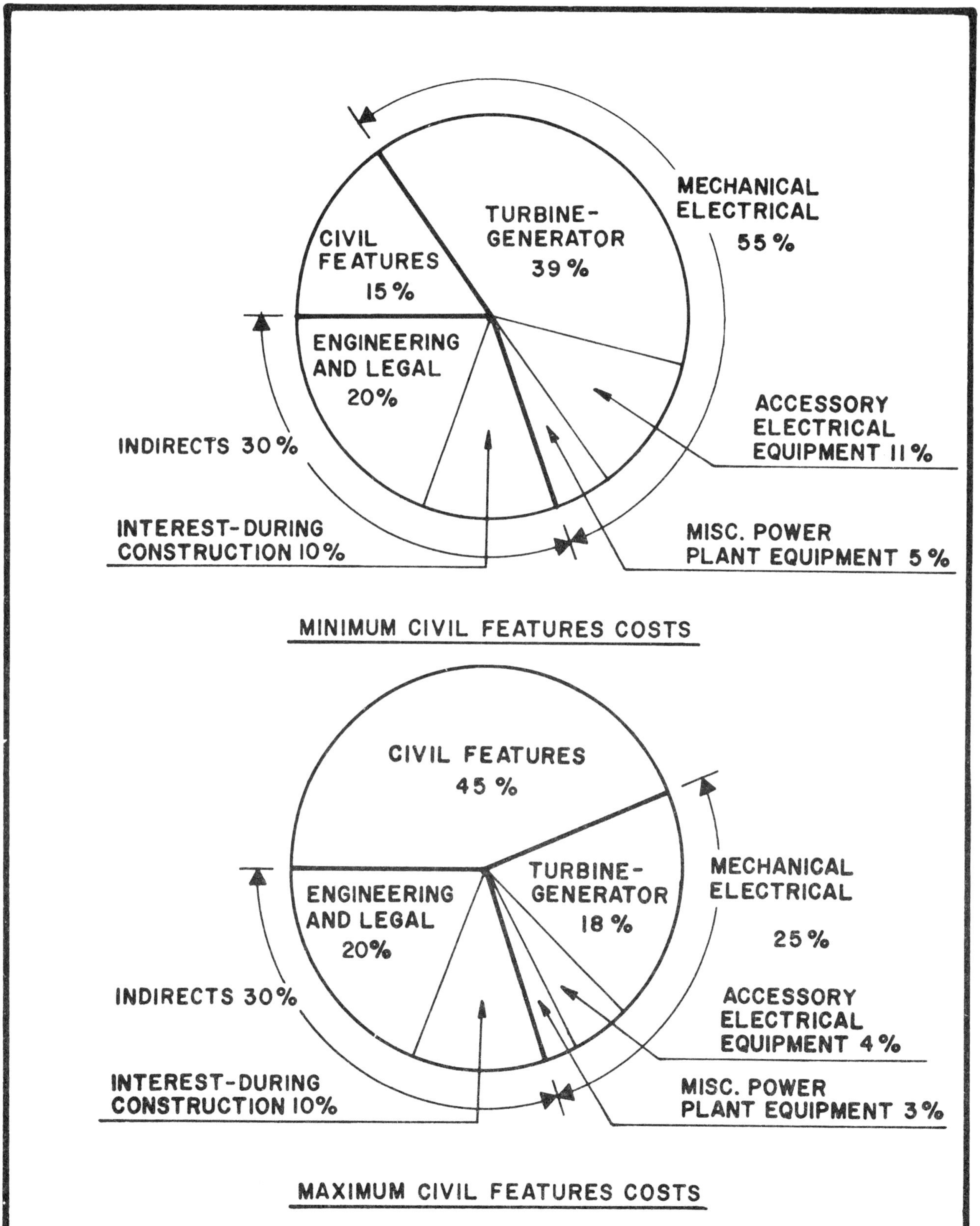

Figure 1-1. Range of Civil Features Costs

ments for working space, storage area, and personnel comfort items such as air conditioning, lockers, potable water and showers, from that which would be required for larger installations. Fire protection systems can also be kept to a minimum, and oil handling, filtering and storage can be portable, with permanent facilities being eliminated.

Methodology for Feasibility Determination

Detail Needed for Feasibility Assessment. This volume presents approximate construction and administrative costs that are adequate for a reconnaissance study. Project costs need only to be estimated to within plus-or-minus twenty percent to be considered satisfactory for an assessment of this type and effort should not be expended with the data to obtain a more precise estimate. The costs of civil features which are site specific and require engineering judgement and experience for their evaluation have not been included in this volume. However these areas are noted and with the addition of these costs the costs thus determined will be satisfactory for a feasibility assessment.

Steps for Determining Costs. Figure 1-3 graphically illustrates the steps that should be followed to determine the civil costs for a potential project. It is assumed that some information regarding the plant configuration will have already been determined from the other volumes of this manual. In particular, turbine type and throat diameter, powerhouse type, number of turbine/generator units and their capacity, flow rate, and design head for each unit should be determined prior to the evaluation of the civil costs.

After the above items have been determined, the project area should be analyzed and a tentative plant location selected. From the plant location and other information, the site area, the powerhouse area and the necessary facilities to convey water to the powerhouse and back to the streambed should be determined. Following the steps shown in Figure 1-3, the total civil cost is then determined for either a reconnaissance or feasibility cost assessment dependent upon the input data used.

After the civil cost has been estimated, the plant location selection should be examined to determine if a change in location or orientation could reduce costs. Several alternate locations may be evaluated and the least cost alternative selected.

General Description of Civil Features

Configuration. There is no standard configuration for adding a small hydroelectric installation to an existing

Figure 1-2. Vertical Drop on an Irrigation Canal

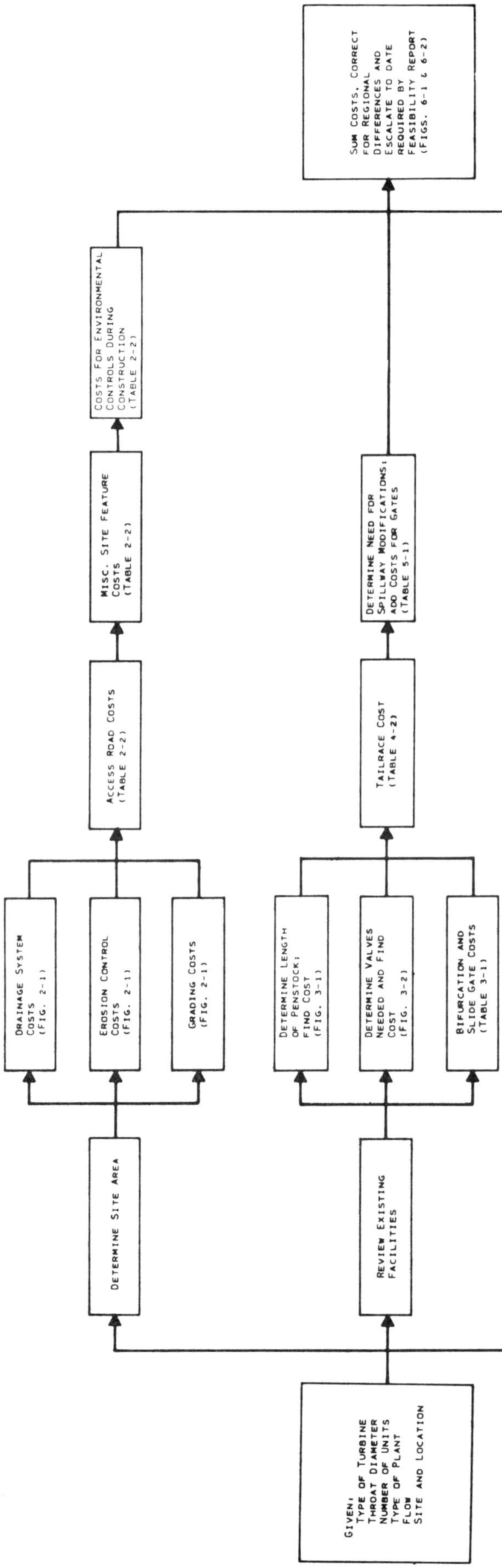

Figure 1-3. Steps for Determining Costs

impoundment. Many design decisions have to be made which are a function of the type of dam, location and type of outlet works, use of impoundment water, and location of the nearest electrical power grid. Nonetheless, the typical design configurations used for the basis of evaluating civil costs in this volume are sufficiently general, so that cost estimates prepared from the data presented should be suitable even for sites that require somewhat unusual configurations.

A potential hydroelectric site will have an existing outlet works, and the use of this existing outlet for the new plant should be made whenever possible, and is often necessary to achieve economic feasibility. This is usually done by constructing a bifurcation in this conduit, one branch becoming the upstream end of the penstock.

Several factors should be considered in the project layout. Ideally, it is best to locate the powerhouse as close to the downstream side of the dam as possible, minimizing the length of penstock. The site should have a minimum overburden, allowing the powerhouse to be founded on rock. The general site should be above a once in a hundred year flood stage. The switchyard site should also be above this flood stage and located for an easy electrical access to the power grid.

Intake. The intakes for small hydroelectric installations are generally already existing as part of the impoundment structure. Occasionally, it may be necessary to construct a new intake. This volume does not include intake construction costs. If a new intake is required, it should be placed high enough to prevent silt from being deposited on the intake floor. A high intake decreases the distance gates must be lifted and simplifies the task of cleaning the trash racks. Intakes must have trash racks and either slide gates or stop logs (depending on the project design) for shutting off the flow of water. The intake opening is generally rectangular or square. The flow passageway must have a transition constructed to match the shape of the penstock or tunnel to which the intake structure is connected. This passageway requires careful design. The water velocity through the trash racks should be realtively low and then gradually increase to the tunnel or penstock velocity without having the flow subjected to sharply converging surfaces or other features creating turbulence.

The trash racks should be designed for a differential pressure of twenty feet of water. The size of each trash rack is dependent on the size of the lifting facilities that are to be provided. The gate which is used for emergency closure should be placed in the intake structure at a point where the water velociy does not exceed ten feet per second. The gates may have fixed wheels and anti-corrosion, anti-friction bearings. Care must be exercised in the selection of bearings and gate material if the gate is to be stored in a submerged position. Stop logs or a bulkhead gate may be provided upstream of the emergency closure gate for proper maintenance of the emergency closure gate tracks or slots. However, it is not normally possible to position all the stop logs unless the water velocity in the passageway is near zero.

Water Passage. Often, the existing waterway from the impoundment to the downstream channel may be adapted for use in the small hydroelectric installation. The water passageway to the turbine should be as smooth and direct as possible. Any bends should be sufficiently upstream of the turbine to allow streamline flow at the turbine entrance. If a proposed water passageway would appear to deliver water to the turbine in an asymmetrical pattern, it is prudent to make a model study for determining the expected flow conditions (Davis and Sorensen 1969 Section 22). Design modifications can be made as a result of a model study and the model study costs are minor when compared to the potential savings in project costs.

Provisions must be made to permit venting of the tunnel and/or penstock downstream of the emergency closure gate. This prevents high negative pressures in the penstock if the gate is closed with the unit in operation.

Powerhouse. There are generally three main areas in a powerhouse; an area for the turbine/generator, a maintenance or erection area, and a service area.

The main area, housing the turbine/generator, is normally the central area around which the service and erection areas are positioned. In multiple unit hydroelectric installations, the service areas may be either at one end of the plant or grouped around each unit. The arrangement will depend on the site characteristics. A similar determination must be made for the erection areas.

Within the turbine/generator area, it is good practice to have walls at least ten feet from the turbine generator on those sides from which access for maintenance purposes is required. The ceiling heights for any interior area must be carefully coordinated with the height of the equipment to be located or removed during normal maintenance or replacement periods. Methods for removal, and clearances required for the replacement of any part of the main generating unit and its supporting equipment must be given consideration in the powerhouse design.

The area required for the erection area is normally determined by providing an area for each individual part which may be removed during an overhaul period. Vertical clearance requirements should be determined by consideration of not only the turbine/generator equipment but also the switchyard equipment. Depending on the location of the main service facilities, it may at times be necessary to disassemble the main transformer or remove the transformer bushings for removal from the plant area.

Space for service requirements is normally minor in small hydroelectric installations. Frequently, a separate service area building can be constructed at a saving in project cost. Many of the service activities can be accomplished in this separate building, where it is often easier to maintain an acceptable working environment, away from the noise and heat of the turbine/generator. However, some area must always be set aside within the main powerhouse structure for the service equipment required by the generating unit.

Tailrace. A tailrace is necessary for the proper operation of a hydroelectric plant. The purpose of the tailrace is to convey the water leaving the power plant back to the stream channel. Depending on the site characteristics, the tailrace can vary from a short, unlined excavation to a long, concrete-lined channel. The tailrace is also constructed to maintain the water surface elevation at a level specified by the turbine manufacturer. This is usually accomplished by adjustment of the tailrace profile to approximate a weir-type structure. Finally if necessary, the tailrace can be designed to dissipate any excess energy of the water leaving the power plant to prevent erosion of the mainstream banks.

Switchyard. Usually the location of the switchyard is the result of an economic balance between construction costs and operating energy losses. See Volume V for switchyard siting details.

Limitation of Data

The data provided within this volume regarding cost and dimensions was obtained from manufacturers, federal agencies, engineering consultants, and contractors. The data was analyzed and factored to represent reasonable costs to be used for the intended purpose of this volume.

There are many factors which can cause the construction costs to vary, some of which are material availability, labor market, and site conditions. Material shortages and construction site remoteness may increase the costs. Unusual labor market conditions, shortage of skilled craftsmen or low labor productivity, subsistance payments and portal to portal pay may also increase the construction costs. Judgement must enter into the cost analysis process for these types of cost increases. The costs given in Sections 2, 3, and 4 are for a July, 1978 cost level.

SECTION 2
SITE PREPARATION DESIGN AND COST GUIDELINES

General

Site preparation for a small hydroelectric development involves the modification of the existing terrain and results in changes in both the topography (cuts and fills), and in the natural or existing drainage pattern. This section describes the items that need to be considered in the evaluation of the site preparation activity. Both the technical design and the costs are considered.

Drainage and Erosion Control

The construction of a new hydroelectric facility usually involves changes in both the topography and the drainage patterns of the project area, which in turn may result in the accumulation, at specific locations, of excessive surface and/or subsurface water. Removal of the excess water is the main objective of a drainage design. Drainage design varies from project to project, and cannot be generalized as to the best method to be used. However, a combination of proper grading plus a system of collection points (catch basins) is generally the most effective method for removal of surface water. Removal of ground water requires the design of an underground drainage system, which will include a network of subdrains for the collection of subsurface water. The subdrain network would be connected to a main collector or the surface water collection system.

Proper grading should prevent accumulation of water at any location within the project area. However, if water flows over the side slopes of cuts or fills, erosion can become a problem. The effect of the water that flows directly over the slope can be minimized by sodding or terracing. If, because of the nature of the cut or fill, none of these solutions is applicable, it is often possible to divert water by means of a ditch (in cut) or a berm (in fill) along the top of the slope, with a pipe spillway arrangement at specific locations for the discharge of surface runoff over the slope.

The costs for grading, drainage collection and erosion control systems are shown on Figure 2-1. Utilizing the area requiring grading, the construction cost is estimated using the grading curve on Figure 2-1.

The drainage cost for any of the site area or parking area subject to ground water and thus requiring the installation of a network of subsurface drains and catch basins, is estimated using the drainage system from Figure 2-1.

Finally, the construction cost for erosion control is estimated using Figure 2-1. This construction cost is estimated using the erosion control cost curve and the slope area that is being considered. Drainage systems, with erosion control costs are not usually significant cost items on small hydroelectric installations.

Access Roads

Access to the project area is an important feature of project planning, both for construction and for operation. Existing impoundments are provided with access to locations where existing structures are located (i.e., dam, intake, spillway, energy dissipator). When planning a hydroelectric addition at an existing impoundment, use can be made of existing access to serve the new facility if appropriate, or new access can be developed as required. An existing access road may require upgrading before being used for construction access. In either case, since hydroelectric developments often involve the transportation of large and heavy pieces of equipment, certain minimum standards for access roads need to be set. Standards for access roads are given in Table 2-1.

TABLE 2-1
ACCESS ROAD DESIGN STANDARDS

Design Speed	40 mph
Minimum Width	10 feet (one lane)
Maximum Grade	10 percent
Minimum Curve Radius	50 feet
Minimum Sight Distance	400 feet

Bridges on existing roads may be restrictive as to the size and weight of equipment that can be transported across them, and could result in additional handling and equipment assembly costs. Any new bridges which may be required should be designed to adequately accomodate future construction and equipment loads.

The estimated construction cost per mile of new paved access road of single lane width is $125,000, and of two lane width is $250,000. This cost is based on a two-inch asphalt concrete pavement, a four-inch subbase and a four-inch base. A single lane unpaved access road has a construction cost of $75,000 per mile. If an existing road requires upgrading, a cost of $50,000 per mile should be used. For a single span access road bridge, constructed of standard prestressed I girders, use $50 per square foot. This bridge cost includes excavation, substructure on piles and superstructure (Table 2-2).

Parking and Miscellaneous Site Features

Site preparation for small hydroelectric installations involves the design of various related features, such as parking areas, equipment erection area, fencing, and landscaping. Depending on the size of the project, the

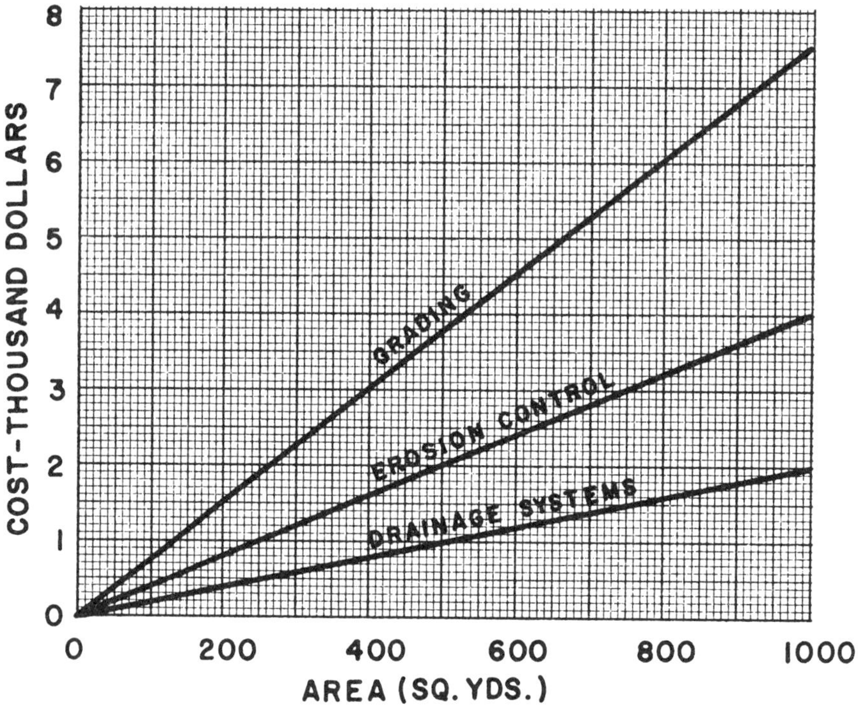

NOTES:

1. Drainage Systems include surface and subsurface systems
2. Erosion Control includes seeding, terracing, dikes, trenches, and pipe spillways
3. Cost Base July 1978
4. Use Site Area for Grading and Drainage
5. Use Area with Slopes Greater than 4:1 for Erosion Control

Figure 2-1. Grading, Drainage and Erosion Control Costs

equipment erection area may be converted into a parking area after all equipment installation is completed. Whether one area serves both purposes, or a different area is assigned for each purpose, the main consideration in the layout of the facility is the relative location of each with respect to the area to be served. The erection area must be located so that the equipment may be moved easily to the installed location. Consideration should also be given to the dimensions of the facility, which will depend on the expected use (number and type of vehicles to be parked, and size of equipment to be erected). The cost for paving the parking and equipment erection area with two inch asphalt concrete pavement and four inch each of base and subbase is approximately $7 per square yard (Table 2-2).

Fencing is provided to protect the project facilities from vandalism and the public from accidents. Normally the cost for fencing is not an important cost item and will not materially influence the final project cost. However, if for some unusual siting conditions it is required to fence off a much larger area than normal, then the cost for the usual chain link type fence eight feet high with a one foot extension arm may be based on a unit cost of $16 per lineal foot. (Table 2-2)

Preserving the natural characteristics of the project area is of importance. Consequently, the area should be landscaped in an attempt to restore the original vegetative condition. The approximate landscaping cost for seeding, planting and fertilizing is $2800 per acre. (Table 2-2)

Environmental Controls During Construction

Environmental problems associated with small hydroelectric projects during construction are generally minor. However, they involve the following types of events:

1. Removal of vegetation, disposal of spoil and change of the land form by grading to provide access roads and level areas for the powerhouse, switchyard and parking areas.
2. Noise and dust created by construction activities including blasting. These disturb recreation areas which may be near the site.
3. Temporary disturbance of the stream caused by building in the streambed, which may result in temporary increase in stream turbidity. Construction may also require an interruption to releases, which could affect aquatic wildlife and downstream users.
4. The long-term commitment of land and part of the streambed for project facilities, thereby preempting use of the area as "wildlife habitat".

Costs associated with mitigation of the above effects are generally insignificant. Damping for dust control, reseeding of vegetation, spacing of blasting to avoid disturbance if recreation areas are nearby, along with other necessary measures, would generally amount to an additional estimated project cost of $10,000 (Table 2-2). Depending on the design of the existing outlet works, cost increases might also result where the releases from the reservoir must be maintained during the construction period.

TABLE 2-2
SITE PREPARATION COSTS
(Cost Base July 1978)

Access Roads		
	Paved Single Lane	$125,000/mile
	Paved Two Lanes	$250,000/mile
	Unpaved Single Lane	$ 75,000/mile
	Single Span Bridge	$ 50/ft^2
Parking and Miscellaneous Site Features		
	Parking Lot Paving	$ 7/yd^2
	Fencing	$16/ft
	Landscaping	$2,800/Acre
Environmental Controls During Construction		
	Noise, Dust and Stream Turbidity Control	$10,000

SECTION 3
HYDRAULIC FACILITIES DESIGN AND COST GUIDELINES

General

Existing impoundments are provided with hydraulic facilities for the normal operation of the project. This may involve water releases for domestic use, irrigation use, fish life, stream flow maintenance or flood control. These hydraulic facilities are normally the following: intake, tunnel or outlet works conduit, valves, energy dissipator, flumes, canals, or penstocks. If a new hydroelectric facility is to be added, additions to or modifications of the existing hydraulic facilities and other related works will be necessary.

Intake

Intake structures at existing impoundments can generally be used for small hydroelectric additions, provided they meet the criteria required for power intakes. Presently, because of new environmental controls, criteria for intakes may have changed from those that were in effect at the time existing impoundments were designed.

For protection of the hydroelectric equipment, there should be the capability to stop the flow of water through the intake structure during emergencies. Most existing impoundments have an emergency closure system, either at the intake itself or at some point along the outlet works conduit under the dam (valve chamber). However, if the existing facility does not have the capability to shut off the flow of water under emergency conditions, an emergency closure system should be included in the design of the hydroelectric facilities. This closure system is usually located just downstream of the intake. The closure device is usually a vertical slidegate that can be remote control operated under a power failure condition; for example, a gravity operated slide gate, or a hydraulic cylinder operated gate provided with an accumulator system.

There should always be a minimum of two closure devices upstream of the hydraulic turbine. One would be the intake gate and the second is usually a valve on the turbine inlet. The cost of the turbine valve is given in Volume V where it is included in the turbine cost. The cost of a slide gate may be estimated at $20,000. (Table 3-1)

TABLE 3-1
MISCELLANEOUS COSTS
(Cost Base July 1978)

Cost Item	Cost
Slide Gate	$20,000
Tunnel, 7 ft. Diameter Steel Lined	$ 1,300 pl ft.
Penstock Bifurcations	
Flow Less Than 200 cfs	$ 5,000
Flow 200 cfs to 600 cfs	$10,000
Flow Over 600 cfs	$20,000

Tunnels and Penstocks

Many types of pressurized tunnels and conduit systems have been developed and used on various hydroelectric applications.

The design and preliminary costs of these facilities are discussed below. Most existing impoundments already have a water conveyance facility in service. If possible, use of the existing tunnel or conduit for the proposed power generation facility should be made. The cost of a new tunnel or conduit will often make a proposed project infeasible. However, new conduit facilities are often required.

If a new tunnel is determined to be necessary, a cost of $1300 per linear foot (Table 3-1) may be assumed for a feasibility cost assessment. This cost is for a seven foot diameter steel-lined tunnel. As this is the minimum diameter that can be achieved with standard boring equipment, no cost savings can be realized by specifying a smaller tunnel.

Penstocks are pressurized, low-friction water conveyance conduits which carry water from the lower end of the existing impoundment outlet works or the tunnel exit portal to the powerhouse. Penstock design is a complicated process involving aspects of economics, turbine regulation requirements, plant siting and materials (Bier, 1966). These items will be presented briefly in this volume in order to permit a better understanding of the costs presented, in case modifications are required. A single figure to estimate penstock costs is presented at the end of the following discussion. Penstocks can be constructed of wood, concrete or steel. Steel is usually the preferred material and costs will be given for only this material. Penstock design must consider the stresses caused by internal pressure (static head plus water hammer), external pressure, temperature, erection and installation.

The elevation of the hydraulic turbine with respect to the surface elevation of the impoundment determines the static head in the penstock. Additionally, a pressure wave, which is termed water hammer, is produced whenever there is an increase or decrease in the penstock velocity (Davis and Sorensen, 1969, Section 27). Water hammer adds to the internal pressure on the penstock. Minimizing the additional head due to water hammer is a design consideration. Using a penstock design velocity of ten feet per second for a small hydroelectric installation rather than the higher velocities often used, will minimize the water hammer effects. One major cause of water hammer is the action of the turbine wicket gates, which can operate over their full range in a matter of seconds, thereby stopping the flow and causing large water hammer effects. The closure time may be increased so that a significant water hammer is not produced. A turbine shutoff valve normally requires minutes for its operating cycle and may not produce a significant water hammer. On hydroelectric installations with short penstocks, the water hammer is often not a major design consideration as it is on hydroelectric projects having long penstocks (Davis and Sorensen, 1969, Section 28). A surge chamber may be constructed to reduce the effects of water hammer. For those projects having long tunnels and/or penstocks, a surge chamber offers other advantages which are discussed later in this Section.

External pressure exists on a penstock whenever it is buried or the internal pressure is below atmospheric pressure. If a penstock is placed in a tunnel and backfilled with concrete, then the design must include stiffener rings or special internal supports during the erection phase. For the penstocks buried in earth, external reinforcing rings are usually required. Even for penstocks placed above ground, it is often required to add external stiffener rings if there is any possibility, including operator error, of having the internal pressure substantially below atmospheric pressure.

A penstock held firmly at each end or bend point will have longitudinal stresses caused by temperature changes. Penstock design must include this consideration or have expansion joints provided. The maximum temperature differential generally occurs when the penstock is unwatered.

Beam stresses occur in the penstock pipe whenever it is placed on supports. These stresses are a function of the distance between supports.This distance is often assumed to be fifty feet for preliminary design then modified in final design. At the support points, it is generally necessary to locally reinforce the penstock, especially when it is relatively thin, as is often the case for small hydroelectric installations.

The final consideration in the penstock design is the stresses caused during the handling and erection phase. A minimum handling thickness, as a function of penstock diameter, is required and is usually the governing factor in penstock design for small hydroelectric power plants as defined in this Volume. Minimum handling thicknesses, over a range of penstock diameters, are shown below in Table 3-2.

Figure 3-1 shows the installed penstock costs for a small hydroelectric installation. The costs include fabrication and erection, including supports, for a steel penstock. The costs are based on a penstock velocity of ten feet per second and a pipe thickness equal to the minimum thickness required for handling. A corrosion allowance, that is an additional thickness, has not been used, as it is normal to coat the interior of the penstock,

TABLE 3-2
Steel Penstock Minimum Thickness

Thickness Inches	Maximum Diameter Inches
0.125	30
0.1875	55
0.25	80
0.3125	105
0.375	130
0.4375	155
0.5	180

usually with coal tar enamel, to decrease the hydraulic friction losses and the effects of corrosion.

If the static head on the penstock is greater than 225 feet, it is possible that the minimum handling thickness will no longer govern the penstock design and the penstock costs will be higher than given in Figure 3-1.

Figure 3-1 costs are based on a relatively flat penstock gradient. If the penstock is placed on a gradient of more than fifteen degrees with respect to the horizontal, then corrections must be made as noted on Figure 3-1.

The unit cost of the penstock is determined from the Figure 3-1 cost curve and the rated turbine flow obtained from Volume V. The total overall penstock cost is the product of the penstock length times the unit cost.

Final penstock design may indicate there could be an overall saving in owning and operating costs by increasing the penstock velocity above the value of 10 feet per second used for Figure 3-1. It is not necessary to consider this possibility in a feasibility analysis.

One specialized type of penstock should be mentioned. Many existing dams have concrete outlet conduits which are designed for unpressurized conditions. When downstream controls are provided, conduit pressures will increase accordingly and measures must be taken to protect the existing outlet conduit structure against failure under the higher pressures. The most effective and least costly method to increase the pressure capacity of the existing outlet structure is to install a relatively thin, welded steel liner inside the existing concrete outlet conduit. This liner may be fabricated in place when the tunnel is in an unwatered condition. The annulus between the liner and the existing concrete conduit is then backfilled with concrete. The cost for this type of modification may be estimated from Figure 3-1, using the unit cost increased by fifty percent.

Valves, Gates, Outlet Works and Other Hydraulic Works

Depending on the project configuration selected, various other hydraulic equipment will be needed for the operation of a small hydroelectric installation. The nature and cost of this additional equipment is presented below.

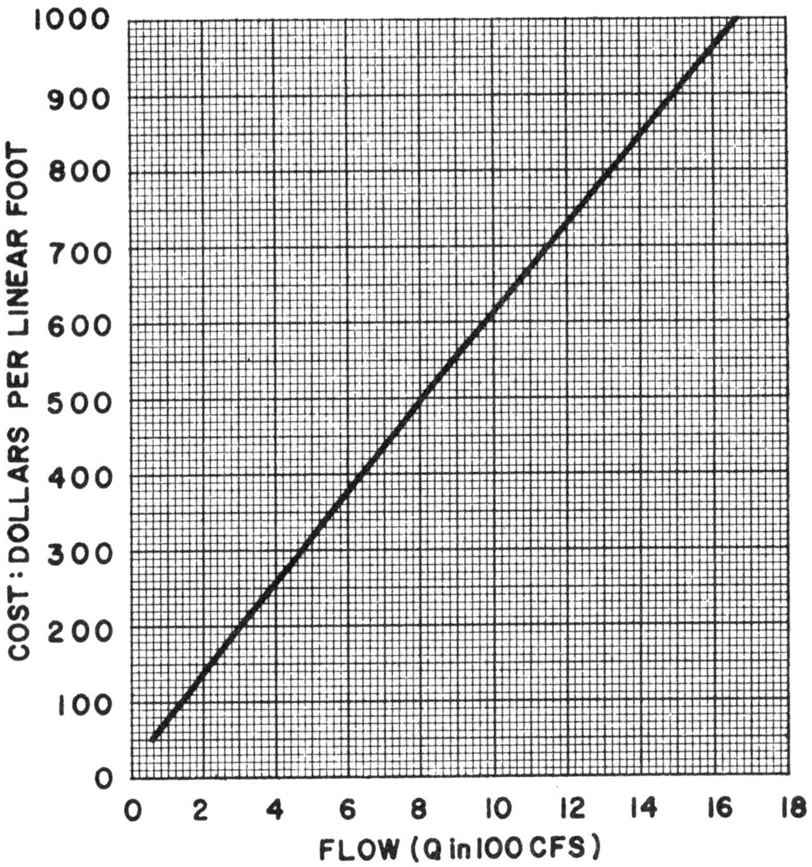

NOTES:

1. Cost Base July 1978
2. Penstock Velocity 10 feet per second
3. Steel Cost at $1.10 per pound
4. Based on Minimum Handling Thickness and Maximum Head of 225 feet
5. If Penstock Gradient is over 15°, add 1% of total for each degree over 15°
6. Valves and Bifurcations not included

Figure 3-1. Installed Penstock Costs

Energy dissipation valves such as the "Howell-Bunger" and the hollow jet are typically used to bypass water when the power house is inoperative. The Howell-Bunger valve is a fixed cone, movable cylinder type valve. Due to lack of streamlining of the cylinder and cone the Howell-Bunger is excessively noisy but serves as an effective energy dissipater. The hollow-jet valve, developed by the United States Bureau of Reclamation, is essentially a movable needle valve with streamlined needle and housing and is thus much quieter. Both valves are available for conduit sizes up to 96 inches. Costs for either valve may be estimated from Figure 3-2.

Butterfly valves up to 12 feet in diameter are used almost exclusively to open or close flow into the turbine spiral case. Butterfly valves are not normally used as flow control valves due to the stresses and flow patterns imposed when they are partially open. It is not necessary to include a butterfly valve in the cost estimate, as the turbine inlet valve is included in the turbine cost (see Volume V). However, in accordance with the requirement that two closure devices are needed on the conduit upstream of the turbine, a butterfly valve may occasionally be required. The cost of a butterfly valve may be estimated from Figure 3-2.

As discussed earlier in this Section, the rapid closure of either turbine valve or turbine gates may cause a water hammer and increase the penstock water pressure which under extreme conditions can cause the conduit to rupture. A surge chamber, placed either at the upper end of the penstock or near the tunnel outlet, reduces the effects of the water hammer. Effects of water hammer may also be minimized by the use of a pressure relief valve connected to the turbine spiral case. If there is a relatively long tunnel ahead of the penstock, a surge chamber at the upstream end of the penstock may also assist in supplying water to the turbine during start-up. Normally, however, a small hydroelectric installation will have a relatively short tunnel and penstock, and a surge chamber will not be required. No costs are presented for either the surge chamber or the pressure relief valve as they must be specially designed for each site, and are seldom used for small hydroelectric projects as defined by this volume.

A bifurcation splits a single flow conduit into a pair of conduits. A bifurcation is used if a single penstock conveys water to two turbines or if a bypass must be provided off the main penstock. Multiple bifurcations are often used. Bifurcation costs are usually estimated by calculating the weight of the bifurcation and multiplying by a cost per pound for steel. For preliminary estimates, however, an approximate cost of $5,000 to $20,000 per bifurcation is used. (Table 3-1)

There are no other facilities for hydraulic conveyance that need to be considered in a feasibility assessment.

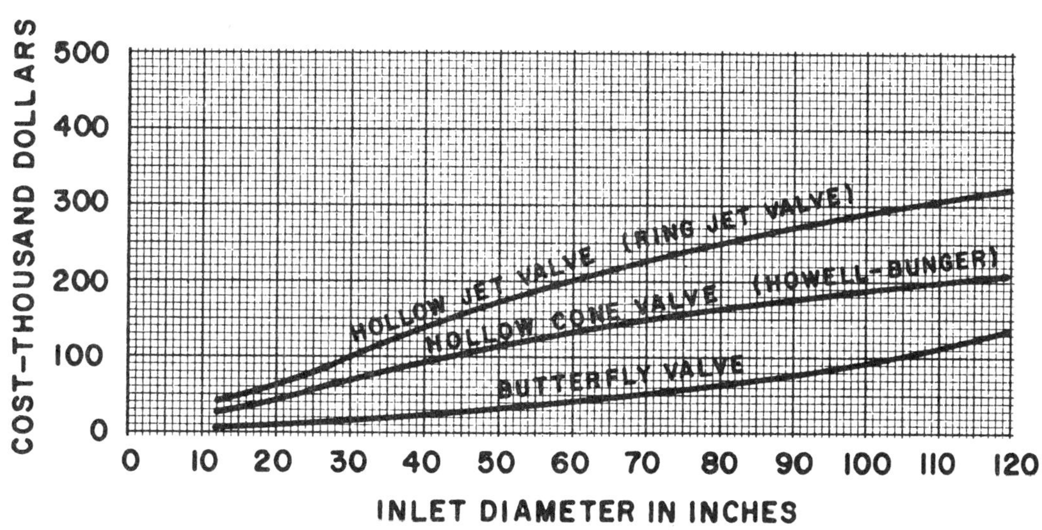

NOTES:
1. Cost Base July 1978
2. Maximum Head 100 ft.

Figure 3-2. Costs for Butterfly, Hollow Jet, and Hollow Cone Valves

SECTION 4
POWERHOUSE AND APPURTENANT FACILITIES SELECTION AND COST GUIDELINES

General

The distinguishing feature between indoor, semi-outdoor, and outdoor hydroelectric plants is basically the type of weather protection afforded the generator and erection area. The main equipment items which may be placed either indoor or outdoors would be the generator, generator breaker, power transformer and crane. Cost data is presented for each turbine configuration.

Indoor Plant. An indoor plant has the erection area and the main equipment items, with the possible exception of the power transformer, within an enclosed building structure. Most small power transformers are air-cooled. Placing the transformer inside a building not only increases the fire hazard but also increases the demands on the cooling system unless an unusually large, well-ventilated area is provided. Consequently, on most contemporary small hydroelectric installation designs, the power transformer is seldom placed indoors, even for an indoor type plant.

With an indoor type plant it is necessary to furnish a bridge-type crane for handling the generator rotor, because portable cranes of the size required are generally not adaptable to indoor plant use. Also, indoor powerhouses can be readily adapted for air-cooling of the generator whereas air-cooling of a generator can be difficult in an outdoor type of plant.

Semi-Outdoor Plant. A semi-outdoor plant has the main generating unit fully enclosed by a building structure. The main lifting equipment, generally a gantry type crane, is located on the powerhouse enclosure roof and the equipment is handled through hatches. Generally the erection area is outdoors. This type of installation is not commonly used and therefore, no costs are presented.

Outdoor Plant. An outdoor plant has a weather-proof housing over the generator with water-cooling coils located within this generator housing. The erection area is outdoors and, accordingly, any major overhaul work requiring dismantling of the generator can only be done during dry weather unless portable shelters are provided. The power transformer is outdoors. Depending on the relative location of the generator with respect to the switchyard, the generator breaker may also be outdoors. If a permanent crane is required, it will probably be of the traveling gantry type. However, portable or mobile cranes can be used to an advantage on outdoor-type small hydroelectric plants.

Location and Setting

The site selected for the hydroelectric installation should be one that maximizes the potential power and minimizes project costs. The maximum power is developed by decreasing the length of water conduit to the turbine while still obtaining the highest vertical fall in the water.

The stream channel conditions downstream of the dam, the accessibility of the site for construction and future maintenance, the foundation conditions and location of the impoundment spillway must all be considered in the site selection. To a lesser degree, the available area for the switchyard enters into the selection of the powerhouse site.

The civil features having the highest cost will be the powerhouse structure, including the excavation, and the waterways (penstock, valves, gates and outlet works). The structure cost is a function of the type of turbine, physical size of the turbine and type of plant (indoor or outdoor). If there is a possibility that more than one type of turbine may have to be considered for economic feasibility, as an example, a Francis turbine or a tube turbine, then estimated project costs may have to be made using each type of turbine. However, a practical approach for a site having a multiple choice of turbines would be to assume one particular type of turbine and, if the site is feasible, the final selection of turbine type can be made during the initial final design period.

The powerhouse civil construction costs may be determined using the cost curves included in this Section. Selection of the turbine best suited to a particular site should be made on the basis of the information presented in Volume V. The curves present the cost as a function of either a principal turbine dimension or the turbine generator nameplate rating. The civil costs are for an outdoor type of plant unless otherwise noted in the following sections describing the turbine and its characteristics. The civil construction costs include the construction of a reinforced concrete powerhouse structure complete with all miscellaneous steel work, and all other civil features required. These cost curve figures, in addition, indicate the area required by the powerhouse structure. This area is used, by following the procedure given in the Excavation Cost Section, in determining the powerhouse excavation costs.

The costs and areas shown in the figures are for single unit powerhouses. For multiple unit installations, these costs should be multiplied by the number of units.

The setting of the actual elevation of the turbine is made based on data given in Volume V. The turbine elevation determines the depth of excavation required for the powerhouse structure.

Structures for Alternate Turbine/Generator Configurations

Tube Turbine. A tube turbine can be efficiently located to become part of the existing outlet works and/or to be adjacent to the existing impoundment. This type is easily adapted to a canal installation. Normally, the generator will be housed within a building. However, it is feasible to have the major erection or overhaul area outdoors. Refer to Volume V for the characteristics of this type turbine. Refer to Figure 4-2 for the civil costs for tube turbine powerhouses. These costs are based on an indoor type plant. Figure 4-1 shows the installation of a tube turbine on a canal drop.

Bulb Turbine and Rim Turbine. The possible configurations for either the bulb or rim turbine are similar to those that are appropriate for the tube turbine. As the turbine and generator for the bulb-type unit are in the water passage, the enclosed structure above the unit is relatively small, unless the erection and maintenance areas are enclosed. Normally, for units less than five MW capacity, these types are not as economical as the tube type, despite the smaller powerhouse. Refer to Volume V for characteristics of this type of turbine. Refer to Figure 4-3 for the civil costs, which are based on an outdoor type plant. Figure 4-4 shows the installation of a bulb turbine.

Fixed and Moveable Blade Propeller Turbine. The propeller turbine can be efficiently located to become part of the existing outlet works and/or to be adjacent to the impoundment. As with tube turbine, propeller turbine installations can be easily adapted to canal drop sites.

A propeller turbine is adaptable to either an indoor or outdoor installation. Refer to Volume V for the characteristics of this type of turbine. Refer to Figure 4-5 for the civil costs, which are based on an outdoor type plant.

Figure 4-1. Tube Turbine Installation, Allis-Chalmers, 420 kW, 16.5 ft. head, 300 cfs, Turnip Power Plant (Courtesy Imperial Irrigation District, Imperial, California).

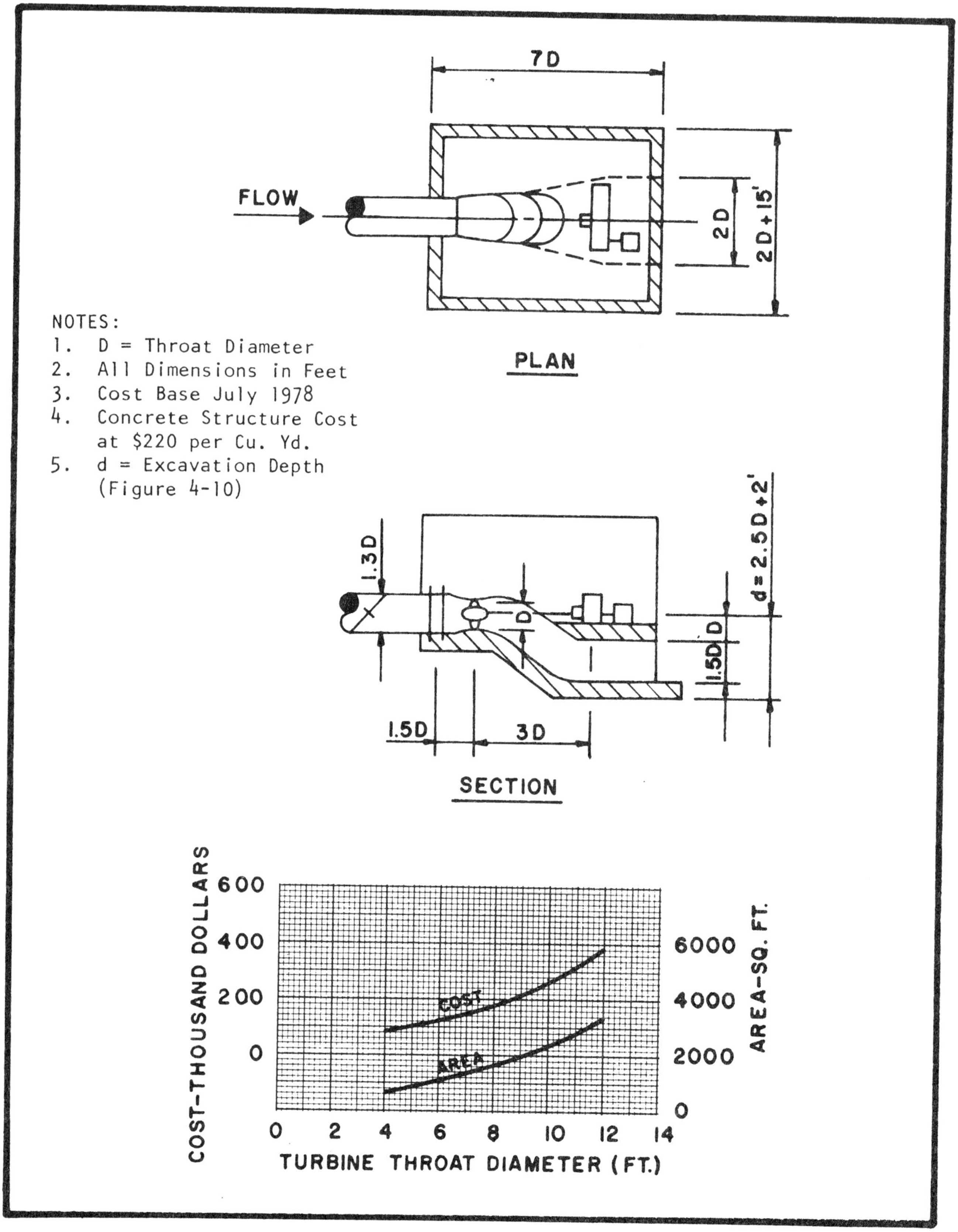

Figure 4-2. Tube Turbine Powerhouse Civil Cost and Area

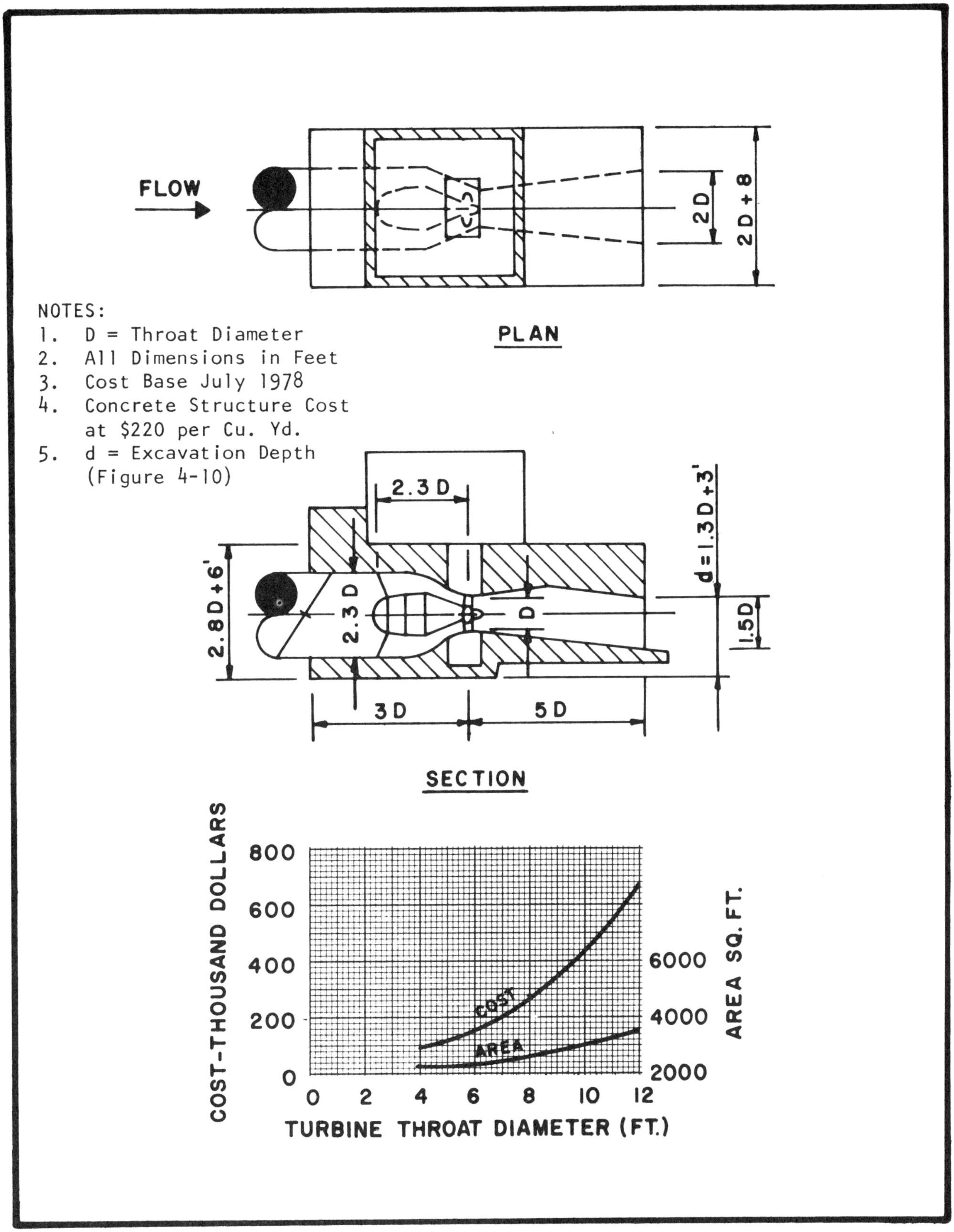

Figure 4-3. Bulb Turbine and Rim Turbine Powerhouse Civil Cost and Area

Francis Turbine. Flow into a Francis turbine is normally conveyed through a penstock. An area must be available downstream from the impoundment to accomodate the larger site requirements of a Francis turbine. This type of turbine can be used either in an indoor or outdoor plant, depending on site conditions. The usual installation is one having a vertical turbine/generator shaft. Refer to Volume V for characteristics of this type of turbine. Refer to Figure 4-6 for the civil costs, which are based on an outdoor type plant.

For very small turbines, those having throat diameters less than 48 inches, there may be a cost advantage in using a Francis type with a horizontal shaft. The arrangement of penstock, discharge and generator can be simpler than those for a vertical shaft unit. Refer to Volume V for the characteristics of this type of machine. Refer to Figure 4-7 for the civil costs which are based on an indoor type plant.

Cross Flow Turbine. The Cross Flow turbine can be used for either a penstock or flume installation. Normally this type of unit is placed indoors. The required erection and maintenance area is minimal. Refer to Volume V for the characteristics of this type of turbine. Refer to Figure 4-8 for the civil costs, which are based on an indoor type plant.

Propeller Turbine - Flume Configuration. A propeller turbine may be used in a flume or canal at an existing drop or vertical discontinuity in the flume or canal. This configuration may be used for either an indoor or outdoor type of plant, depending on the site conditions. Penstocks are not used with this type of configuration. Refer to Volume V for characteristics of this configuration. Refer to Figure 4-9 for the civil costs, which are based on an outdoor type plant.

Impulse Turbine. The impulse turbine wheel is limited in its use on low head, small hydroelectric installations. Refer to Volume V. No civil costs are given for impulse turbine installations.

Figure 4-4. Bulb Turbine Kleinmunchen Power Plant, Austria (Courtesy VOEST-ALPINE AG).

Excavation and Foundation for Powerhouse

Excavation. At small hydroelectric sites, excavation is necessary to correctly set the turbine elevation with respect to the tailwater elevation. The method for determination of the turbine setting elevation is given in Volume V.

The excavation cost may be approximated as a function of the powerhouse area and the maximum depth of excavation. Figure 4-10 shows the relationship of the total cost of excavation to the powerhouse area and the maximum excavation depth. This cost curve for small hydroelectric installations was developed using the following assumptions:

1. Excavation would be done to full depth to a distance of five feet outside the powerhouse perimeter and all side slopes would be on a 45 degree angle.

2. The total volume of excavation would be one-half common excavation and one-half rock excavation.

3. The unit excavation costs assumed were two dollars per cubic yard for common excavation and ten dollars per cubic yard for rock excavation. These can be typical unit costs where the construction haul is normal.

The approximate powerhouse area requirements for each type of turbine are indicated in Figures 4-1, 4-3, and 4-5 to 4-9 inclusive. These figures also show the depth required by the turbine in terms of a turbine parameter. Knowing the turbine size or parameter from Volume V, and the powerhouse area and depth from the above figures, use Figure 4-10 to determine the powerhouse excavation cost.

It is required that the powerhouse structure be placed on sound material in order to develop full resistance to shearing and sliding. Any weathered material and material shattered by blasting must be removed prior to concrete placement. To insure proper foundation conditions, it may be necessary to excavate to a depth greater than that indicated by Figures 4-1, 4-3 and 4-5 to 4-7. Often it is necessary to make a sample boring at a proposed site to determine the below grade foundation conditions. An unusual condition might be cause to select a

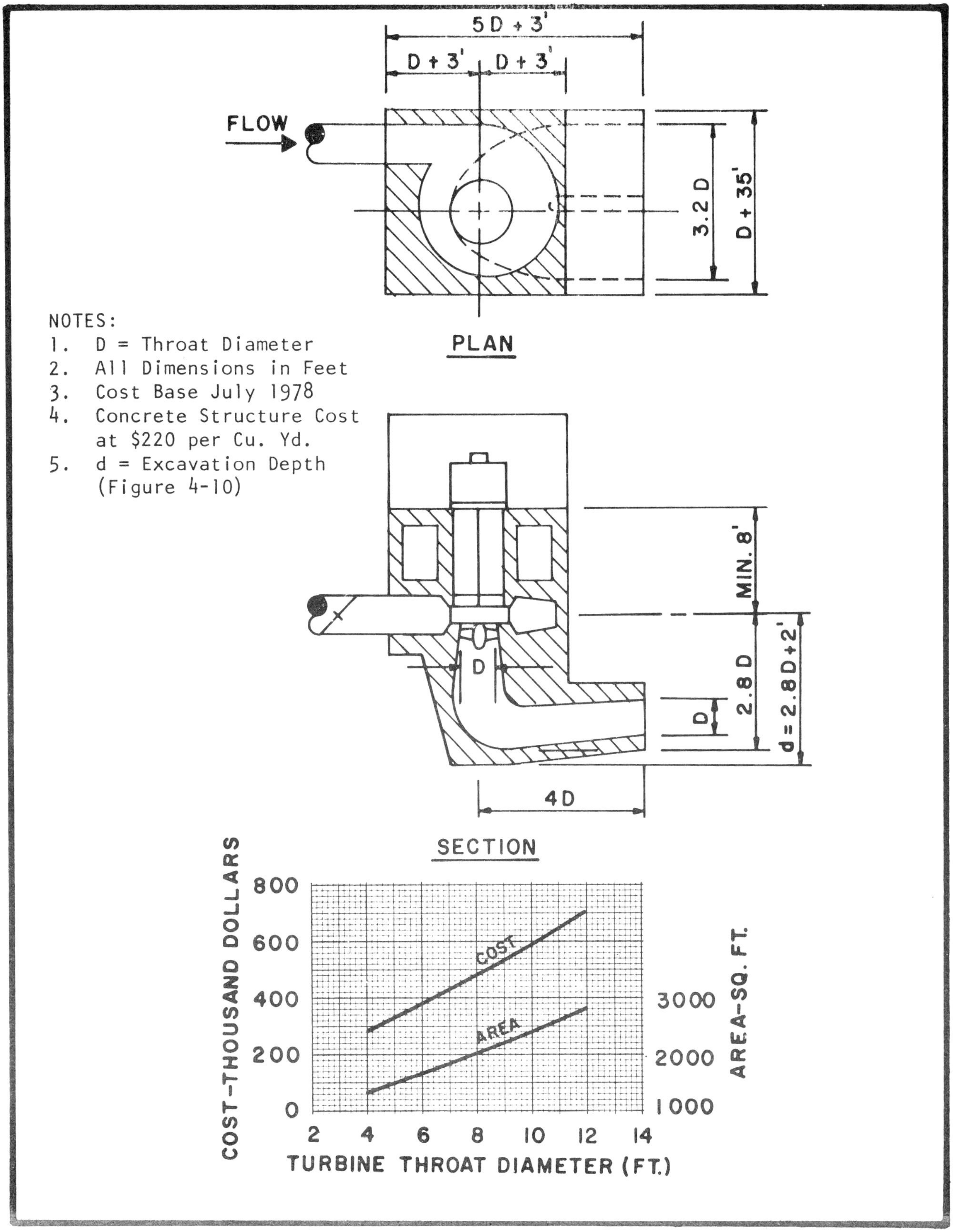

Figure 4-5. Propeller Turbine Powerhouse Civil Cost and Area

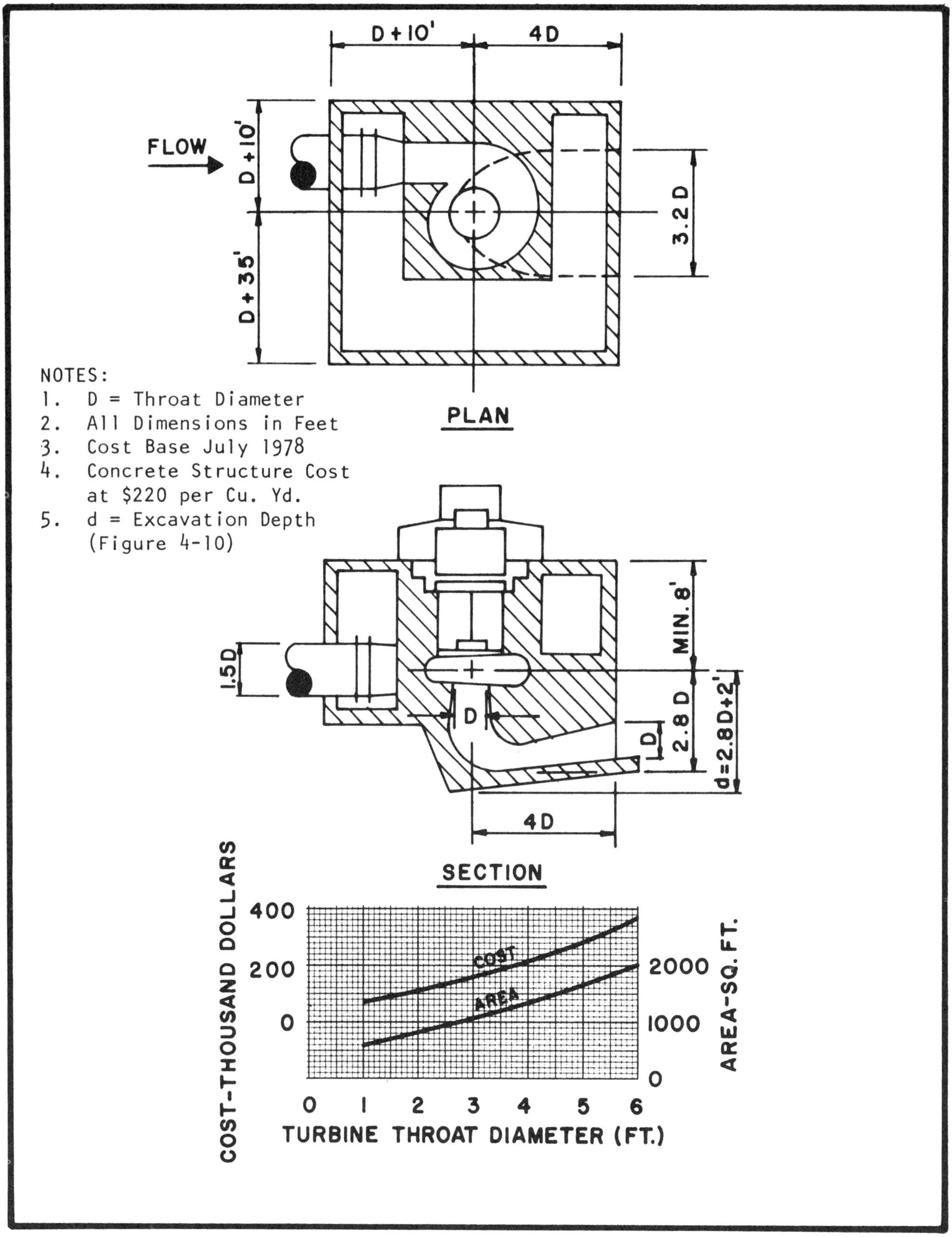

Figure 4-6. Francis Turbine Powerhouse Civil Cost and Area

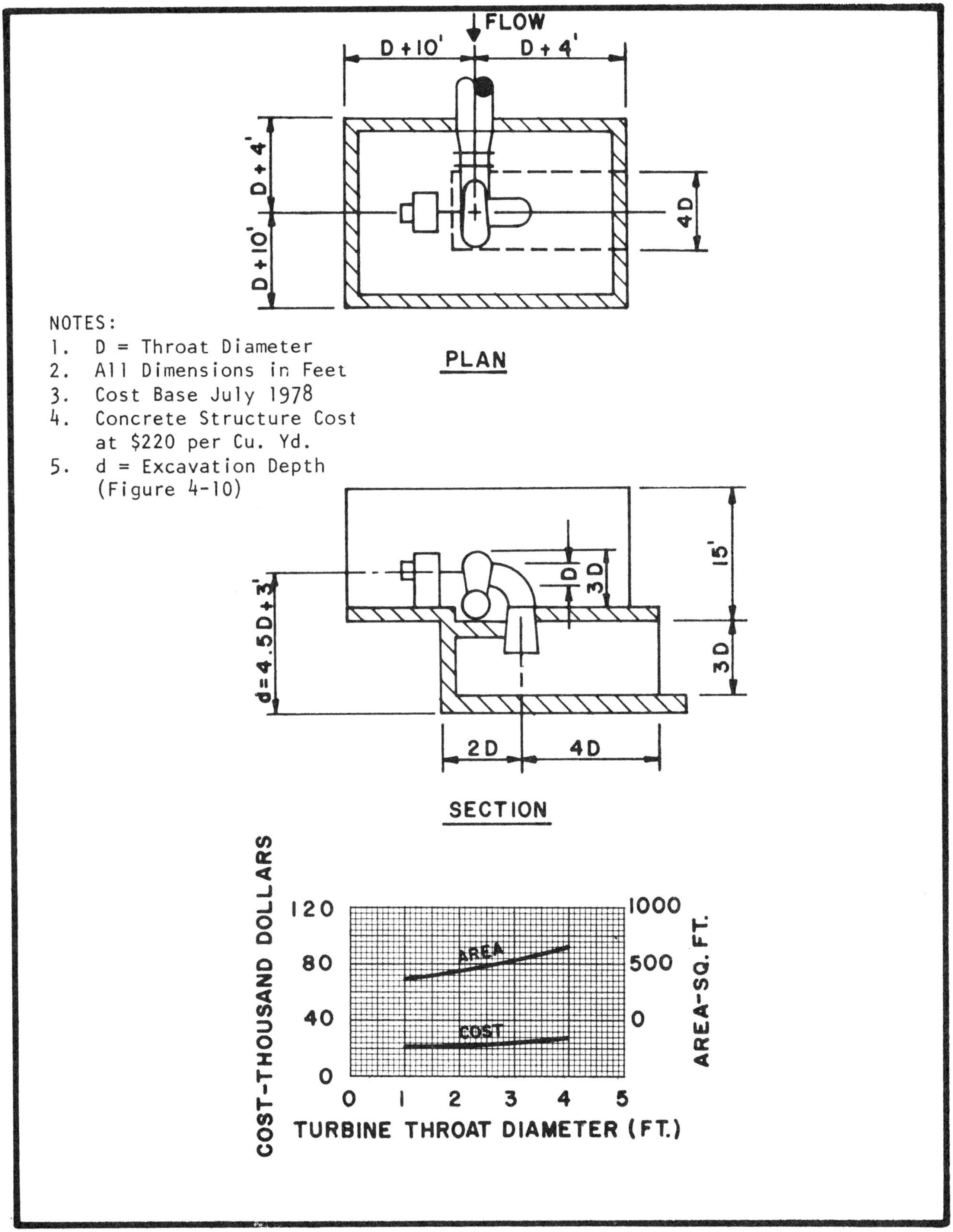

Figure 4-7. Horizontal Francis Turbine Powerhouse Civil Cost and Area

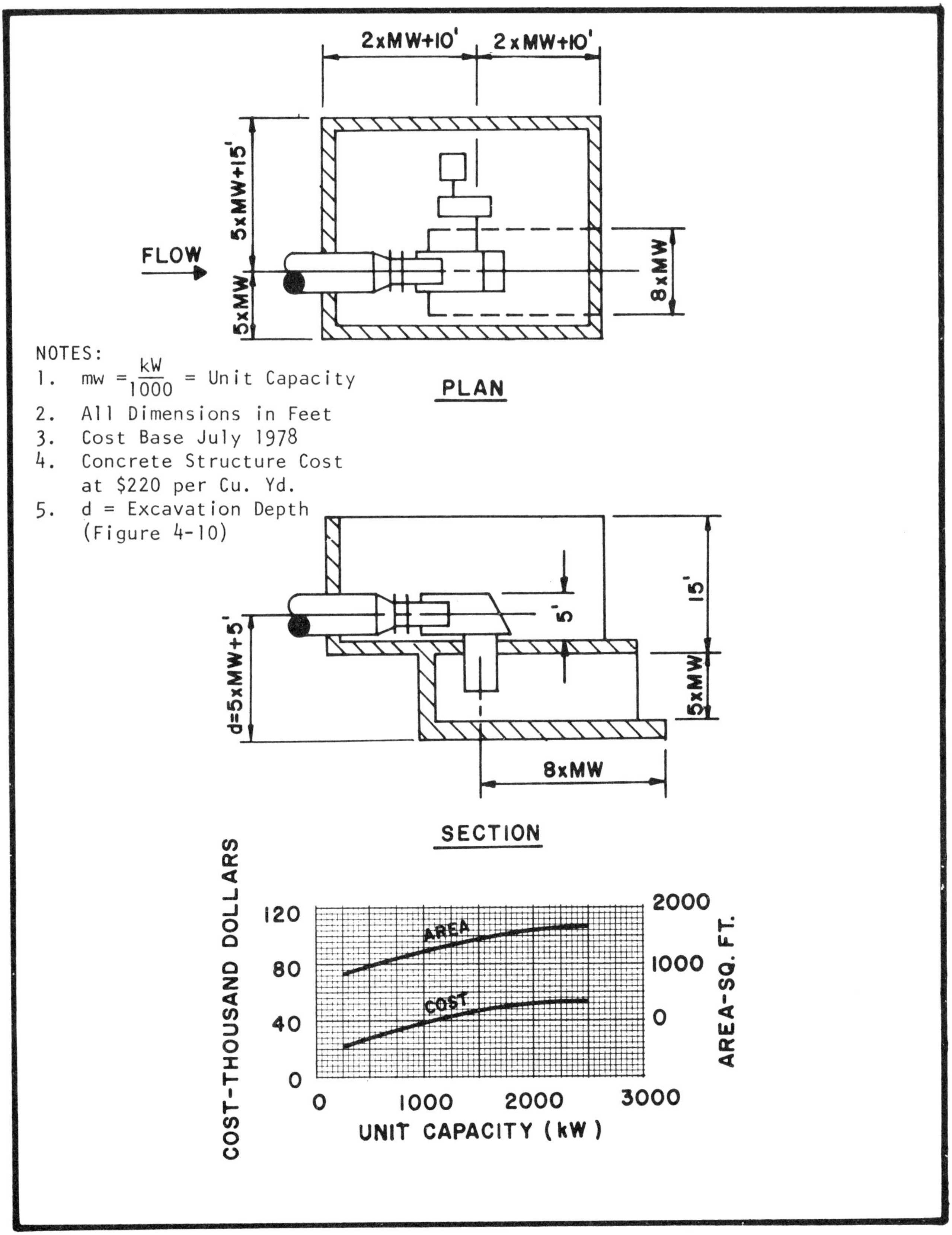

Figure 4-8. Cross Flow Turbine Powerhouse Civil Cost and Area

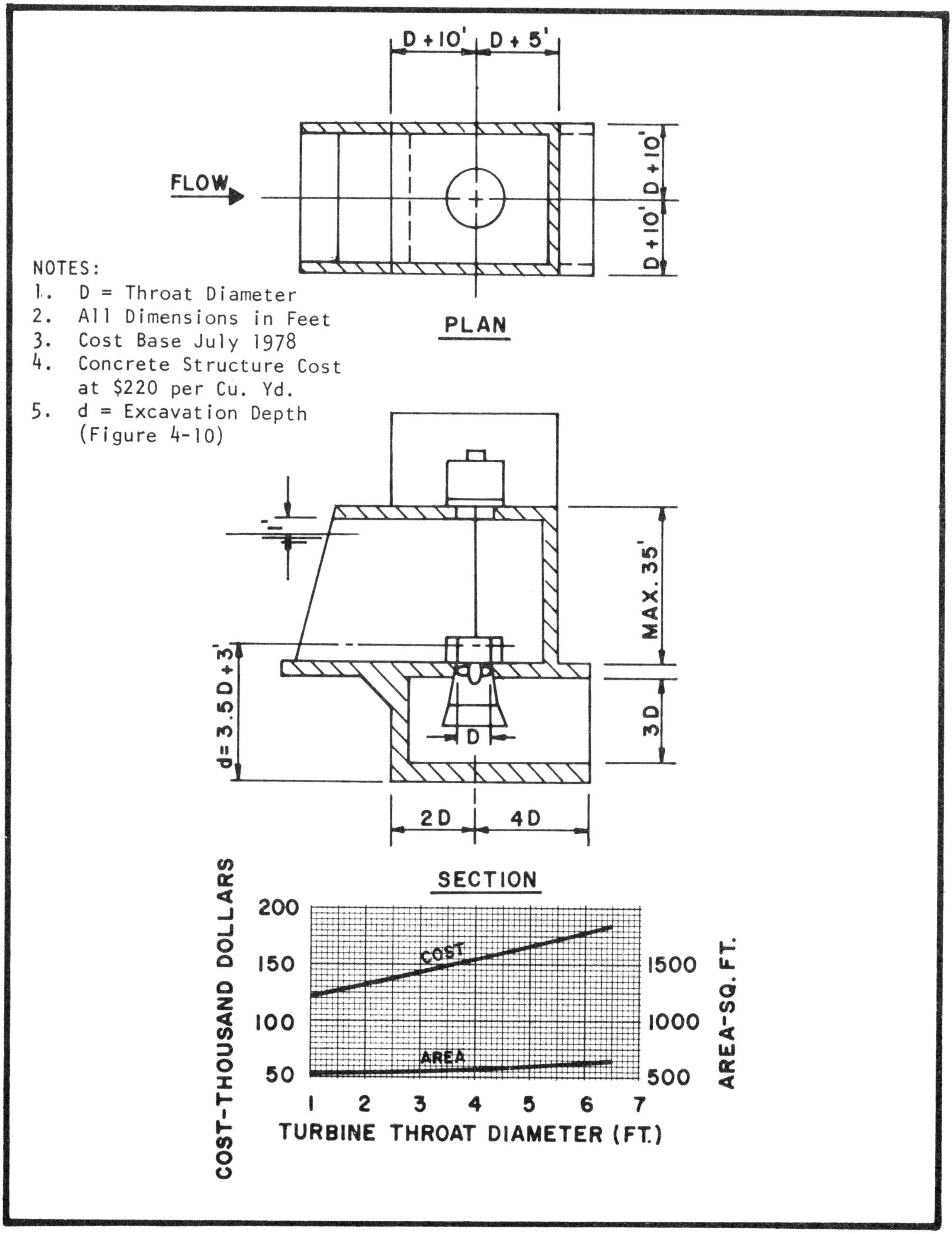

Figure 4-9. Flume Type Installation Powerhouse Civil Cost and Area

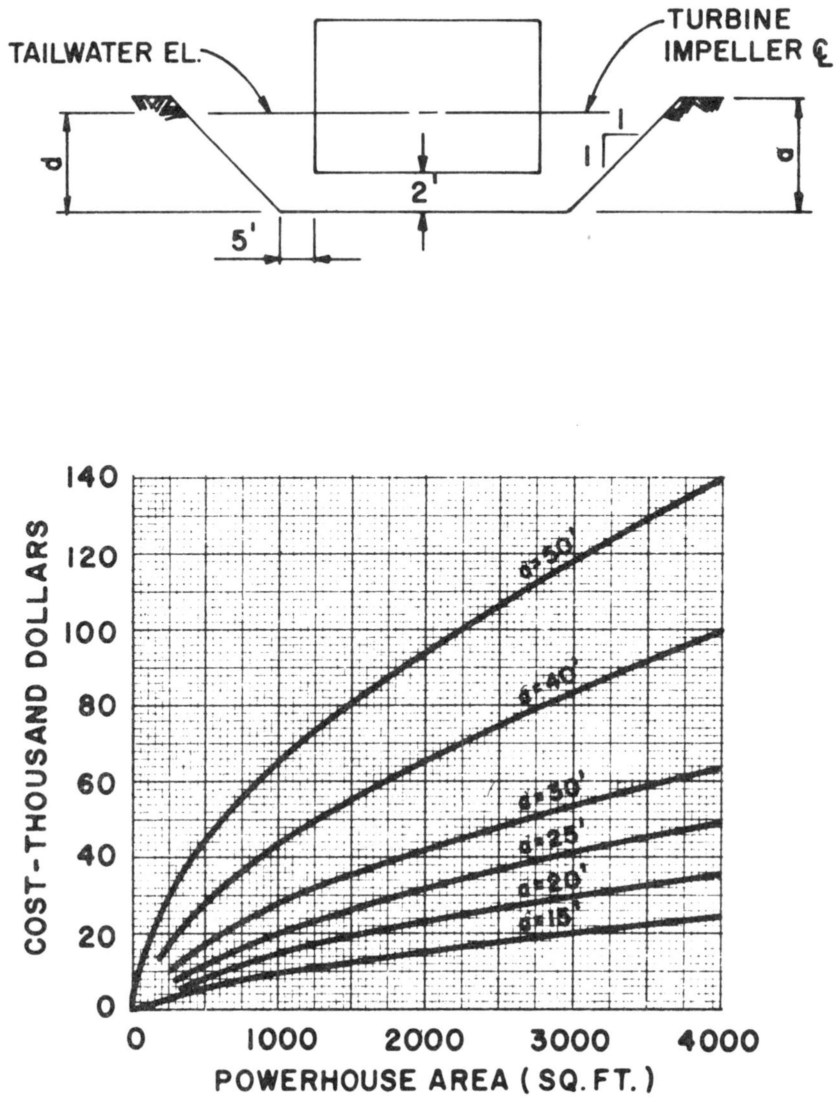

NOTES:

1. Cost Base July 1978
2. Excavation Costs: 50% common at $2 per Cu. Yd. and 50% rock at $10 per Cu. Yd.
3. The absolute value of "a", depth of excavation, is generally more than the value of "d" shown in Figures 4-1 to 4-7, as the original ground elevation is usually higher than the design tailwater elevation.

Figure 4-10. Powerhouse Excavation Costs

deeper depth curve on Figure 4-10 than that indicated by the turbine parameter. However, it is not necessary for a reconnaissance evaluation to assess this possibility. A feasibility evaluation will have to evaluate the requirements of using a deeper depth curve.

Some sites may require the construction of a cofferdam to protect the construction site. This protection may be either in the form of sheet piling or a dike and rip-rap. Construction dewatering and care and handling of the stream facilities normally will be required for the excavated area. As these costs are unique to the site and soil conditions, the costs for flood protection and dewatering facilities are not included in Figure 4-10. An evaluation should be made on a site specific basis for the costs for these items. Protection of the construction site could total ten percent of the total civil features cost.

Foundation and Stability. The cost of the powerhouse foundation should be considered. It is difficult to accurately estimate the extent and cost of required foundation work without some detailed soil information. On the basis that the reconnaissance assessment will be made without the benefit of a soils report, some allowance should be made in the total estimated cost for possible additional foundation work. This additional work would primarily include cut-off walls and drain systems.

For foundation assessments, there are three basic types of power plant sites, each with different foundation requirements and associated costs. The first type of site has the power plant in, or as a part of, an existing structure. With this type of site, there would be little or no foundation work required, and therefore no additional foundation costs. The second type of site has the power plant as a part of a new water retaining structure or dam. With this installation, there is a head difference across the structure which presents the potential for subsurface flows below the structure, causing uplift and a possible overturning moment. The additional cost for added excavation or structural work can be substantial, and would vary considerably with the specific site conditions. A nominal cost of $100,000 (Table 4-1) may be assumed for this case. The third type of installation has the power plant separate from the dam, with the water conveyed to it through an enclosed conveyance, so that water pressures are not a problem. Although there will be some foundation costs for this type of installation, depending on the soil types and the site topography, the cost would normally be much less than for the previous case. A nominal cost of $20,000 (Table 4-1) may be assumed.

TABLE 4-1
Foundation Costs
(Cost Base July 1978)

Type of Site	Cost
Addition to Existing Structure	$ 0
New Structure Below Impoundment	$ 20,000
New Structure Acting as Impoundment	$100,000

Figure 4-11 shows various methods of stabilizing the powerhouse structure. There must be enough mass in the powerhouse structure and its contents that will prohibit the structure from floating if there is a hydraulic uplift. If sufficient mass is not available then lips are provided at the base of the structure which effectively allows an earth mass to be added to the powerhouse mass weight for overcoming floatation. Adequate drainage around the powerhouse subgrade will decrease the hydraulic uplift. However, provisions may have to be provided to offset the tendency of the tail water to produce a hydraulic uplift. Full penstock pressure against the closed turbine valve or wicket gates will produce a hydraulic thrust which produces an overturning moment on the powerhouse structure. The overturning moment must be resisted by either deadweight mass, anchors, increasing the size of the powerhouse base or any combination of these features. The hydraulic thrust has to be resisted by either a shear key or anchor bolts.

Tailrace Improvements and Costs. The main function of the tailrace is to maintain a minimum tailwater elevation below the power plant and to keep the draft tube submerged. All turbines, with the exception of the impulse turbine, require that the tailwater be maintained above a minimum elevation to minimize the effects of cavitation. It is also important to keep the draft tube submerged, even when there is no flow in the downstream channel or tailrace, in order to improve the turbine start up conditions. This is normally accomplished by excavating the channel immediately downstream of the power plant to maintain a pool of sufficient depth to keep the draft tube covered, and by including a control structure, such as a weir, to maintain the pool at a minimum elevation. Also, a section of new channel might be necessary to connect the new installation with the existing stream channel.

The major portion of the tailrace cost is in the cost for the required excavation, with some additional cost for concrete channel lining, a concrete sill or weir, and riprap. The amount of excavation required depends on the elevation of the turbine spiral case and the length, width and depth of the channel required to return the plant discharge to the existing stream channel. The cost for the tailrace is predominately proportional to the tailrace length, but there is also a fixed cost to cover the excavation immediately downstream of the draft tube. The tailrace cost may be estimated as $15,000 plus $200 per linear foot. (Table 4-2)

Switchyards. Economic studies are usually required to determine the location of the power transformer, circuit breaker and other items of electrical equipment specified in Volume V which may be placed in the

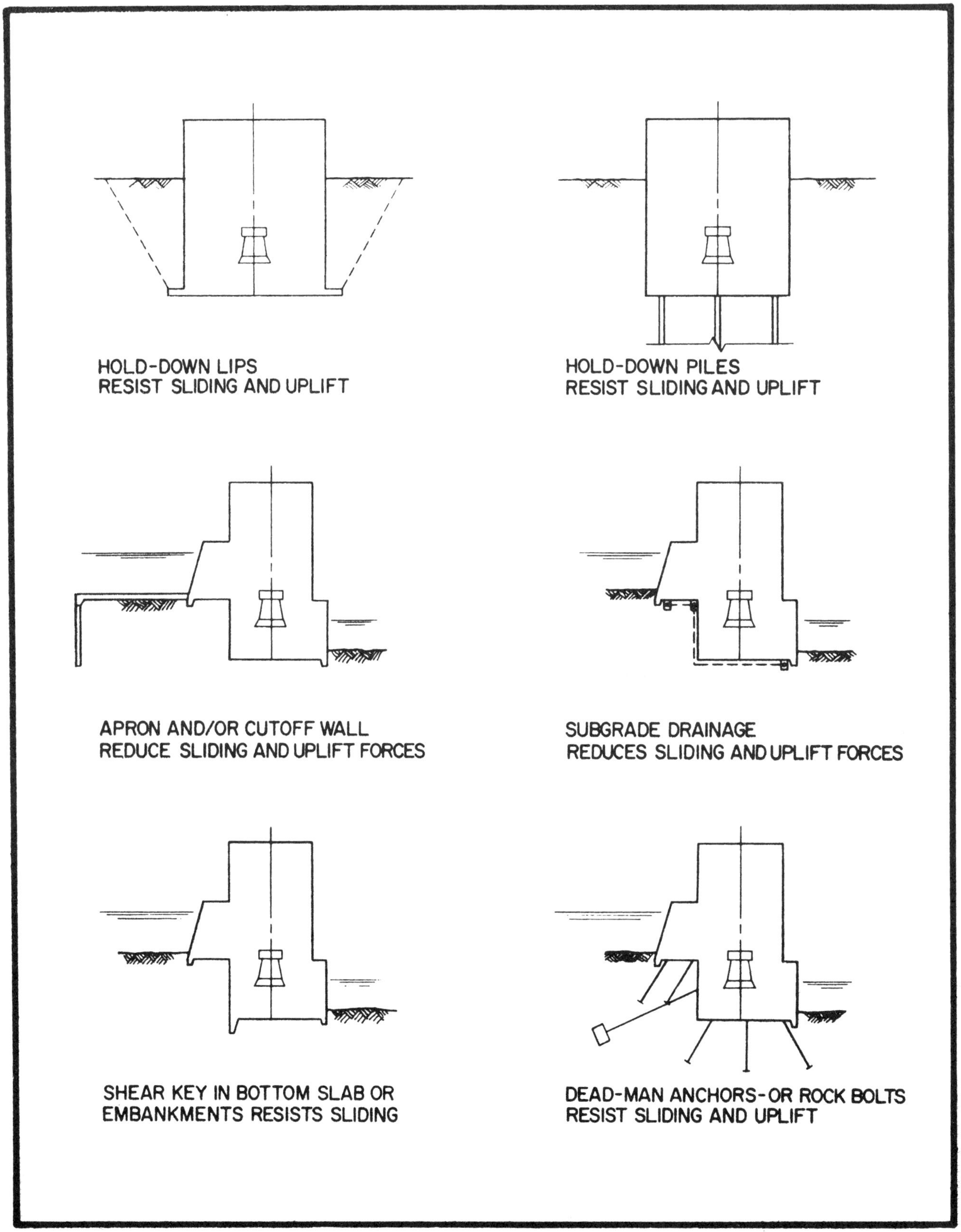

Figure 4-11. Methods for Obtaining Stability Against Sliding, Uplift and Overturning

TABLE 4-2
Tailrace Cost
(Cost base July 1978)

Fixed Cost	$15,000
Proportional Cost	$200 per linear foot

switchyard, and to determine the method of routing the electrical power conductors from the generator to the initial point of the transmission line for connecting to the power grid. Normally, for small hydroelectric installations, the power transformer is located within the switchyard. Accordingly, the switchyard should be placed as close as practical to the generator to minimize the electrical losses and length of the generator conductors. Further, the switchyard site must be above the flood elevation and placed where any possible water spray will not effect the high voltage equipment. Using the generator rating from Volume V, the switchyard civil costs may be determined by the use of Figure 4-12.

Costs shown in Figure 4-12 include a normal amount of grading and fencing costs. If the switchyard site is sufficiently remote from the powerhouse structure, and where more than a normal amount of grading may be required, the extra grading costs can be determined by applying the parameters of Table 2-2 and Figure 2-1. Except for extremely unusual site conditions, any of these increases will not be significant project costs and need not be considered.

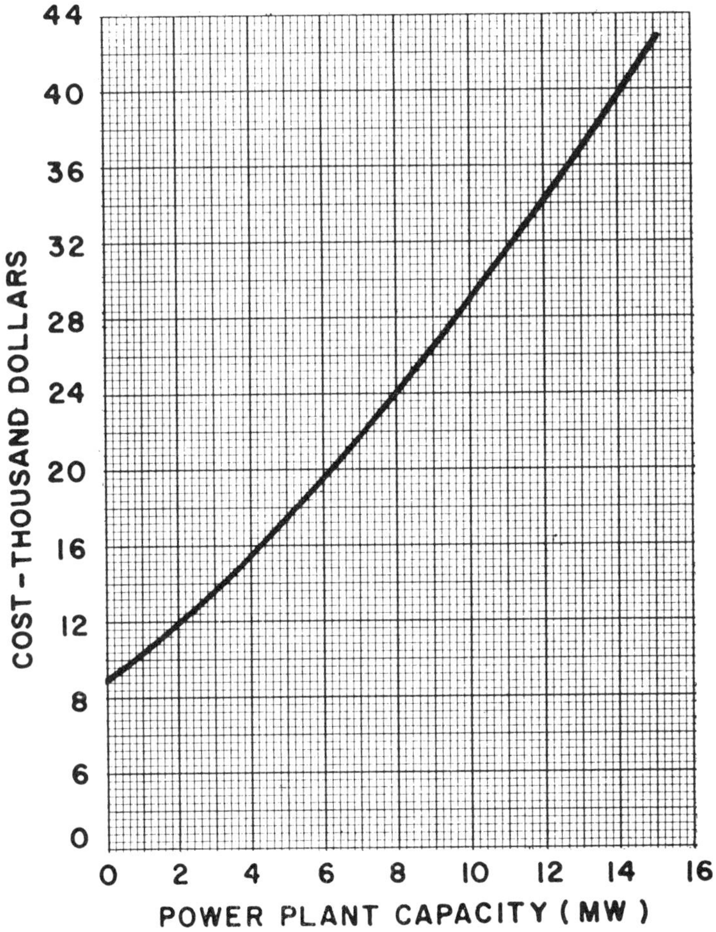

NOTE: Cost Base July 1978

Figure 4-12. Switchyard Civil Costs

SECTION 5
SPECIAL NEEDS FOR POWER ADDITIONS TO DAMS

General

It may be found that, to use an existing impoundment for developing hydroelectric power, the project will not be feasible unless existing civil features are utilized with a minimum of modification costs. Unusual design elements can often be used to simplify this utilization. With the inclusion of an unusual design element, a marginal project may become feasible. The resulting design, however, would not be one that would have been followed if the original impoundment had included a hydroelectric plant.

This Section will describe some innovative designs to stimulate thought of possible solutions to the foregoing problem.

Innovative Design Possibilities

The following items describe several unusual designs that may be considered:

1. Concrete gravity section dams are often constructed in narrow canyons with no apparent location for a powerhouse. Normally, an outlet works through the dam exists which can be easily connected to a turbine. To avoid a large excavation in the canyon well, a powerhouse can be constructed at the downstream toe of the dam by extending the spillway lip downstream. The extension of the spillway floor would form the roof on the proposed powerplant. Access to the powerplant could be developed along one side of the spillway.

2. Low concrete gravity dams with spillway gates are often constructed in congested areas with no space on either abutment for a powerhouse. A powerplant could possibly be constructed by the conversion of several of the spillway bays into a powerplant. The modifications would include the extension downstream of the spillway piers and construction of a back wall with a large gate to form a forebay for the powerplant. The turbine and draft tube would be placed in the extended spillway. The generator would be placed on a deck level with the top of the spillway piers. The turbine-generator shaft would be encased in a hollow pier between the apron and generator floor. The gates in the downstream wall of the powerhouse would be opened only to allow the spillway to pass flood flows.

3. An old abandoned powerhouse may now have historical importance. As modern turbine generator units have a smaller overall size than earlier units of the same rating, a new powerplant can often be constructed inside the existing structure. The original facade of the structure can be left intact. Demolition costs will be saved and environmental problems may be avoided.

4. Often, more than one conduit penetrates an existing dam. However, the conduit diameters are sometimes small and limit the generation of power. By joining two of the existing conduits together, a larger turbine/ generator can be installed which would produce enough generation capacity to make the project feasible.

5. The effective head on the turbine can be increased with a decrease in the tail water elevation. This can be accomplished by excavating the tail race to a lower depth and joining the tail race of the powerplant to the existing stream farther downstream from the dam.

6. Use can sometimes be made of inflatable rubber or fabric bags, placed on the spillway crest to raise the reservoir water level which increases the head on the turbine. The increased head increases the power output of the powerplant. The storage capacity of the reservoir is also increased which could result in an increase in energy production for the unit. At a predetermined increase in elevation the inflated bags would automatically deflate and the capacity of the spillway would not be changed. Costs are given in this Section for a similar design which uses bascule gates to increase the effective head on the turbine.

Spillway Modification

Frequently, the feasibility of a particular hydroelectric site can be enhanced by increasing the available head. The most practical method for raising the water surface elevation at an existing dam is to add bascule gates to the spillway crest to allow a higher water surface elevation. The bascule gates offer several advantages over other types for this application. Because the bascule gates rotate about their base and can be controlled from one end, they don't require a superstructure or intermediate supports. This results in a lower cost than other types of gates. With the gate in the lowered position (fully open), the face of the gate is almost flush with the spillway crest, so there is little change in the original discharge rating for the spillway. The bascule gates can be automatically controlled to maintain a predetermined water surface elevation within close tolerances. For gates up to 10 feet high, the cost for the gate and the complete operating mechanism is about $5,000 per foot of length (Table 5-1).

The function of the spillway is to protect a dam from being overtopped during a design flood. In adding bascule gates, or any similar device, to the spillway crest it is mandatory that the device allow the spillway crest to have its normally rated capacity.

TABLE 5-1
Bascule Gate Costs
(Cost Base July 1978)

Cost Item	Cost
Bascule Gate (maximum 10 foot height)	$5,000 per linear foot

Site Planning and Facilities Arrangements

The location and arrangement of the powerhouse and related facilities can pose a problem when adapting a hydroelectric facility to an existing dam. The facilities which are normally included in each site are the access road, parking area, switchyard, maintenance building, and powerhouse. The greatest restrictions on the locating of facilities are caused by natural obstacles such as cliffs, canyons, and the stream channel. In situations where the site is confined, space can be saved by such methods as placing the switchyard on top of the powerhouse or on a platform over the tailrace or by combining the maintenance building and switchyard with the powerhouse. In a very narrow canyon, where access roads would be too expensive or impossible to build, it might be necessary to provide access to the powerhouse with elevators, cranes, or cableways.

SECTION 6
SUMMARY AND FEASIBILITY STUDY COST GUIDELINES

General

This section describes the method of updating the civil features costs, as presented in this volume, from the July 1978 base to the date required for either the reconnaissance or feasibility study. Two indices are presented. The first, which is for escalation, is the United States Bureau of Reclamation index of project component costs. The second is a correction for site location and reflects the variation of construction costs within the continental United States. The first index is given on Figure 6-1 and the second is given on Figure 6-2.

The method for obtaining indirect costs, which include engineering, construction management, and the operation and maintenance and insurance is also provided in this section. The method and the percentage presented is the same as used in Volume V.

Escalation

The United States Bureau of Reclamation publishes on a quarterly basis the cost indices for thirty-four construction items that are common to irrigation and hydro projects. These are published in *Engineering News Record* (1977-1978) and are applicable to the eighteen western states. Four of these indices, which are considered the most significant, are included on Figure 6-1 for the last six years. By future indices, as they are published, the four curves in Figure 6-1 may be extended beyond July 1978 and extrapolated, if necessary, to the date required for the feasibility assessment. The construction item for which an escalated cost is required must be considered to be represented by one of the four classifications in Figure 6-1.

The escalated construction cost is obtained by determining the index number on the extended curve for the required date. The ratio of the July 1978 index to the index for the date used in the feasibility assessment is the multiplier by which the July 1978 base cost is multiplied to obtain the escalated cost.

Regional Cost Correction

A regional cost adjustment is made on the final cost after all the individual costs have been escalated. Figure 6-2 shows the regional cost variation. The cost base used in the preceding Sections represent a regional cost value of one. If the construction site is in a region having a cost value other than one, as shown by the Figure 6-2, then this different regional value is used as a multiplier to correct the total escalated cost for any regional cost difference.

Manpower Allocation for Studies

Personnel required to perform the analysis described must have civil engineering experience in hydroelectric plant design and project engineering. The studies should be directed by a Senior Civil Engineer having broad experience in this field. The majority of the study will be prepared by a Civil Engineer with less experience. The remainder of the work, including layout drawings and quantity takeoffs, will be done by a Designer. A total of ten man-days of effort should be allocated to prepare the feasibility study and cost estimate. A reconnaissance study and cost estimate will require about five man-days of effort. The allocation of time will be approximately ten percent for the Senior Civil Engineer, sixty percent for the Civil Engineer and thirty percent for the Designer. The civil engineering cost of this study is two percent of the total civil features cost.

Contingency

A contingency allowance is added to the escalated and regionally corrected construction costs to cover unknown and omitted items which would normally be included in a more detailed cost estimate. Contingencies also include an allowance for possible cost increases due to unforeseen conditions. Contingencies are normally estimated as 20 percent of the construction cost.

Engineering, Construction Management and Other Costs

Once the escalated and regionally corrected construction cost has been determined, it is necessary to estimate the engineering, construction management and administration costs, sometimes referred to as development or indirect costs. These costs include expenditures for feasibility study, license and permit applications, preliminary and final design, construction management, and administration. A multiplier of 20 percent should be applied to the total final construction cost, including contingencies, to estimate these development costs.

For a more detailed breakdown of these development costs the following percentages, applied to the final construction costs plus contingencies, may be used:

Feasibility Study	2%
License and Permit Applications	2%
Preliminary Design	3%
Final Design	6.5%
Construction Management	5.5%
Administration	1%

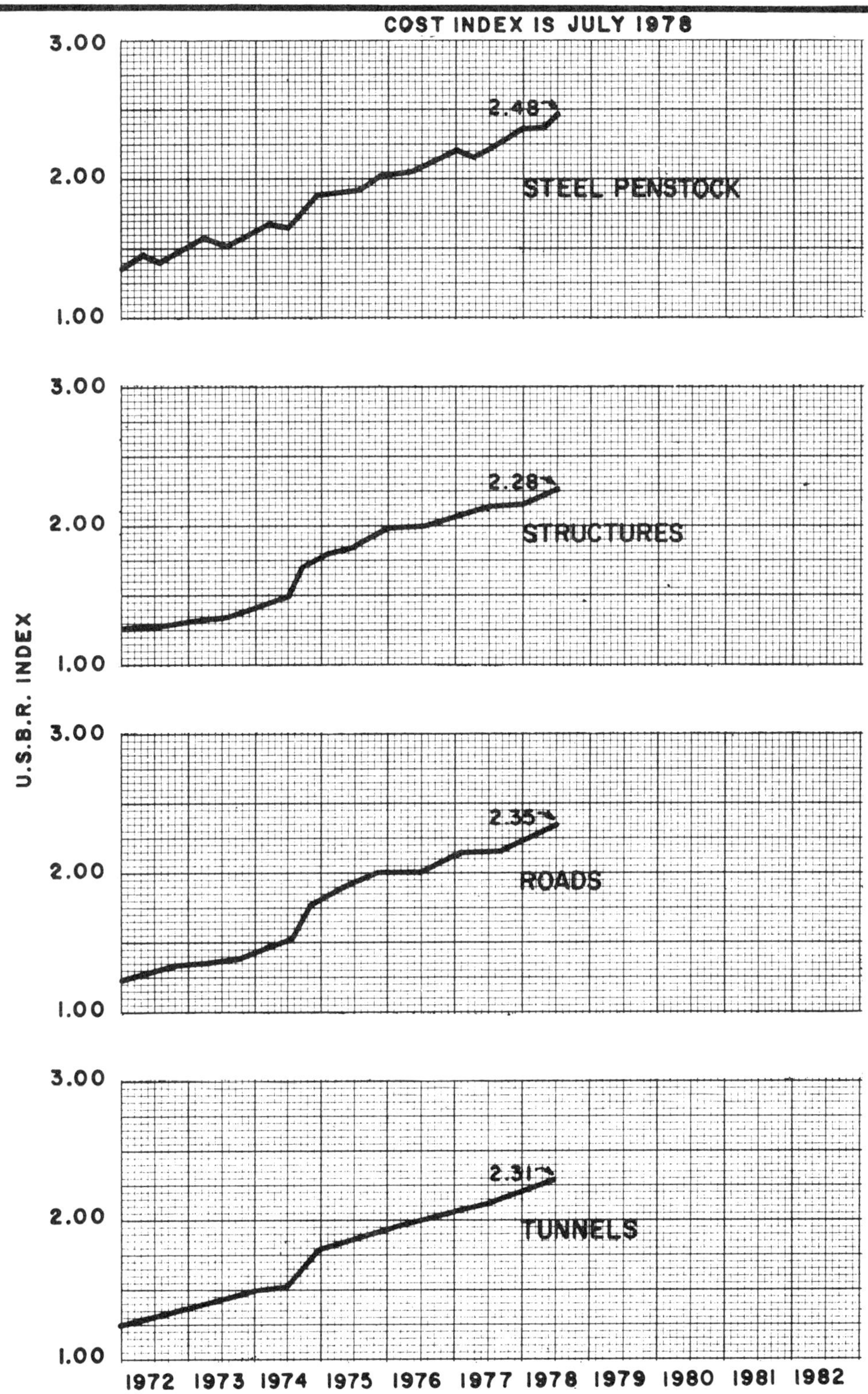

NOTE:

1. United States Bureau of Reclamation Construction Cost indices for the 18 Western States for Concrete Lined Tunnels, Primary Roads, Hydroelectric Power Plant Structures (Rein. Conc. Structures and Improv.), and Steel Penstocks.

Figure 6-1. Historical Cost Indices

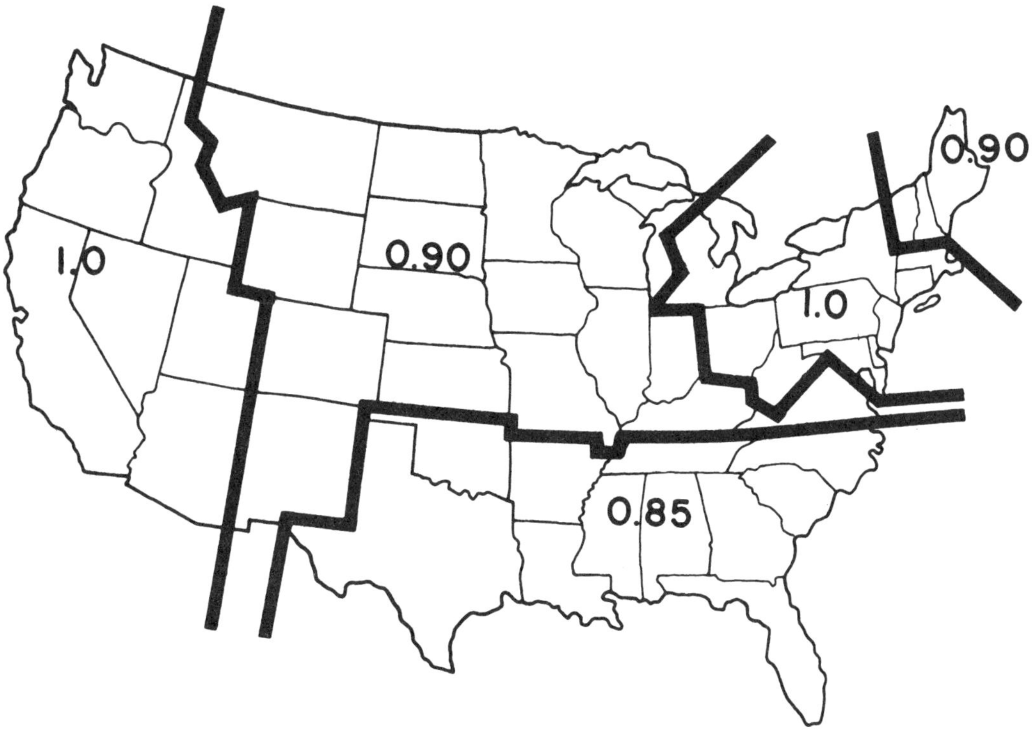

NOTE: San Francisco, California Base = 1.0, July 1978

Figure 6-2. Construction Cost Variation in the United States

The above percentage are for the civil feature costs only, hence the multipliers should be applied only to the costs of this volume. Not included in the above development costs are interest during construction, legal fees and financing fees. These omitted costs will be covered in Volume II which describes economic and financial considerations.

Operation and Maintenance Costs

General. Operation and maintenance costs for small hydroelectric plants are difficult to forecast accurately. The costs are directly related to the site and the owner's capability to perform the operation and maintenance function. The amounts which are suggested to be used in this report are based on those published by the U.S. Bureau of Reclamation and are updated to reflect recent experience.

Operation and maintenance costs as described herein, include the items listed below.

Insurance. The government is basically a self-insurer, however, for a commercial installation, coverage is required for fire and storm damage, vandalism, property damage and public liability.

Routine Maintenance and Operation. An amount must be budgeted to cover the costs of manpower, wages, services, equipment and parts utilized in the normal operation and maintenance of the hydroelectric plant.

General Expenses. The final portion of operation and maintenance costs are made up of those expenditures for administration fees and other miscellaneous costs required during project operation.

Operation and Maintenance Cost. The cost of operation and maintenance expenses can be estimated by multiplying the investment cost for the powerplant, including contingencies and development costs, by 1.2 percent. The resulting amount will be the estimated cost for operation and maintenance of the hydroelectric plant for the first year of operation. The operation and maintenance costs will increase with time, corresponding to inflationary trends. The current annual increase for operation and maintenance costs is taken to be 6-1/2 percent.

There are two final comments to be observed in determining the operation and maintenance costs of hydroelectric plant facilities. First, the total annual costs for operation and maintenance (from Volume V and VI) should never be estimated below a certain minimum amount, approximately $20,000 in 1978 dollars. Second, the multiplier given previously, 1.2 percent, should be used only if the owner can integrate the operation of the small hydropower facility with other related operations. If the operating entity will operate and maintain only the small hydroelectric facility under consideration, a multiplier of 2 to 4 percent should be used to determine annual O&M costs.

Cost Summary Sheet

Completing the Cost Summary Sheet, shown as Exhibit I, provides a method for determining the civil cost to be used in the feasibility assessment estimate. The account numbers used in Exhibit I are those designated by the Federal Energy Regulatory Commission for hydroelectric development. Although provision has been made for a civil costs contingency item, it is normal to include this with the other contingencies as an overall project cost. Refer to Volume II.

REFERENCES

Bier, P.J., *Welded Steel Penstocks design and construction, Engineering Monographs, No. 3,* United States Department of Interior, Bureau of Reclamation, 1966.

Davis, Calvin Victor, and Sorensen, Kenneth E., *Handbook of Applied Hydraulics,* Third Edition, 1969, McGraw-Hill Book Company, Section 22, Pg. 22-66 & 22-67, Hydraulic Design Factors; Section 27, Water Hammer Pg. 27-1; Section 28, Surge Tanks Pg. 28-1.

Engineering News-Record, McGraw-Hill's Construction Weekly, McGraw-Hill Inc., McGraw-Hill Building, 1221 Avenue of the Americas, New York, N.Y. 10020; December 22, 1977 Page 101; March 23, 1978 page 97; June 22, 1978 page 109; September 21, 1978 page 121.

EXHIBIT I
COST SUMMARY SHEET

PROJECT ____________________ PLANT CAPACITY __________ MW

JOB NO. ____________________ AVG. ANNUAL ENERGY __________ MWH

DATE ____________ BY ____________

FERC ACCT. NO.	DESCRIPTION	COST*	ESCALATION FACTOR	ESCALATED COST	TOTAL
331	STRUCTURES & IMPROVEMENTS				
	SITE				
.11	DRAINAGE SYSTEM				
.12	EROSION CONTROL				
.13	FINAL GRADING				
.14	ACCESS ROAD				
.15	PARKING & MISC. SITE FEATURES				
.16	ENVIRONMENTAL CONSTRUCTION CONTROLS				
.17					
.18					
	POWERHOUSE				
.21	STRUCTURAL				
.22	EXCAVATION				
.23	FOUNDATION				
.24	SWITCHYARD				
.25					
.26					
	TOTAL, ACCOUNT 331				

* COST BASE JULY 1978

FERC Acct. No.	Description	Cost*	Escalation Factor	Escalated Cost	Total	
332	RESERVOIRS, DAMS & WATERWAYS					
.01	Penstock					
.02	Valves					
.03	Bifurcation & Slide Gates					
.04	Tailrace					
.05						
.06						
	Total, Account 332					
	TOTAL CIVIL COSTS					
	Contingencies					
	Regional Correction Factor					
	CORRECTED CIVIL COSTS					
	Engineering, Construction Management and Other Costs					
	GRAND TOTAL					
	* Cost Base, July 1978					

GLOSSARY

Abbreviations

alternating current	ac	head in feet	H
barrel (42 gallons)	bbl	Hertz	Hz
benefit-cost ratio	B/C	horsepower	hp
British thermal units	Btu	kilovolt	kV
cents	¢	kilovolt-ampere	kVA
cubic feet	ft^3	kilowatt	kW
cubic feet per second	cfs	kilowatt-hours	kWh
cubic yard	cu yd	megavolt ampere	MVA
direct current	dc	megawatt	MW
dollars	$	megawatt-hours	MWh
efficiency in percent	E	percent	%
feet	ft	pound	lb
flow in cfs	Q	pounds per square inch	psi
gigawatt	GW	revolutions per minute	r/min
gravitational constant	g	square yards	sq yd

ALTERNATING CURRENT (ac)—an electric current that reverses its direction of flow periodically as contrasted to direct current.

ANADROMOUS FISH—fish, such as salmon, which ascend rivers from the sea at certain seasons to spawn.

AVERAGE LOAD—the hypothetical constant load over a specified time period that would produce the same energy as the actual load would produce for the same period.

BENEFIT-COST RATIO (B/C)—the ratio of the present value of the benefit stream to the present value of the project cost stream computed for comparable price level assumptions.

BENEFITS (ECONOMIC)—the increase in economic value produced by the hydropower addition project, typically represented as a time stream of value produced by the generation of hydroelectric power. In small hydro projects this is often limited for analysis purposes to the stream of costs that would be representative of the least costly alternative source of equivalent power.

BRITISH THERMAL UNIT (Btu)—the quantity of heat energy required to raise the temperature of 1 pound of water 1 degree Fahrenheit, at sea level.

BUS—an electrical conductor which serves as a common connection for two or more electrical circuits. A bus may be in the form of rigid bars, either circular or rectangular in cross section, or in form of stranded-conductor overhead cables held under tension.

BUSBAR—an electrical conductor in the form of rigid bars, located in switchyard or power plants, serving as a common connection for two or more electrical circuits.

CAPACITOR—a dielectric device which momentarily absorbs and stores electrical energy.

CAPACITY—the maximum power output or load for which a turbine-generator, station, or system is rated.

CAPACITY VALUE—that part of the market value of electric power which is assigned to dependable capacity.

CAPITAL RECOVERY FACTOR—a mathematics of finance value used to convert a lump sum amount to an equivalent uniform annual stream of values.

CIRCUIT BREAKER—a switch that automatically opens an electric circuit carrying power when an abnormal condition occurs.

COSTS (ECONOMIC)—the stream of value required to produce the hydro electric power. In small hydro projects this is often limited to the management and construction cost required to develop the power plant, and the administration, operations, maintenance and replacement costs required to continue the power plant in service.

COST OF SERVICE—cost of producing electric energy at the point of ownership transfer.

CRITICAL STREAMFLOW—the amount of streamflow available for hydroelectric power generation during the most adverse streamflow period.

CRITICAL DRAWDOWN PERIOD—the time period between maximum pool drawdown and the previous occurrence of full pool.

DEMAND—see LOAD.

DEBT SERVICE—principle and interest payments on the debt used to finance the project.

DEPENDABLE CAPACITY—the load carrying ability of a hydropower plant under adverse hydrologic conditions for the time interval and period specified of a particular system load.

DIRECT CURRENT (dc)—electricity that flows continuously in one direction as contrasted with alternating current.

ENERGY—the capacity for performing work. The electrical energy term generally used is kilowatt-hours and represents power (kilowatts) operating for some time period (hours).

ENERGY VALUE—that part of the market value of electric power which is assigned to energy generated.

ELECTRIC RATE SCHEDULE—a statement of the terms and conditions governing the sale of electric service to a particular class of customers.

FEASIBILITY STUDY—an investigation performed to formulate a hydropower project and definitively assess its desirability for implementation.

FEDERAL ENERGY REGULATORY COMMISSION (FERC)—an agency in the Department of Energy which licenses non-Federal hydropower projects and regulates interstate transfer of electric energy. Formerly the Federal Power Commission (FPC).

FIRM ENERGY—the energy generation ability of a hydropower plant under adverse hydrologic conditions for the time interval and period specified of a particular system load.

FORCE MAJEURE—an event or effect that cannot be reasonably anticipated or controlled.

FORCED OUTAGE—the shutting down of a generating unit for emergency reasons.

FORCED OUTAGE RATE—the percent of scheduled generating time a unit is unable to generate because of forced outages due to mechanical, electrical or another failure.

FOSSIL FUELS—refers to coal, oil, and natural gas.

GENERATOR—a machine which converts mechanical energy into electric energy.

GIGAWATT (GW)—one million kilowatts.

GRAVITATIONAL CONSTANT (g)—the rate of acceleration of gravity, approximately 32.2 feet per second per second.

HEAD, GROSS (H)—the difference in elevation between the headwater surface above and the tailwater surface below a hydroelectric power plant, under specified conditions.

HERTZ (Hz)—cycles per second.

HYDROELECTRIC PLANT or HYDROPOWER PLANT—an electric power plant in which the turbine-generators are driven by falling water.

INSTALLED CAPACITY—the total of the capacities shown on the nameplates of the generating units in a hydropower plant.

INTERCONNECTION—a transmission line joining two or more power systems through which power produced by one can be used by the other.

KILOVOLT (kV)—one thousand volts.

KILOWATT (kW)—one thousand watts.

KILOWATT-HOUR (kWh)—the amount of electrical energy involved with a one kilowatt demand over a period of one hour. It is equivalent to 3,413 Btu of heat energy.

LOAD—the amount of power needed to be delivered at a given point on an electric system.

LOAD CURVE—a curve showing power (kilowatts) supplied, plotted against time of occurrence, and illustrating the varying magnitude of the load during the period covered.

LOAD FACTOR—the ratio of the average load during a designated period to the peak or maximum load occurring in that period.

LOW HEAD HYDROPOWER—hydropower that operates with a head of 20 meters (66 feet) or less.

(AT) MARKET VALUE—the value of power at the load center as measured by the cost of producing and delivering equivalent alternative power to the market.

MEGAWATT (MW)—one thousand kilowatts.

MEGAWATT-HOURS (MWh)—one thousand kilowatt-hours.

MINIMUM REVENUE REQUIREMENT—funds required to pay all costs incurred by a project.

MULTIPURPOSE RIVER BASIN PROGRAM—programs for the development of rivers with dams and related structures which serve more than one purpose, such as - hydroelectric power, irrigation, water supply, water quality control, and fish and wildlife enhancement.

NUCLEAR ENERGY—energy produced largely in the form of heat during nuclear reactions, which, with conventional generating equipment can be transfered into electric energy.

NUCLEAR POWER—power released from the heat of nuclear reactions, which is converted to electric power by a turbine-generator unit.

OUTAGE—the period in which a generating unit, transmission line, or other facility, is out of service.

(IN) PARALLEL—several units whose AC frequencies are exactly equal, operating in synchronism as part of the same electric system.

PEAKING CAPACITY—that part of a system's capacity which is operated during the hours of highest power demand.

PEAK LOAD—the maximum load in a stated period of time.

PLANT FACTOR—ratio of the average load to the installed capacity of the plant, expressed as an annual percentage.

PONDAGE—the amount of water stored behind a hydroelectric dam of relatively small storage capacity used for daily or weekly regulation of the flow of a river.

POWER (ELECTRIC)—the rate of generation or use of electric energy, usually measured in kilowatts.

POWER FACTOR—the percentage ratio of the amount of power, measured in kilowatts, used by a consuming electric facility to the apparent power measured in kilovolt-amperes.

POWER POOL—two or more electric systems which are interconnected and coordinated to a greater or lesser degree to supply, in the most economical manner, electric power for their combined loads.

PREFERENCE CUSTOMERS—publicly-owned systems and nonprofit cooperatives which by law have preference over investor-owned systems for the purchase of power from Federal projects.

PROJECT SPONSOR—the entity controlling the small hydro site and promoting construction of the facility.

PUMPED STORAGE—an arrangement whereby electric power is generated during peak load periods by using water previously pumped into a storage reservoir during off-peak periods.

RATE OF RETURN ON INVESTMENT—the interest rate at which the present worth of annual benefits equals the present worth of annual costs.

RECONNAISSANCE STUDY—a preliminary feasibility study designed to ascertain whether a feasibility study is warranted.

SECONDARY ENERGY—all hydroelectric energy other than FIRM ENERGY.

SERVICE OUTAGE—the shut-down of a generating unit, transmission line or other facility for inspection, maintenance, or repair.

SMALL HYDROPOWER—hydropower installations that are 15,000 KW (15 MW) or less in capacity.

SPINNING RESERVE—generating units operating at no load or at partial load with excess capacity readily available to support additional load.

STEAM-ELECTRIC PLANT—a plant in which the prime movers (turbines) connected to the generators are driven by steam.

SURPLUS POWER—generating capacity which is not needed on the system at the time it is available.

SYSTEM, ELECTRIC—the physically connected generation, transmission, distribution, and other facilities operated as an integral unit under one control, management or operating supervision.

THERMAL PLANT—a generating plant which uses heat to produce electricity. Such plants may burn coal, gas, oil, or use nuclear energy to produce thermal energy.

THERMAL POLLUTION—rise in temperature of water such as that resulting from heat released by a thermal plant to the cooling water when the effects on other uses of the water are detrimental.

TRANSFORMER—an electromagnetic device for changing the voltage of alternating current electricity.

TRANSMISSION—the act or process of transporting electric energy in bulk.

TURBINE—the part of a generating unit which is spun by the force of water or steam to drive an electric generator. The turbine usually consists of a series of curved vanes or blades on a central spindle.

TURBINE-GENERATOR—a rotary-type unit consisting of a turbine and an electric generator. (See TURBINE & GENERATOR)

VERTICALLY INTEGRATED SYSTEM—refers to power systems which combine generation, transmission, and distribution functions.

VOLTAGE OF A CIRCUIT—the electric potential difference between conductors or conductors to ground, usually expressed in volts or kilovolts.

WATT—the rate of energy transfer equivalent to one ampere under a pressure of one volt at unity power factor.

WHEELING—transportation of electricity by a utility over its lines for another utility; also includes the receipt from and delivery to another system of like amounts but not necessarily the same energy.

Part 3
Equipment for Small Scale Hydroelectric Power Plants

The information in this part is from *Proceedings—Small/Low-Head Hydropower PRDA-1706 Contractors' Symposium*, held in Albany, New York, prepared for the United States Department of Energy, Idaho Operations Office under DOE Contract DE-AM07-76ID01570, May 8-10, 1979. The material used is from a paper, "Available Low Head Hydro Technology," presented by W.W. Wayne, Jr. of Stone & Webster Engineering Corporation.

ABSTRACT

The present status of equipment technology for low head hydroelectric developments is summarized. Included are descriptions and data on available types of turbogenerating equipment and information on small packaged units and standard designs. Approximate cost information suitable for guidance in making preliminary assessments of potential low head hydro sites is given.

INTRODUCTION

Historically, the economic feasibility of low head hydro developments has been adversely affected by two features, namely: (1) the relatively large fluctuations in head caused by changing flow rates, which affects turbine efficiency, and (2) the high turbine discharge required to produce each kilowatt of power, which results in large size machinery and relatively high costs per kilowatt of installed capacity.

Since World War II, significant steps have been taken to develop more efficient turbogenerating equipment for low head hydro plants. These include the TUBE unit, STRAFLO unit, and bulb unit designs. And, more recently, manufacturers have focused their attention on the development of standarized designs and small packaged units, with the intent of further reducing overall equipment costs. This paper summarizes the present status of hydro equipment technology.

SMALL LOW HEAD FRANCIS AND PROPELLER TURBINES IN TRADITIONAL ARRANGEMENTS

Francis and propeller type turbines have long been the backbone of the hydroelectric power industry, with an established record of dependable and efficient service. They are particularly suitable for low head installations. They also are adaptable to a wide variety of settings and arrangements, including such specific features as: vertical or horizontal shaft; open flume, pressure casing, or concrete semi-spiral casing; conical or elbow type draft tube; adjustable or fixed blade propeller runners; and, single or double Francis runners. Maximum turbine efficiencies from 92 to 94 percent are commonly obtained for units of modern design.

The Kaplan turbine is a propeller turbine with adjustable runner blades which are coordinated with the wicket gates to obtain the best blade-gate relationship for most efficient operation. This arrangement eliminates the rapid drop in efficiency which occurs at part load and overload with fixed blade propellers, and enhances the economic attractiveness of such equipment.

One of the most widely used low head turbines in past years was the vertical shaft Francis type, such as the Samson turbine manufactured by The James Leffel & Company of Springfield, Ohio. It can be used with either an open flume or pressure flume setting, as shown in Figure 1. Leffel's standard designs cover 13 sizes with runner diameter ranging from 17 to 74 inches for use in an open flume setting. The 74 inch size at 10 feet head is rated 350 hp at full gate and at 26 feet, which is the maximum design head, the output would be 1,466 hp. Corresponding discharges would be 450 cfs and 614 cfs, respectively. For use in the pressure case setting, Leffel's standard sizes go up only to 56 inches maximum runner diameter. The maximum output for this size would be 200 hp at 10 feet head and 839 hp at the maximum 26 feet head. The corresponding discharges would be 218 cfs and 352 cfs, respectively.

The expected performance curve for a Leffel 56 inch Samson type turbine operating at 130 rpm under a net head of 20 feet is shown in Figure 2.

In addition to Leffel, there are a number of other long established turbine manufacturers who still retain in their files proven designs of small size low head turbines which could be readily updated to suit modern fabrication techniques and the higher strength materials currently available. Such firms include Allis-Chalmers Corporation of York, Pennsylvania, Barber Hydraulic Turbine Ltd. of Port Colburne, Ontario, and Dominion Engineering Works of Montreal.

PACKAGED UNITS

Introduction

In this paper, a "packaged unit" is defined as a hydro turbogenerator set or unit, completely assembled in the manufacturer's shop and shipped as a unit mounted on its base, or in its housing. At the site, it would require little more than bolting in place, connecting the water source to the intake, and making the necessary external electrical connections. Shipping and handling capabilities would set the limit on maximum size of a packaged unit.

The chief advantages of packaged units are minimizing field installation labor generally, and eliminating the need in the field of skilled mechanics with hydroelectric machinery installation experience. With a good set of instructions, relatively inexperienced personnel should be able to set up the unit and put it in operation.

Barber Hydraulic Turbine Ltd. of Canada

This company has developed a small hydro generating station standard packaged unit which can be transported by plane to remote sites in northern Canada and installed with a minimum of labor. These units, which are called "Canada Frontier Low Head Mini-Hydel Stations," will come from the factory complete in a steel housing ready to bolt in place on a field prepared foundation. Only the lower part of the steel draft tube would be shipped separately and would require bolting during site installation. The general arrangement of such a station is shown in Figure 3.

The Mini-Hydel Stations will have 59 inches diameter (1500 mm) vertical shaft Francis runners and will be designed for heads ranging from 10 feet to 25 feet. The power output will range from about 100 to 600 kW, depending on the head. Because the units would be for isolated use, they will have synchronous generators. The latter will be driven through a gear box at 1200 rpm. Oil pumps and governors would be belt driven. Maximum turbine efficiency is expected to range from 85 to 89 percent, depending on the head at the particular site. The expected turbine performance for a unit operating at a head of 25 feet is shown in Figure 4.

Barber has received an order for the first Mini-Hydel Station from the Ontario Hydro Commission. This unit, which is scheduled to be installed during the summer of 1979, will be primarily for developmental experience and demonstration. It will have an induction generator and will be connected to the local electric system. To minimize civil costs, water will flow to the turbine through a 6 foot diameter diphon constructed over the top of an existing dam. The differential head will be approximately 12 feet. The location is at Wasdell Falls at the north end of Lake Couchiching about 90 miles north of Toronto.

Barber believes if it goes to a model without wicket gates, that a considerably bigger runner could be used, possibly doubling the power output without increasing the overall size of the package. However, some sort of intake gate or flow control valve would have to be provided for shutting down the unit.

Barber believes there may be applications for the Mini-Hydel Station in the United States as new, or replacement, units for old hydro plants, or for installation at dams where there are presently no power facilities. The lower cost per kW of a modified larger capacity model without wicket gates and with an asynchronous generator would be of interest wherever the unit could readily be tied into an existing electric utility system.

Other Manufacturers

While not all inclusive, a listing of other manufacturers that can offer packaged units follows:

Ossberger-Turbinen Fabrik
8832 Weissenburg, Bayern
Otto-Rieder Str. 7 - Postfach 425
Germany

Called Hydro-Light
Outputs 1.5 to 6.3 kW
Heads 4 through 10 m
Discharge 40 to 80 l/s

Barber Hydraulic Turbine, Ltd.
Suite 210
65 Queen St. West
Toronto, Ontario
Canada M5H 2M5

Called Micro-Hydel
Outputs 10 to 50 kW
Heads up to 35 ft

The James Leffel and Company
Springfield, Ohio 45501

Called Hoppes Hydro-electric Unit
Outputs 0.5 to 10 kW
Heads 8 through 25 ft
Discharge 1.2 to 16.3 cfs

Gilbert Gilkes & Gordon Ltd.
Kendal, Cumbria
England LA9 7BZ

or its agent,

Border Electric Company
Blaine, Washington 98230

Called Hydec Hydro Electric Generating Set
Output up to approximately 35 kW

Wind and Water Power
P.O. Box 49
Harrisville, NH 03450

Called Packaged Unit
Outputs 20 to 300 kW
Heads 3 to 20 m
Discharge 250 to 3000 liters/s

BULB UNITS

General Design Criteria

The term "bulb unit" refers to an axial flow unit which has the generator mounted inside a steel bulb located within the powerhouse water passage. The principal advantages of a horizontal shaft bulb unit, when compared to a conventional vertical shaft propeller turbine with external generator mounted overhead, are as follows:

1. Flow through the turbine is more favorable since the substantial directional changes in the spiral casing and draft tube are eliminated.

2. Maximum turbine efficiencies are generally from 1 to 2 percent higher for the equivalent size unit.

3. The axial flow results in greater power output for the same runner diameter and head.

4. The length, width, and height of the powerhouse structure can be reduced in most cases.

5. Normally, less excavation is required for the powerhouse.

6. Formwork for the water passages is simpler and less costly.

7. Although the total cost of the turbogenerating equipment and accessories is about the same, overall structural savings from 10 to 15 percent are usually obtained.

8. The higher unit efficiency increases the annual power output and operating revenues.

Many different design configurations have been investigated by bulb turbine manufacturers and the results to date indicate the best arrangement is to place the generator upstream of the turbine, rather than downstream. Placing the generator bulb downstream seriously affects the draft tube efficiency and can cause cavitation and vibration problems.

Because of the very low speed of bulb units, directly connected generators are physically large and relatively costly. To alleviate this problem, small to medium capacity machines have been, on occasion, provided with geared type speed increasers. Such units usually have one large vertical access shaft through which all equipment components can be either installed, or removed for repairs.

With the increasing demand for higher capacity units and modern advances in generator technology, the trend is now towards directly driven generators which are built as large as practicable to obtain additional flywheel effect, improved cooling, and higher efficiencies. This design generally requires two access shafts, one serving the upstream generator compartment, and the other serving the downstream turbine compartment. The access shafts are limited in size but can be used for personnel access, oil and water piping, generator leads, and control cables. Such components as bearings, coolers, pumps, and valves normally can be installed, or withdrawn, through these shafts. Larger hatches, or openings, in the powerhouse substructure must be provided for the installation and/or removal of the main turbine and generator assemblies.

The large capacity bulb generators manufactured by Alsthom of France generally are designed to operate under from 2 to 3 atsmospheres of pressure, so as to improve the heat transfer capability and obtain more effective cooling. Pressurized operation may, for some units, only be required above 75 percent load. Since the maximum diameter of the generator housing (bulb shell) must not exceed the turbine runner diameter by more than about 10 percent, the use of pressurized cooling is a significant item, since it minimizes both the generator diameter and length for a desired power output.

A very important consideration in the mechanical design of a bulb unit is the number, type, and size of the main shaft journal bearings. It is essential for the larger size units that the journal bearings be a self-aligning type, with adequate provisions for lubrication, particularly during startup and shutdown. Whenever possible, it is preferable to adopt a two bearing design, since the distribution of loads can be more accurately determined with that arrangement. However, for extremely large size units, three bearings may be required to keep within allowable stress limits.

Figure 5 shows a comparison of the expected turbine efficiency for a bulb unit, with adjustable blades and wicket gates with an equivalent vertical shaft Kaplan unit. It will be noted that the peak efficiency of the bulb unit is about 1-1/2 percent higher. These curves are based on turbines with a 3000 mm (119 inches) runner diameter.

Bulb Units for Rock Island Project

The largest capacity bulb units in the world are now being installed in the new second powerhouse at the Rock Island Dam on the Columbia River near Wenatchee, Washington. Each of the eight units can produce 51,300 kW at a net head of 39.7 feet and a speed of 85.7 rpm, giving a total plant capacity of 410,400 kW. The turbogenerators, electric governors, and other accessory equipment are being furnished by Alsthom-Neyrpic of France.

The bulb turbines have a runner diameter of 7400 mm (291 inches) and are equipped with stainless steel adjustable runner blades, wicket gates, steel intake and draft tube liners, and an exterior wicket gate operating ring.

The horizontal shaft bulb generators are directly connected to the turbines and are rated 54 MVA, 0.95 pf, 6.9 kV, 3 phase, 60 Hz, 85.7 rpm. They are equipped with forged steel shafts, thrust and journal bearings, pressurized air cooling systems with heat exchangers, voltage regulators, brushless exciters, and other necessary appurtenances.

A cross-section of the new powerhouse is shown in Figure 6 and further details of the bulb unit setting are shown in Figure 7. Construction photographs of the powerhouse structures and equipment are contained in Figures 8 through 12.

As of December 1, 1978, four units had been placed in commercial operation by the project owner, Public Utility District No. 1 of Chelan County, Washington. All eights units are expected to be in commercial operation by July 1979.

OSSBERGER CROSS-FLOW TURBINES

General Description

The Ossberger cross-flow turbine, which is a low speed unit suitable for both low and medium head installations, is a radial, impulse-type turbine with partial air admission. The intake water, which is distributed to the periphery of a hollow cylindrical shaped runner by adjustable guide vanes, flows through the blades, across the inside of the runner, and out through the blades, on the other side, as indicated in Figure 13. An exploded view showing the relationship of the various parts of an Ossberger turbine is presented in Figure 14.

If operating conditions require, the guide vanes can be divided into two separately controlled sections, as indicated in Figure 15. Normally, for most installations, the lengths of the two guide vane sections would be in the ratio 1:2. The smaller section would be used when the water supply is low, the larger section would be used for medium flows, and both sections would be used together for larger flows.

Francis and propeller type turbines usually have peak efficiencies higher than Ossberger turbines. However, with divided guide vane sections, as outlined above, the efficiency curve of an Ossberger turbine is flat over a wide range of flowand head conditions. This is illustrated in Figure 15. Guaranteed maximum efficiency is 83 percent but actual maximum efficiency at the best gate position has, in many cases, exceeded 85 percent.

The Ossberger runner consists of curved blades fitted and welded into the end discs, a longitudinal shaft through the center, and intermediate rings to support relatively long runner blades. A typical large size runner is shown in Figure 16.

The bearings for Ossberger runners are standard self-aligning roller bearings which normally require greasing only once or twice each year.

The steel intake transition section and the steel draft tube, except on very small units, would be shipped separately and bolted to the main turbine assembly at the site. An adjustable air inlet valve in the casing above the draft tube automatically controls the amount of vacuum and, thereby, the water level in the draft tube.

Generators, speed increasing gear boxes, roller bearings, and couplings are standard items furnished with the turbines, but produced by other manufacturers. For Asynchronous generating units, Ossberger regulators are furnished to control turbine output, usually as a function of head pond water level. For synchronous units, control is by a relatively simple type of hydraulic governor produced by a governor manufacturer, such as Woodward Governor Company.

Manufacturing Experience

Ossberger Turbinen Fabrik of Weissenburg, Germany, is the only manufacturer of cross-flow turbines and F.W.E. Stapenhorst Inc., of Pointe Claire, Quebec, Canada, is the sole representative for Ossberger in North America. In the past 25 years, hundreds of Ossberger units have been furnished to countries world wide. Many of the units are for small isolated power systems in remote areas. A substantial portion of the units, have capacities less than 100 hp. The largest capacity units, installed in 1975, are rated 1360 hp each at 98 feet head and are located at the Taltson River Project of the Northern Canada Power Commission in Alberta. Design heads have varied from as little as 6 feet for a 21 hp unit to as much as 562 feet for a 250 hp unit.

The unit which is largest in physical size was installed in 1977 at a small paper mill in the vicinity of Stuttgart, Germany. It is rated 276 hp at 11.8 feet head and discharges 247.2 cfs. The runner is 1250 mm (49.2 inches) in diameter and 129.9 inches long. This unit is shown in the upper photograph of Figure 16. Ossberger does not intend to manufacture a larger diameter runner until the 1250 mm machine is fully integrated into its production program and several years of operating experience have been achieved.

Proposed U.S. Installation

F.W.E. Stapenhorst, Inc., has acquired the existing Goodyear Lake Dam and power plant in New York State and recently received a Federal Energy Regulatory Commission license to install and operate two Ossberger turbines capable of producing 1000 hp each under a head of 30 feet. The power output will be used by a private utility company which previously owned this plant and decided to abandon it because of high operating costs. Since the station is to be unattended, the new units will be arranged for fully automatic operation and remote control. Special equipment will be provided for controlling forebay water levels.

STRAFLO TURBINES WITH RIM TYPE GENERATORS

Introduction

STRAFLO is the proprietary name given by Escher Wyss, Ltd., of Zurich, Switzerland to its axial flow turbine which has a generator rotor mounted on the periphery of the turbine runner blades. The concept is not new and 73 units of this type, all built by Escher Wyss, have been operating successfully for many years. These units are relatively small (1 to 1.9 MW), design heads are low (8.0 to 9.2 m) and all have fixed blade propeller type runners. Escher Wyss has recently developed new bearings and seals which should make it feasible to build units of this type in considerably larger sizes and for higher heads. Turbine runners with adjustable blades should also be possible. Escher Wyss anticipates that STRAFLO units will be suitable for the same head range which vertical Kaplan turbines and bulb turbines are normally considered.

Existing Installations

The 73 existing STRAFLO units are installed in 14 power stations located on three rivers, the Lech and Iller in Germany, and the Saalach in Austria. The runner diameters vary from 1950 mm to 2100 mm rated flows are from 20 to 25 m^3/S, rated heads are from 8.0 to 9.2 m, and rated outputs from 1.0 to 1.9 MW. These units all have adjustable guide vanes. However, since the vanes do not fully close, each unit has a butterfly valve on the upstream side for stopping the flow when the unit is shut down. Typical details of these units are shown in Figure 17. The units may be installed with either a horizontal shaft, as in the Lech River plants, or a sloping shaft, as in the Iller River plants.

Since the outer rim of the turbine runner supports the generator rotor, a sealing system is required to prevent penstock water from entering the generator area. The complete generator, including stator and rotor, is entirely outside the water passage and thus easily accessible for maintenance. Cooling poses no unusual problems, since the generator is in the open. The greater inertia of the larger diameter rotor, as compared to a bulb unit, is a distinct advantage in terms of operating stability. With the generator outside the water passage, the "bulb" of the STRAFLO unit is only large enough to accommodate the runner journal and thrust bearings, and the inner wicket gate bearings.

The compactness of the STRAFLO unit is probably its most important feature, when compared to bulb and tubular units, and makes it possible to achieve cost savings in the civil works, powerhouse crane, and accessory equipment. This fact becomes evident when comparing the size of structures and crane span required for a STRAFLO unit versus that needed for an equivalent bulb unit, as shown in Figure 18. The compact nature of the STRAFLO unit also enables it to be installed inside a concrete spillway dam, as illustrated by Figures 19 and 20. Such an arrangement eliminates the need for a separate powerhouse structure, with resulting reductions in project costs.

Planned Installations

In cooperation with the City of Zurich, Switzerland, Escher Wyss has designed a demonstration STRAFLO unit to be installed in the existing Hoengg plant. It will have adjustable runner blades, the new type hydrostatic seals, and the new hydrostatic rotor rim bearings. The runner diameter will be 3000 mm, head 4 m, output 1.5 MW, and discharge 50 m^3/s. Escher Wyss recently received an order to proceed with preparation of shop drawings and manufacture of this machine. The scheduled date for commissioning of this unit is December 1980.

Escher Wyss is designing and furnishing key elements of seven STRAFLO units for the Lixhe and Andenne power stations in Belgium. ACEC of Charleroi, Belgium, under license from Escher Wyss, has a contract from Socolie for these units and is to do the major portion of the manufacturing work. The units will have fixed blade runners, conventional hydrodynamic central bearings, and lip seals. Data describing these units follow:

	Andenne	Lixhe
Number of units	3	4
Runner Dia, mm	3,550	3,550
Rated Head, m	4.9	7.4
Rated Discharge, Each m^3/s	77	87.8
Rated Output, Each MW	3.21	5.53
Max. Output, Each MW	3.5	5.85
Rated Speed, rpm	107.1	120
Runaway Speed, rpm	245	281
Centerline of Runner Below Min. Tailwater, m	1.7	3.55
Distance Between Turbine Axes, m	8.35	8.35

Application Ranges of STRAFLO Units

The application of the alternative seal and bearing designs which can now be provided are shown in Figure 21, based on fixed blade runner designs.

TUBE TURBINES

The term TUBE is a trademark of Allis-Chalmers Corporation of York, Pennsylvania, and refers to an axial flow turbine with a generator located outside the water passage and connected to the turbine either by a direct drive shaft, or through a geared type speed increaser.

Allis-Chalmers has pioneered in the development of this type of unit and is now in a position to offer standardized packaged designs based on horizontal shaft units and 10 different runner diameters ranging from 750 mm to 3000 mm. Unit outputs will vary from 50 kW to 5000 kW, depending upon runner size and operating head.

TUBE turbines are propeller type units which provide the highest feasible operating speed and maximum power output for a given runner size. This permits use of a smaller powerhouse structure in terms of both area and height, thus reducing excavation and other civil costs. Unit spacing is only about 2/3 that required for a conventional vertical shaft propeller turbine of the same rating.

The packaged design now includes an intake butterfly valve for flow control, either a fixed blade or adjustable blade turbine runner, fixed guide vanes, hollow turbine shaft with guide and thrust bearings, a hydraulic operator for blade control (if required), a geared type speed increaser, a synchronous or induction type generator with fan cooling, static excitation and voltage regulation equipment, electrical controls, and low voltage switchgear. Figure 22 is a photograph of the first standardized unit to be manufactured (late 1978) and shows the various components associated with this 1500 mm diameter turbine and 1000 kW generator.

Figures 23 and 24 show details of two sloping shaft custom designed TUBE units which were installed in the Dolby Plant in Maine. These are adjustable blade turbines rated 5680 hp each at 48 feet head, and directly connected to induction generators.

Figure 25 is a TUBE unit sizing chart showing the potential power output for different runner sizes and operating heads. Five blade runners are used for the higher heads and four blades for the lower heads. Figure 26 shows the expected TUBE turbine efficiency curves for different blade and gate configurations.

COSTS OF TURBOGENERATORS AND ACCESSORIES FOR LOW HEAD HYDRO DEVELOPMENTS

As mentioned previously, the available operating head has a significant effect on the physical size and cost of hydraulic turbines. This is due to the fact that the volume of water needed to produce a specified amount of power varies inversely with the head. For example, when the head changes from 10 m to 5 m, twice as much water is needed to produce the same power output.

Valid cost data for hydraulic turbines, generators, and accessory equipment is difficult to obtain since there are so many design variables involved. However, based on various pricing information recently obtained, the author has developed approximate equipment cost curves which can be used as a guide in making very preliminary assessments of a potential hydroelectric installation.

The cost curves in Figure 27 show the relationship between the installed cost per kW of capacity and the rated net head in meters for several of the different types of turbogenerators covered in this paper. An explanation of each of the four curves follows.

Curve A shows the approximate installed cost of cross-flow turbine units, including generator, speed increaser, governor, flywheel, exciter, and other required accessories. The costs are based on unit capacities in the range of 200 to 400 kW.

Curve B shows the approximate installed cost of axial flow propeller turbines with rim type synchronous generators (STRAFLO units), including fixed runner blades, adjustable wicket gates, governor, exciter, and other normal accessories. The costs are based on unit capacities in the range of 1500 to 7000 kW and a runner diameter of 3000 mm.

Curve C shows the approximate installed cost of small TUBE units, including adjustable blade propeller turbine, fixed guide vanes, synchronous generator, thrust bearing, governor, excitation equipment, and butterfly valve for intake flow control. The costs are based on unit capacities in the range of 50 to 5000 kW, and runner diameters from 750 to 3000 mm.

Curve D shows the approximate installed cost of small vertical shaft Francis units of the Samson type in an open flume setting, including speed increaser and a horizontal shaft synchronous generator, exciter, governor, and usual accessories. The costs are based on unit capacities ranging from 200 to 600 kW and a runner diameter of 1423 mm. While the open flume setting results in simpler structures and somewhat lower equipment costs, there is some reduction in the overall unit efficiency which affects the annual power production and value.

When selecting the type of turbogenerating equipment to be used for a proposed low head hydroelectric development, many aspects must be considered, with equipment cost being only one of them. The evaluations should give due consideration to such other features as: existing site conditions, variation of flows and heads, overall unit efficiency, operating experience with similar equipment, annual operation and maintenance costs, total powerhouse cost (structures and equipment), and any special plant operating requirements. It is also important to note that the cost of foreign manufactured equipment, expressed in terms of U.S. dollars, can change drastically in a short time due to changes in the exchange rate and differing inflationary trends. The results can be either beneficial or detrimental, insofar as the purchaser is concerned.

TOTAL POWER COSTS FOR LOW HEAD HYDRO PLANTS AT EXISTING DAMS

The writer has prepared additional cost curves to give a very quick and rough estimation of total power costs for a potential hydro development at an existing low head dam. The curves in Figure 28 show the total investment cost per kW of installed capacity in 1979 dollars and the average annual cost per kWh of energy produced for different plant capacity factors and different rated heads.

The total investment costs are based on standard TUBE units and include allowances for an at-site switchyard, engineering, construction supervision, owners administrative and legal expenses, contingency, interest during construction, and project licensing costs. It also is assumed that development is by a tax exempt agency using revenue bond financing.

ACKNOWLEDGEMENT

Much of the information and data contained in this paper were obtained from a report prepared by Stone & Webster Engineering Corporation for the U.S. Department of Energy which is entitled "Report on Turbogenerating Equipment for Low Head Hydroelectric Developments" and dated April 1978.

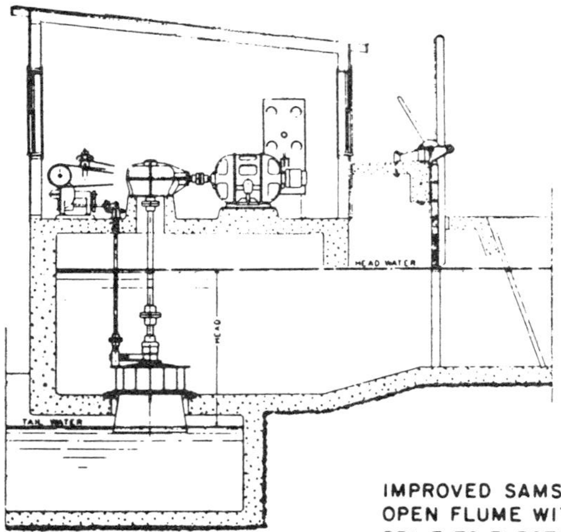

IMPROVED SAMSON TURBINE IN OPEN FLUME WITH SPEED INCREASER DRIVE TO ELECTRIC GENERATOR.

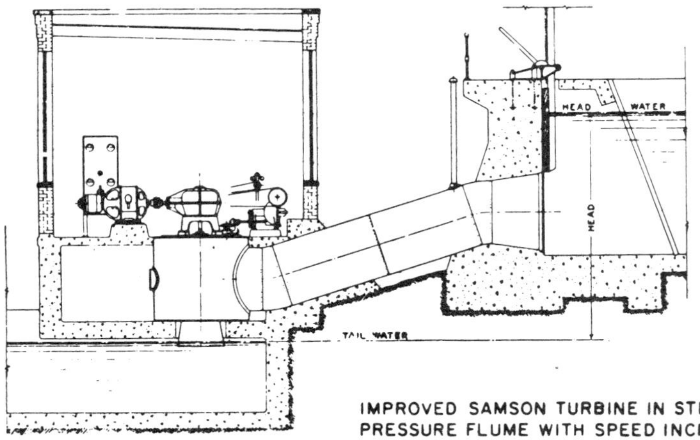

IMPROVED SAMSON TURBINE IN STEEL PRESSURE FLUME WITH SPEED INCREASER DRIVE TO ELECTRIC GENERATOR.

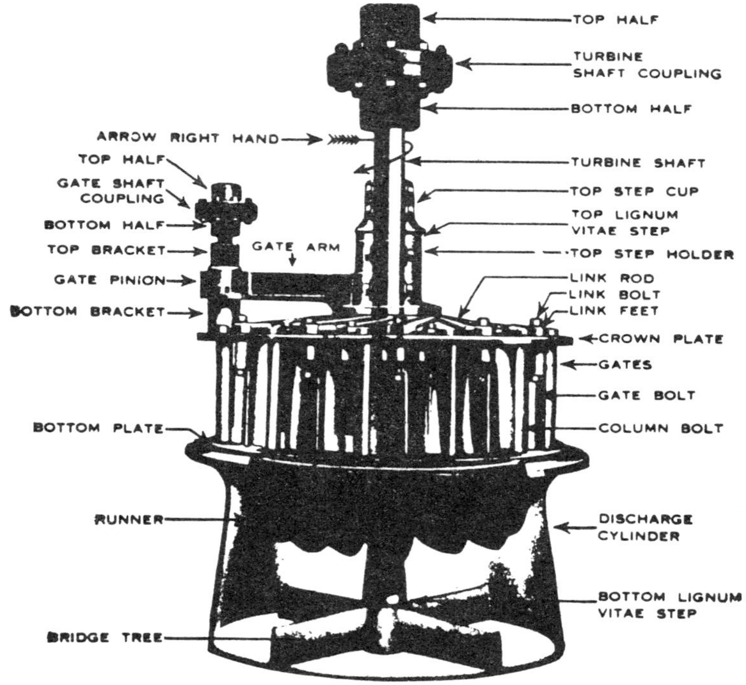

FIGURE I
THE JAMES LEFFEL & CO.
VERTICAL SAMSON TURBINES

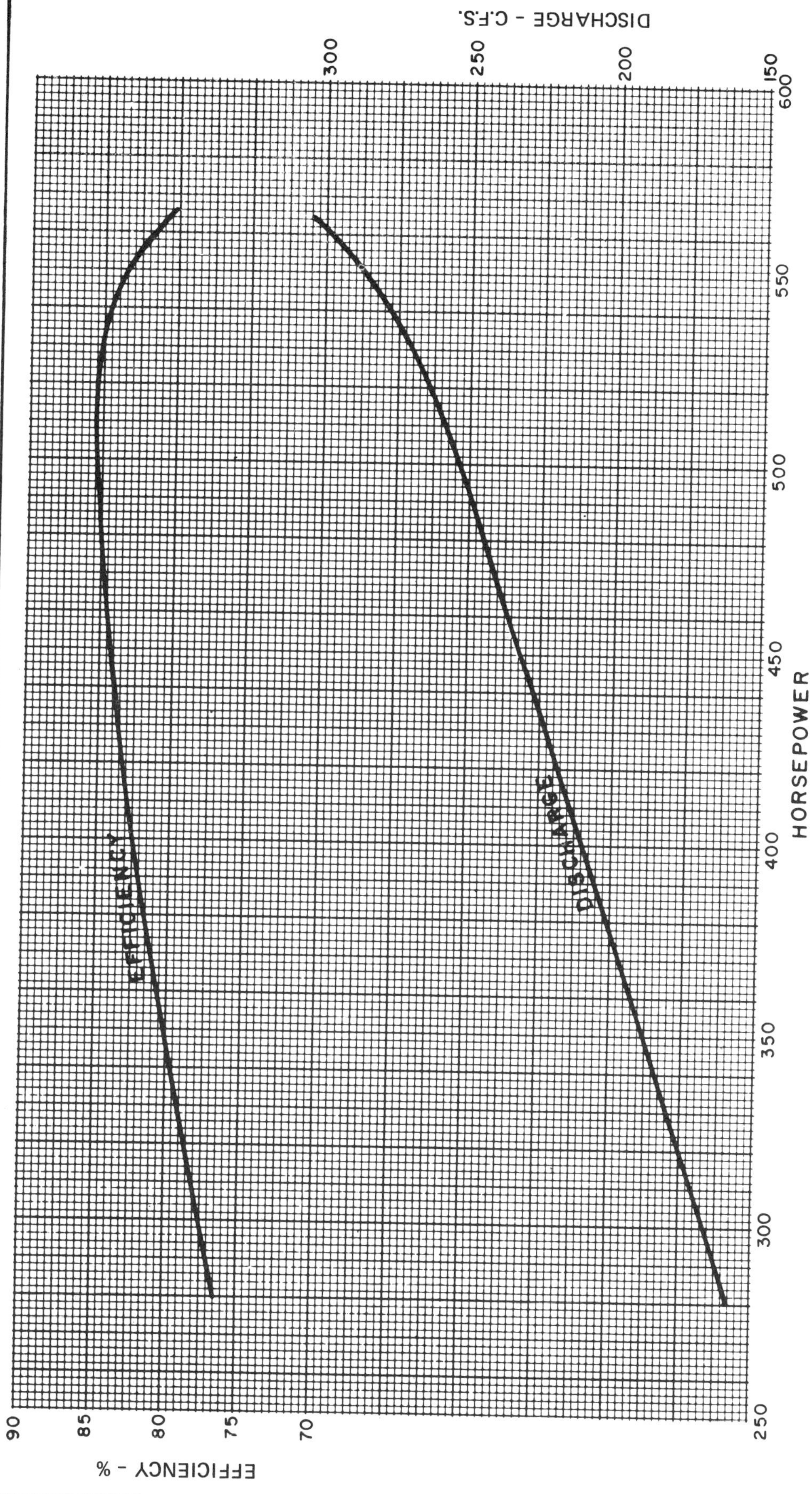

FIGURE 2
EXPECTED PERFORMANCE
56 IN DIA SAMSON TURBINE
130 RPM AND 20 FT HEAD
THE JAMES LEFFEL CO.

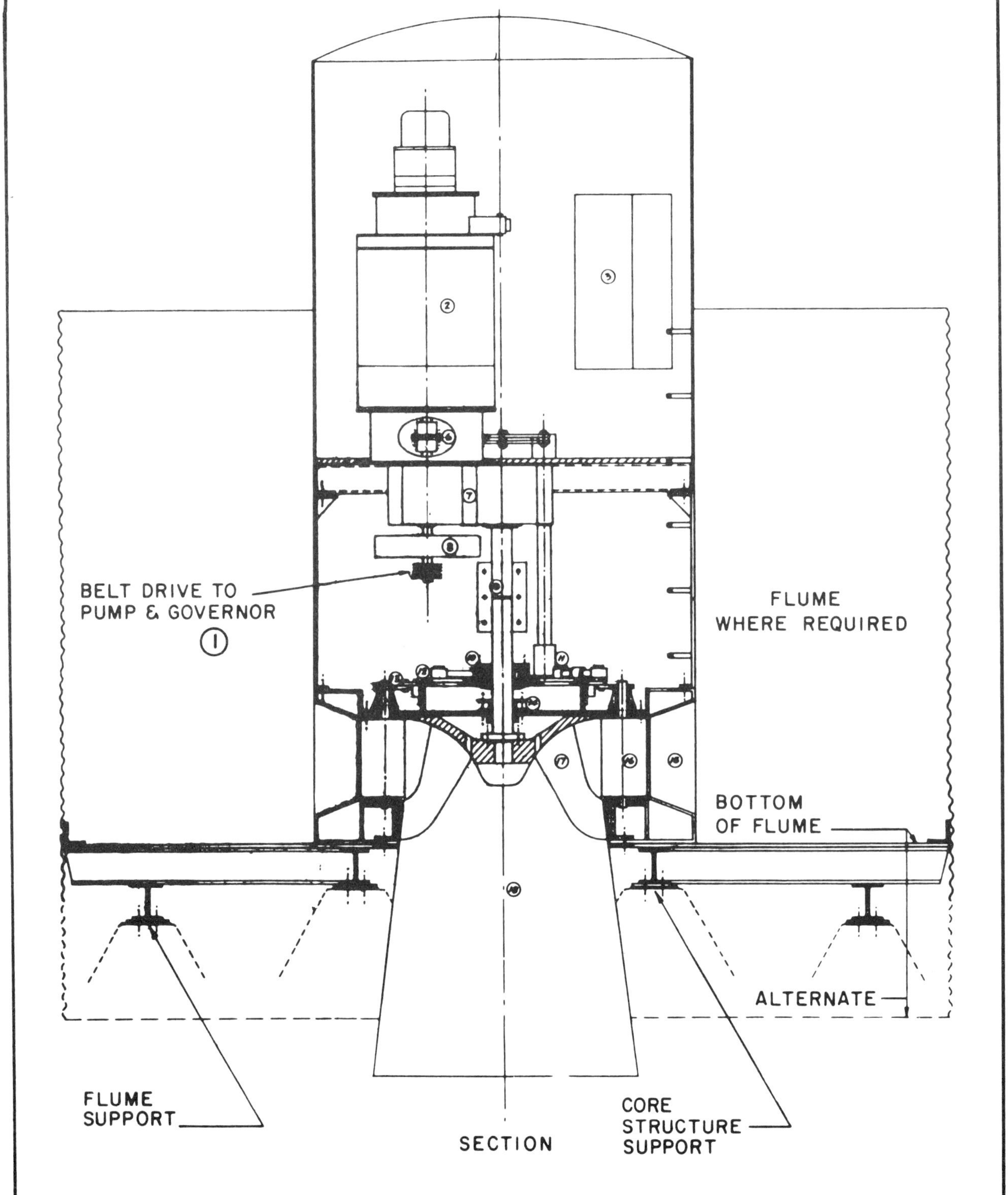

1. GOVERNOR
2. GENERATOR
3. CONTROL BOX
4. —
5. —
6. SPLIT COUPLING
7. GEAR BOX
8. FLYWHEEL
9. RUNNER SHAFT & CMPRSN. COUPLING
10. TURBINE GUIDE BEARING
11. GATE SHAFT & PULL ROD
12. REGULATING RING
13. LINK
14. PACKING GLAND
15. STAYVANE
16. WICKET GATE
17. RUNNER
18. DRAFT TUBE

FIGURE 3
MINI-HYDEL STATION
CANADA FRONTIER WATER & POWER LTD

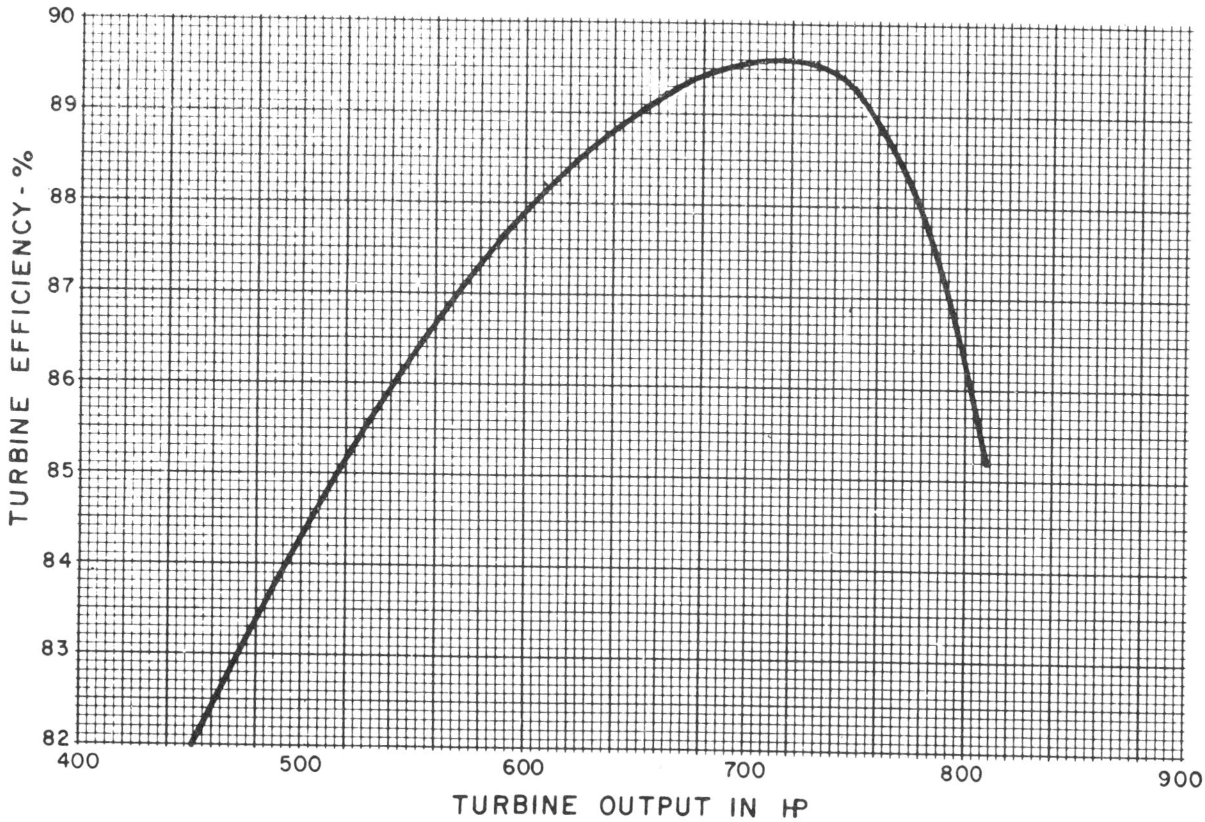

NOTE: CURVE BASED ON MODEL TEST DATA STEPPED-UP BY MOODY FORMULA

FIGURE 4

EXPECTED PERFORMANCE
MINI-HYDEL STATION
1500 MM RUNNER DIA.
RATED HEAD = 25 FT.

CANADA FRONTIER
WATER & POWER LTD.

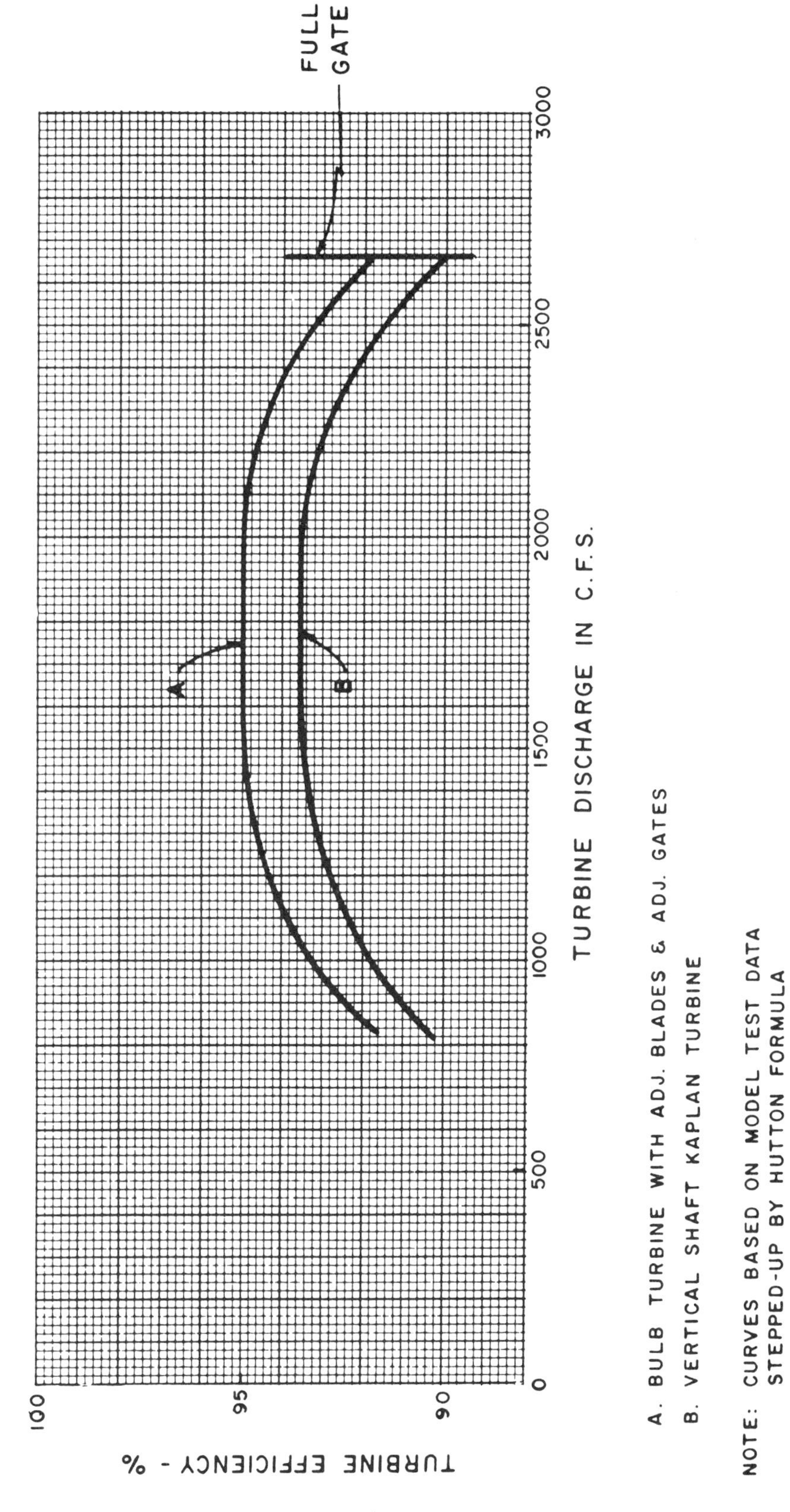

FIGURE 5
COMPARISON OF TURBINE EFFICIENCY
BULB UNIT VS KAPLAN UNIT
3000 MM DIA. RUNNER - HEAD = 10M

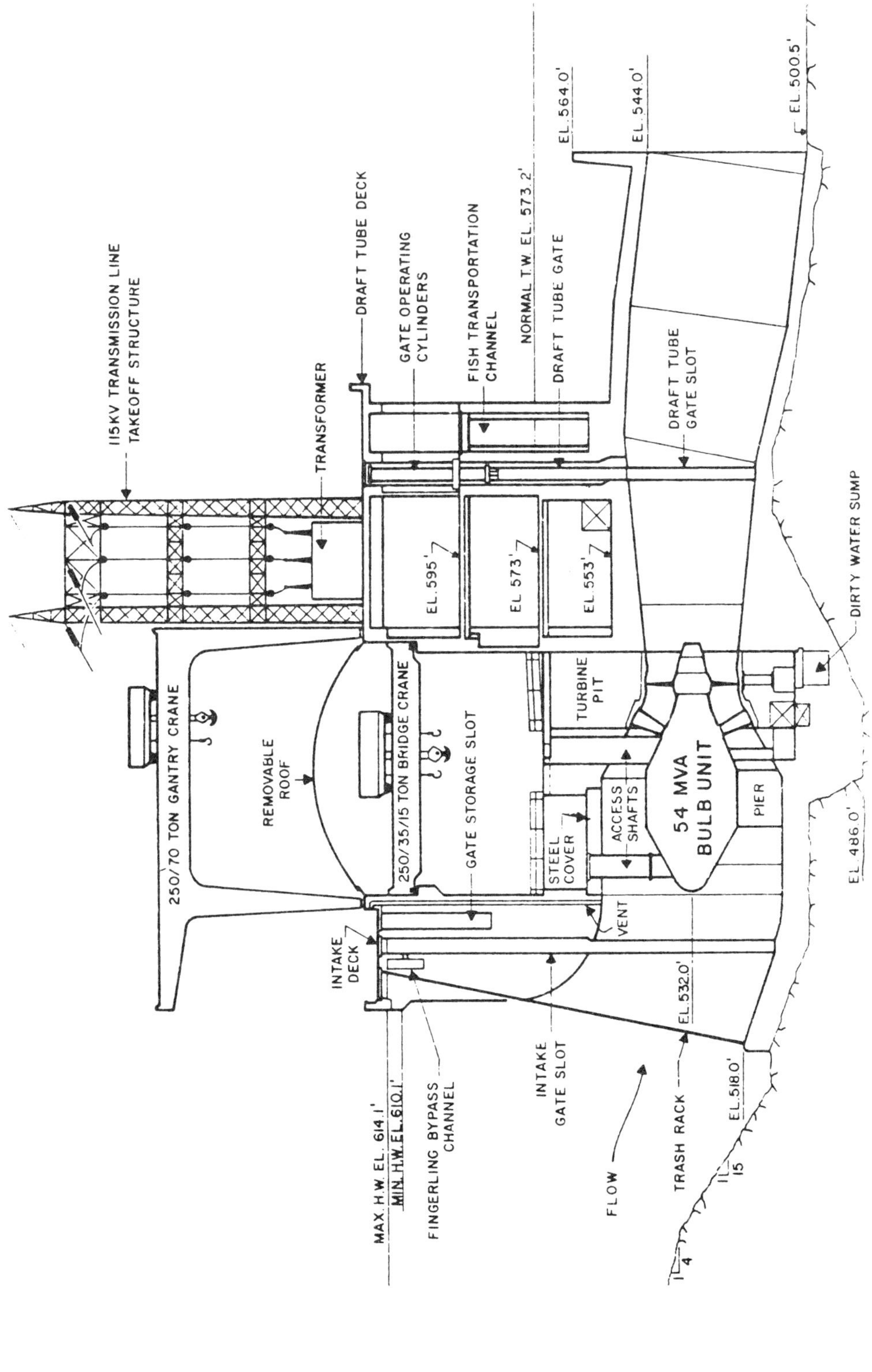

FIGURE 6
SECOND POWERHOUSE CROSS SECTION

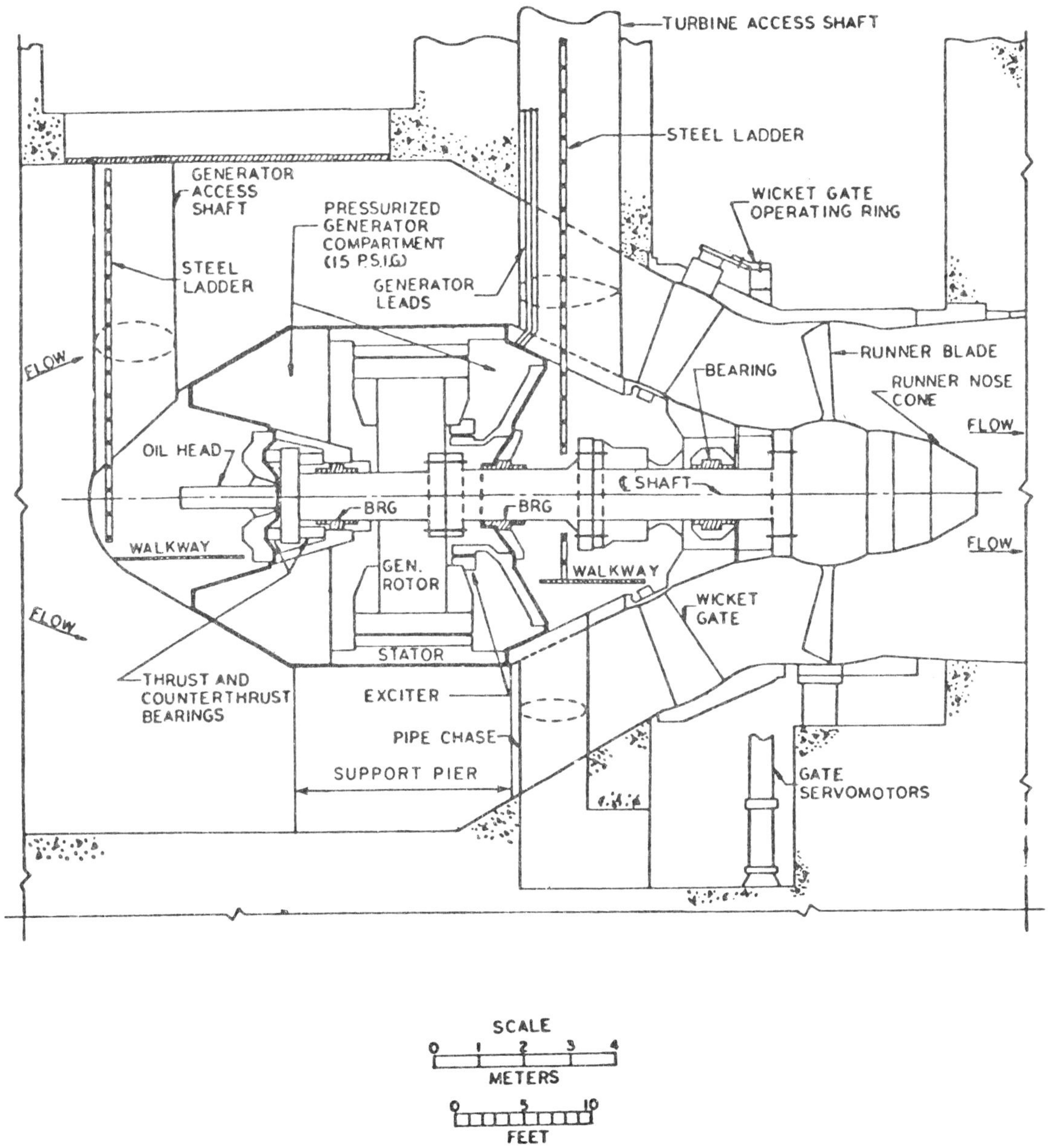

FIGURE 7

SECTION AT CENTERLINE OF BULB UNIT, ROCK ISLAND

FIGURE 8

LOWERING GENERATOR ROTOR DURING BULB UNIT ASSEMBLY

FIGURE 9

ROTATING COMPLETED GENERATOR ASSEMBLY TO PERMIT LOWERING THROUGH HATCH AND ERECTION IN WATER PASSAGE BELOW

VIEW FROM UNIT 8 INTAKE SHOWING GENERATOR ASSEMBLY RESTING ON SUPPORT PEDESTAL AND PIERS PRIOR TO MOVING AGAINST STAYRING.

PUBLIC UTILITY DISTRICT NO. 1 OF CHELAN COUNTY, WASHINGTON

Stone & Webster Engineering Corporation, Agents

FIGURE 10

July 27, 1977

VIEW FROM DOWNSTREAM COFFERDAM AT CELL 9 SHOWING PROGRESS AT UNITS 1 THROUGH 8

PUBLIC UTILITY DISTRICT NO. 1 OF CHELAN COUNTY, WASHINGTON

Stone & Webster Engineering Corporation, Agents

FIGURE 11 September 26, 1977

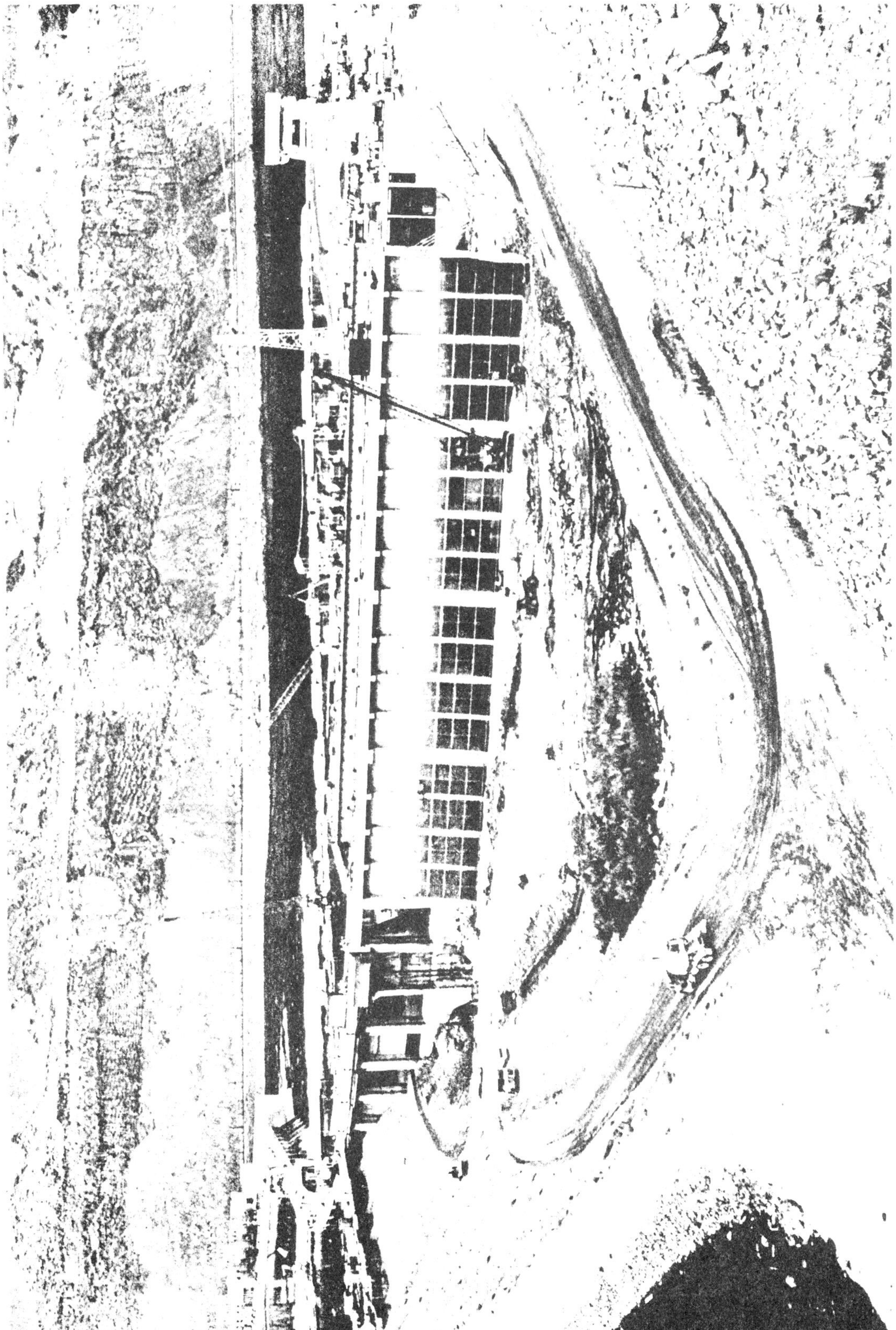

VIEW FROM TOP OF RIGHT BANK SHOWING INTAKE STRUCTURE
AT UNITS 1 THROUGH 8

PUBLIC UTILITY DISTRICT NO. 1 OF CHELAN COUNTY, WASHINGTON

Stone & Webster Engineering Corporation, Agents

FIGURE 12

August 24, 1977

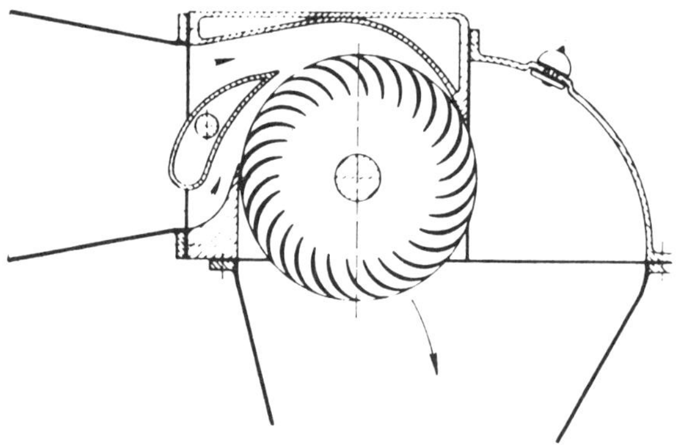

FLOW PATTERN HORIZONTAL ADMISSION

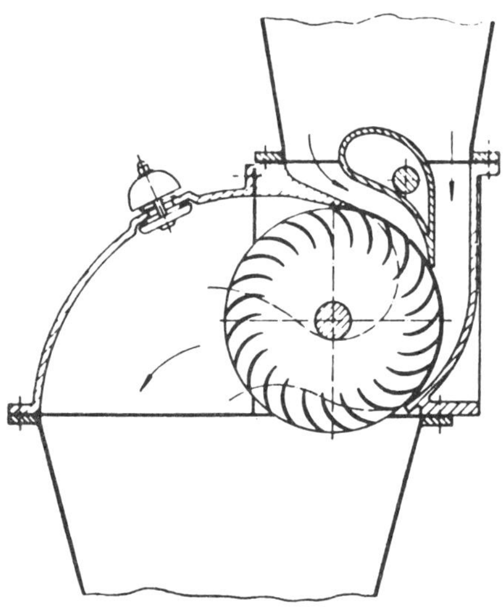

FLOW PATTERN VERTICAL ADMISSION

FIGURE 13
PRINCIPLE OF THE OSSBERGER CROSS FLOW TURBINE

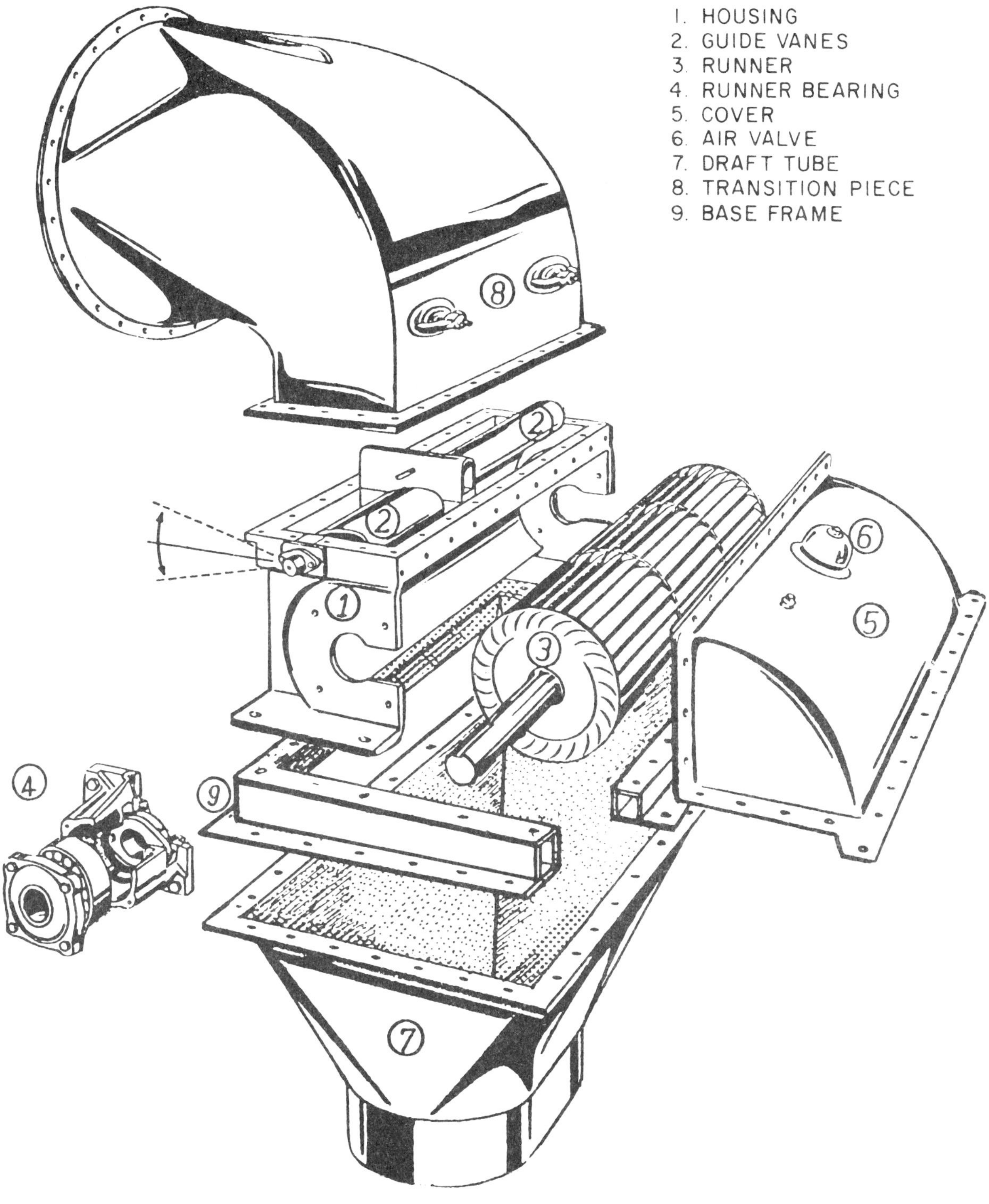

FIGURE 14

VIEW OF TYPICAL OSSBERGER TURBINE ASSEMBLY

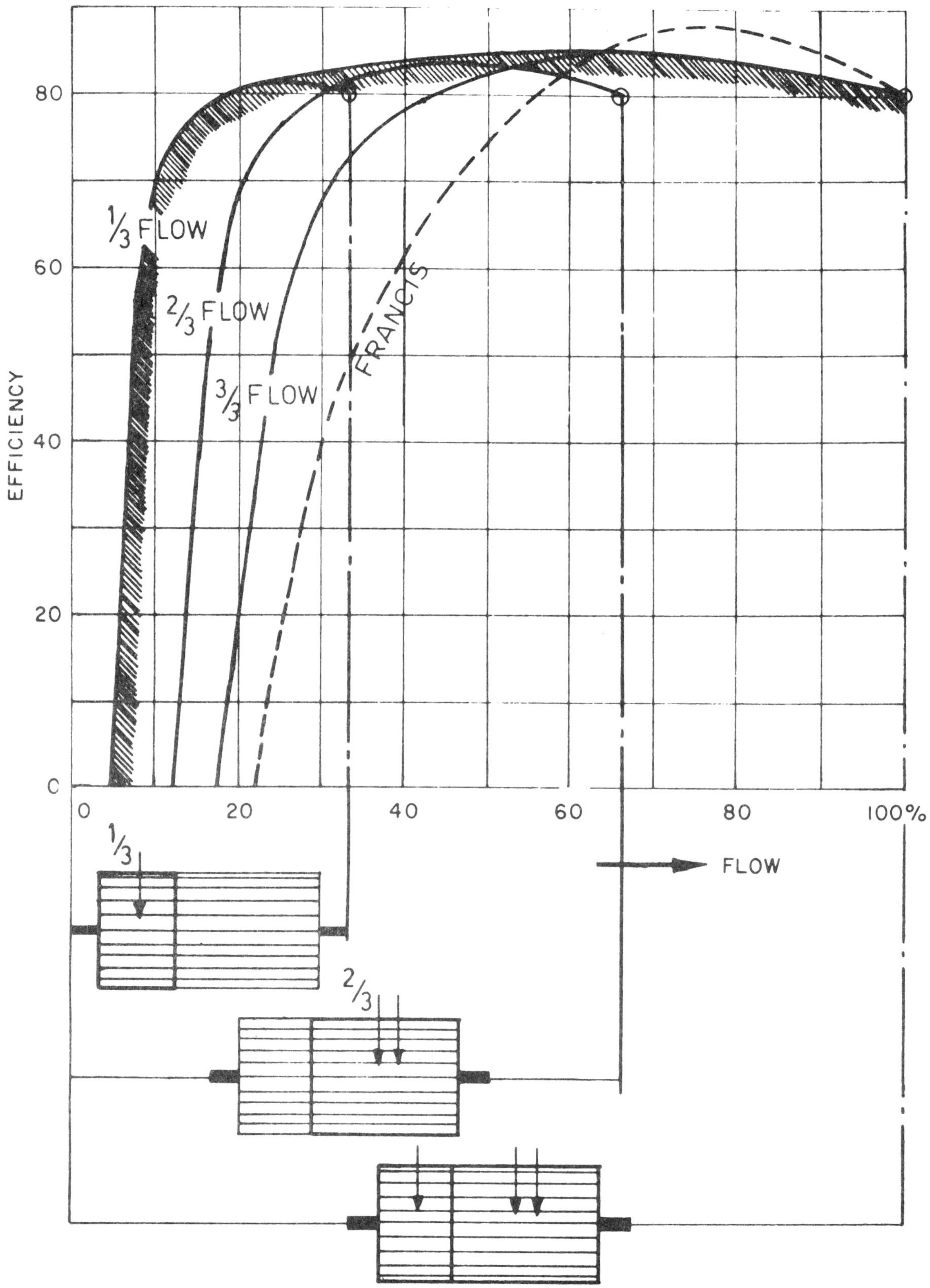

FIGURE 15

TYPICAL EFFICIENCIES
OSSBERGER VS FRANCIS TURBINES

OSSBERGER UNIT
WISSMAN MILL, GERMANY
276 HP, 248 cfs,
11'-9" Hd

AN OSSBERGER RUNNER

FIGURE 16

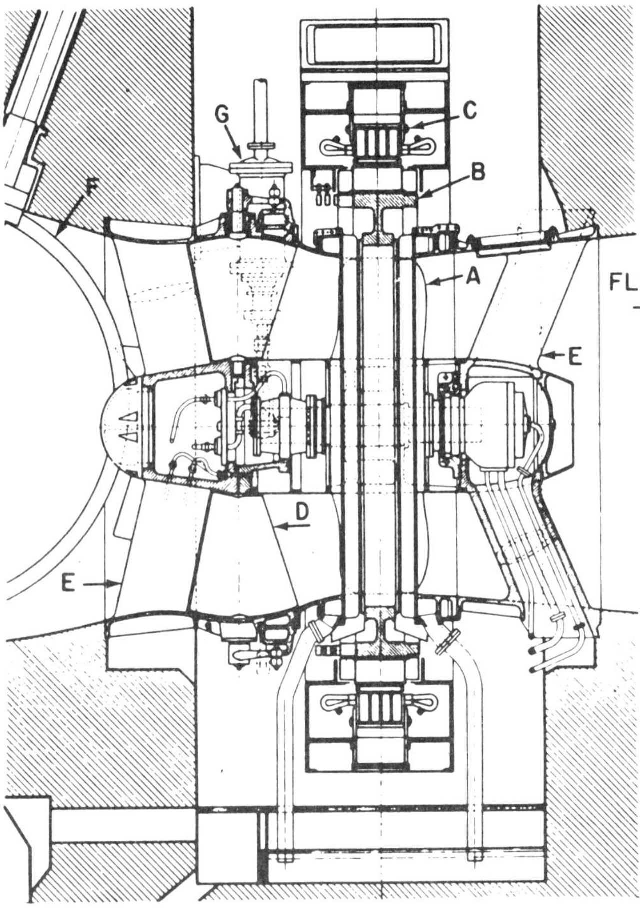

KEY

A. RUNNER BLADES
B. GENERATOR ROTOR
C. GENERATOR STATOR
D. WICKET GATES
E. STAY (SUPPORT) VANES
F. BUTTERFLY VALVE
G. GATE SERVOMOTOR

STRAFLO UNIT

TYPICAL FOR PLANTS ON ILLER, LECH AND SAALACH RIVERS

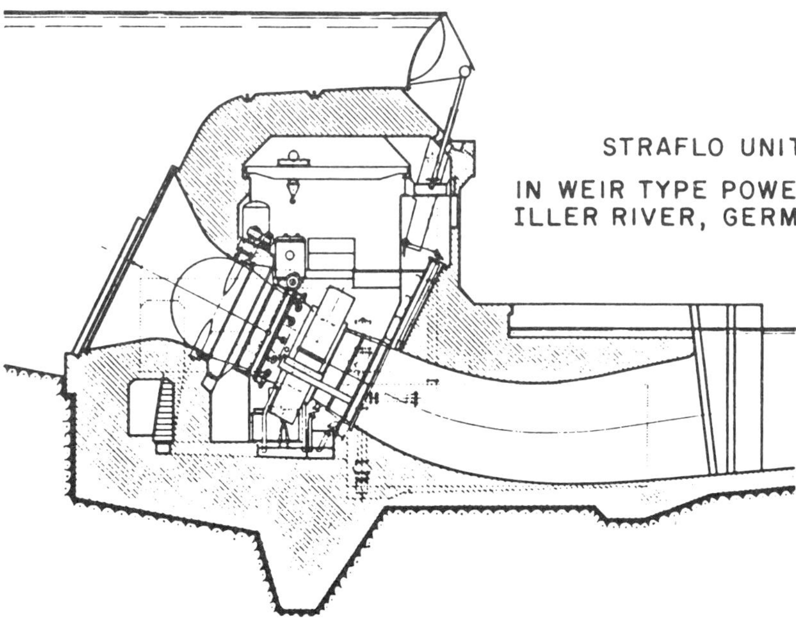

STRAFLO UNIT

IN WEIR TYPE POWERHOUSE
ILLER RIVER, GERMANY

FIGURE 17

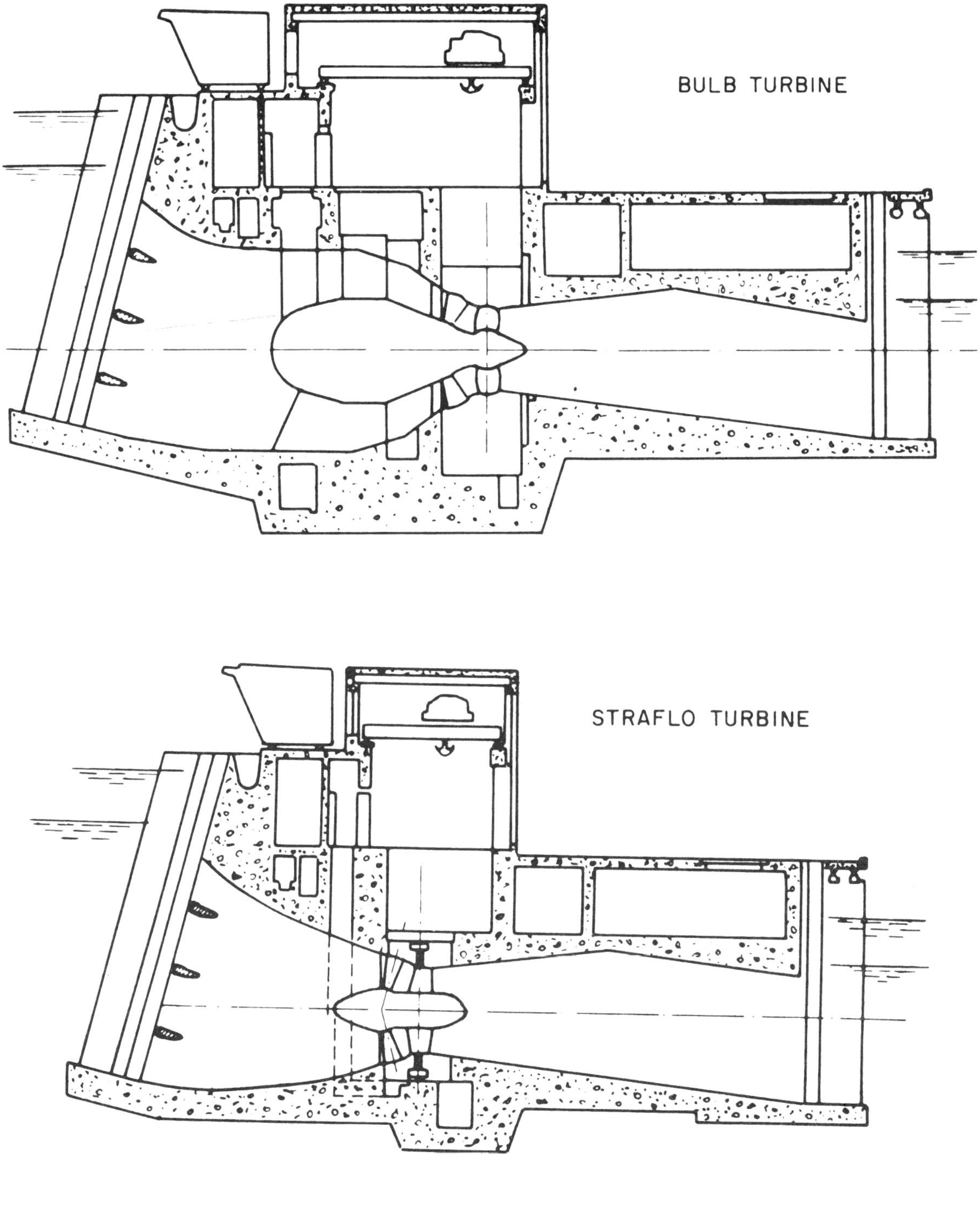

FIGURE 18

COMPARISON OF STRUCTURES REQUIRED FOR STRAFLO VS BULB TURBINES WITH SAME OUTPUT AND HEAD

POWERHOUSE IS LOCATED BENEATH SPILLWAY

FIGURE 19
STRAFLO UNITS
ILLER RIVER, GERMANY

FROM RIGHT TO LEFT:

(1) INTAKE SECTION
(2) GATE OPERATING RING
(3) WICKET GATE ARMS
(4) GENERATOR HOUSING
(5) DISCHARGE SECTION

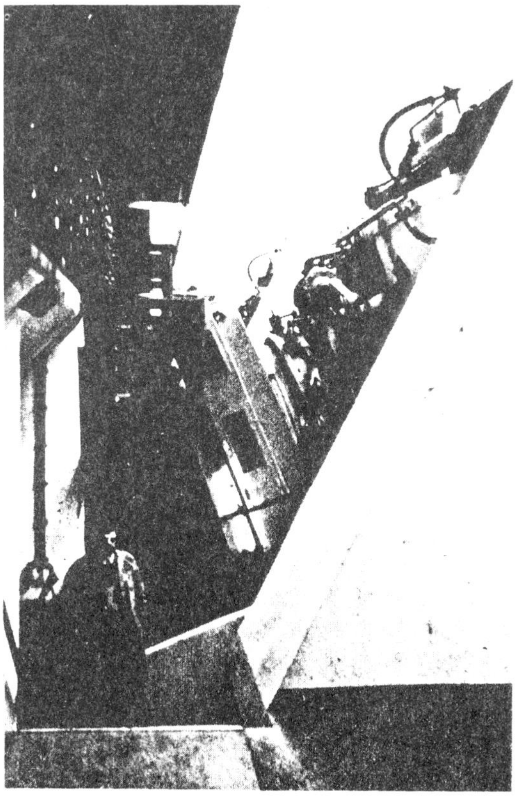

FIGURE 20

STRAFLO UNITS
ILLER RIVER, GERMANY

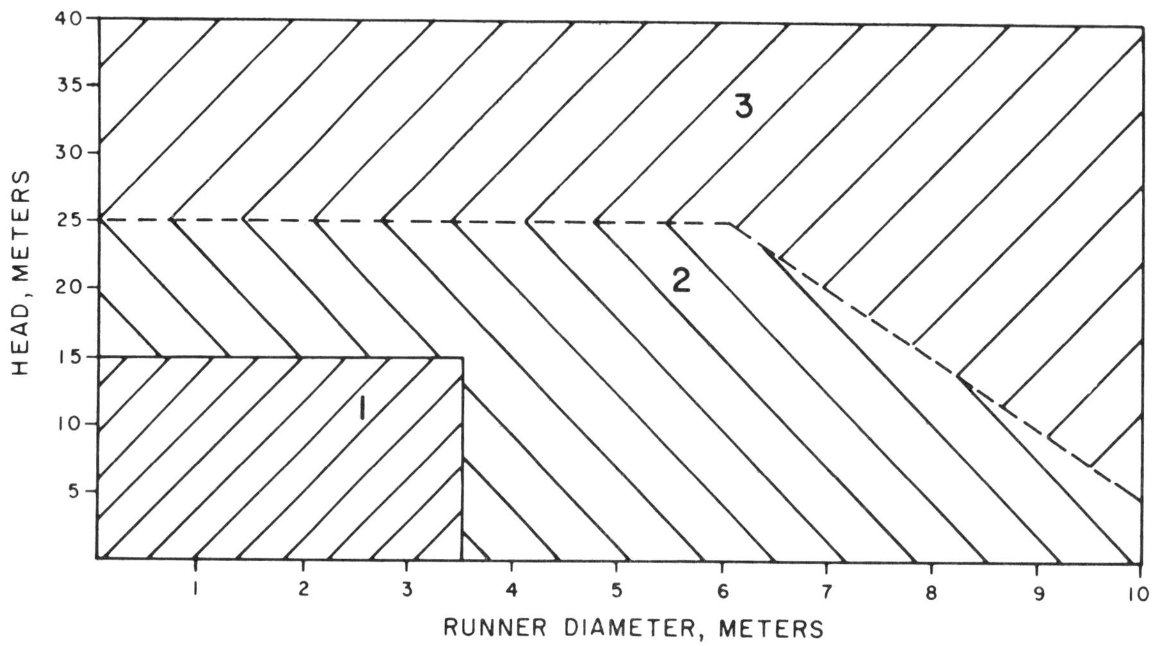

1. HYDRODYNAMIC CENTRAL BEARINGS, LIP SEALS.
2. HYDRODYNAMIC CENTRAL BEARINGS, HYDROSTATIC SEALS.
3. HYDROSTATIC OUTER RIM BEARINGS, HYDROSTATIC SEALS.

NOTE:

THE LIMITS SHOWN ARE SUBJECT TO REVISION AFTER THE COMPLETION OF ESCHER WYSS VIABILITY STUDIES

FIGURE 21

STRAFLO FIXED BLADE UNITS APPLICATION RANGES OF ALTERNATIVE SEAL AND BEARING DESIGNS

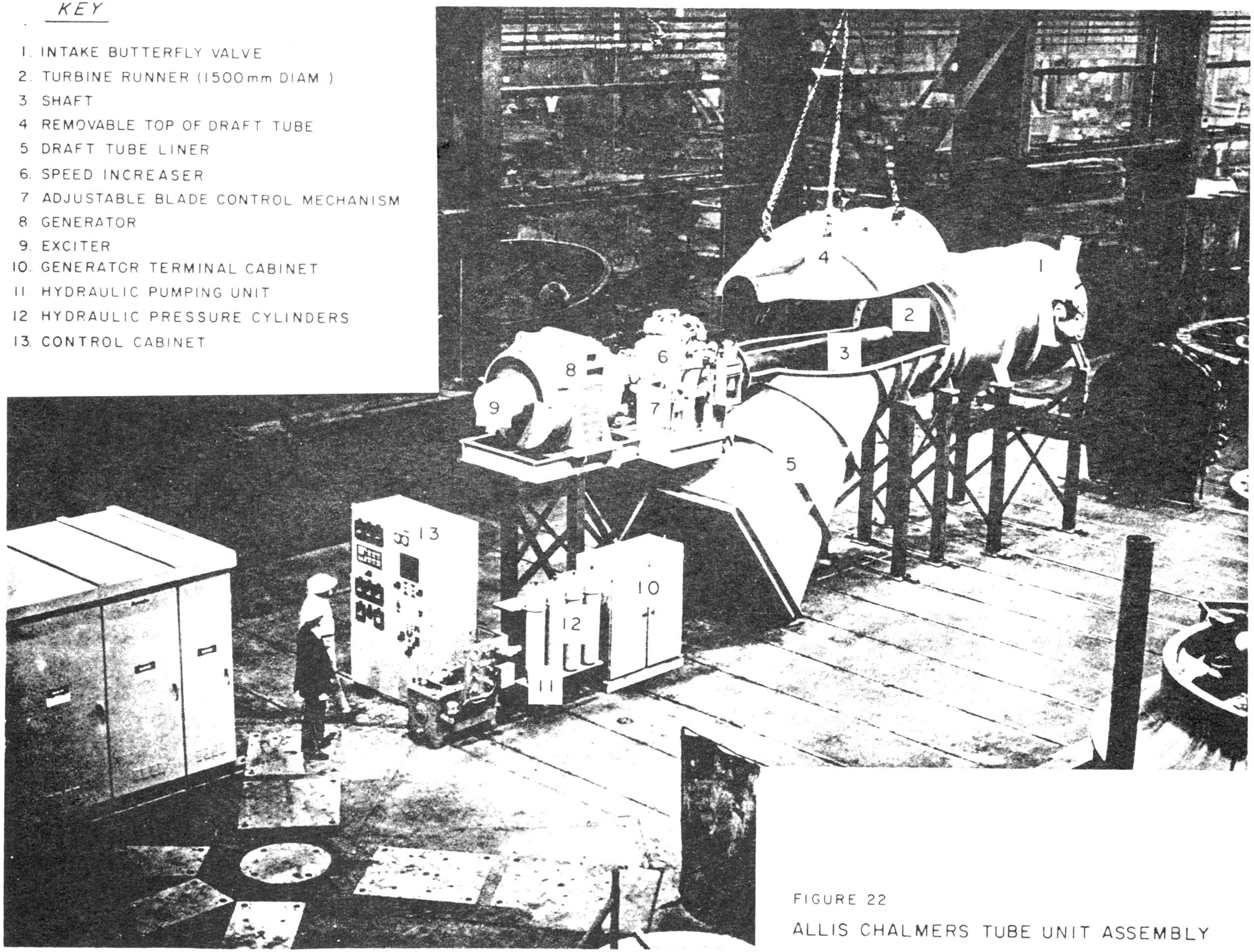

FIGURE 22
ALLIS CHALMERS TUBE UNIT ASSEMBLY

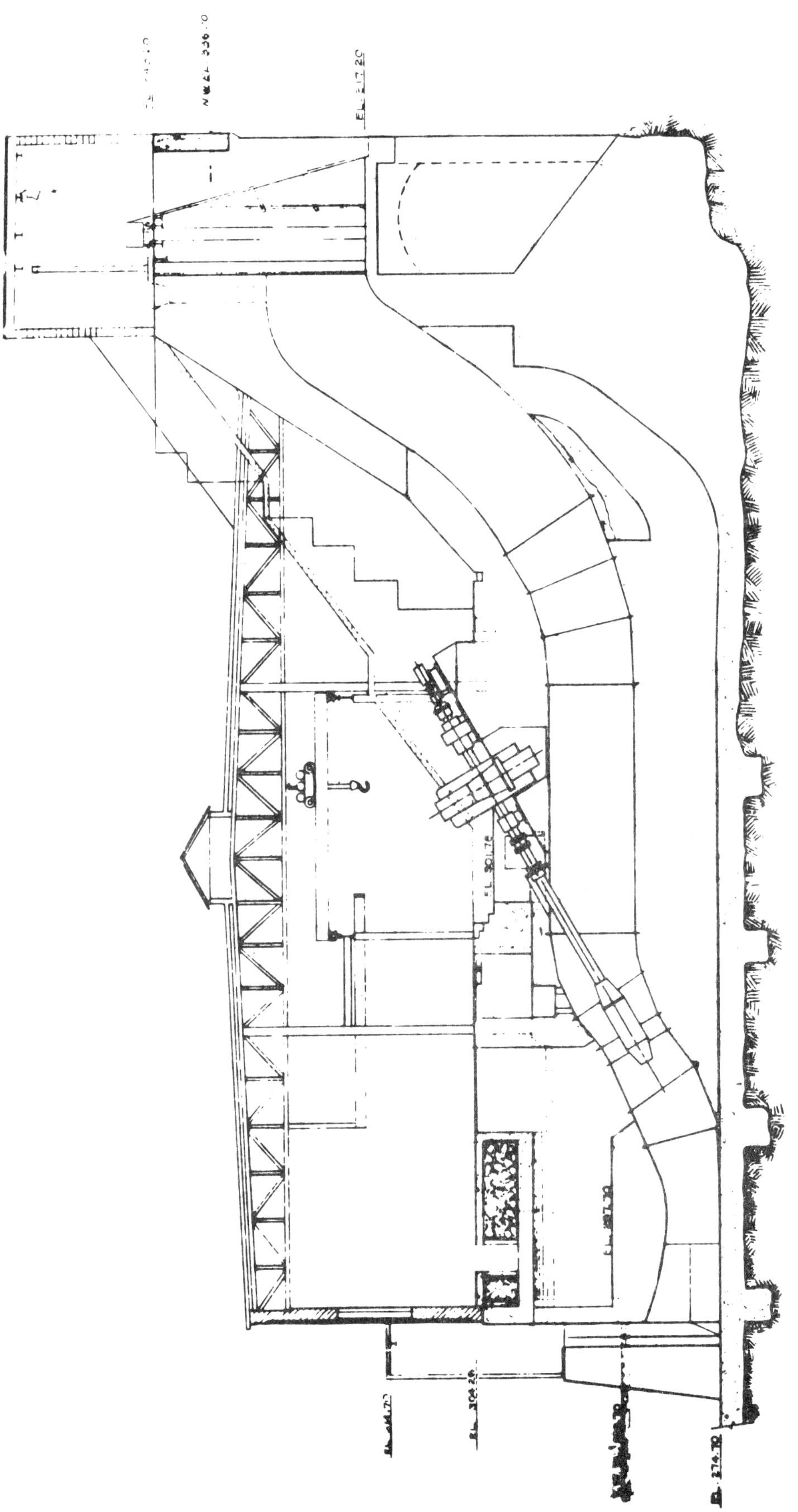

FIGURE 23
ALLIS CHALMERS TUBE UNITS
MODERNIZATION OF DOLBY PLANT
COURTESY OF GREAT NORTHERN PAPER CO

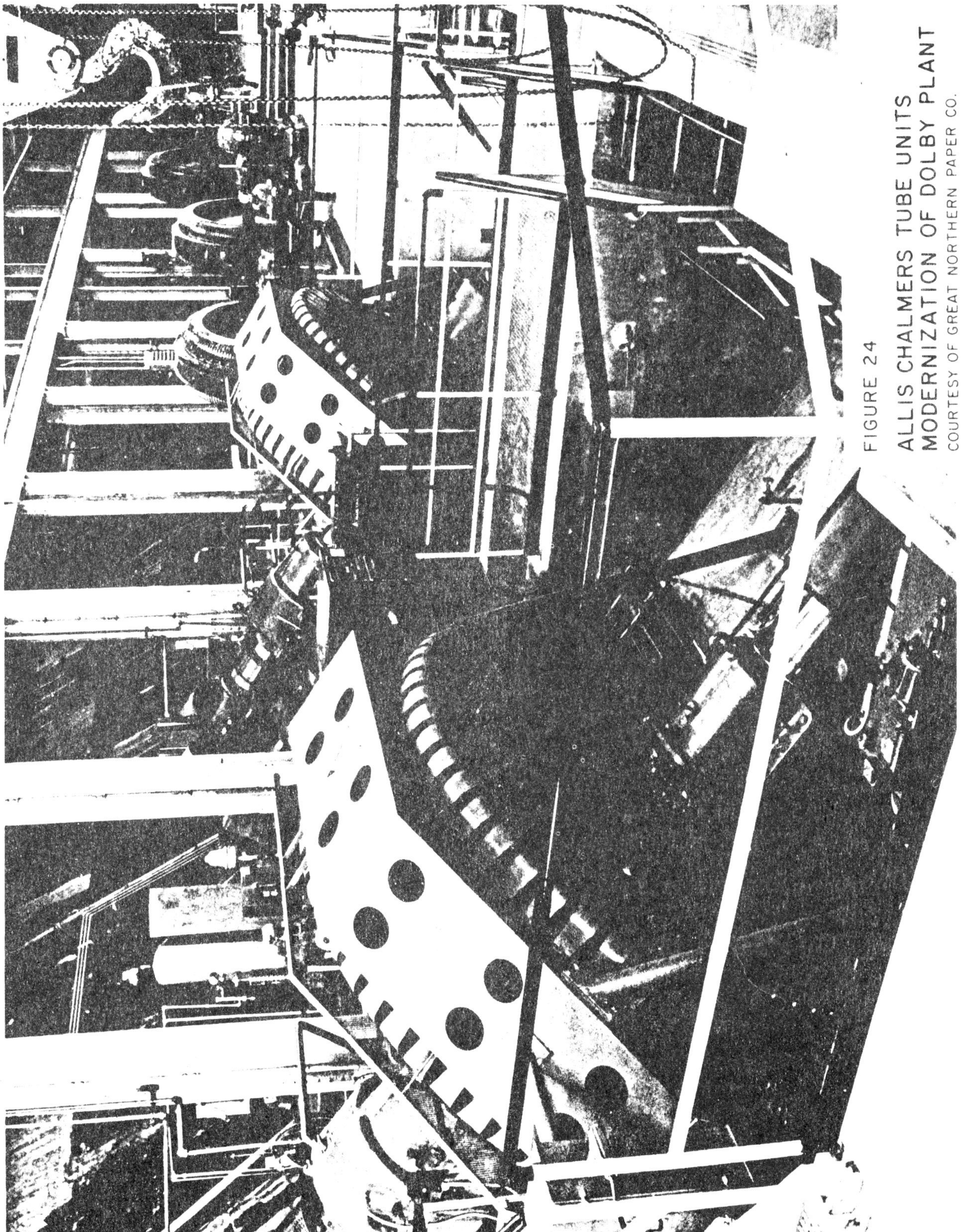

FIGURE 24

ALLIS CHALMERS TUBE UNITS
MODERNIZATION OF DOLBY PLANT

COURTESY OF GREAT NORTHERN PAPER CO.

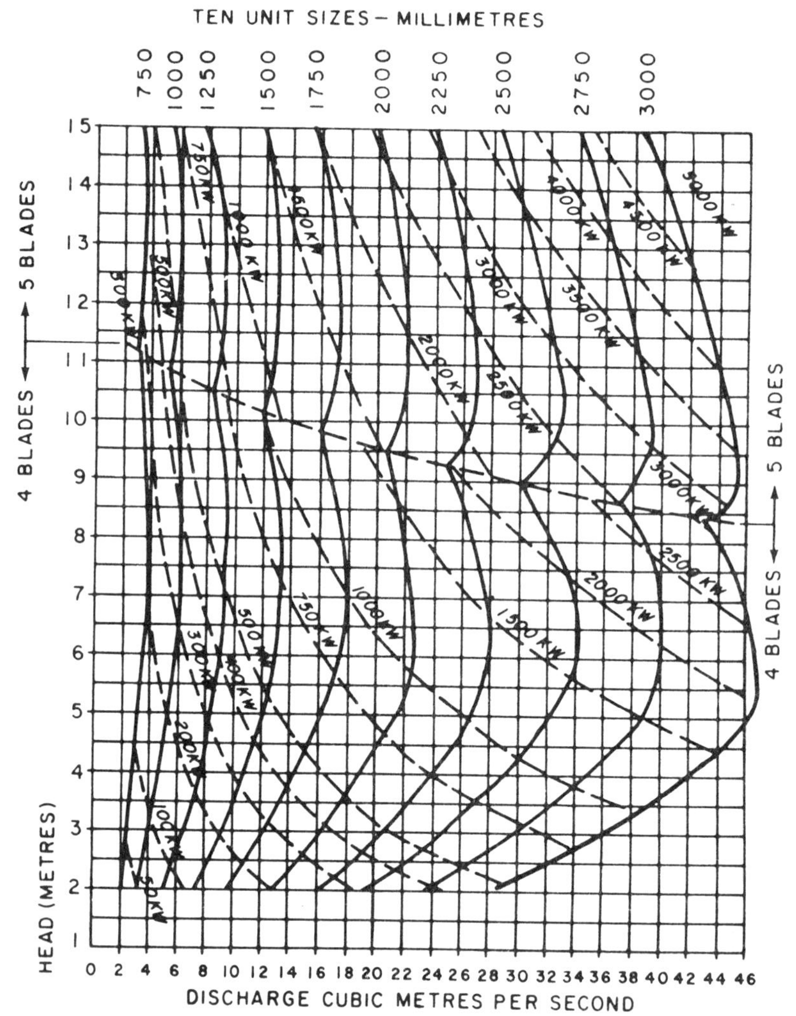

NOTE: OUTPUT BASED ON UNIT CENTERLINE SETTING OF ONE-HALF (1/2) RUNNER DIAMETER (DIA/2) ABOVE TAILWATER. CENTERLINE OF UNIT AT ELEVATION 150m (500FT.) ABOVE SEA LEVEL.

FIGURE 25

ALLIS-CHALMERS STANDARD TUBE UNITS SIZING CHART

FROM BULLETIN: 54B10241-01

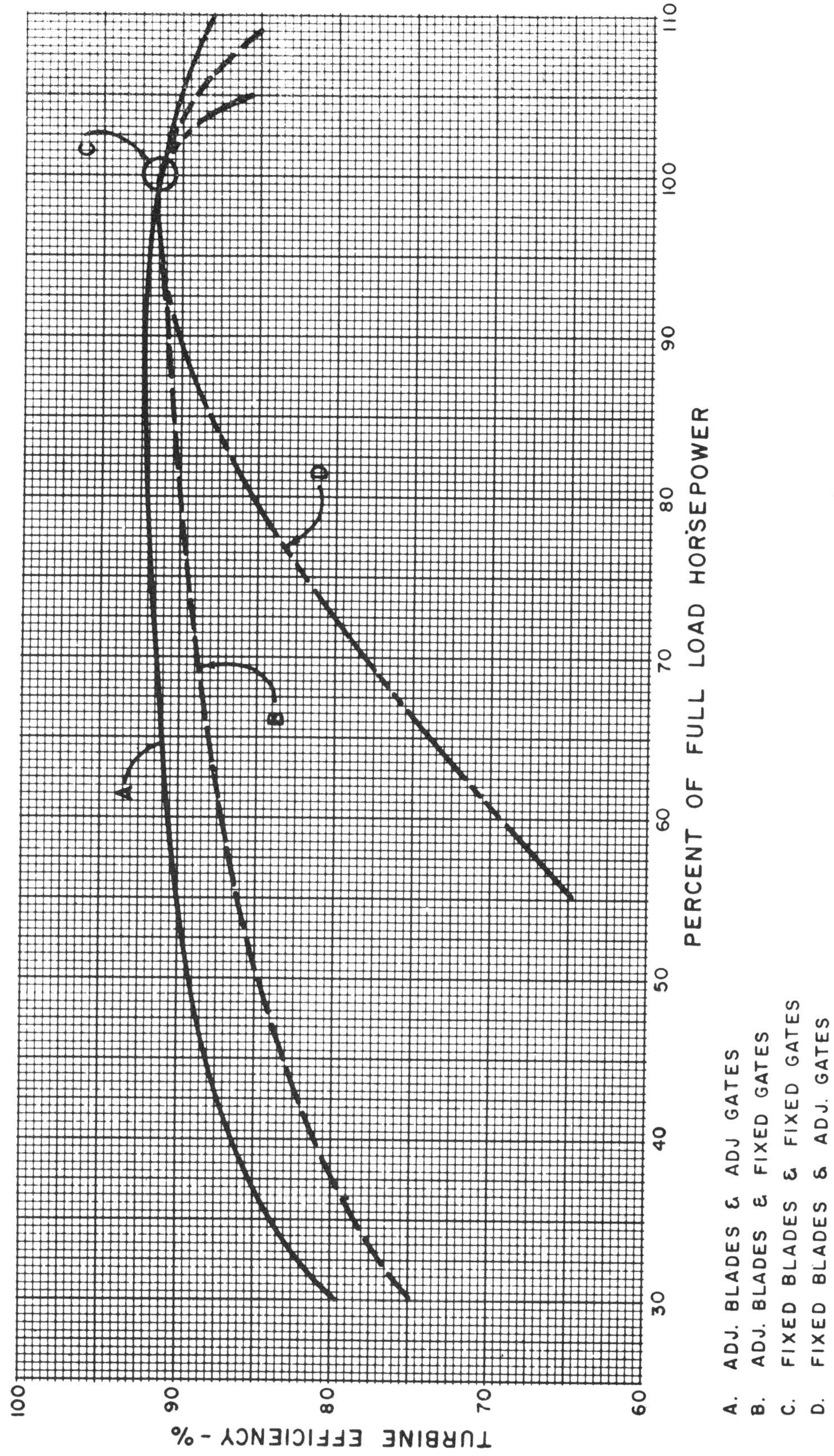

FIGURE 26
EXPECTED EFFICIENCY CURVES
ALLIS-CHALMERS TUBE TURBINES
WITH 3000 MM DIA. RUNNERS

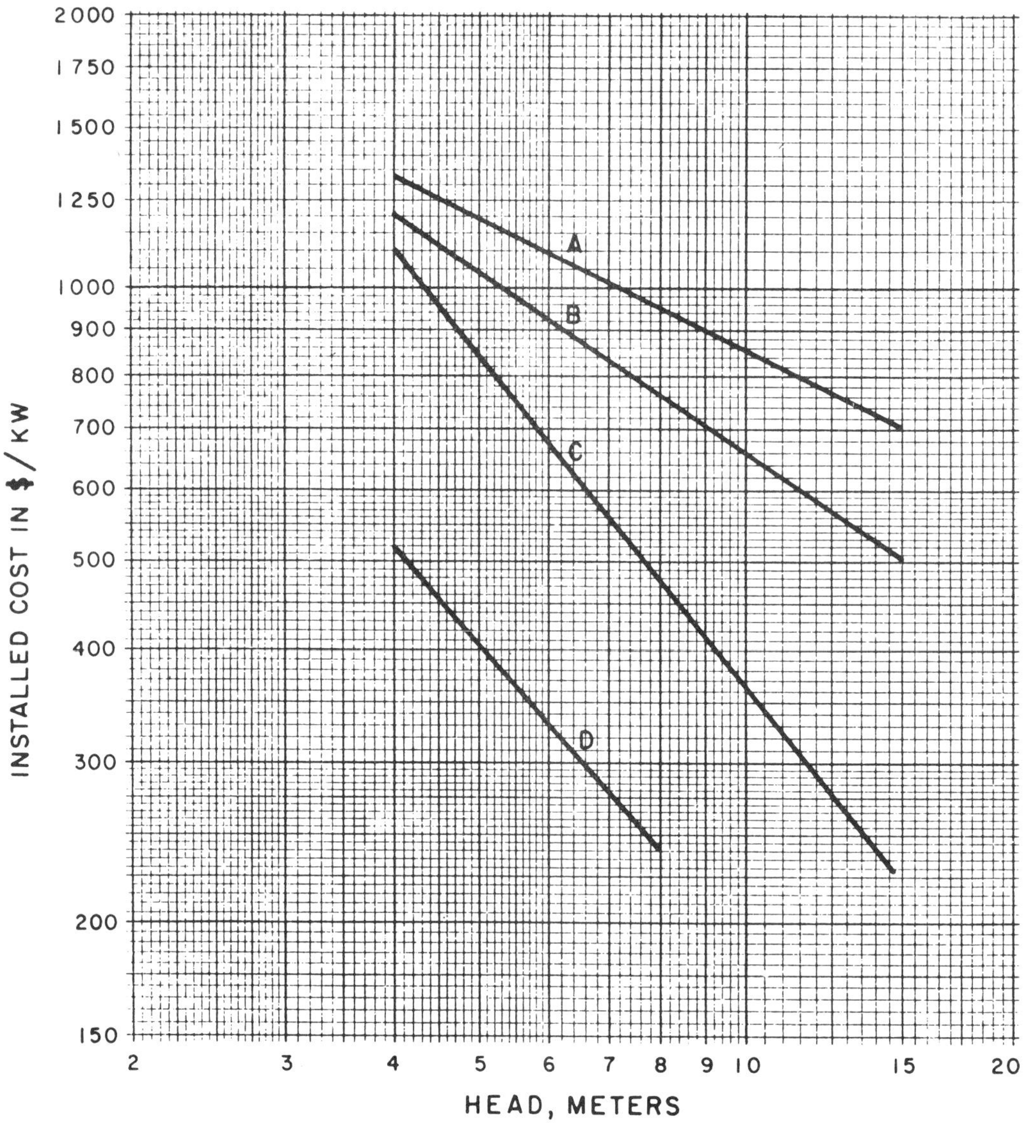

A. CROSS-FLOW (200 TO 350 KW)
B. STRAFLO (1800 TO 6700 KW)
C. TUBE (50 TO 5,000 KW)
D. VERTICAL FRANCIS SAMSON TURBINE WITH OPEN FLUME SETTING (200 TO 600 KW)

FIGURE 27

INSTALLED COST OF SMALL LOWHEAD TURBO-GENERATOR EQUIPMENT

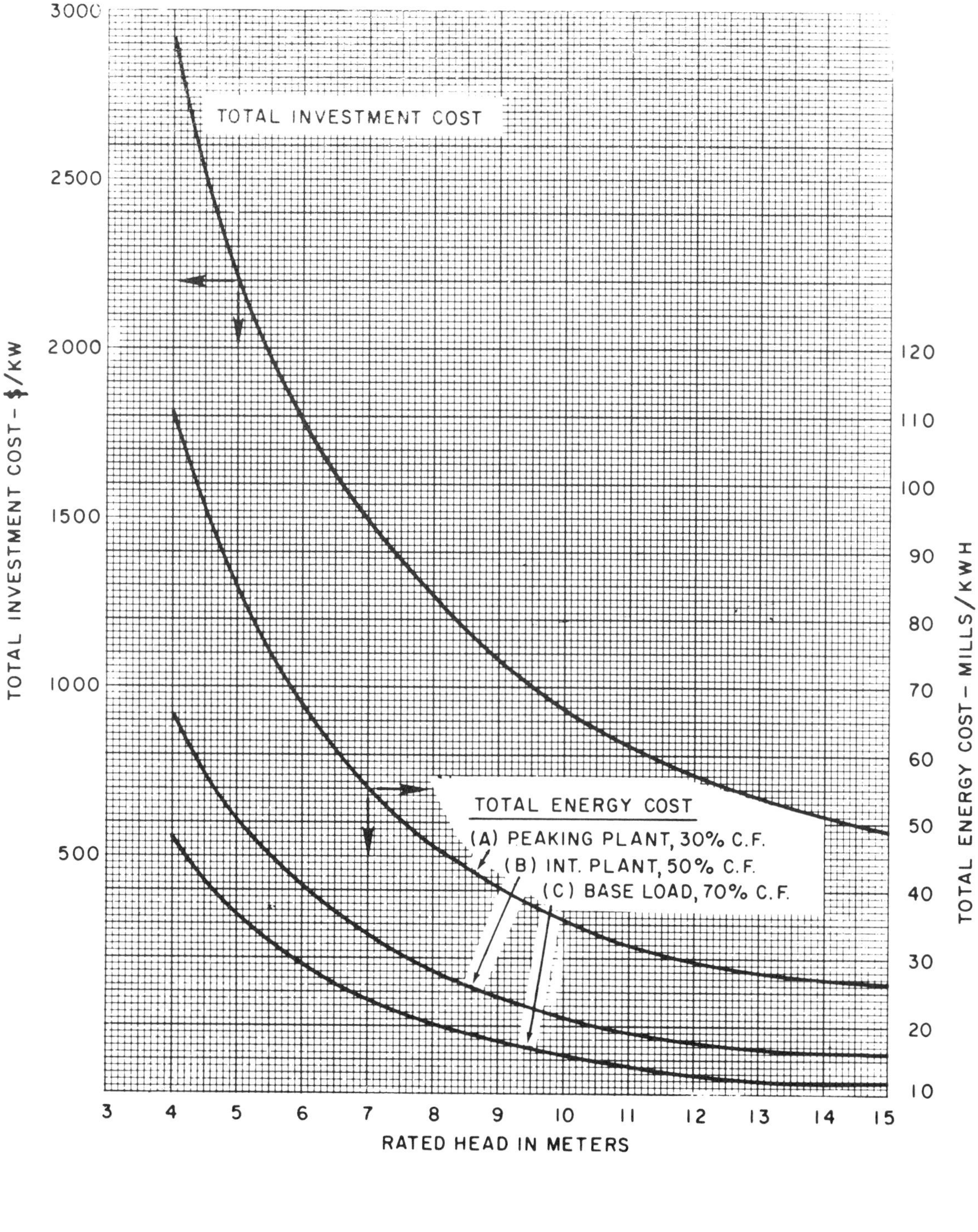

FIGURE 28

TOTAL POWER COSTS, NEW PLANT AT EXISTING DAM BASED ON STANDARD TUBE UNITS AND TAX EXEMPT BONDS

Part 4
Micro Hydropower

The information in this part is from *Micro-Hydro Power: Reviewing an Old Concept* by R. Alward, S. Eisenbart, and J. Volkman of the Technical Research Staff of The National Center for Appropriate Technology, prepared for the U.S. Department of Energy under DOE Contract ET-78-S-07-1752, January 1979.

ACKNOWLEDGEMENTS

This information package was prepared by members of the Technical Research Staff for The National Center for Appropriate Technology, Butte, Montana, under a small grant from the United States Department of Energy.

The material for the contents of this booklet was researched, compiled and edited by Ron Alward, Sherry Eisenbart and John Volkman. Hans Haumberger did the illustrations and assisted in the final lay-out.

Other staff members at NCAT assisted in the work leading to the final report. These include Bob Corbett, Lea Anne Dumezich and Noel Nedved.

Bill Delp of Independent Power Developers in Montana and Doug Smith from Hanover, New Hampshire, provided invaluable assistance in the preparation of materials. Some of the information contained herein derives from their considerable experience in the field of micro-hydro power. In addition, some of the graphics have been modeled after those in the Independent Power Developers information brochure entitled "Hydroelectric Power".

INTRODUCTION

Today there is a great deal of public interest in renewable energy sources such as solar, wind, tides, flowing water and biomass for producing power for home, shop and farm appliances. Many of the technologies for converting these renewable sources into useful power have been with us for centuries and are now once again receiving widespread attention. The generation of power from flowing and falling water is no exception. In fact, it is one of the world's oldest and most common energy technologies. But right now hydro power means big dams and large-scale generation facilities to most people in North America. On a world-wide basis, however, small-scale, environmentally benign, mechanical and electrical hydro power systems are much more common. These are the systems for individual homes, farm and shop use and generally have power outputs less than 100 Kilowatts. For convenience in terminology, this scale of hydro power is referred to as micro-hydro.

This information package has been prepared to respond to an increasing number of requests for information on micro-hydro systems. It is designed to introduce the reader to all aspects of micro-hydro, from first consideration of the idea through to production of power. We have attempted to include as much information as necessary to get you started in the process and to assist you at each step along the way. We are stressing your involvement in the development of your hydro power system, so a major part of the contents of this brochure only serve to give you guidelines. They do not, and cannot, provide detailed descriptions of every stage of the process, since micro-hydro development is very site specific. Each town, municipality and state has differing regulations, and your personal motivation and economic situation are probably very different from the next reader.

One of the particular problems we are addressing is where you can go to get adequate information that will meet your needs -- whether you want to do it yourself or get someone else to do it for you. The Extension Service of the Department of Agriculture used to be the place to go for this information back in the 1920's. They thought of it as part of their job to help rural dwellers with their energy problems. But R.E.A. took care of that, putting most farm and home sites on a regional and eventually national electricity grid. That need, though, has come around again. Now a lot of the older information is outdated. Newer equipment has come on line; different manufacturers and suppliers exist. Nevertheless, today if you go to any major library, including technical and engineering libraries, very likely the best resources in the field are still going to be books published prior to the 1930's. That is because of the lack of interest in micro-scale hydro systems in North America since that time. Some up-to-date written information does exist, but for the most part it is fragmented. There are also a number of people around the world who know a lot about small and micro-hydro systems, but it is difficult for the average person to find out who these people are and where they can be contacted. This information package contains a resource directory, which is an attempt to put you in contact with the literature, plans, people and companies appropriate to your needs.

DECISION TREE

Are you interested in using micro-scale hydro power to generate electrical or mechanical power? To cut your energy bills? To become more, or completely, independent of the local utility? To become more responsible for and responsive to the technologies you use? Whatever the reasons, you have probably already given it some serious thought, but may not know how to go about determining the feasibility of using hydro power, or who manufactures and sells the equipment, or if a permit to develop it is necessary, and where and from whom it can be obtained. What are the problems found along the way and where do you go for help when a problem is encountered?

We have attempted, in this section, to evolve a sort of "decision tree" to help you find out if you are able to, or want to, and how to, develop your site for micro-scale hydro power. We have tried to indicate points of access to and exit from this decision tree. And we have tried to indicate at what point and how these decisions are made.

As you follow through this section, each step will lead you farther along the way, or eliminate you from the process. However, we are interested in leading you to eventual success, so don't get discouraged if the going gets rough. Once you become involved you will realize that this is not like operating a car; you don't simply turn the key on. You're in a rediscovered field and you are a pioneer in its development. So although things may be in a bit of disarray, and you feel yourself getting bogged down in a water rights issue or in licensing problems, remember, that in most situations with the right attitude you can reduce or eliminate the red tape.

There are several avenues to final success and all of them depend on you and the degree of your involvement in the process. You may follow every step through on the do-it-yourself model, including building your own equipment and hand-holding every official through permit formalities, or you may hire an attorney or consultant and pay to have it installed. Or, you may try numerous combinations of these two.

What we want to do here is to give you enough information so that you are able to make that first decision as to whether it is worthwhile pursuing the subject any further. And if it is worthwhile, the following pages provide you with guidelines for each step along the way.

We will put cautions and stop-flags where we think the limits are, and we'll indicate to you how rough the estimates may be. But one thing we want to stress is: *If it looks as though there is any possibility at all, go on to the next step!*

If you come out of this process still with us, then go talk to the equipment suppliers, buy the equipment, or make it, and install it!

STEP 1. OBTAIN ACCESS TO LAND WITH RUNNING WATER.

The running water may be a spring or a permanent stream (or intermittent if seasonal energy use is considered). If you own the land, go on to Step 2. If you are thinking of purchasing the land, or leasing it for long term, then go to Steps 2 through 4 to determine if it is worth your while to obtain the land.

STEP 2. DETERMINE IF THE RESOURCE MEETS YOUR LIFESTYLE ENERGY REQUIREMENTS.

If there is not enough energy available in your stream to come close to what you feel are your needs, then there is not much sense in going any further. On the other hand, your water resource may contain an excessive amount of developable power and you may only want to use a part of it. In order to determine the potential power you have available in the stream, you will have to quantify the head and flow rate.

To determine the particular head and flow rate of your stream, turn to the section, Determining the Hydro Potential of Your Site, and pay special attention to the minimum flow considerations that are indicated. In order to find the power output potential from a typical hydro system installed in your stream, go to the nomograph on Page 21 with your flow rate and head measurements. You now have an idea of how much power you can get from your water resource.

Next, you need to determine how much power you require. If you are considering mechanical shaft power as your end use, e.g., for sawing, grinding, or whatever, then you need to know the power requirements of your machinery. If you want to use the energy available in the water for generating electrical power for domestic, shop or farm consumption, then you have to determine your electrical power needs. The best way to do this is to look at your current electricity bills and get an idea of the number of kilowatt hours you are using per month. Remember, if you are using electrical resistance heating, your bills will indicate a significantly higher electricity consumption during the winter months than during the summer. Another way to determine your electrical power needs is to turn to the table of Typical Household Appliance Loads on Page 420. This is a table of estimated average monthly power consumptions for appliances listed. Now compare your power needs with what is available from your water resource. If your monthly requirements are greater than the hydroelectric system will generate in a month, see where you can reduce consumption to try to match the available power. If your output is greater than the demand, you may have surplus power for other end uses such as space heating, operating a small electric kiln, selling to the utility, etc.

It is important to note that there is a head and a flow rate below which there is currently no economic advantage of trying to obtain electrical power. These minimum heads and flow rates are difficult to specify because combinations of high values of one with low values of the other can give some useful power. For practical purposes in micro-scale hydroelectric systems, any head less than 10 feet is probably going to be uneconomical to develop. Similarly, 10 gallons per minute can be considered the lower limit to the flow rate. However, 10 gallons per minute at 10 feet of head is not going to give any usable electrical power. The following examples will indicate some minimum energy situations:

* A flow rate of 10 gallons per minute at 100 feet of head will give about 100 watts of useful power -- enough to light a 100-watt light bulb continuously.

* A flow rate of 100 gallons per minute at 10 feet of head will also deliver about 100 watts of useful power.

STEP 3. DETERMINE IF YOU CAN USE THE RESOURCE.

Do you have a right to use the water or does that right reside elsewhere? Remember, in many states and particularly in the East, water rights do not necessarily transfer with land title. You will have to investigate this before you proceed any further.

Next, if there is a dam nearby or on your property, can you use it (e.g., buy or lease it)? Can you get a right-of-way for needed pipeline across adjacent properties? Can you get building permits if they are required, such as for a small powerhouse, dam or diversion?

Can you obtain the necessary licenses and permits for the installation (see the Section on Regulatory Conflicts)? There are numerous licensing and permitting procedures you may have to go through. A lot depends on which State you are located in, whether or not the water rises and falls on your land, or just passes through, whether it comes from, goes to, or is in National Forest or Indian Tribal lands.

The water use permit process can be long and drawn out, so just investigate the feasibility at this point. If it looks at all possible, then go on to Step 4.

STEP 4. CALCULATE THE COSTS OF THE SYSTEM.

Determine who the equipment suppliers are for your type of water resource and find out the costs of their equipment. Most equipment suppliers will be able to give you a rough estimate if you can supply them with the following information:

* Head

* Usable flow rate

* Length of pipe required from take-off to generator location, or location of dam with respect to generator location.

* Power demand -- quantity, and what used for.

* Whether you want AC or DC and what you want to do with any surplus power.

Include the costs of piping, dam repair or construction (if needed), laying the pipe, electrical set-up and constructing a small powerhouse to enclose the turbine, generator and electrics.

There are peripheral advantages to a hydro system. These include using the water for fire suppression, domestic water supply and irrigation. Can the inclusion of any or all of these offset other costs you currently have or expect to have? How do these affect the cost figure associated with the hydro system?

The next question that arises is: Are the above costs reasonable?

This question can only be answered by knowing your own reasons for getting involved with micro-hydro. Your non-economic reasons (e.g., environmental, energy independence) may outweigh any other considerations, but if you are like most of us, economics helps play a deciding role. To determine if you can afford the money outlay to put in a hydro system, you will have to look at the cost of alternatives, Step 5, and if and how the system can be financed, Step 6.

STEP 5. THE COST OF ALTERNATIVES TO MICRO-HYDRO SYSTEMS.

Are you currently connected to the utility lines? How much are you paying for your electricity and would it be advantageous, economically, for you to disconnect and supply all your own power? You can determine the economic advantage to you, if any, by doing a calculation similar to that in the section on Economics.

If your site is isolated and the nearest utility lines are a mile or more away, you might find that the cost of installing a small hydro power system is very competitive with the cost of line extension. On the other hand, if you must use some commercial power, continuously or seasonally, in some cases the so-called "minimum rate charge" may eliminate a significant cost saving even if you do install your own micro-hydro system. It is wise to consider all these things.

STEP 6. METHODS OF FINANCING A MICRO-HYDRO SYSTEM.

The capital costs associated with micro-hydro power systems are somewhat high. Typically they can run between $750 and $1500 per kilowatt of installed capacity. Some imported units can cost up to $2000 per kilowatt. Your method of financing the project is going to determine what you will actually pay per kilowatt hour of power used. The section on Economics will help you in making this calculation.

You can keep the costs down by doing a lot of the work yourself. You can build and install some of the needed components (all of them if you are so inclined and have the time). However, be forewarned -- if you do most of it yourself, such things as building the various components, this might close off the traditional sources of financing. Loans are made on the basis of guaranteed collateral, and this collateral often relates to the item in question. So, if the item does not have proven value to the financier, that is, if it cannot be repossessed and have a guaranteed resale value comparable to the prorated value of the loan, then it may not be financable.

STEP 7. START THE PERMITTING PROCESS.

If you have successfully arrived at this point, then now is the time to proceed with obtaining the necessary permits. There is a lot of red tape and hassle along the way -- but persevere! If you approach it with a positive attitude, then you should be successful. Request help from the various officials concerned -- do not go in demanding everything. See the section on Regulatory Conflicts for further information.

STEP 8. BUILD OR BUY THE EQUIPMENT.

Some people will be interested in making or refurbishing as much of their own equipment as possible. They won't be going to the supplier or manufacturer for the whole package, but will want to search out plans and specifications from various sources. A list of these appears in the annotated bibliography under Plans and Specifications.

For those who want to buy their equipment, a List of Manufacturers and Suppliers is included in the section of the same title.

Whatever approach you take, it is important to be sure of the quality of the materials and equipment you buy. There is some junk on the market. This is a young industry, at least in its modern rebirth, so you will probably encounter some unscrupulous entrepreneurs selling basically untested or worthless equipment. If you are looking to make your system cost effective, you may be attracted to the lowest cost equipment. This may or may not be the approach to take in your particular case. The best way to ensure product quality is to ask the supplier for a list of previous customers in your area. Go to one of these sites, talk to the users, and get their opinions on the equipment.

BUY OR OBTAIN GOOD QUALITY PIPE! Do not use seconds. Use gate valves in your system. A good gate valve takes long enough to shut off the water passage that it creates very little water hammer effect. The pressure wave (water hammer) caused by closing off a high pressure line too quickly can cause severe damage to the pipe. DO NOT USE BALL VALVES. A good trash screen at the water intake is also vital to continued system operation.

If you have to build or rehabilitate a dam for your hydro system, go see an engineer and get some advice. Dams can be hazards (usually minor in the case of micro-hydro systems) and their soundness should be ensured.

The powerhouse you build need only be large enough to house the turbine, generator, electrics and battery storage (if any). Minimum requirements are that it be weatherproof and have a dry floor. Nothing sophisticated is required.

Household-size hydro projects are usually most economically developed as DC-to-AC systems. However, if you are tying in with an existing utility grid, synchronous generating systems can often be installed for less per kilowatt.

STEP 9. INSTALL AND CHECK OUT THE EQUIPMENT.

During installation and equipment check-out, it is important that you follow all the manufacturer's instructions. Do not take any shortcuts! Remember, that reputable manufacturers have been in the business for awhile and they know their equipment. If you encounter any problems, contact the manufacturer. He will deal with them. It is much to the manufacturer's benefit to see that the system goes in and functions well.

If you are installing a DC system, install the turbine-generator set as close to the use point as possible. This will keep electrical transmission line losses to a minimum.

DETERMINING THE HYDRO POTENTIAL OF YOUR SITE

* FLOW MEASUREMENT

* HEAD MEASUREMENT

* POWER CALCULATION

To determine the hydro potential of the water flowing from your spring or in your stream, you must know both the flow rate of the water and the head through which the water can usefully fall. The flow rate is the quantity of water, usually measured in gallons or cubic feet, flowing past a point in a given time. Typical flow rate units are gallons per minute (gpm) and cubic feet per minute (cfm). The head is the vertical height in feet from the headwater (in the case of a dam) or the point where the water enters the intake pipe (where no dam exists), to where the water leaves the turbine housing.

FLOW MEASUREMENT

In order to adequately assess the minimum continuous power output to be expected from your hydro unit, you will have to determine the minimum quantity of water that will pass through the system. For this reason, it is important to know both the minimum flow rate of your stream or spring and what portion of this flow you can use for power generation. The percentage of the minimum flow you temporarily divert for power generation becomes a consideration when you are addressing fisheries (fish movement up and down the stream) and questions of aesthetics. One manufacturer suggests that only 25% of the dry season flow be used for power generation. This, of course, depends upon your particular case -- whether you are using a run-of-the-river system or stored pondage, or if your stream is high head with no evident fish life.

If you are already familiar with the stream's seasonal variations, then you can limit flow measurements to the few months surrounding the driest, or lowest flow, period. However, if you don't know approximately when the flow is the lowest, you will have to make at least once monthly, but preferably bi-weekly, flow measurements throughout the complete year.

Once your flow data has been compiled, you are in a position to begin some calculations. Was this a dry year or a wet year? To determine this, you will have to get further information, usually from the water resources people in your area. They will have several years' precipitation and snow pack data available, if not for your immediate location then for some nearby major drainage basin. You will have to take a look at this information to see how your measured year fits into the pattern. Once you have determined whether your year was dry, typical or wet, then make the necessary corrections to your data to determine the minimum expected flow rates for your stream.

There is something further which should be noted. If your hydro system will be producing electricity for a household, it will in many instances be a DC-to-AC conversion system, so you will be concerned only with minimum flows. A good flow sampling through the dry season (assuming you know what the dry season is) is usually adequate. However, if you are considering a system considerably larger than for a single household, then you will likely be looking at direct AC systems. You might want to do a little bit more with load projections, particularly with respect to what can be done with the energy at the time of year it is available. This will require some feeling for the maximum and mean stream flows as well as the minimum. In addition, if your system requires a dam, then it becomes vital to know maximum stream flows in order to adequately size spillways for bypassing excess water in order to prevent damage to your installation.

How to measure stream or spring flow:

1. For small mountain streams or for springs, temporarily dam up the water and divert the entire flow into a container of known size. Carefully time the number of seconds it takes to fill this container.

   For example:

   > If it takes 40 seconds to fill a 55 gallon barrel, the flow rate is 1.375 gallons per second, or 82.5 gallons per minute, or 11 cubic feet per minute.

2. For larger streams, the float method can be used. If done carefully, and repeated several times, it can give results accurate enough for most calculations. In order to use this method, you need to know the cross-sectional area of the stream and the stream velocity.

   The cross-sectional area should be determined at some easily measured spot in the stream, preferably in the middle of a straight run of the stream. Measure the width (w) of the stream in feet. Then, using a stick, measure the depth in feet at equal intervals across the width of the stream (see Figure A). Record the depth at each interval and calculate the average depth.

   For example:

   With a stream cross-section as in Fig. A,

   d = depth

   d1 = 1.0 feet
   d2 = 1.3 feet
   d3 = 1.2 feet
   d4 = 1.8 feet
   d5 = 1.0 feet
   d6 = 0.8 feet
   d7 = 1.1 feet
   d8 = 1.8 feet
   d9 = 1.3 feet
   d10 = 0.7 feet

   Total = 12 feet

   $$\text{Average } d = \frac{12}{10} = 1.2 \text{ feet}$$

Next, multiply the width (w) by the average depth (d) to get the cross-sectional area (A) of the stream in square feet.

For example:

> In the above stream, say the width at the point of making the depth measurements was 8 feet, then the cross-sectional area (A) is
>
> $$A = w \times d$$
>
> $$A = 8 \text{ feet} \times 1.2 \text{ feet}$$
>
> $$A = 9.6 \text{ square feet}$$

The stream velocity can be determined by choosing a straight stretch of water at least 30 feet long with the sides approximately parallel and the bed unobstructed by rocks, branches or other obstacles. Mark off two points, say 30 feet apart, along the stream. On a windless day, place a float upstream of the first marker, in midstream. A pop bottle, partially filled with small stones so that it rides with its neck out of water, is a good float. Carefully time the number of seconds it takes the float to pass from the first marker to the second. Repeat this process several times and average the results.

For example:

> The average time for a float to travel between two markers placed 30 feet apart is 15 seconds. The velocity of the float is thus
>
> $$\frac{30 \text{ feet}}{15 \text{ sec.}} = 2 \text{ feet/second}$$

This float velocity does not, however, represent the velocity of all the water in the stream. The water at the sides and bottom of the stream flows less quickly than that at the center or near the top due to stream bed friction and channel roughness. A correction factor, depending on the roughness or smoothness of the stream bed, is usually included to give an estimated average stream velocity. This correction factor can vary from 0.6 for a rocky hill stream, to 0.85 for a stream with very smooth bed and sides.

For example:

> Taking the float velocity computed above, the stream velocity (V) for a fairly rough hill stream is
>
> $$V = 2 \text{ feet/second} \times 0.65 = 1.3 \text{ feet/second}$$
>
> or $V = 78$ feet per minute.

The flow rate of the stream can now be calculated by multiplying the cross-sectional area of the stream (A) by the stream velocity (V).

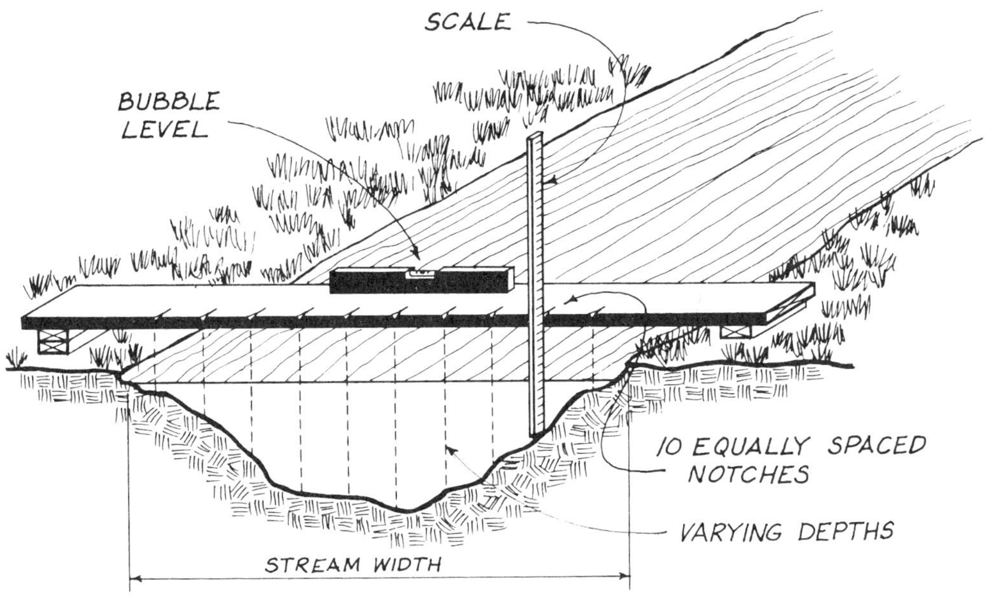

FIGURE A

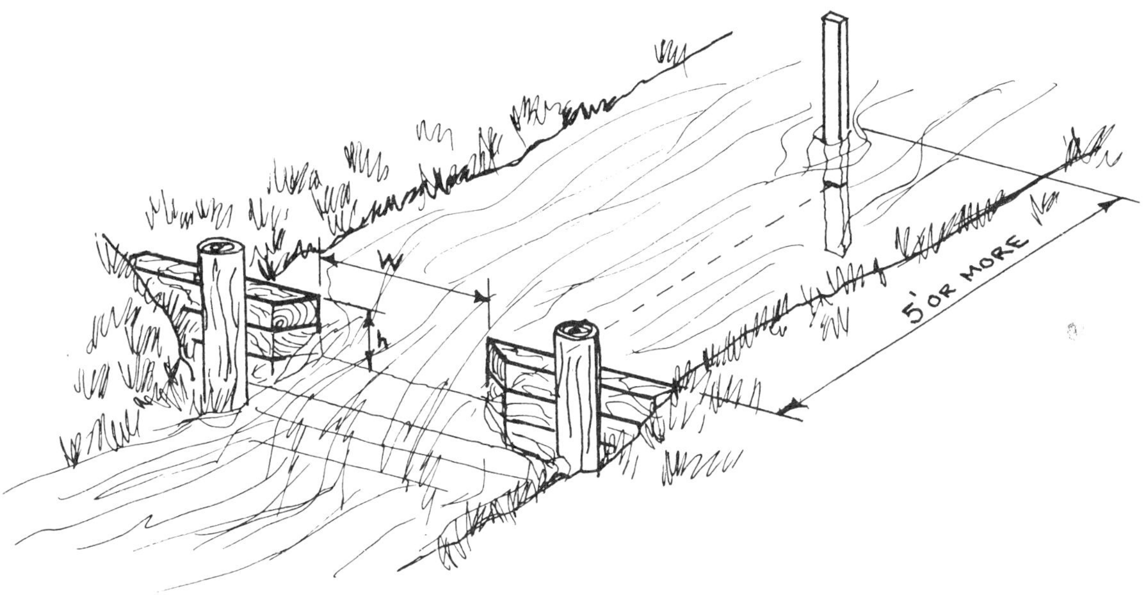

FIGURE B

For example:

Flow = A X V

Flow = 9.6 sq ft X 78 ft/min

Flow = 748.8 cubic feet/min.

Now, depending on what portion of the stream flow you can or want to use, you can now determine the usable flow. Simply multiply the stream flow rate you have just calculated by the portion of the flow you will be using.

For example:

If you will only be using 25% of the minimum stream flow, and the stream flow you have determined above is 748.8 cubic feet per minute, then the usable flow is

Usable Flow = 748.8 cfm X 0.25

Usable Flow = 187.2 cfm

3. There is a third method for determining stream flow. This is called the Weir Method. This method is accurate and can be used to measure the flow rate of any stream. It is particularly advantageous for flow measurements in shallow streams where a weighted float would have difficulty floating freely. However, it is also a more complicated technique for measuring flow.

Essentially, a temporary dam structure is built across the stream perpendicular to the flow, with a rectangular notch or spillway of controlled proportions located in the center section. This notch has to be large enough to take the maximum flow of the stream during the period of measurement, so make some rough estimate of the stream flow prior to building the Weir. The notch width (W) should be at least three times its height (h) and the lower edge should be perfectly level. The lower edge and the vertical sides of the notch should be beveled with the sharp edge upstream. The whole structure can be best built out of timber with all edges and the bottom sealed with clay, earth and sandbags to prevent any leakage. A typical Weir is illustrated in Figure B.

In order to measure the flow of water over the Weir, you have to set up a simple depth gauge. This is done by driving a stake in the stream bed at least 5 feet upstream from the Weir, until a pre-set mark in the stake is precisely level with the bottom edge of the notch. The depth of water on this stake, above the pre-set mark, will indicate the flow rate of water over the Weir. You will need to refer to a "Weir Table" in order to determine this flow rate. A typical "Weir Table" follows.

WEIR TABLE

Depth on Stake in inches	c.f.m. per inch of notch width	Depth on stake in inches	c.f.m. per inch of notch width
1	0.40	12.5	17.78
1.25	0.55	12.75	18.32
1.5	0.74	13	18.87
1.75	0.93	13.25	19.42
2	1.14	13.5	19.97
2.25	1.36	13.75	20.52
2.5	1.59	14	21.09
2.75	1.83	14.25	21.65
3	2.09	14.5	22.22
3.25	2.36	14.75	22.70
3.5	2.63	15	23.38
3.75	2.92	15.25	23.97
4	3.22	15.5	24.56
4.25	3.52	15.75	25.16
4.5	3.83	16	25.76
4.75	4.16	16.25	26.36
5	4.50	16.5	26.97
5.25	4.84	16.75	27.58
5.5	5.18	17	28.20
5.75	5.54	17.25	28.82
6	5.90	17.5	29.45
6.25	6.28	17.75	30.08
6.5	6.65	18	30.70
6.75	7.05	18.25	31.34
7	7.44	18.5	31.98
7.25	7.84	18.75	32.63
7.5	8.25	19	33.29
7.75	8.66	19.25	33.94
8	9.10	19.5	34.60
8.25	9.52	19.75	35.27
8.5	9.96	20	35.94
8.75	10.40	20.25	36.60
9	10.86	20.5	37.28
9.25	11.31	20.75	37.96
9.5	11.77	21	38.65
9.75	12.23	21.25	39.34
10	12.71	21.5	40.04
10.25	13.19	21.75	40.73
10.5	13.67	22	41.43
10.75	14.16	22.25	42.13
11	14.67	22.5	42.84
11.25	15.18	22.75	43.56
11.5	15.67	23	44.28
11.75	16.20	23.25	45.00
12	16.73	23.5	45.71
12.25	17.26	23.75	46.43
		24	47.18

To use the table, determine the depth of water in inches over the pre-set stake mark. Take this value to the Weir Table and read off the flow rate in cubic feet per minute per inch of notch width. Multiply this volume flow rate by the width, in inches, of your Weir notch. This will give you the stream flow rate in cubic feet per minute.

For example:

> On a particular stream you have built a Weir with a notch width of 30 inches. The depth of the water on the stake above the pre-set marking is 6¼ inches. On the Weir Table, read opposite 6.25 inches to the flow rate of 6.28 cfm per inch of notch width. The flow rate of the total stream is then
>
> 6.28 cfm X 30 in.
>
> or 188.4 cfm

When you have the Weir in place, you can easily take readings at your convenience. If you are going to use the Weir for any extended period of time, it is important to frequently ensure the watertightness of the sides and bottom.

HEAD MEASUREMENT

The greater the vertical distance that water falls, the more potentially useful power there is available in the water.

How to measure head:

1. For high head systems, detailed topographical maps of your area may give some indication of the vertical height difference between proposed intake and tailwater levels. However, the accuracy of map reading is limited so this technique should only be used for very preliminary estimations.

2. For those who are acquainted with photographic surveying techniques, this method can give fairly accurate results. Pictures taken in the field can be developed and the elevations scaled on the photographs. But caution -- this is not a method for amateurs. Photographic surveying requires some skill and training.

3. Pocket altimeters can give you preliminary estimations of the elevation difference between intake and tailwater locations on proposed high head systems. However, the accuracy of measurement is not suitable for any serious calculations.

   There are some larger portable altimeters on the market, which tend to be very expensive, that can enable you to make elevation measurements within an accuracy of a couple of feet. These instruments are suitable for engineering calculations; however, they do not give as precise measurements as do the following three methods.

4. Any good surveyor can be hired to determine the head for you. Ask the surveyor to simply find the vertical distance between your water source, or proposed intake location, and the proposed location of the power plant. Hiring a surveyor is going to be somewhat expensive, so if this is your only alternative, you want to be reasonably sure that you intend to carry through with the project. If your head is less than 25 feet, you need very precise measurements, so a surveyor may be advisable.

5. If you know how to use standard surveying equipment, for example a transit or a surveyor's level and leveling rod, borrow or rent the appropriate pieces and get a friend to help you make the necessary measurements.

6. Another technique involves a do-it-yourself approach. The equipment required is a carpenter's level, some sort of stand to raise the level a few feet off the ground, and a tape measure. You may or may not use

someone else to assist you. The method is described below in relation to the following diagram:

a) Set the level on the stand; make sure the level is horizontal (level) and that its upper edge is either at the same elevation as the water source, or a known vertical distance above the water surface.

b) Sight along the upper edge of the level to a spot on a nearby object (tree, rock, building) that is further down the hill and which can be reached for measuring.

c) Note this precise spot on the object and mark it (Point A in the diagram).

d) Move your level and stand down the hill slope and set it up again so that this time the upper edge of the level is at some Point B below Point A on the first object, as in the drawing. Mark this Point B and measure and record the vertical distance A to B. Now sight along the upper edge of the level in the opposite direction to another object that is further down the hill.

e) Repeat this procedure until you end up at the same elevation as the proposed power plant site.

f) If more than one set-up was required, add all the vertical distances A-B. If your first set-up was above the water surface, subtract the vertical distance between the water surface and the upper edge of the level from the sum of the vertical distances. You now have the total head.

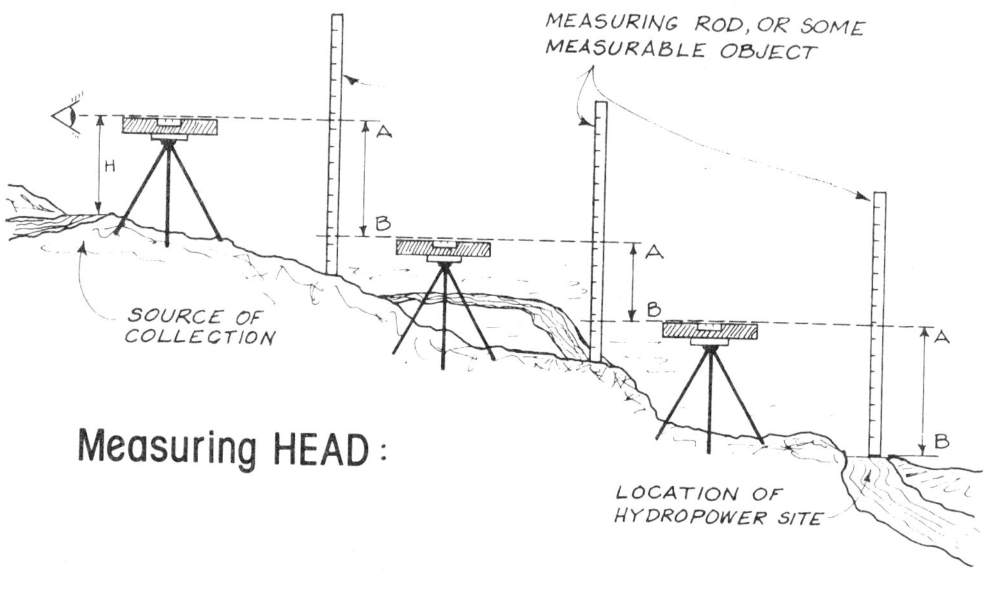

FIGURE C

In undertaking the above measurements there are a few things you should remember:

-- You do not need to be concerned with horizontal distances for head determination.

-- Every time you re-set the level, its upper edge should be at precisely the same level as Point B (sight back to check).

-- You need not travel in a straight line.

Once you have the total, or gross, head, there are various losses to be considered before you can make any theoretical power calculations. The net head is required for these calculations.

NET HEAD = GROSS HEAD - LOSSES

Now, losses occur for several reasons. Whenever water flows through a pipe there are friction losses. These friction losses are greater for increased flow rates and also greater for smaller pipe diameters. Elbows and bends in the pipe will also increase friction losses. PVC pipe offers low friction loss, rarely exceeding 8% of the gross head. Good steel pipe has twice the friction loss as PVC. Iron, asbestos and concrete pipe all cause higher losses. Typical PVC pipe losses are indicated in the nomograph on Page 22 . Any reputable pipe manufacturer or supplier will be able to supply you with the pipe size and friction loss information for your particular flow conditions.

Other losses that might occur in your hydraulic system depend on the type of turbine or wheel that you will be using. For example, in an impulse turbine there is a slight head loss due to the vertical distance between the nozzle jet and the tailwater. Overshot water wheels also suffer some inherent loss since the wheel has to run free of, and therefore slightly above, the tailwater. On the other hand, crossflow, Francis and propeller turbines with draft tubes have almost no inherent head losses.

POWER CALCULATION

Once you have determined the usable flow rate and the net head for your particular site, you are in a position to calculate the amount of power you can expect. You will first have to calculate the theoretically available power, assuming that 100% of the power available in the water can be usefully converted.

The theoretical power available (P_{th}) is given by the following equation:

$$P_{th} = 62.4\ Q \times h$$

where Q = usable flow in cfm

h = net head, in feet

62.4 = density of water in lbs/cubic foot

This gives P_{th} in foot-pounds per minute.

To convert this to horsepower, the equation becomes:

$$P_{th} = \frac{62.4\ Q \times h}{33{,}000}$$

$$= \frac{Q \times h}{529}$$

To convert to Kilowatts, the same equation becomes:

$$P_{th} = \frac{62.4\ Q \times h}{33{,}000} \times 0.745$$

$$= \frac{Q \times h}{709}$$

For example:

Using a flow rate (Q) of 5 cfm and a net head (h) of 100 feet, the theoretical power, in Kilowatts, is

$$P_{th} = \frac{Q \times h}{709}$$

$$= \frac{5 \times 100}{709}$$

$$= 0.7 \text{ KW}$$

$$= 700 \text{ Watts}$$

The theoretical power available represents more power than you will get out of your equipment. All of the machinery and other equipment that you use to convert the power available in the flowing water to mechanical shaft or electrical power are less than 100% efficient. As an indication, typical efficiencies of water wheels and turbines are listed below. More precise figures are available from the manufacturers.

Typical Efficiency Ranges for Small Water Wheels and Turbines

Prime Mover		Efficiency Range
Water Wheels	- Undershot	25 - 45%
	- Breast	35 - 65%
	- Poncelet	40 - 60%
	- Overshot	60 - 75%
Turbines	- Reaction	80%
	- Impulse	80 - 85%
	- Crossflow	60 - 80%

To transmit the power from water wheel or turbine to a generator, alternator, or some mechanical system also entails losses. Belt drives are 95 to 97% efficient for each belt; gear boxes 95% and higher; alternators and generators 80%. If you use second hand equipment, you can adjust the efficiency rating down slightly.

Typical overall efficiencies for electrical generation systems can vary from 50 to 70%, with the higher overall efficiencies occurring in the high head, high speed impulse turbines. Overall efficiencies of systems using water wheels are usually well under 50%.

For example:

> Using our previous figures, the theoretical power available (P_{th}) was 700 Watts. If an impulse turbine, with efficiency of 80%, is used with a single belt drive to an 80% efficient alternator, then the useful power (P) is
>
> $$P = 700 \text{ Watts} \times 0.8 \text{ (turbine)} \times 0.95 \text{ (belt drive)} \times 0.8 \text{ (alternator)}$$
>
> $$P = 426 \text{ Watts}$$

In summary, here are the steps to follow to determine the hydro potential of your site:

1. Measure the water flow rate using one of the following:

    -- timed container filling method,
    -- float method, or
    -- Weir method.

2. Determine the usable flow -- that portion of the stream flow you can use.

3. Measure the total or gross head, preferably with:

    -- surveyor's equipment or
    -- carpenter's level and stand.

4. Determine the net head by subtracting friction and other losses from the gross head.

5. Calculate the theoretical power available using either

    $$P_{th} = \frac{Q \times h}{529} \text{ horsepower} \quad \underline{\text{or}} \quad P_{th} = \frac{Q \times h}{709} \text{ Kilowatts}$$

    where Q = usable flow rate in cubic feet per minute
    h = net head, in feet.

6. Calculate the useful power available by multiplying P_{th} by the efficiency of each piece of machinery linked into the system between and including the water wheel or turbine and the unit giving out the useful power.

NOMOGRAPH TO DETERMINE TYPICAL OUTPUT POWER FROM A MICRO-HYDRO SYSTEM

(Assumed system efficiency of 53%)

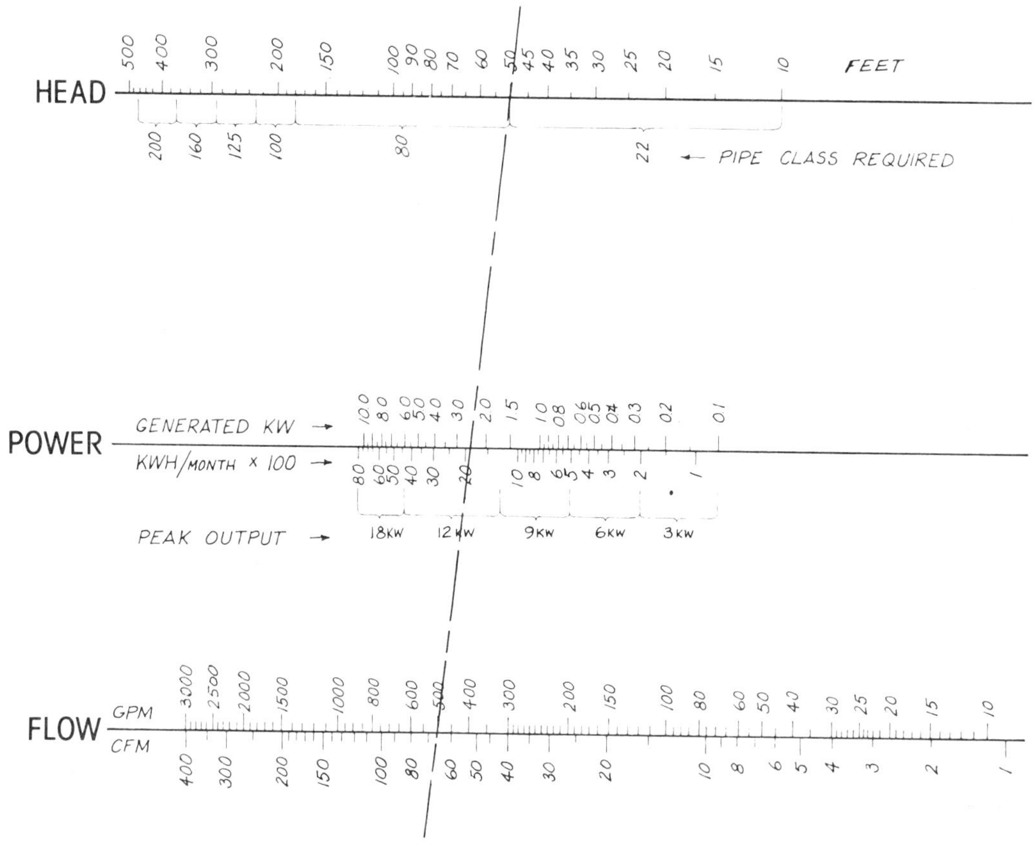

Adapted from: Independent Power Developers' brochure "Hydroelectric Power".

Once you know the usable flow rate of your stream and the net head, then you can use a nomograph similar to the one above to determine the power you can expect from your turbine. For example, suppose you have a usable water flow rate of 500 gpm through a net head of 50 feet. To determine the power you can expect from the turbine, locate the flow rate, 500 gpm on the FLOW line, and the head, 50 feet, on the HEAD line. Join these two points with a straight line. The point where this line cuts the POWER line is the power output of your turbine. The POWER line gives three pieces of information. The continuous power output, in generated kilowatts, is 2.5. This means that your turbine will be putting out a constant 2.5 KW if the water flow rate conditions stay the same as assumed. At a continuous power output of 2.5 KW, you can expect to produce nearly 2000 kilowatt hours per month, as indicated on the first scale on the left side of the POWER line. If you are using a DC system, feeding into storage batteries, then your system can have a peak power output of 12 KW, as shown by the brackets.

NOMOGRAPH TO DETERMINE LOSSES DUE TO FRICTION IN PVC PIPE

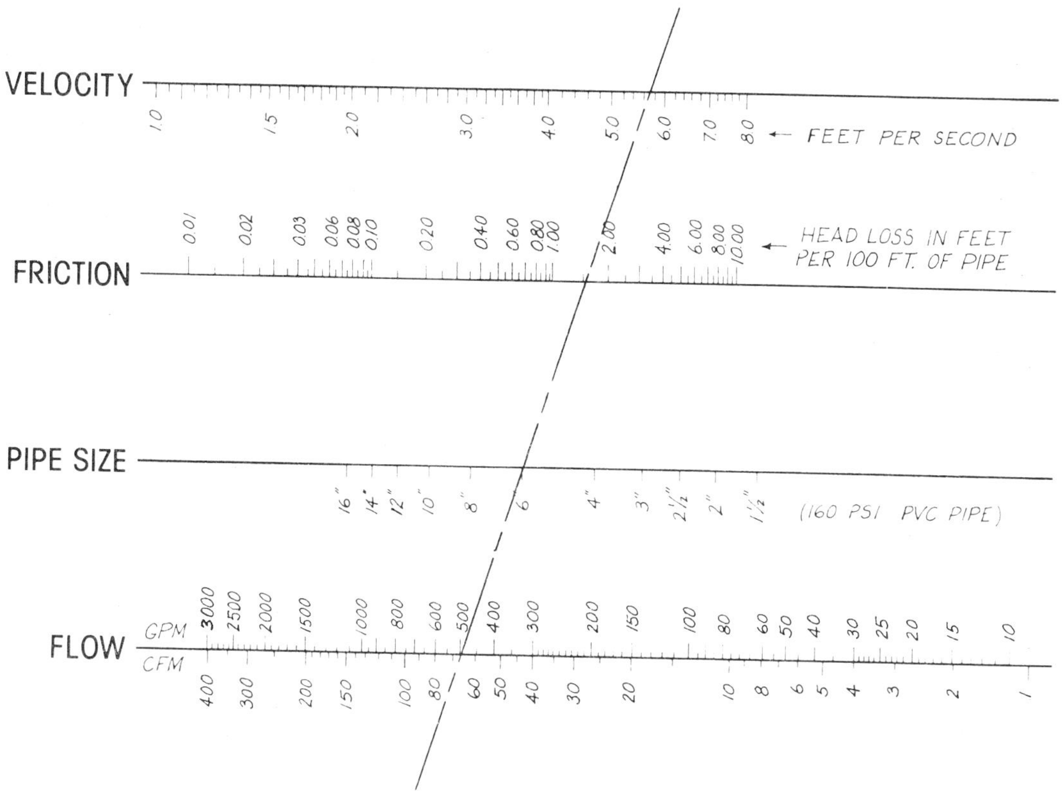

Adapted from: Independent Power Developers' brochure "Hydroelectric Power".

Sample calculation to illustrate use of nomograph:

Suppose you have a water flow rate of 500 gallons per minute through a 6-inch diameter PVC pipe. To determine the head loss due to friction, locate the flow rate value, 500 gpm, on the FLOW line, and the pipe diameter, 6 inches, on the PIPE SIZE line. Using a straight edge, draw a straight line through these two points and continue it to cut the FRICTION line. The FRICTION line is cut at a value of 1.50, which means that for the pipe size and flow rate given, there is a head loss due to friction of about 1.50 feet for each 100 feet of pipe length used in your system. An extension of this same straight line to intersect the VELOCITY line indicates that the water is flowing through the pipe at about 5.8 feet per second.

TYPICAL HOUSEHOLD APPLIANCE LOADS

APPLIANCE	POWER (WATTS)	AVG. HOURS USE/MO.	TOTAL POWER CONSUMP. KW HR/MO.
Blender	600	3	2
Car Block Heater	450	300	135
Clock	2	720	1
Clothes Dryer	4600	19	87
Coffee Maker	600-900	12	7-11
Electric Blanket	200	80	16
Fan (kitchen)	250	30	8
Freezer (chest, 15 cu ft)	350	240	84
Hair Dryer (hand-held)	400	5	2
Hi-Fi (tube type)	115	120	14
Hi-Fi (solid state)	30	120	4
Iron	1100	12	13
Light (60-Watt)	60	120	7
Light (100-Watt)	100	90	9
Lights (4 extra, 75-Watt)	225	120	27
Light (fluorescent, 4')	50	240	12
Mixer	124	6	1
Radio (tube type)	80	120	10
Radio (solid state)	50	120	6
Refrig. (standard, 14 cu ft)	300	200	60
Refrig. (frost free, 14 cu ft)	360	500	180
Sewing Machine	100	10	1
Toaster	1150	4	5
TV (black & white)	255	120	31
TV (color)	350	120	42
Washing Machine	700	12	8
Water Heater (40-gal)	4500	87	392
Vacuum Cleaner	750	10	8
* * * * * * * * * * *			
Shop Equipment:			
Water Pump (½ hp)	460	44	20
Shop Drill (¼", 1/6 hp)	250	2	.5
Skill Saw (1 hp)	1000	6	6
Table Saw (1 hp)	1000	4	4
Lathe (½ hp)	460	2	1

EQUIPMENT

* TURBINES
* WATER WHEELS
* SIZING THE SYSTEM
* POWER GENERATION & STORAGE
* LOAD CONTROL & GOVERNORS
* OTHER EQUIPMENT

TURBINES

A water turbine is basically a device that converts the energy in falling water into rotating mechanical energy. This energy, available in a rotating shaft, may either be used directly to operate mill and grinding equipment or hooked to a generator to produce electricity. Water turbines take many shapes and have been used around the world for centuries. Although this section will deal primarily with the generation of electricity, references on mechanical applications have been included for those who are interested.

The energy potential that is available at a hydro site is the result of the combination of the "head" and "flow" that is present. It is possible to produce a given amount of power with high head and low flow, low head and high flow, or any combination in between. Sites generally with less than 60 feet of head are referred to as "low head" and those with over 60 feet are "high head." It is also generally assumed that a head of one meter is the absolute minimum needed to develop hydro power, but less than 10 feet (approx. 3 meters) is probably uneconomical to develop.

Because low head projects require large amounts of water and thus physically large equipment, the resulting unit cost of power generated will be higher than for high head sites with similar power potential. For this reason, most micro-hydro systems being developed in the United States are of the high head type. This is not, however, to say that low head sites are uniformly uneconomical. There are a great many large-scale low head systems operating around the world.

Water turbines are generally very efficient in converting the energy available in the water resource into mechanical and electrical energy. In fact, efficiencies of 70-85% are not uncommon. Some turbine designs are intended to operate at a specific flow and their efficiency suffers greatly with any variation, while others are more flexible. Also, because of operating characteristics, turbines are usually designed for a specific application and output. You will generally gain nothing by installing a unit that is larger than your water resource is capable of driving. It is important that you have a good understanding of both your energy requirements and characteristics of the water resource, so that the proper equipment can be selected.

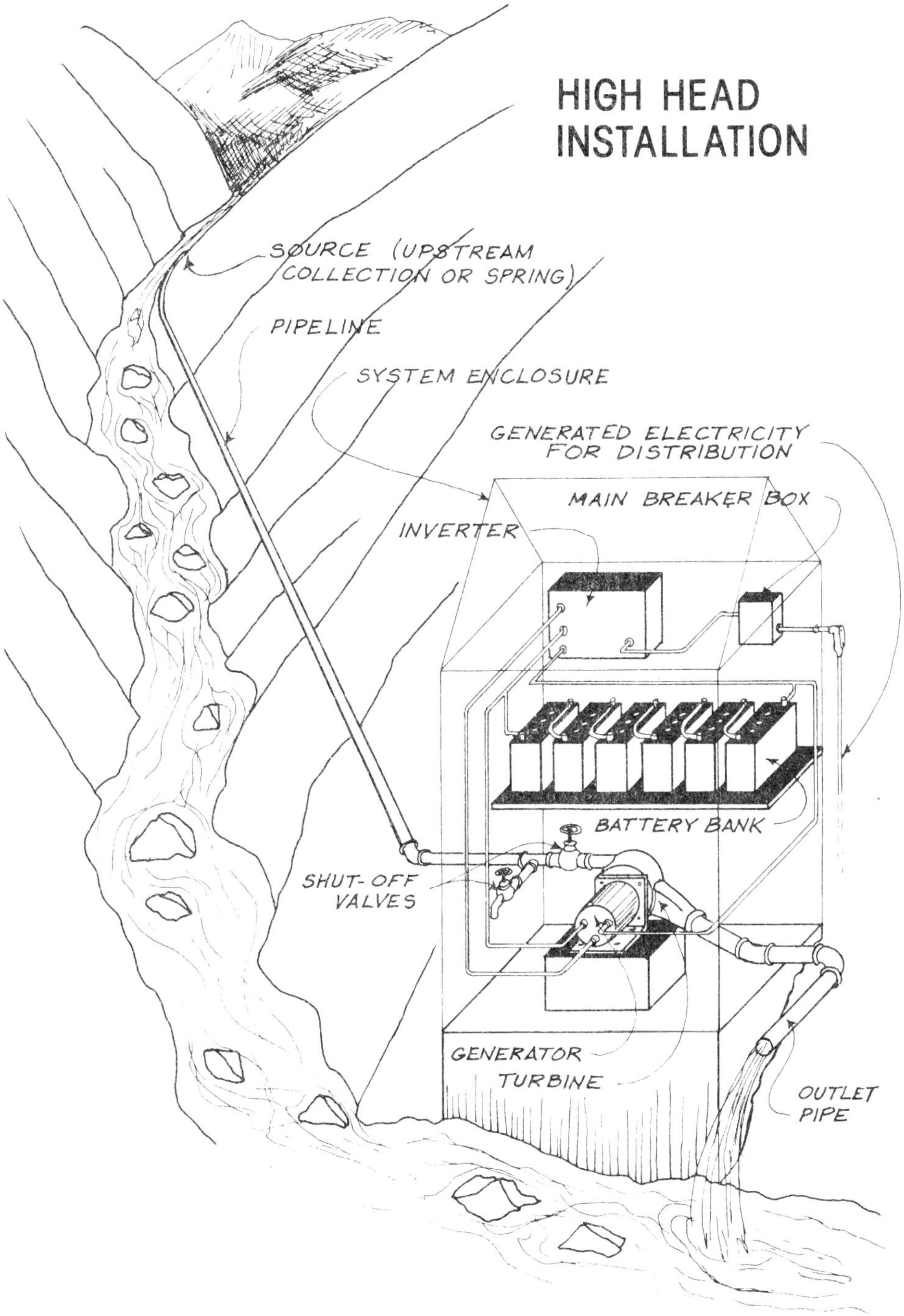

Adapted from: Independent Power Developers' brochure "Hydroelectric Power".

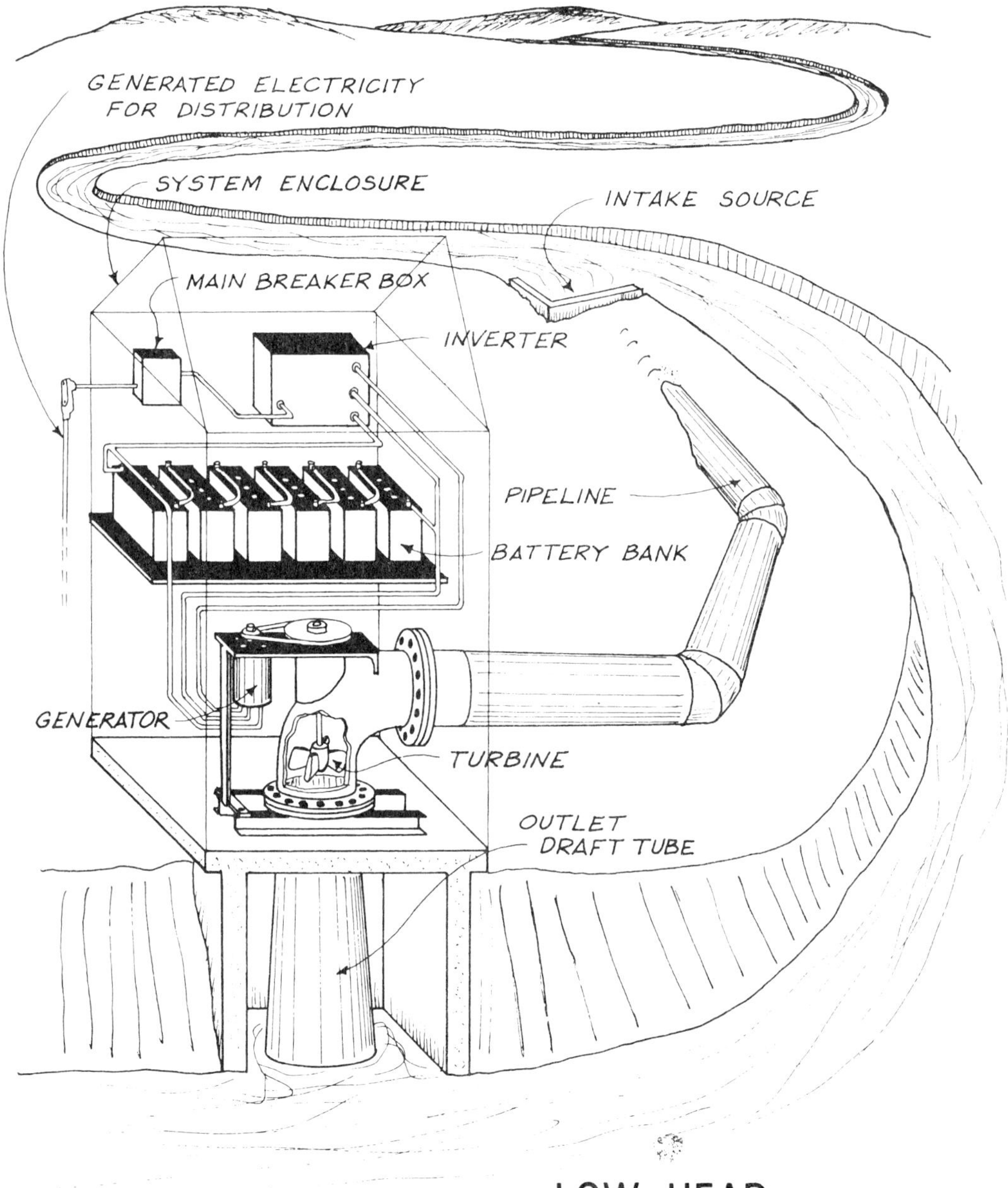

LOW HEAD INSTALLATION

Adapted from: Independent Power Developers' brochure "Hydroelectric Power".

Impulse Units In General

Impulse turbines generally use the velocity of the water to move the runner, rather than pressure, as is in the case with reaction and propeller designs. Also, because of this, they generally discharge to atmospheric pressure, there is no suction on the down side of the turbine, and the water simply falls out the bottom of the turbine housing after hitting the runner. Impulse units are generally the simplest of all common turbine designs and are widely used in micro-hydro applications.

The Pelton Wheel:

In general terms, a Pelton Wheel is a disc with paddles or buckets attached to the outside edge. The water passes through a nozzle and strikes the paddles one at a time, causing the wheel to spin. The buckets are shaped so that the water stream is split in half and caused to change direction, heading back in the opposite direction to the original water stream for greatest efficiency. The shape and smoothness of the buckets is important. Because the power developed by a Pelton Wheel is largely dependent on the velocity of the water, it is well suited for high head/low flow installations. Operating efficiencies in the 80% range are common and micro units using the Pelton Wheel are produced by several firms in North America.

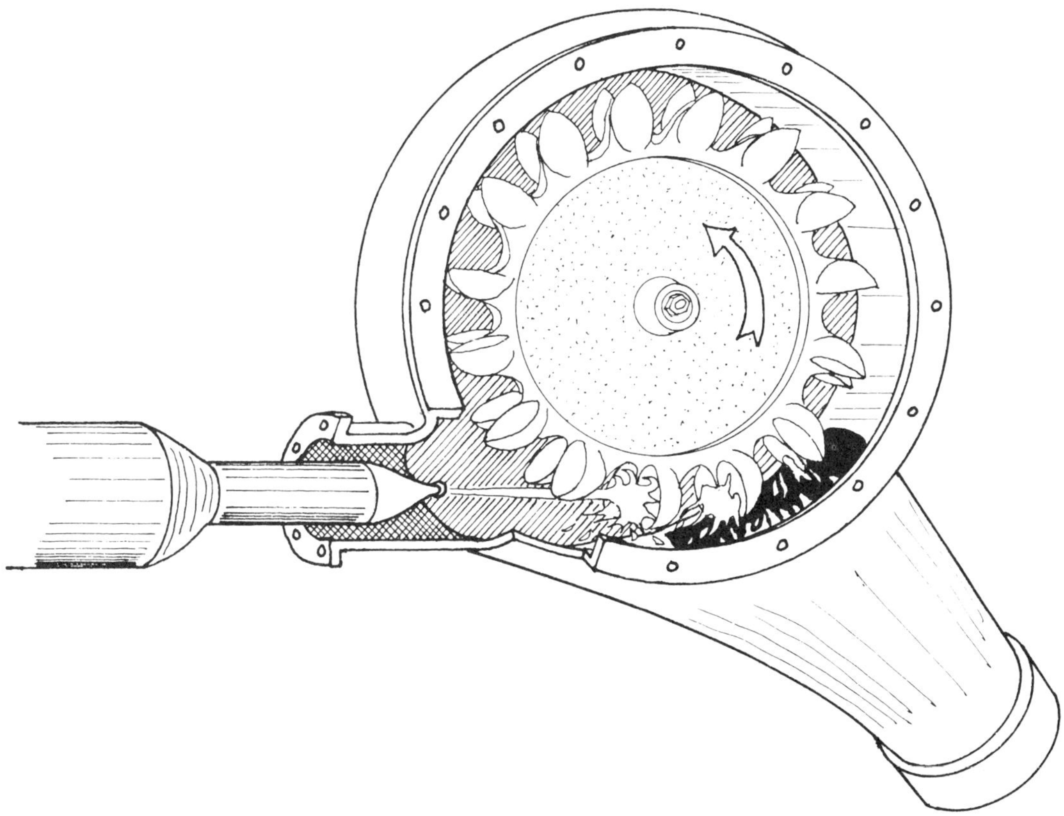

The Turgo Impulse Wheel:

The Turgo unit is a variation and perhaps improvement on the Pelton. It is made exclusively by Gilkes of England. The Turgo runner is a cast wheel whose shape generally resembles a fan blade that is closed on the outer edge. The water stream is applied to one side, goes across the blades and exits on the other side. The stated advantages of this design are that power equivalent to the Pelton can be produced with a smaller wheel at higher speed. Like the Pelton, it is possible to use more than one water jet on a single wheel in situations where relatively lower head and higher flow are present. Also as with the Pelton, the wheel itself is made in relatively few sizes and different nozzle sizes are used to match the equipment to the site conditions.

The Crossflow Turbine:

A crossflow runner is drum-shaped with the blades fixed radially along the outer edge. The unit, open in the center, resembles a "squirrel cage" blower. When looked at from the end as though it were a clock face, the water enters at 9 o'clock, crosses across the center and exits at 4 o'clock; thus the name crossflow. Most commercially available crossflows are made by Ossberger of West Germany or by someone else under their license. Because of its design, the crossflow is said to largely be self-cleaning and is well suited to low head applications. Ossberger has, in fact, installed them successfully in situations with only 39 inches (1 meter) of head. The Ossberger crossflow uses a metering vane at the intake side and maintains high efficiency over a wide range of flow rates. Because the runner and housing fit fairly close, a draft tube is used on the down side of the turbine, allowing some flexibility in installation relative to turbine placement and tail water level. The crossflow is used widely around the world, although it is less common in the United States.

Crossflow Turbine

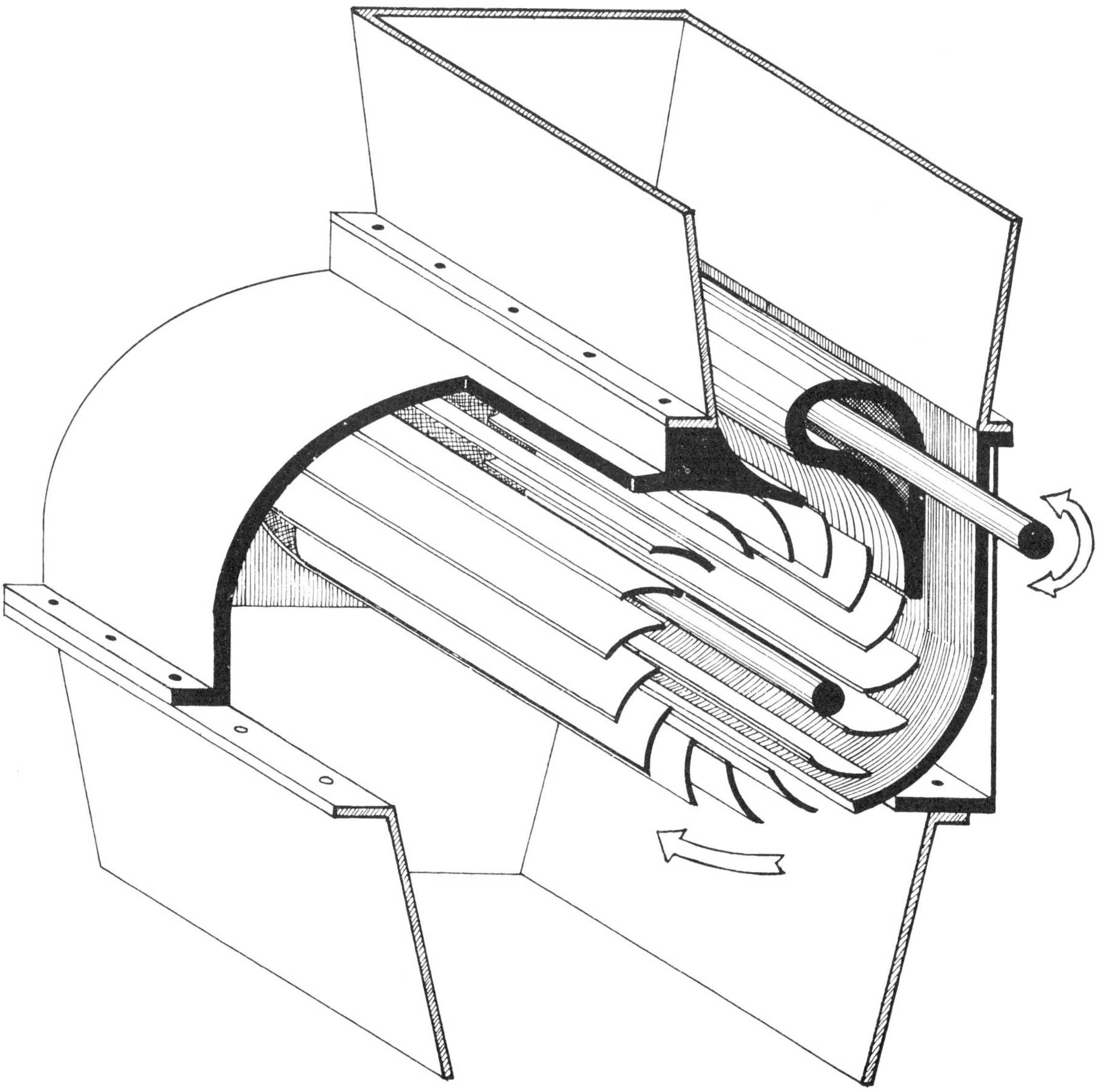

Reaction Turbines in General

Reaction turbines, while doing the same thing as impulse designs, work on a different principle. The runner is placed directly in the water stream and power is developed by water flowing over the blades rather than striking each individually. Reaction turbines use pressure rather than velocity. The function is more like that of a centrifugal water pump running in its reverse mode. Reaction units tend to be very efficient at specific designed-for situations and their efficiency falls sharply with any variation. The Kaplan design is intended to address that problem. Reaction units are usually the ones used in very large installations.

The Francis Turbine:

Francis units are typically installed in very large hydro power developments. The water is introduced just above the runner and all around it and then falls through, causing it to spin. Francis units are designed very specifically for their intended installation, use a complicated valve system, and thus are not generally used in micro-hydro applications.

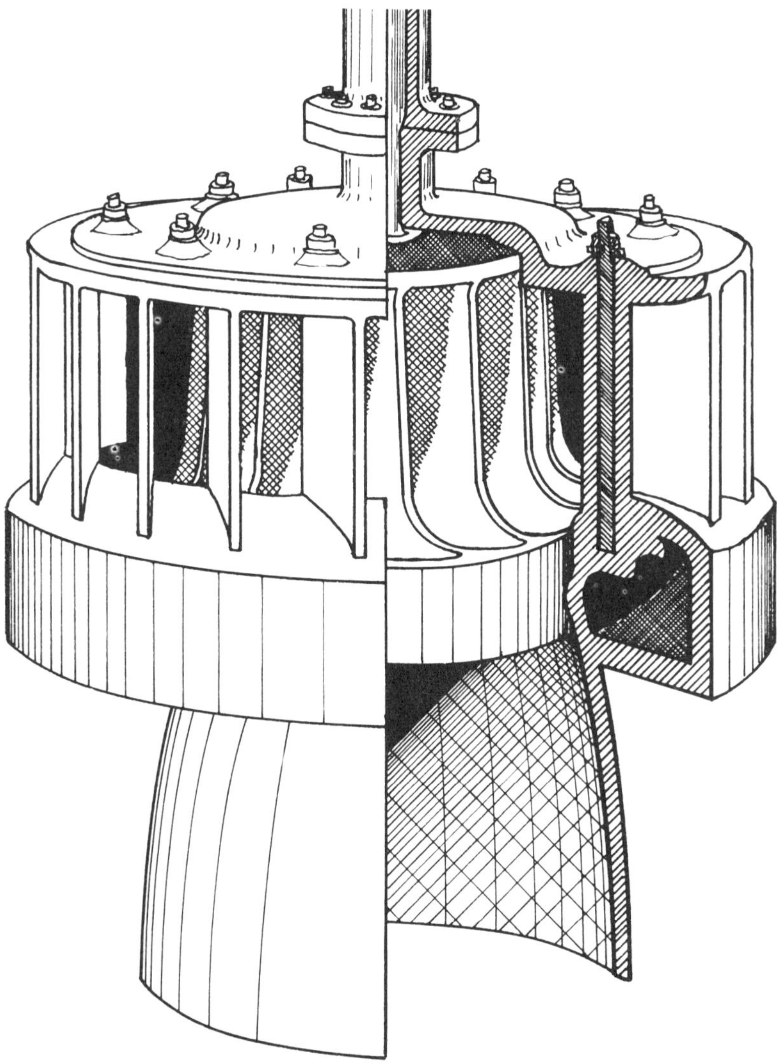

Propeller Turbines:

This design resembles a boat propeller running in a tube and operates on a similar principle. As with the Francis, the water contacts all of the blades constantly and it is thus imperative that the pressure in the cross section of the pipe be uniform. If, for example, the unit were operating horizontally and the pressure at the top of the tube were less than at the bottom, the runner would be out of balance. There are several variations on how the principle is applied, i.e., the *bulb turbine* in which the turbine and generator are a sealed unit placed directly in the water stream; the *straflow* in which the generator is attached directly to the perimeter of the turbine; the *tube turbine* in which the penstock bends just before or after the runner, allowing a straight line connection with the generator, which is located outside of the pipe. The *Kaplan*, a variation on all of these, has adjustable blades on the propeller to allow for variations in flow rates. Depending on site conditions, the unit can be installed in any position from horizontal to vertical. The most common reaction turbine design used in micro-hydro is the tube, sometimes with the Kaplan blade arrangement.

WATER WHEELS

Water wheels are the traditional technologies used for converting the energy in flowing and falling water into useful mechanical power. They are usually large diameter, slow turning wheels that are most suited to generating mechanical power. This mechanical power can be efficiently used for milling and grinding processes, operating saws, lathes, drill presses and pumps. Or, it can be inefficiently but still effectively used for generating electricity after the rotational shaft speed has been geared up to some suitably higher rpm. Water wheels offer some advantages over higher speed turbines. First, it is quite possible for the individual to construct his/her own water wheel with minimum workshop facilities. Some plans and specifications for water wheel construction are listed in the Bibliography. Water wheels can offer high torque and are thus capable of driving heavy, slow turning mechanical equipment. They will operate in situations of large water flow rate variations and they require minimal maintenance and repair. In addition, trash racks and screens are usually not required, since most water wheels can operate with dirt, stones and leaves entrained with the water. The disadvantages of water wheels are several. They are significantly less efficient than higher speed turbines, they operate at slow speed, and they are bulky. As well, in many areas of North America it is necessary to house them in fairly large structures or otherwise provide some protection to avoid freeze-up if year-round operation is required.

Undershot Wheel:

This is the most basic of water wheels. Water passing under the wheel strikes the blades or paddles, causing the wheel to rotate. It can operate on a minimum of one foot of head (producing almost no power at this head), and has a low efficiency. The optimum head range is 6 to 15 feet, with minimum wheel diameter being about 15 feet.

Poncelet Wheel:

This is an adaptation of the undershot wheel, in which the blades are curved to provide more efficient water interaction with the wheel. This wheel makes use of the velocity of the water which has been held back and forced through a narrow opening. Minimum diameter of a Poncelet Wheel is about 14 feet and they usually operate best with heads of 7 feet or less. Efficiencies are higher than for the undershot wheel.

These wheels require a breastworks of concrete fitted close to the rim of the wheel in order to help retain water in the buckets. The close clearances necessitate the use of trash racks to keep stones and wood from entering the system and causing damage.

Breast Wheel:

Water enters the breast wheel below the top of the wheel and is kept in the wheel buckets by a close fitting breastworks until it discharges at or near the lowest point on the wheel. Breast wheels operate best with heads less than 10 feet, and wheel diameters usually range between the head and three times the head. High breast wheels (water entering above the center shaft) have efficiencies that can approach 65%. Low breast wheels (water entering below the center shaft) usually have efficiencies between 35 and 40%.

Breast wheels require somewhat complicated, curved breastworks. In addition, the buckets have to be ventilated to allow air to escape to the next higher bucket as each bucket fills. The close tolerances of the breastworks give the same disadvantages in terms of water-borne debris as for the Poncelet wheel. The complicated construction techniques and the lower efficiencies, particularly in the low breast wheels, usually make other types of water wheels more attractive.

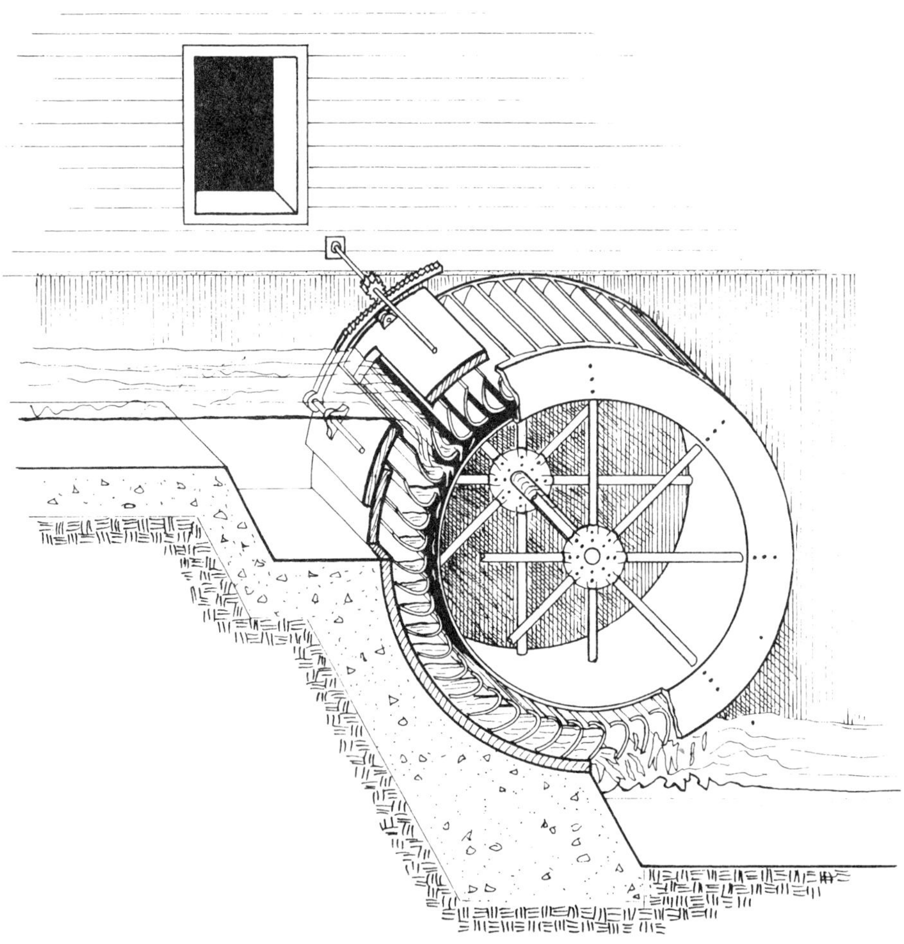

Overshot Wheel:

Water is supplied by a nearly horizontal chute to the top of the water wheel. The weight of the water in the rim buckets causes the wheel to turn. The entering water usually strikes the buckets with a velocity somewhat greater than the rim speed so as not to be struck by the back of the buckets and be splashed off the wheel. The water supply, and thus the power output, is controlled by a hand-operated sluice gate. Overshot wheels are generally the most efficient of the water wheels. They can operate on any head above 10 feet. Today, the upper limit on the head is around 30 feet because of the cost of constructing a wheel of that diameter.

SIZING THE SYSTEM

It is important early in your project to examine your power needs and the characteristics of that need. There are two separate but related questions here: Total Consumption and Peak Consumption.

THE TOTAL CONSUMPTION IS THE NUMBER OF KILOWATT HOURS USED IN A GIVEN PERIOD OF TIME, MOST COMMONLY KWH PER MONTH.

If you are presently getting your power from the utility company, you need only check a number of past bills to get this information. Keep in mind that most household consumption varies by the month, season, unusual changes in weather, or a change in the number of household members. For example, taking consumption figures for the months of June, July and August may not reflect true annual averages. If you are building or buying a new home, you can estimate your power consumption from the Table on Typical Household Appliance Loads. If more appliance energy consumption information is required, you can find this in the following books:

Energy Primer　　　　Other Homes and Garbage

The need for accuracy in developing these calculations is real, as over or undersizing the system will result in a system that is either unnecessarily expensive or does not meet your needs.

THE PEAK CONSUMPTION IS THE MAXIMUM AMOUNT OF ELECTRICAL ENERGY NEEDED AT ANY ONE TIME.

This second vital piece of information can further be explained in that if all of the electrical items in the house were operating at once, the resulting power need would be the peak. Do not confuse this with total consumption, as the situation could exist where a system could meet one need but not the other. In micro systems the peak will be more likely to cause you problems than will the total consumption. As with utility power systems, situations with high peak load relative to the average load will be less efficient and more costly than situations where consumption is more uniform. When analyzing your situation, it may be advantageous or necessary to adjust lifestyles to fit the system rather than buy a system that can cover a high peak consumptive pattern.

If you have access to the utility grid, it will be possible to purchase power needed, but not available from your hydro system, from the power company. It may also be possible to sell your surplus power back to the utility company, although that situation will depend on local policy of the power company and the regulatory agencies.

POWER GENERATION & STORAGE

Mechanical energy is converted to electricity by a generator, similar to the one found in your car. The electricity can be either Direct Current (DC) or Alternating Current (AC). Most, but not all, household appliances run on AC, 110 volts, 60 cycles, and that will be the most useful type of electricity to have.

If you intend to be completely independent from the power grid, a *synchronous generator* is used. AC systems in North America operate at a frequency of 60 cycles per second and any variation from that will affect the accuracy of clocks, phonographs and the like. In order to generate power at this frequency, the speed of the synchronous generator must be very constant and a *governor* is used to control water flow and thus turbine speed. If you have access to utility power, you can use an *asynchronous generator* with no governor, since it is self-regulating. In either case, the generator must run at its design speed and thus some kind of speed increaser is needed between the turbine and generator.

Another option is to generate direct current and either use it as is or convert it to AC when needed through the use of an *inverter*. A DC-to-AC system has several advantages, especially in very small systems (less than 5-KW). The excess power generated by a DC system can easily be stored in batteries, thereby extending the system's peak capacity. DC generators are not speed-sensitive and no governor is needed. Thus, a small DC system will generally cost less and serve needs better than a comparable AC system because a small direct AC unit often cannot cover peak needs. Battery storage systems in hydro units generally work better than those in wind power systems because the hydro generator is always putting some power back into the battery set except in the most extreme cases. This means that a deep discharge condition, common with wind systems, is very rare. Deep discharge is a common cause of battery failure. The storage function does limit the size of a DC system, as batteries become unwieldly and very costly in systems over 6-KW.

With a DC system, it is possible to use only DC appliances, although this is obviously somewhat restrictive and may be expensive if all of your current appliances run on AC. It would, however, save the cost of a DC/AC inverter. For those seeking a lowest cost option, an all 12-volt DC system is possible, as there are many appliances available from recreational vehicle suppliers.

LOAD CONTROL & GOVERNORS

The *governor*, a device that regulates turbine speed through water flow on synchronous generators, is available from a variety of sources. It is reliable and accurate, but does introduce additional cost and maintenance factors.

In systems with no storage capability, some sort of *load control* is needed. Hydro units that have access to the power grid may consider the grid as a sort of storage function, i.e., putting your excess into the grid and buying it back when demand exceeds supply. This concept is currently being used in some areas, but wider use is still open to questions. Otherwise, something must be done with excess power and that is the function of load control units. Such units usually dump excess energy as heat and that heat can often be put to good use such as supplementary space or water heating. When demand begins to approach system capacity, it is possible to use a device that turns key equipment on and off on a priority basis. Non-essential uses are turned off and critical needs are met.

OTHER EQUIPMENT

Dams/Impoundments:

Although many micro-hydro units operate without dams, it may, in some situations, be necessary to build such items into the system. If a dam is required, depending on where you live, you may have to get a registered engineer to design it or at least approve your plans. Good dam plans and construction techniques are described in a number of books, among them:

- Cloud Burst
- Design of Small Dams
- Small Earth Dams
- Ponds for Water Supply and Recreation

Powerlines:

If the hydro unit is any distance from the house, you will obviously need powerlines. Keep in mind that the longer the distance, the larger the wires that will be needed to avoid unreasonable power loss. This is especially true of 12-volt systems which require very large wire to avoid excessive losses.

ECONOMICS

Any analysis of the economics of a micro-hydroelectric system must begin with a determination of the system's true total cost. An investment of this size requires that your calculations be as accurate as possible. This cost calculation must include not only the turbine and generator, but also any pipe, cable, buildings, dam, civil engineering work, permits, legal work and the like. You then have to consider the other options available for electric power and determine their costs. An important consideration that might help the economics of developing a micro-hydro system is the other uses that can be made of the water resource. Fire suppression, domestic water supply and irrigation might involve additional expenses if they had to be developed separately. But because there is very little extra cost to develop these along with a hydro system, they should be figured into your analysis. One way to do this is to add the costs of developing alternative methods of fire suppression, water supply and irrigation to the cost of obtaining electrical power other than by hydro

One characteristic of many renewable energy resources is that the "front end costs" are high although the lifecycle costs may be very competitive with conventional energy sources. Small-scale hydro, for example, is fairly expensive to purchase, but except for small maintenance costs the system should provide "free" energy for 10 years or longer. The economic analysis of hydro must include such lifecycle costs. For those who are developing a home some distance from existing powerlines, hook-up costs must be considered when looking at the cost of alternative choices. In many areas of the United States, power companies charge by distance to hook homesteads up to powerlines; sometimes as much as \$1.50 to \$2.00 per foot. A person living 1 mile from powerlines could thus spend as much as \$10,000 to hook-up, thus making the hydro option very attractive.

The cost for hydroelectric systems is generally in the range of \$750 to \$1500 per kilowatt. Variations of cost will depend on any site work that may be needed, the unit's size (cost per kilowatt will go down as output goes up), and head (high head units are generally cheaper than low head units of equal output).

In the event that your site has access to conventional power, and the water has no other uses, it may be difficult to justify a micro-hydro unit on economic terms unless you are willing to take a very long range view. There are a great many factors that will affect your analysis and any such calculations must be done on a case-by-case basis. For example:

* Will you base calculations on life of the hydro unit, which may be 20 years or more, or some shorter period?

* Are you faced with large hook-up charges from the power company?

* Are tax credits or grants available from the State or Federal Government?

* Is your taxable income sufficient to take advantage of the deductible interest payments on loans?

* How long do you expect to remain at this residence and will the hydro system have market value when you want to sell?

* Is there any opportunity to sell or barter power with neighbors or the electric company?

* How much of the installation and maintenance work can you do yourself?

For a great many people the ideas of self-reliant and self-sufficient lifestyles are very appealing, irregardless of economic considerations. However, for most people it is important to separate these ideological considerations from the economic ones. There are a great many people for whom micro-hydro presents a viable alternative and economic arguments are sufficient.

A SAMPLE ECONOMIC ANALYSIS

One realistic way to analyze your electric power options is to figure the total cost for each over some period of time and then compute the costs in current dollars. The following is such an analysis, computed at three typical interest rates. The following assumptions have been made:

* The hydro system is a 6 KW, DC-to-AC battery storage unit.

* The total cost for the system is $6,000.

* Total maintenance is $1358 over 15 years ($50 per year with 8% annual cost increases).

* The site is near existing powerlines and no hook-up charges are required.

* Utility power costs start at 4¢ per kilowatt hour and increase by 10% per year.

* The hydro unit is financed by a 15-year, 10% loan.

* Average monthly consumption is 500 Kilowatt hours.

* The hydro unit is worth $1800 (30% of purchase price) at the end of the 15-year period.

* No tax deductions or credits are used.

The figures shown below reflect the cost for each option in current dollars. The computations were done at three interest rates. It may be helpful to view the differences shown as "profit" or "loss" resulting from having selected one option over another. The actual numbers reflect the investment needed today to cover the cost over 15 years. Were you to invest the amounts shown below at the corresponding interest rates, the revenue generated (principal and interest) would be sufficient to cover the cost of the 6 KW hydro system, including maintenance:

5% interest	7.5% interest	10% interest
$9517.54	$8260.37	$7258.44

If we then subtract the salvage value of the system in 15 years, the resulting figures will show the cost of the system at present prices:

5% interest	7.5% interest	10% interest
$8651.71	$7652.03	$6827.53

If the same approach is taken with the cost of utility-supplied power, the amounts shown must be invested at different interest rates in order to pay for the power over the 15-year period:

5% interest	7.5% interest	10% interest
$5086.80	$4249.25	$3600.00

Thus, the cost difference between the two options is shown below:

5% interest	7.5% interest	10% interest
($3564.91)	($3402.78)	($3227.53)

As can be seen in this example, the hydro option will be somewhat more expensive over the 15-year period, assuming all of the conditions mentioned above. A change in any of the assumptions can significantly affect the economics one way or the other. For example, if the site were one mile from the existing powerline, a hook-up charge would result. Assuming a cost of $1.50 per foot for the new line, the cost of utility power would increase by $7920.00. Figuring this into the above equation, the hydro system would offer a profit of:

5% interest	7.5% interest	10% interest
$4355.09	$4517.22	$4692.47

The example shows that for many people, micro-hydro is a viable option; the economics depends heavily on individual conditions. There is also the question of how highly you value independence from traditional energy sources. For some, economic conditions permitting, a small additional cost may be worth the option of unplugging from the power grid.

Another way of looking at the costs associated with installing, operating and maintaining a micro-hydro system is to determine how much all of this will cost you in terms of money spent per kilowatt hour of electricity generated and used. Independent Power Developers have come up with some graphs that can assist you in this estimation. These are reproduced on the following pages. The graphs are set up for various amortization periods, operation and maintenance costs and interest rates.

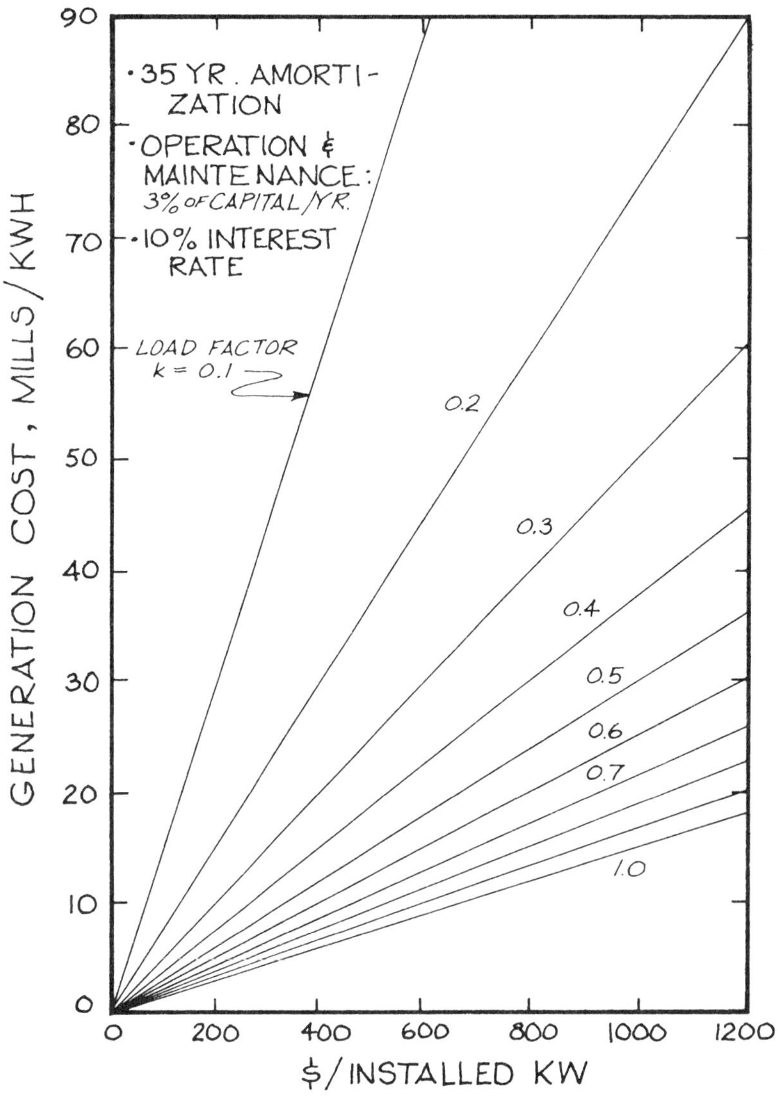

The horizontal axis of each graph is the cost of the micro-hydro installation, given in dollars per installed kilowatt. The vertical axis is the cost of the electrical power generated in mills per kilowatt hour (10 mills equal one cent). The family of lines emanating from the lower left in each graph represents lines of constant load factors. Load factor is simply the average power output of the generating equipment, divided by its rated power output.

What the load factor lines mean, in essence, is that the more you are able to use your electrical power, up to the rated capacity of the system, the closer your system load factor comes to 1.0 and the less per kilowatt hour you spend on electricity. Micro-hydroelectric systems normally have a load factor range between 0.6 and 1.0.

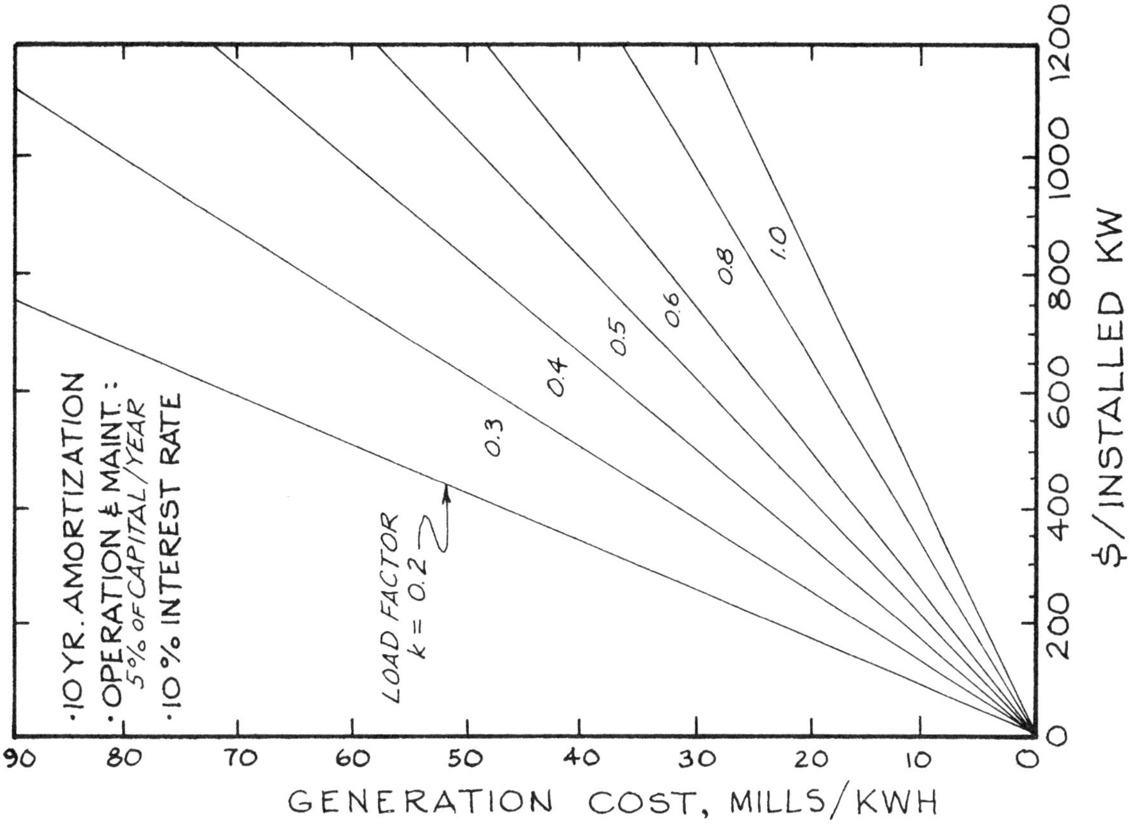

·10 YR. AMORTIZATION
·OPERATION & MAINT.: 5% OF CAPITAL/YEAR
·10% INTEREST RATE
LOAD FACTOR k = 0.2
0.3
0.4
0.5
0.6
0.8
1.0
GENERATION COST, MILLS/KWH
$/INSTALLED KW

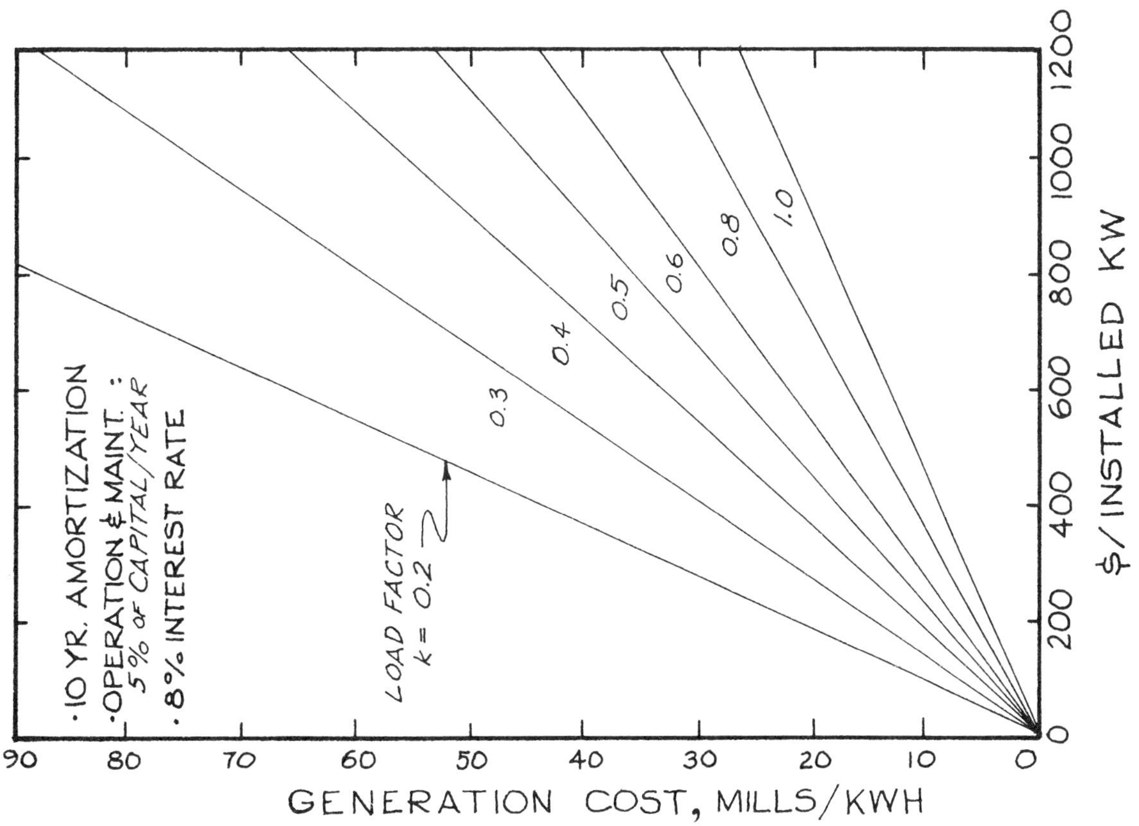

·10 YR. AMORTIZATION
·OPERATION & MAINT.: 5% OF CAPITAL/YEAR
·8% INTEREST RATE
LOAD FACTOR k = 0.2
0.3
0.4
0.5
0.6
0.8
1.0
GENERATION COST, MILLS/KWH
$/INSTALLED KW

SOURCES FOR FINANCIAL ASSISTANCE

At the time of this writing, there are several potential sources for financial assistance usable in micro-hydro. This area is rapidly expanding and new programs are likely to be developed.

DOE Small Hydro Program

In 1978 Congress produced legislation that will encourage the development of small-scale hydro (up to 15 MW) on existing dams. The legislation as written will allow $10 million for feasibility studies and $100 million for construction. It apparently was not intended for individual systems, although they are not specifically excluded. The program may be useful for projects of a neighborhood or larger scale.

DOE Small Grants Program

The Small Grants Program, designed to encourage the demonstration of appropriate technologies in general, is administered through regional Department of Energy offices. The regional offices have been allowed some flexibility in their administration, and checking with the office is your area is recommended.

State Grants/Tax Credits

Many states now offer tax credits and/or grants designed to encourage the development of renewable resources. Many of these programs specifically reflect solar, but taken in its widest interpretation, wind and hydro in the broader sense are both solar energies. Check the legislation in your state to see what interpretation is used. Keep in mind that grant programs are often very competitive. In the case of the tax credit, you are generally allowed to deduct some percentage of the system's total cost from your gross income. Be sure to run a quick calculation using your own figures to determine the program's effect on the real cost of your system.

Conventional Bank Loans

One possible source of money is a loan from your local bank. The interest you pay is deductible, but again be sure to calculate real effects in your own situation. People that function in a relatively cash-free economy may derive little benefit from tax deductions. Also, you may encounter some resistance from the bank, as they may be uncomfortable about loaning money on something they don't understand. The development of solar heating systems encountered a similar problem some years ago. With hydro, this is likely to be less of a problem in the future. Also, of course be sure that you are able to afford the monthly payments.

REGULATORY CONFLICTS

A key part of any micro-hydro development program is a study of various institutional and/or legal barriers that may be present and what can or must be done to deal with them. There are a wide range of potential problems in this area, as well as a range of solutions. There may also be some problems for which there are no solutions, or solutions that are too costly to allow the project to proceed. It is best to identify those problems very early in the project time frame and begin what action is needed.

The first step is to determine the ownership of water rights for the resource being considered. Law pertaining to water rights varies greatly across the country and a close examination of regulations in your area is important. Generally, you should not assume that because you own the land you also own the water that runs across it. In many states, it is possible to buy land without water rights, as well as buying water rights without owning the land along it. A property owner downstream may own water upstream from that property, or the water may be owned by several downstream users, with each allowed to take some percentage of the flow. In many states the fact that your hydro system will not consume water but only use its energy makes no difference; you must have legal right to use the water. The water right for a particular site may go back a great many years and researching old records may be necessary to determine the true holder of those rights.

Much of the state law regulating water usage was written relative to its irrigation and domestic use, and thus is not directly applicable to hydroelectric systems. This may require new legal decisions in some cases and could mean some hesitancy on the part of regulatory agency personnel. *Do not be discouraged and do not give up because your first response is "no".*

A second step is to secure a water use permit, which, in some cases, may be required from both state and federal agencies. In some states, having legal right to the water is all that is required. The other situation is when an additional permit is required, referring to the specific use intended. Often the actual installation work must begin within some time period or the permit expires and must be reapplied for. Here again, most water use laws are oriented toward agricultural uses and questions may arise relative to their interpretation in the micro-hydro context.

A federal use permit is generally required when the resource or its development involves federal land, be that Forest Service, Bureau of Land Management, or whatever. An important point to remember is that your use is beneficial and non-consumptive. This is different from agricultural or industrial uses which may be beneficial but are also consumptive. Be certain to mention that your intended use is beneficial and non-consumptive, as it should help in the permit process.

The Federal Energy Regulatory Commission (FERC) is generally the agency that issues licenses for projects intended to generate electricity, including hydro. Because most recent hydro projects have tended to be very large developments, the FERC licensing process has developed into a lengthy effort. In April of 1978, reflecting the new interest in small-scale hydro development, a new short form was developed. The short form, which covers systems up to 1500 KW, is only two pages long.

The Army Corps of Engineers has jurisdiction over all navigable streams in the United States, meaning that any changes made to stream banks, beds or flows must get Corps approval. A question that arises often in this context is that of the definition of "navigable river." One definition relates to the lumber industry -- a stream that can float a log of a certain size is navigable. Another line of reasoning, and in fact legal ruling, takes the position that because even the smallest creek eventually ends up in larger (navigable) streams, they all fall under Corps jurisdiction. The question is somewhat complicated by the fact that interpretations of this type are left to regional Corps offices, and differences do exist.

A second question relates to the level of stream modification that requires Corps approval. Obviously, the placement of a dam constitutes significant modification, but the installation of a small run-of-the-river unit may not. Again, the interpretation of regulations in this area is the role of the regional Corps office. In any event, it is best to develop the hydro site with the least stream alteration possible. The Army Corps of Engineers is organized by major river basin or drainage areas, and a determination of what basin you are in is the first step in locating the appropriate Corps office.

In those situations where a dam is necessary, a permit will be required from the Corps of Engineers and may be required from state and local agencies as well. There is often some minimum size in terms of dam height and acres of water behind it, and units smaller may be exempt from the permit process. There may also be annual fees, either fixed or variable amount reflecting dam size. These fees are generally modest and are intended to cover the cost of inspection programs. Some states may require that construction plans be done by a registered engineer, which may be a good idea for those individuals who lack engineering skills, but will also raise the cost of the project. Generally speaking, the inclusion of a dam in your hydro project will complicate the effort substantially and thus should be avoided if possible.

A final area for regulatory concern is environmental impact. This may include changes in stream flow, blockage of fish movement because of the dam, flooding of some areas above the dam, and the like. The state agency that is most likely to be involved is the Department of Natural Resources, Fish and Wildlife, Conservation or something similar. There may also be federal agencies with jurisdiction, depending on local conditions. In the case of dams, you may be required to install a fish ladder (a water stairway that allows the fish to move upstream past the dam) and depending on the size and type needed, they can be very expensive. In some states, people intent on redeveloping the power potential of an existing dam have been required to put in a fish ladder even though there had never been one before. Some regulations require that public hearings be held on the proposed development and from a strictly administrative standpoint, these can be very time-consuming. As mentioned above, when planning your hydro project, *keep stream disruption to an absolute minimum*, and things will go much easier and quicker.

Although it is generally assumed that the power generated by a micro-hydro system will be used by the individual or family that owns it, there are efforts around the country to enter the electric generation business on a small scale. For example, it is now possible to purchase a control unit that when your demand is greater than the system's output, the balance is purchased from the power grid (assuming you have access to utility power). When, however, your system is producing more than you need, the control unit feeds the excess back into the grid and in effect winds the electric meter backwards. This type of arrangement makes the power grid a sort of storage battery and may allow the possibility of installing a smaller, less expensive system. It may also affect the income side of the economic analysis by allowing the sale of small amounts of power to the grid. Systems of this type are operating in a few isolated areas and the question is being decided in several others, but generally, widespread usage is somewhere in the future.

For those people who have more generating capacity than they need, it may be feasible to sell or barter the power to a nearby neighbor. In some states the Public Utility Commission will enter the regulatory picture, while in other areas very small projects will be ignored. Also, just as many local governments operate power companies, it may be possible for several families living near a hydro resource to form some kind of power co-op to build and operate the system jointly. Here again, the State Department of Commerce may have jurisdiction and you may encounter opposition from the local power company. Electric companies sometimes interpret their monopolistic license as meaning that no one else can generate power within their service area.

Micro-scale power co-ops and similar arrangements are still very much in their infancy, but may be a viable option for some people.

By way of summary, some general guidelines for approaching the regulatory issue may be valuable. First, develop a clear idea of what you want to do, what the site looks like and how the hydro unit will fit into it. Anticipate questions and develop sound answers. Be able to explain the project clearly to people who will hear of micro-hydro for the first time. Because this is a very new field, you are likely to get a range of fairly negative responses to your first inquiries. Do not be discouraged, but attempt to determine the reason behind his/her response, do what is necessary to correct the situation and continue on.

It will be wise to correspond in writing, make copies and keep a file on all that has transpired. You may want to consult an attorney when dealing with obtuse legal language.

A final word of caution is that you should start checking out these regulatory questions very early, before you buy a lot of equipment or otherwise invest a lot of money. Micro-hydro projects frequently consume one year from first considering the idea until the first kilowatt is generated.

CAUTIONS AND SUGGESTIONS FOR THE DO-IT-YOURSELFER AND THE SELF-INSTALLER

The assumption in this section is that you have successfully arrived at Step 8 in the Decision Tree. Now you are entering the final design and equipment selection stage. The following list of considerations will make installation much easier and avoid a lot of troubles at some future date.

In the final design stage, ensure that you:

1. Consider your stream bed loading conditions. Silt and rocks coming down the stream, particularly during periods of high run-off, can cause intake clogging or even destruction of the intake pipes.

2. Size the pipe so that it is capable of handling the volume flow rate that you require. Any responsible pipe supplier can give you the correct size for the flow conditions you expect.

3. Route the pipeline, from intake to the turbine, so that it contains the minimum number of bends. Do not use 45° (or greater) elbows in the pipeline. Otherwise, there will be too much strain on the pipe and excessive friction losses.

4. Keep a downhill slope in the pipe at all times (except for the initial siphon intake, if used) in order to avoid air locks and silt deposit.

5. Do not let the water velocity in PVC pipe get much above 5 feet per second. Above this line velocity there are other design considerations coming into play that the do-it-yourselfer is not usually prepared to deal with. One of these considerations is thrust-blocking wherever the pipe bends.

6. Size the pipe in order to maintain about 5 feet/second line velocity in order to avoid excessive ice build-up in the pipe. If your line velocity is much less than this and you are installing the system in an area where winters are severe, then consider insulating or burying the line.

7. Consider installing a water by-pass before the turbine in case the water may be needed for fire control.

8. Locate the DC turbine and generator adjacent to the point of use. This is important in order to keep electrical transmission lines as short as possible so that line losses are kept to a minimum.

9. Plan to install the system in warmer weather, and not under freezing conditions if at all possible. You are dealing with a liquid that freezes at certain times of the year, and if it does so while you are working with it, it could be dangerous to the equipment and to yourself.

When buying or otherwise acquiring your equipment you should:

1. Be sure you deal with a reputable supplier. There is some junk around, so buyer beware!

2. Expect delays in getting quotes and deliveries from equipment suppliers, since none of them are currently very big and they are usually quite busy.

3. Obtain pipe that is rated and classed for the pressure rating it's in. Don't buy seconds.

4. Ensure that you have a good trash control system for the intake. A screen mesh should be used that has openings smaller than the minimum nozzle diameter which leads into the turbine. This way the only solid particles that can come down the pipe will be small enough to pass through the nozzle without clogging it.

During installation:

1. Be sure you follow the manufacturer's or supplier's instructions and suggestions.

2. Slow down and do things properly. You are dealing with mechanical and electrical systems. They can cause you a lot of grief unless treated with respect.

3. Watch for rocks, and be careful placing them, when burying PVC pipe.

4. Use gate valves wherever valving is necessary. Other kinds of valving allow you to turn off the water too quickly, causing dangerous water hammer effects.

5. Use standard house wiring procedures with the electrical hook-up. Go to your local bookstore and pick up the appropriate electrical do-it-yourself book. Or, if you feel unsure of your own capability, go to your local electrician and get him/her to help you.

Once the system is operational, be very sure that you:

Close valves, when you have to, slowly. Take a little time and screw them closed gradually. Closing a valve too quickly can cause a shock wave (a high pressure wave) to move through the water in the pipe. It is going to have some detrimental effect somewhere along the length of that pipe. This damage is going to vary depending on the length of the pipe and the velocity of the water at the point of valve closure. In many cases it will crack the pipe and you will see a slow leak, or again it could cause a serious rupture in the pipeline.

MANUFACTURERS AND SUPPLIERS

Prime Movers

Independent Power Developers, Inc. Route 3, Box 285 Sandpoint, Idaho 83864	Pelton and Propeller Units, Complete Systems
The James Leffel Company Springfield, Ohio 45501	Francis/Propeller/Hoppes Units
Gilbert, Gilkes & Gordon, Ltd. Westmoreland, England LA9 7BZ	Wide range of turbines from 10 KW to multi-megawatt, Turgo and Kendal
Small Hydroelectric Systems P.O. Box 124 Custer, Washington 98240	Pelton, with power range 5 to 25 KW for heads from 50 to 350 feet.
Ossberger Turbinenfabrik D-8832 Weissenberg Pastfach 425 Bayern, West Germany	Crossflow (Mitchell or Banki type) turbines of 1 to 1000 KW.
Escher Wyss, Ltd. CH-8023 Zurich, Switzerland	400 KW mini-straflow
Barata Metal Works and Engineering PT J.L. Ngagel (109) Surabaya, Indonesia	18 KW (plus or minus) Francis and Crossflow
Jyoti, Ltd. Industrial Area PO Chemical Industries R.C. Dutt Road Baroda 390 003, India	Francis and Turgo sets from 5 to 25 KW, plus some larger units
Officiene Buehler Canton Ticino, Switzerland	---
Westward Mouldings, Ltd. Greenhill Works, Delaware Road Gunnislake, Cornwall, England	Fiberglass Water Wheels
Campbell Water Wheel Company 420 South 42nd Street Philadelphia, Pennsylvania 19104	Water Wheels
Drees & Co. GMBH 4760 Werl/West. Postfach 43 West Germany	Medium to large turbines

G&A Associates 223 Katonah Avenue Katonah, New York 10536	Francis units
Maschinenfabrik Kossler GMBH A-3151 St. Polten St. Georgen, Austria	Small turbine sets of 12 to 1250 KVA rating, plus variety of larger machines
Karlstads Mekaniska Weskstad Fack S-681 01 Kristinehamn, Sweden	Range of medium sized horizontal axis propeller turbines of 50 to 1800 KW
Niagara Water Wheels, Ltd. 706 E. Main Street Welland, Ontario L3B 3Y4, Canada	Four models of propeller turbines with power in range of 20 to 250 KW
AB Bofors Nohab S-46101 Trollhatten, Sweden	Propeller turbines from 100 to 2000 KW
Barber Hydraulic Turbines, Ltd. Barber Point, P.O. Box 340 Port Colborne, Ontario L3K 5W1, Canada	Propeller and Francis turbines
Elektro GMBH St. Gallerstrasse 27 Winter Thur, Switzerland 8400	Francis, miniature turbines in range 50 to 2000 watts
Canyon Industries 5346 Mosquito Lake Road Deming, Washington 98244	Francis, miniature turbine set offering 50 to 750 watts
Briau SA BP 43 37009 Tours Cedex, France	Francis/Propeller turbine sets to 50 KW
Northern Water Power Co. P.O. Box 49 Harrisville, New Hampshire 03450	Axial flow propeller turbines with output range 20 to 250 KW
Land & Leisure Services Priory Land St. Thomas Launceston Cornwall, England	Propeller turbines
Alaska Wind and Water Power P.O. Box G Chigiak, Alaska 99567	Pelton turbines
Pumps, Pipe and Power Kingston Village Austin, Nevada 89310	Pelton turbines

Bell Hydroelectric 3 Leatherstocking Street Cooperstown, New York 13326	Crossflow turbines
Balaju Yantra Shala (P) Ltd. Balaju, Katmandu, Nepal	Crossflow turbines
Maine Hydroelectric Development Groups Goose River, Maine	Belfast turbines
Allis Chalmers Hydro Turbine Division P.O. Box 712 York, Pennsylvania 17405	Large turbines

Miscellaneous Equipment Suppliers

Windworks Box 329, Route 3 Mukwonago, Wisconsin 53149	Gemini Inverter
Lima Electric Company, Inc. 200 East Chapman Road Box 918 Lima, Ohio 45802	AC Alternator
Woodward Governor Company 5001 N. 2nd Street Rockford, Illinois 61101	Mechanical Governor
Natural Power, Inc. New Boston, New Hampshire 03070	Governor

SOURCES OF PROFESSIONAL SERVICES

P.W. Agnew Department of Mechanical Engineering The University Glasgow, Scotland	Research & Development
Dry Buck Ranch P.O. Box 5 Banks, Idaho 83602	Consulting
Guy Immega Lacqueti Island British, Columbia, Canada	Consultant/Author
Intermediate Technology Group, Ltd. 9 King Street London, England WC2E 8HN	Research & Development
Low Impact Technology 34 Martin Street South Melbourne, Victoria Australia	Research & Development
David Master P.O. Box 302 Milford, New Hampshire 03055	Consultant
Niagara Water Wheels, Ltd. Box 326, Bridge Station Niagara Falls, New York 14305	Design/Installation
Mike Harper, P.E. P.O. Box 21 Peterborough, New Hampshire 03458	Consulting Construction Installation
Doug Smith P.O. Box 43 Hanover, New Hampshire 03759	Consultant
Tientsin Elector-Driving Research Institute Tientsin, China	Research & Development
Bill Delp c/o Independent Power Developers Route 3, Box 285 Sandpoint, Idaho 83864	Consulting Fabrication Installation

BIBLIOGRAPHY

THE BANKI WATER TURBINE

Prepared and
Published by: School of Engineering
Oregon State University
219 Covell Hall
Corvallis, Oregon USA 97331

This is a short pamphlet which is basically a translation of a German paper by D. Banki written in 1917. It includes a good section on both the theory of the Banki crossflow turbine as well as runner construction details. Some test data is given on a typical turbine. This paper is only recommended for those serious enough to want to build their own. The approach is quite scientific.

CLOUDBURST

Edited by Vic Marks
Published by: Cloudburst Press, Ltd. (1973)
Mayne Island
British Columbia, Canada VON 2J0

This book contains an excellent 30-page section on micro-hydro development. It goes through the standard techniques of measuring head and flow, gives good descriptions of do-it-yourself dam building, and tells you how to build an overshot water wheel and a crossflow turbine. There is also a fairly good section on water wheel design, describing particular features of overshot, breast and Poncelet wheel buckets.

A DESIGN MANUAL FOR WATER WHEELS

By W.G. Ovens
Published by: VITA (1975)
3706 Rhode Island Avenue
Mt. Rainier, Maryland 20822

This is a do-it-yourself type booklet intended for use specifically in developing countries. However, it has obvious application in North America as well. It deals specifically with design and construction details of an overshot water wheel for mechanical power.

ENERGY PRIMER

Prepared and Published by: Portola Institute (1978 ed.)
558 Santa Cruz Avenue
Menlo Park, California 94025

This comprehensive information catalogue focuses on small-scale renewable energy systems, among which is water. It devotes 22 pages to basic descriptions of most aspects of harnessing water power. Included are sections on water rights, environmental considerations, measuring available water power, building dams, water wheels and turbines. It gives some construction details on crossflow turbines and the Pelton wheel. Contained also is a list of over 60 water turbine manufacturers around the world.

A HANDBOOK OF HOMEMADE POWER

By "Mother Earth News"
Published by: Bantam Books (1974)

This book is available in most bookstores. It includes only a brief section on hydro power, including plans for a small water wheel.

HARNESSING WATER POWER FOR HOME ENERGY

By Dermot McGuigan
Published by: Garden Way Publishing Co. (1978)
Charlotte, Vermont 05445

This is an excellent book describing all manner of material related to small and micro-scale hydro. It gives a number of examples of installations of the various types of water wheels and turbines in the United Kingdom and the United States. Manufacturers are listed along with their products and outputs. Equipment costs are often included. It contains a good bibliography.

HYDRO POWER ENGINEERING

By J.J. Doland
Published by: Roland Press Co. (1954)
New York, New York

This is a good book if you are an engineer and have a background in hydraulics. It is good on theory and describes practical, but larger, hydro systems than would be of interest to the homeowner.

LOW-COST DEVELOPMENT OF SMALL WATER-POWER SITES

By H.W. Hamm
Published by: VITA (1967)
3706 Rhode Island Avenue
Mt. Rainier, Maryland 20822

This 43-page booklet gives a lot of good information for every step in the process of developing small-scale hydro power sites. Descriptions are included of water wheels, a small 12" diameter crossflow turbine and the Pelton wheel. Small earth dam construction is also covered.

OTHER HOMES AND GARBAGE

By J. Leckie, G. Masters, H. Whitehouse and L. Young
Published by: Sierra Club Books (1975)
San Francisco, California

This book gives a brief 12-page introduction to micro-hydro power. It describes techniques of measuring water flow, gives simple dam construction, and illustrates the basic types of water wheels and turbines. It is not worth buying this book just for the water power section.

PRACTICAL WATER POWER ENGINEERING

By W.T. Taylor
Published by: Van Nostrand, USA (1925)
Crosby Lockwood, UK (1925)

This is not necessarily a practical book, particularly when it comes to construction and installation of water power systems. However, it does cover, in detail, information on rainfall and run-off, important considerations in determining your stream capacity. Contains good sections on site selection, reservoir and canals and electrical power transmission.

PRODUCING YOUR OWN POWER

Edited by Carol Stoner
Published by: Rodale Press, Inc.
Emmaus, Pennsylvania

This book deals with a variety of renewable energy systems and includes a 40-page section on water power. It includes an adequate section on measuring head and flow and calculating power available. Included is a five-page piece on determining channel, pipe and other head losses. Small earth and rock dams are described, as well as typical water wheel and turbine characteristics and uses. There is also a section dealing with the VITA hydraulic ram.

USE OF WEIRS & FLUMES IN STREAM GAUGING: TECHNICAL NOTE NO. 117

Published by: World Meteorological Organization (1971)
Publications Center
P.O. Box 483
New York, New York 10016

Describes, in very much detail, techniques for making an accurate assessment of stream water flow rates.

WATER MEASUREMENT MANUAL

By Bureau of Reclamation (1967)
Available from: Superintendent of Documents
U.S. Government Printing Office
Washington, D.C. 20402

This book gives detailed descriptions of a wide variety of techniques for measuring flow rates with Weirs, flumes, gates, pipes and orifices. It is somewhat broader in scope than the World Meteorological Organization book, but no more useful for the micro-scale hydro developer.

DESIGN OF SMALL DAMS

By Department of the Interior (1973)
Available from: Superintendent of Documents
U.S. Government Printing Office
Washington, D.C. 20402

This is an excellent book! It is 816 pages long and fairly expensive, but it may well be worth your while. It describes medium and large-size earth fill dams -- including all details from site selection, soil sampling, design considerations and construction techniques. Included are some excellent sections on environmental impacts and code and regulation requirements.

PONDS FOR WATER SUPPLY AND RECREATION: AGRICULTURAL HANDBOOK NO. 387

By Soil Conservation Service, U.S. Department of Agriculture
Available from: Superintendent of Documents
U.S. Government Printing Office
Washington, D.C. 20402

This book does include most of the information you need to know concerning small earth-filled dams. It contains good sections on water needs assessment, site selection, drainage areas required for various pond sizes, spillway requirements, and construction techniques. This book is not directly applicable to small-scale dams, since it deals with large farm ponds built with some heavy equipment; nevertheless, much of the information is easily transferable.

SMALL EARTH DAMS: CIRCULAR 467

By L.N. Brown
Published by: California Agricultural Extension (1965)
90 University Hall
University of California
Berkeley, California 94720

This circular describes site selection for small dams, construction, maintenance and management details and has an interesting section on permits and regulations. It is generally applicable to small earth dams anywhere, but is specifically related to California experiences.

Do-It-Yourself Plans and Specifications for water wheel and turbine construction are included in the following publications:

A DESIGN MANUAL FOR WATER WHEELS	Overshot Wheel
HANDBOOK OF HOMEMADE POWER	5-Foot Diameter Wheel
CLOUDBURST	Overshot, Undershot and Crossflow
ENERGY PRIMER	Overshot, Crossflow and Pelton
LOW-COST DEVELOPMENT OF SMALL WATERPOWER SITES	Crossflow
THE BANKI-WATER TURBINE	Crossflow
WATER POWER HYDRAULIC ENGINEERING, 1899	Water Wheels

Reprinted in Alternative Sources of Energy, No. 14
Route 2, Box 90-A
Milaca, MN 56353

GAS TURBINES FOR AUTOMOTIVE USE 1980

Edited by John P. O'Brien

Energy Technology Review No. 54

The gas turbine engine has been the object of extensive R&D programs, and could be an interesting radical alternative to the conventional piston engine.

In the past, gas turbines have had difficulty competing in the automotive field in spite of lighter weight, low emissions and absence of a cooling system, because passenger cars operate at varying speeds which is contrary to the turbine principle. Turbines also operate at higher temperatures.

The purpose of this book is to describe in concise form the most recent technological achievements in this field which might well result in a new and better form of passenger transportation.

A summarized table of contents follows here with **some chapter headings and examples of subtitles.**

ISBN 0-8155-0786-0 **342 pages**

HEAT EXCHANGE FLUIDS AND TECHNIQUES 1979

by M.W. Ranney

Energy Technology Review No. 50
Chemical Technology Review No. 143

The use of heat exchange fluids (also named heat transfer fluids) is essential to many manufacturing and energy generating as well as energy consuming processes.

Sophisticated designs of the heat exchange apparatus, together with a reduction in size, have made it possible to apply energy-saving engineering feats to building and home construction, often in connection with solar heating and cooling.

The transfer media used in such apparatus must be liquid over a wide temperature range and have a low vapor pressure. They must also be stable over extended periods of time, often at temperatures in the order of 300°C.

This book gives a large selection of such fluids and describes their synthesis, properties and applications. Many details on the construction of heat exchangers are furnished also. A summarized table of contents follows here with chapter headings and **examples** of subtitles. Numbers in ( ) are numbers of topics.

ISBN 0-8155-0778-X **392 pages**

INDUSTRIAL AND INSTITUTIONAL WASTE HEAT RECOVERY 1979

Edited by P.G. Stecher

Energy Technology Review No. 37

Here is an excellent guidebook to productive management of discarded or waste heat and a guide to technology for institutional integrated energy/utility systems.

The first part of the book discusses industrial waste heat recovery. Simplified examples are explored, e.g., the use of medium- to high-temperature exhaust gases to preheat combustion air for gas turbines using regenerators. Effective waste heat management including its measurement technology and economics are discussed, with emphasis on energy conservation to increase profits.

What are the benefits and costs? How does a manager evaluate equipment? Answers to these questions—and many more—are furnished in this book. Commercial firms of all kinds can benefit from the financial guidelines given to determine the feasibility of installing a waste heat recovery system.

Of great value are the descriptions of 14 successful installations in U.S. industries such as glassmaking. Design specifications for recovery devices, cost studies, and resultant energy savings combine to form a complete case study.

There is a chapter devoted to equipment to aid the reader in assessing the merits of various heat transfer devices. Instrumentation techniques for measuring the essential parameters of a waste heat recovery system are elaborated. Engineering data for analyzing waste heat problems and a list of referral organizations are also provided.

Architect-engineers, as well as industrial and power plant owners and suppliers, will find the integrated utility system (IUS) described in the second part a promising way to conserve energy. By combining the utilities of electrical power, water, sewage treatment and waste disposal previously purchased from outside sources into one within an institution, the wastes from each individual system are used to furnish another needed utility. Power generation yields energy for heating and cooling; solid waste is burned for heat content; sewage effluent is reused for equipment make-up water and irrigation.

A partial and condensed table of contents with chapter headings and subtitles follows:

ISBN 0-8155-0738-0 **362 pages**

SOLAR CELLS FOR PHOTOVOLTAIC GENERATION OF ELECTRICITY 1979
MATERIALS, DEVICES AND APPLICATIONS

by Marshall Sittig

Energy Technology Review No.48

The direct conversion of sunlight into electricity by the use of semiconductor devices has an intriguing potential. Quiet and reliable, they can convert sunlight directly into electricity without the use of moving parts. Considerable emphasis in this book is on obtaining high current densities and the resultant economics.

Originally the high cost of producing photovoltaic cells did confine them to relatively exotic uses—powering space craft, for example. But silicon, the prime raw material, is cheap and abundant, and the technology for mass production, especially of the so-called ribbon cells, is available, as described in this book.

More research is needed to improve the conversion efficiency from sunlight to electrical current, but the way is also shown clearly in this book, e.g., by integrating finished solar cells into whole systems suitable for large scale applications.

Once the critical cost figures are reached, the market for solar cells should grow dramatically. Installation of whole banks of these cells on individual homes and office buildings, together with large arrays as regional electrical generating stations, should provide electricity at rates competitive with those now provided by utilities to many commercial and residential customers in the future.

The scope of this book is shown by the following table of contents, which presents chapter headings and subtitles. The numbers in parentheses indicate the number of processes per topic.

ISBN 0-8155-0765-8 **350 pages**